Usha Rani Palaniswamy
Kodiveri Muniyappa Palaniswamy

Handbook of Statistics for Teaching and Research in Plant and Crop Science

Handbook of Statistics for Teaching and Research in Plant and Crop Science

FOOD PRODUCTS PRESS®
Crop Science
Amarjit S. Basra, PhD
Editor in Chief

Plant-Derived Antimycotics: Current Trends and Future Prospects edited by Mahendra Rai and Donatella Mares

Concise Encyclopedia of Temperate Tree Fruit edited by Tara Auxt Baugher and Suman Singha

Landscape Agroecology by Paul A Wojkowski

Concise Encylcopedia of Plant Pathology by P. Vidhyaskdaran

Molecular Genetics and Breeding of Forest Trees edited by Sandeep Kumar and Matthias Fladung

Testing of Genetically Modified Organisms in Foods edited by Farid E. Ahmed

Fungal Disease Resistance in Plant: Biochemistry, Molecular Biology, and Genetic Engineering edited by Zamir K. Punja

Plant Functional Genomics edited by Dario Leister

Immunology in Plant Health and Its Impact on Food Safety by P. Narayanasamy

Abiotic Stresses: Plant Resistance Through Breeding and Molecular Approaches edited by M. Ashraf and P. J. C. Harris

Teaching in the Sciences: Learner-Centered Approaches edited by Catherine McLoughlin and Acram Taji

Handbook of Industrial Crops edited by V. L. Chopra and K. V. Peter

Durum Wheat Breeding: Current Approaches and Future Strategies edited by Conxita Royo, Miloudi M. Nachit, Natale Di Fonzo, José Luis Araus, Wolfgang H. Pfeiffer, and Gustavo A. Slafer

Handbook of Statistics for Teaching and Research in Plant and Crop Science by Usha Rani Palaniswamy and Kodiveri Muniyappa Palaniswamy

Handbook of Microbial Fertilizers edited by M. K. Rai

Eating and Healing: Traditional Foods As Medicine edited by Andrea Pieroni and Lisa Leimar Price

Handbook of Plant Virology edited by Jawaid A. Khan and Jeanne Dijkstra

Physiology of Crop Production by N. K. Fageria, V. C. Baligar, and R. B. Clark

Plant Conservation Genetics edited by Robert J. Henry

Introduction to Fruit Crops by Mark Rieger

Sourcebook for Intergenerational Therapeutic Horticulture: Bringing Elders and Children Together by Jean M. Larson and Mary Hockenberry Meyer

Agriculture Sustainability: Principles, Processes, and Prospects by Saroja Raman

Introduction to Agroecology: Principles and Practice by Paul A. Wojtkowski

Handbook of Molecular Technologies in Crop Disease Management by P. Vidhyasekaran

Handbook of Precision Agriculture: Principles and Applications edited by Ancha Srinivasan

Dictionary of Plant Tissue Culture by Alan C. Cassells and Peter B. Gahan

Handbook of Potato Production, Improvement, and Postharvest Management edited by Jai Gopal and S. M. Paul Khurana

Handbook of Statistics for Teaching and Research in Plant and Crop Science

Usha Rani Palaniswamy
Kodiveri Muniyappa Palaniswamy

Food Products Press®
The Haworth Reference Press™
Imprints of The Haworth Press, Inc.
New York • London • Oxford

For more information on this book or to order, visit
http://www.haworthpress.com/store/product.asp?sku=5256

or call 1-800-HAWORTH (800-429-6784) in the United States and Canada
or (607) 722-5857 outside the United States and Canada

or contact orders@HaworthPress.com

Published by

Food Products Press® and The Haworth Reference Press™, imprints of The Haworth Press, Inc., 10 Alice Street, Binghamton, NY 13904-1580.

Cover design by Jennifer M. Gaska.

Library of Congress Cataloging-in-Publication Data

Palaniswamy, Usha R.
    Handbook of statistics for teaching and research in plant and crop science / Usha Rani Palaniswamy, Kodiveri Muniyappa Palaniswamy.
        p. cm.
    Includes bibliographical references and index.
    ISBN-13: 978-1-56022-292-7 (hc. : alk. paper)
    ISBN-10: 1-56022-292-1 (hc. : alk. paper)
    ISBN-13: 978-1-56022-293-4 (pbk. : alk. paper)
    ISBN-10: 1-56022-293-X (pbk. : alk. paper)
    1. Botany—Research—Statistical methods—Handbooks, manuals, etc. 2. Crop science—Research—Statistical methods—Handbooks, manuals, etc. I. Palaniswamy, Kodiveri Muniyappa. II. Title.

QK51.P35 2005
519.5′02′458—dc22
                                                                                    2004026105

Invocation to Lord Ganesha
vakratu.nDa mahaakaaya koTisuuryasamaprabha.
nirvighnaM kuru me deva sarvakaaryeshhu sarvadaa.
O elephant-headed and large-bodied Lord,
you are radiant as one thousand suns,
I ask for your grace so that all tasks that I start
may be complete successfully without any hindrances.

Dedicated to Indrani,
Meera Devi, Vijayaraghavan, Rajeswari, Gowri Shankar,
Krishna Jayanthi,
Sambandan, Archana, and Madhushree
For their motivation of our work
and enriching of our lives and spirits

ABOUT THE AUTHORS

Usha Palaniswamy, PhD, holds a joint appointment in the School of Allied Health and the Asian American Studies Institute. Her major publications include over forty peer-reviewed journal articles, several book chapters, the monthly newsletter *Purslane,* and the book *A Guide to Medicinal Plants of Asian Origin and Culture.* She is the lead Editor of *Acta Horticultura: Traditional Medicine and Nutraceuticals.* In 2002 she received the Excellence Award for Teaching Innovation from the American Association of University Professors and an official citation from the State of Connecticut general assembly for her innovation in teaching. Dr. Palaniswamy's current research focuses on Asian traditional medical systems, Asian medicinal plants, human issues in horticulture, the design of health education programs for ethnic minority groups, and phytochemicals that promote human health and prevent diseases.

K. M. Palaniswamy, PhD, taught statistical methods, design of experiments, agricultural statistics, sampling techniques, quality control, and biostatistics at both undergraduate and graduate levels for thirty years, as well as serving as faculty and Head of the Department of Physical Sciences at Tamil Nadu Agricultural University, India, for over twenty-five years. He has extensive consultancy experience in applied statistics, including appointments as UN Expert in Agricultural Statistics, Baghdad, and UN Consultant in Agricultural Statistics at Makerere University, Uganda. Dr. Palaniswamy has published over seventy research articles. He held the position of Research Scholar at the International Rice Research Institute in the Philippines and earned his MS (experimental statistics) degree at the University of the Philippines. He was also Senior Biostatistician and Head of the Epidemiology Division at the Regional Cancer Research Institute and Hospital in Madras, India, in the Cancer Registry Scheme of the Indian Council of Medical Research, New Delhi.

CONTENTS

Foreword

An important aspect of experimental research is to draw proper inference from observed data. The tools of statistics are the windows through which scientists can peer for suitable interpretation. That is why statistics is included as a subject and taught to students of both undergraduate and postgraduate levels in all universities. For a proper understanding of the different techniques of statistics used by researchers, good textbooks that explain the basic concepts and tools are essential. Viewed from that angle, this book is good reading material for both researchers and teachers.

The authors have written this book in a simple and lucid manner. The book covers the following commonly applied tools for research data: descriptive statistics, univariate analysis, multivariate analysis, and design of experiments. The authors have explained in detail theory and application aspects with many practical examples drawn from different fields of scientific research. The steps of analysis are well explained and illustrated with examples using actual data. A salient feature of the book is the set of numerous exercises and questions included at the end of each chapter. Those assignments will give students the self-confidence to apply these important tools in any situation.

I congratulate and also express my appreciation for the sincere effort made by Kodiveri M. Palaniswamy and Usha R. Palaniswamy in bringing forth this well-compiled textbook.

I am confident this book will be of very great help to scientists and researchers in the analysis of their experimental data. This book will also be of immense help to teachers of statistics at the undergraduate and postgraduate levels.

C. Ramasamy, PhD
Vice Chancellor
Tamil Nadu Agricultural University
Coimbatore, India

Preface

Writing the preface to a book is more strenuous than writing the book itself. *Handbook of Statistics for Teaching and Research in Plant and Crop Science* is the outcome of lecture notes delivered to undergraduate and graduate students in various universities in India, the Philippines, Sudan, Iraq, Tanzania, and the United States; research papers published and presented in seminars; workshops; and conferences during the past three decades. Special lectures and materials on sample surveys, crop-response studies, design of experiments, crop forecasting and crop estimations, and some other special topics developed to address the needs of statisticians of Asian and African countries training in applied statistics at the Munich Center for Advanced Training in Applied Statistics for Developing Countries of Asia, Africa, and Germany also form part of this book. Experimental data from exhaustive research publications, master's theses, and doctoral dissertations are used widely as real examples whenever possible to improve the understanding of concepts and principles of statistics. Many years' experience in teaching and research in statistics and the requests of students and colleagues for a textbook were the impetus behind this book. Many textbooks on statistics available are mainly based on theory and are not of much help to science-oriented students and researchers as many do not like mathematics. This has been kept in mind while preparing this book. This book is written in a simplified way to create a significant impact on the learning of statistics. Tables and figures are provided to help clarify some of the more difficult concepts. The ideas of a number of statisticians and researchers have been obtained and analyzed, and as many of them as possible have been incorporated for wider use.

The specific objectives of this book are to

1. present the fundamental concepts of important statistical methods and experimental designs to the students and researchers who wish to apply them in their specific problems;
2. motivate students and readers to develop a positive attitude toward learning and enjoying statistics as a subject;
3. serve as a textbook for two semester statistics courses: one semester covering the statistical methods, and the second semester covering experimental designs;

4. serve as a reference book for researchers and instructors in different fields;
5. serve as a textbook not only to students in crop science but also in other fields such as medicine, economics, education, social science, engineering, veterinary, commerce, banking, and others as the statistical principles explained in this book are common to all sciences;
6. facilitate analysis of the experimental data in different ways and to get maximum information from the experimental data;
7. include some of the topics that are not explained in other textbooks which are useful to plant scientists.

The following are the important features of this book:

1. The book can be understood by readers with minimal mathematical background.
2. Derivations of formulas are purposely avoided in most cases.
3. Mathematical symbols are used sparingly and only when absolutely essential.
4. In illustrating the statistical procedures, data from actual experiments have been used.
5. Suitable examples have been inserted to illustrate the principles.
6. At the end of each chapter, exhaustive exercises are provided for the students to work on and practice.
7. The computational procedures are discussed and illustrated with numerical examples taken from real data.
8. Figures have been provided and explained to facilitate the use of statistical tables in the Appendix.

ORGANIZATION

This book consists of two parts. Part I consists of 14 chapters beginning with a brief introduction explaining the various uses of statistics in applied research problems. Descriptive statistics are presented in the beginning to help students to understand the later chapters. In Chapter 1, uses of various tables, graphs, diagrams, and pictorial diagrams for presentation of statistical data are explained. Important mathematical concepts are explained in Chapter 2. In Chapters 4 and 5, various measures of central tendency and dispersion are explained with suitable examples. Chapter 6 deals with normal distribution; its properties and application to practical problems are stressed. Chapter 7 discusses probability which is a prerequisite for courses such as statistics, genetics, medicine, education, sociology, and other fields at undergraduate and graduate levels. Chapter 9 deals with histogram, frequency tables, ogives, and their uses in practical problems. Concepts of population, sample, sampling, sampling distribution of mean, standard error, point estimate, and confidence interval estimate for the population mean are explained with suitable examples in Chapter 10. Chapter 11 deals with tests of hypothesis and tests of significance and their uses in different situations. Chapters 12 and 13 deal with correlation and regression studies, respectively. Computational methods for correlation and regression coefficients, and procedures for fitting of the linear regression equation are provided. Different uses of the Chi square test are given in Chapter 14.

Part II includes 15 chapters. Emphasis is given to the most commonly used designs in plant and crop sciences. Chapter 15 discusses definitions for experimental design and importance of the study of plot size and shape. A detailed procedure for this study is explained with actual data. The structure and technique of the analysis of variance, and the use of F statistics are explained in Chapter 16. In Chapter 17, the basic principles of experimental design, several suggestions to improve the precision of the experiment, and the procedure to fix a sampling unit for the estimation of plant attributes are given. A detailed description of the completely randomized design with layout plan, method of analysis of data and its advantages and disadvantages are given in Chapter 18. Chapter 19 explains the randomized complete block design model; procedures to lay out the design, methods of analysis of the data, and its advantages and disadvantages are explained. The use of contrasts in the comparison of treatments in an experiment is discussed in Chapter 20. In Chapter 21, procedures for tests, namely least significant difference (LSD) test, Duncan's New Multiple Range Test, Dunnett's test, Tukey's test, S test, and Newman-Keuhls test, are explained with suitable numerical examples for comparing individual pairs of means. Chapter 22 deals with Latin square design. Chapter 22 reviews factorial experiments, the layout in different designs, and discussion with main effects and interaction to help clarify this topic. Chapters 23 and 24 discuss the layout of split plot design and strip plot design, respectively, and the procedures for analysis and comparison of treatments for these designs. The principle of confounded design is explained in Chapter 26 with suitable examples. The analysis of covariance technique and its uses are explained in Chapter 27. The importance of data transformation in the analysis of experimental data is explained and illustrated with suitable examples in Chapter 28. The topic of statistical quality control is dealt with in Chapter 29, according to the concept of W. A. Shewart for adoption in scientific laboratories.

This book was completed after several revisions and modifications for use as a textbook and also as a reference. Though we provide many examples to illustrate the principles and procedures based on plant science data, it does not mean that this text is limited to this field alone. We believe that students, instructors, and readers interested in statistics will benefit a great deal from this book.

Acknowledgments

I thank K. M. Palaniswamy, the co-author of this book, for his excellent contributions and the experience that he brings to the world of teaching and research in statistics, and plant and crop sciences. He has continued to be my greatest mentor and a perfect role model in my academic performance, always demanding excellence in my accomplishments. I am personally indebted to Roger Buckley, director of Asian American Studies Institute, University of Connecticut, for his constant encouragement and strong support of my research and teaching efforts. I express my gratitude to Archana Sambandan, University of Connecticut, for help with proofreading of the manuscript and Manjit S. Kang, professor, Quantitative Genetics, Louisiana State University, Baton Rouge, for review of the manuscript before publication. I thank my delightful daughter, Madhushree, for her wonderful smile and responsible behavior at home and school that allowed me to spend time on completing this manuscript.

Usha R. Palaniswamy

First of all, I thank Usha R. Palaniswamy, the co-author of this book, for the continuous inspiration, support, cooperation, and encouragement to complete this unique textbook for the benefit of students, teachers, and researchers. Next, I acknowledge the help rendered by my son, P. Vijay, Department of Surgery, Indiana Medical Center, Indianapolis, and my daughter, Krishna J. Palaniswamy, for their valuable suggestions in the computation and typing and in the preparation of the charts, tables, and figures. I am indeed very grateful to my wife, Indrani, for allowing me to spend time and concentrate on this project during late hours for several years and for all the sacrifices she made in her life in several ways. I am particularly indebted to K. A. Gomez, statistician, International Rice Research Institute (IRRI), the Philippines, for shaping my career as a statistician and under whose guidance I did my graduate studies in experimental statistics. She is my mentor and she is responsible for motivating me to develop a sustainable interest in statistics; I conducted several experiments at IRRI under her able guidance and expertise when I was a research scholar there. I express my thanks to the Tamil Nadu Agricultural University, Coimbatore, India, where I worked as a statistician and conducted several experiments that are used as examples in this book. I am also thankful to the undergraduate and graduate students at Tamil Nadu Agricultural University, India; University of Khartoum, Sudan; Makerere University, Uganda; University of Dar es Salaam, Tanzania; Arab Institute for Training and Research in Statistics, Baghdad, Iraq; and my colleagues who insisted that I write this book. I am thankful to the Indian Agricultural Research Statistics Institute, New Delhi, for giving me training and helping me to analyze my uniformity trial data and other experimental data. My thanks are due to K. Kumaran Kutty, professor and head of the Department of Statistics, Calicut University, India, under whose guidance, support, and encouragement I did my doctoral work in statistics. I am also thankful to Dieter Stentzel, director, Munich Center for Advanced Training in Applied Statistics for Developing Countries of Asia and Africa, Germany, for inviting me to offer special lectures on applied statistics to statisticians of Asia and Africa.

K. M. Palaniswamy

We both are thankful to Dr. C. Ramaswamy, Vice Chancellor, Tamil Nadu Agricultural University, Coimbatore, India, for his support and for writing the foreword for this book.

Introduction

The purpose of this book is to provide working scientists, graduate students, and other researchers who are engaged in different areas of research with sufficient and efficient statistical techniques and procedures for dealing with data collected from their experiments. Few books are available to teach researchers in detail how to run their experiments scientifically, collect data, analyze and draw valid conclusions, and make sound decisions. Much of this book is the result of research work of the authors and lecture notes given to students in statistical courses in various universities. This book will also be useful to professional statisticians whose main interest is applied statistics in biological sciences. For better comprehension, many examples are provided from some of the authors' published articles and from their theses and dissertations. By following the procedures in this book, researchers can develop the ability to convert a mass of data into a form for meaningful analysis. With the current emphasis in interdisciplinary research, an urgent need exists for a book of this type oriented with practical applications in agriculture, plant, and crop sciences. It is written in such a way that it is easily understandable by nonmathematicians. It contains materials sufficient for courses in statistical methods, and for courses in experimental design for students in plant and crop sciences, and others who work in empirical disciplines. Several topics such as field layout of experiments, sampling for plant traits, and plot size and shape are included.

Scientists conduct experiments on plants, human beings, mice, insects, cell cultures, etc., and gather a variety of data, accumulated at a phenomenal rate, which are nothing but numerical data. Such experiments are conducted in agricultural fields, greenhouses under controlled conditions, and laboratories. The data thus collected from their experiments or investigations refer to the actual observations. The deluge of various types of information has precipitated the need for statisticians to process, store, and analyze it. Since decisions are made based on the collected data, the data must be scientific and must meet certain conditions laid out by theoretical considerations. Otherwise, the results can be erroneous and misleading. Every caution must be exercised to obtain data that are as representative as possible. The data are random in nature, i.e., variable that is measured or observed assumes a set of numerical values that are unknown prior to the experimentation or investigation. They further show variability when repeated measurements are made. The crux of the problem faced by scientists is how to deal with variability among the observed values. This variability phenomenon occurs due to unknown causes that are definitely not under the control of the experimenters. Variability occurs even if the experiment is executed very carefully.

There is an old canard, "figures never lie, only statisticians do," and to make this statement false, the data should be collected in a systematic and scientific way. Any numerical number represents a result arrived at under certain conditions. For example, the number 180 carries no information without a context, but if it refers to the blood pressure of a patient, we appreciate that the patient is showing symptoms of hypertension. While collecting data, one has to foresee errors due to (1) measurement equipment or instruments (e.g., weighing scales, color meters); (2) transcription; (3) carelessness and lack of proper instructions while recording observations; and (4) computing errors. These errors are certainly not random and hence one must make every effort to avoid them while collecting data.

Researchers often consult statisticians only after the data collection is complete and all results are in. This is similar to calling a doctor after the patient is dead. The planning phase of experimental design is as important as data collection and data analysis. Valid effects of the treatments are tested while avoiding other extraneous variations in a well-designed experiment. For example, if you want to know whether a new type of fertilizer will give increased yields, you will conduct experiments in terms of an appropriate hypothesis that you formulated is acceptable or whether it is wrong and unacceptable, and consequently has to be rejected. The accuracy of results is highly dependant on the selection of the design. If you select the wrong design, it is quite likely that the decisions you arrive at may be absolutely misleading. If data are not collected scientifically and the results are reported incorrectly, the chances of reaching wrong conclusions are high. Thus, lack of knowledge in statistical science and use of faulty designs result in pointless expenditure, time, and labor consumption. The experimental design has to be formulated following the basic principles of experimental design: replication, randomization, and local control. With appropriate statistical methods scientists can get accurate and reliable results at less cost. In this book, several important experimental designs, which are very useful to researchers, are explained with real examples.

One important feature of this book is that the authors have taken all efforts to keep readers' interest in understanding the subject. Generally, students do not like to study statistics as it is considered dull and unbearably time-consuming. Often, students take a statistics course only to satisfy a requirement. They have to study it because it contains tools for use in research work. Because students cannot learn by mere reading, a number of exer-

cises are provided after each chapter. Sufficient importance has been given
to preliminary topics such as data condensation, tabulation, graphical representations, and tests of significance. Since the emphasis in this book is on
practical applications, the statistical theory is deliberately kept to a minimum. The subject is explained in the simplest nontechnical manner possible. The exercises may vary according to the different disciplines, but the
scientific methods are common to all. Each example is selected to bring out
one or more points. These exercises require a fair amount of computation
and will help students analyze their data with confidence.

Due to differences in syllabi adopted in different universities, some chapters may be inappropriate for some students. Statistics are used in disciplines
such as public health, political science, law, market research, demography,
sociology, anthropology, economics, geology, astronomy, medicine, agriculture, genetics, business administration, industry, education, bio-informatics
(a combination of biotechnology and information technology), etc. No one
statistical theory is applicable to one particular subject. A general theory of
statistics is applicable to any field of study in which observations are recorded. The procedures in this book can be extended to many other fields. It
also provides a solid foundation for advanced training in statistics. To read
and benefit from this book, no pre-requisite is expected or assumed.

Researchers, particularly students, are hesitant to accept nonsignificance
or negative results. In this context, researchers should note that there is no
such thing as a negative result. If the null hypothesis is not rejected, it is positive evidence that there may be no real differences among the treatments
tested. Consult a statistician for the selection of design, choice of treatments, response variable, etc., before the start of an experiment. Note that
large significant effect will not often occur by pure chance. Since mere significance does not report the size of an effect, such effect can be too small to
be of practical importance.

Practical or applied statisticians are not interested in the derivations of
formula. They are mainly concerned with the use of techniques already per-
fected by the mathematical statisticians and approved by specialists. Thus,
the success of an applied statistician depends upon expert knowledge in the
field of application, and not upon training in advanced mathematics. Induc-
tive statistics play a significant role in modern statistics and involve estimation and decision making on the basis of sample data.

Statistics may be defined as a tool for decision making in the light of uncertainty. Decisions are made on the basis of data collected and hence the
ways in which they are collected become extremely important. Experimental designs are used to help reduce the experimental errors in data collection. Scientists use statistics to describe the results of an experiment or investigation. This process is referred to as data analysis or descriptive

statistics. They often observe only a portion of the population, i.e., a sample, and use statistics to answer questions about the whole population. This process is called inferential statistics. Several topics in this book explain the methods of drawing inferences.

Special symbols are introduced as needed. Statistical methods are an important adjuvant to knowing more about the experiments. Basic statistics, which is presented in the first part of the book, will help students to understand the second part of the book. When dealing with examples, the authors have presented all steps in solving to ensure that students will have no difficulty in following the procedures used in arriving at the solutions to the problems.

PART I:
STATISTICAL METHODS

Chapter 1

Tables, Graphs, and Diagrams

The results of experiments or investigations should be presented in such a way that readers understand the inferences arrived at by the investigators. Statistical information can be displayed in many ways. Data can be presented as tables, charts, graphs, diagrams, or pictures. Each method of presentation is discussed here and suitable examples are provided.

TABLES

It can be confusing when data are presented in large numbers in text. Tables are used for communicating detailed information in rows and columns. Actual numerical data can be summarized meaningfully in the form of a table. Any amount of information can be incorporated in the tables. A good table must contain a complete "tell-it all" title, caption, body, and footnote. Usually, the title is placed above the table. The units of measurements for variables are indicated directly below the variable. Use parallel rules to separate column headings from the body of the table and place a rule below the table. The source of the data in the table is set below the rule at the end of the table. Do not use vertical lines in a table. An example of data presented in tabular form appears in Table 1.1 (Palaniswamy et al., 2002).

GRAPHS AND DIAGRAMS

A picture is worth a thousand words and at least ten thousand numbers! When data are presented by a visual method, one can easily grasp their magnitude. Sometimes, diagrams or graphs may be more attractive than a mere table of figures. If the data are presented in diagrams, they may be retained longer in memory than when presented in tabular forms.

Graphs are a visual way to portray the relationship between a set of variables. Graphs show the present and the possible future set of data. They also help us to check the assumptions that are made in the analysis or suggest alternative hypotheses to be investigated or other new aspects. Graphs play a very important role in descriptive statistics. Graphics (graphs that summa-

3

TABLE 1.1. Antioxidant vitamin contents in curry leaves and postharvest storage temperature and treatment (mean concentration per gram dry weight ± SD).

Postharvest storage temperature and treatment	Lutein (µg/g)	α-tocopherol (ng/g)	β-carotene (ng/g)	Chlorophyll (mg/g)
Fresh	27 ± 2.8	592 ± 10.1	511 ± 10.5	27.8 ± 1.2
Oven-dried (60°C)	11 ± 1.9	296 ± 8.2	148 ± 1.2	9.9 ± 1.2
Air-dried (22°C)	10 ± 2.1	515 ± 8.5	357 ± 1.3	18.9 ± 1.1
Frozen (−15°C)	8 ± 2.5	589 ± 11.2	398 ± 1.5	26.9 ± 1.1

Source: Palaniswamy et al., 2002.

rize the data) are extensively used in personal and business computers as they are easier to understand and easier to create than even the best tables. In addition, graphs can add to the aesthetic quality of data presented.

Some Information About Graphs

1. Graphs are intended to give an overview rather than detailed pictures of the data set.
2. The title should be clear and give complete information about the contents of the data.
3. The units must be specified in the vertical and horizontal axes.
4. Too much information in the graph should be avoided.
5. A nonzero origin should be indicated by a zigzag symbol breaking the lower end of the scale to attract the reader's attention.

Advantages of Using Graphs

1. The nature of the data progression can be seen quickly.
2. They help to forecast the future trend.
3. They help to interpret the data.
4. When there are two or more factors, they help us to identify the interaction, if any, among the factors.
5. They help us to form some new ideas about the factors studied.

Limitations of Graphs

1. Data cannot be specifically shown as when incorporated in the tables.
2. Graphs usually take more time to construct than tables.
3. We cannot get the predicted values exactly from the y-axis for different values of independent variables shown in the x-axis.

How to Construct a Graph

In the coordinate system, there are two axes, one is the horizontal axis (x-axis) and the other is the vertical axis (y-axis). The x-axis starts from zero origin. If the range in y values is high, the lowest value is small, and one cannot show all the values in the y-axis from zero value. If the zero value is shown as origin, then the scale is broken by drawing a zigzag symbol in the y-axis between origin and the first value. By doing so, the graph can be shown in a more complete manner. A comparison of Table 1.2 and Figure 1.1 demonstrates this point.

TABLE 1.2. The average per capita availability of milk in India in different years.

Year	Per capita availability of milk (g/d)
1970-1971	112
1996-1997	202
1997-1998	204
1998-1999	211
1999-2000	214

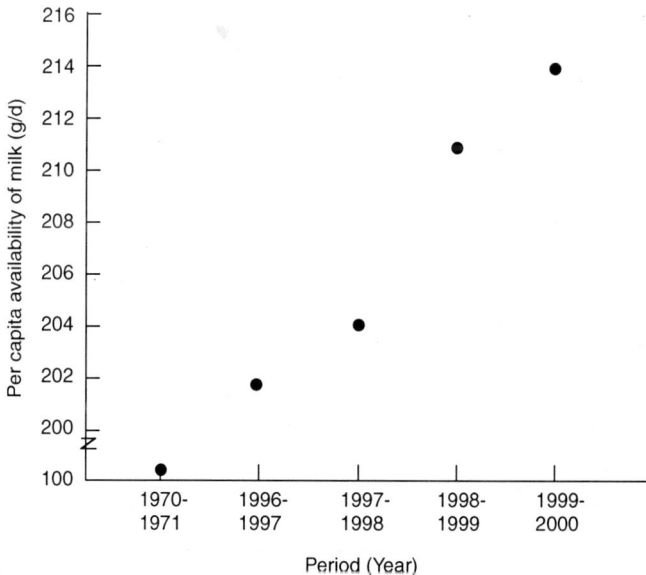

FIGURE 1.1. Graph showing the use of a zigzag symbol for breaking the scale in y-axis. Data represent per capita availability of milk (g/d) in India in different years.

The horizontal distance from the vertical axis to the point is called the *abscissa* and the vertical distance from the horizontal axis to the point is called the *ordinate*. The abscissa and the ordinate are called the *coordinates* of the points. Several forms of graphical representations are used in descriptive statistics. Some of them include the following:

1. Histogram
2. Frequency polygon
3. Frequency curves and ogive curves
4. Bar diagrams
   a. Vertical
   b. Horizontal
   c. Subdivided
        i. Single subdivided
       ii. Multiple subdivided
      iii. Percentage
       iv. Multiple
5. Line diagram
6. Pie diagram
7. Pictogram
8. Scatter diagram
9. Zone graph
10. Graph of time series
    a. One variable
    b. Two variables (same unit) and two different units
    c. More than two variables

Histogram

A histogram is used to represent the frequency distribution of a variable. It is constructed by erecting bars or rectangles by taking class intervals as base of the rectangles whose areas are proportional to class frequency. Along the x-axis, the class intervals of the variable are shown. Along the y-axis, frequencies of the classes are shown. Heights are proportional to the frequencies.

Example

The data on ascorbic acid content (mg/100g) in the 73 chili (*Capsicum annuum* L.) varieties recorded by Usha Rani (1996) can be exhibited in a histogram in a meaningful and understandable manner (see Figure 1.2). The distribution of the data follows.

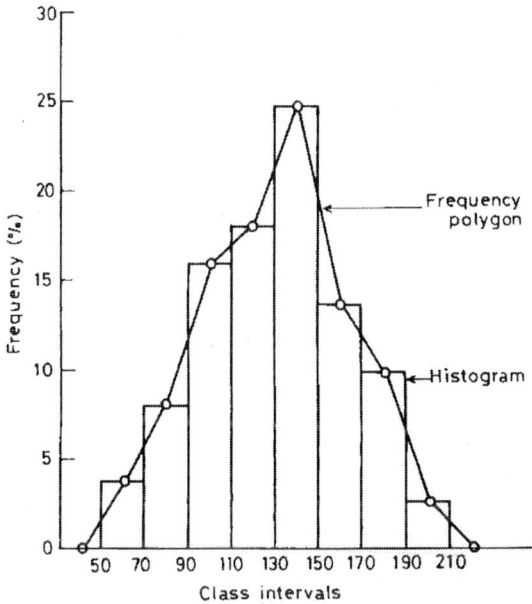

FIGURE 1.2. Histogram and frequency polygon for the distribution of ascorbic acid content in 73 chili (*Capsicum annuum* L.) genotypes. (*Source:* Usha Rani, 1996.)

160.03	118.73	64.43	154.61	92.68	125.26	117.86	133.78
87.95	149.87	183.52	165.92	150.66	143.52	146.65	88.08
98.37	96.13	146.83	97.54	92.09	112.80	168.96	129.51
155.05	108.60	170.10	117.77	174.24	127.03	176.94	
192.10	139.22	191.10	87.04	145.27	119.08	131.10	
127.40	139.94	163.78	102.79	144.18	163.79	131.28	
138.18	110.39	77.34	66.93	175.88	91.46	164.61	
140.30	98.34	145.41	109.70	135.81	178.98	147.52	
152.54	99.01	83.76	138.43	139.51	86.42	154.23	
182.04	106.49	121.74	104.87	120.99	58.73	137.72	

These data are summarized in a frequency table (Table 1.3).

Frequency Polygon

At the top of the bars or rectangles of the histogram, the midpoints of the rectangles are marked and then joined by straight lines to form a frequency

TABLE 1.3. Frequency distribution of the data on ascorbic acid content in the 73 chili (*Capsicum annuum* L.) varieties recorded.

	Distribution	Frequency	Relative frequency	Relative frequency (%)
	50-70	3	0.041	4.1
	70-90	6	0.082	8.2
Class limits	90-110	12	0.165	16.5
	110-130	13	0.178	17.8
	130-150	18	0.247	24.7
	150-170	12	0.164	16.4
	170-190	7	0.096	9.6
	190-210	2	0.027	2.7

Source: Usha Rani, 1996.

polygon; it is customary to add a class at each end of the class intervals. The resulting diagram is called frequency polygon (see Figure 1.2). The advantage of a frequency polygon over a histogram is that in one graph we can exhibit several polygons and compare. Several histograms cannot be shown in a single graph. To show two histograms, two separate graphs should be drawn.

Frequency Curve

Suppose we have several k classes in a frequency distribution. The ith class has the midvalue x_i and frequency f_i. Class intervals of the variable X are shown in the x-axis and frequencies in the y-axis. The points (x_1,y_1), $(x_2,y_2) \ldots (x_k,y_k)$ are marked. Joining the points by a smooth curve results in a frequency curve. This can be shown in the same graph along with a histogram and a frequency polygon (see Figure 1.3).

Ogive Curves: Cumulative Frequency Curves

In frequency curves, the distribution of the variable under investigation is formed in a frequency table, and the cumulative frequencies are computed. Two types of cumulative frequencies are generally depicted. One is called *lesser than* and the other *greater than.* These two types of frequencies are shown in one graph. One is called a *less than ogive curve* and the other *greater than ogive curve.* For computing the less than ogive curve, the frequencies lesser than the upper limit of each class are considered; for

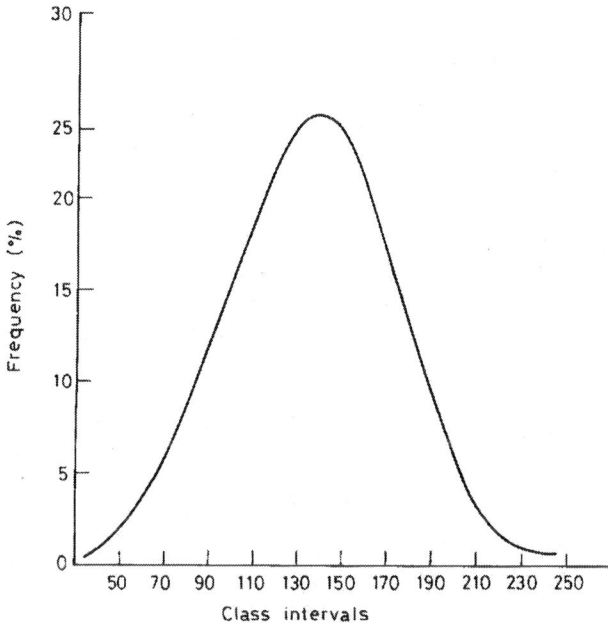

FIGURE 1.3. Diagram showing the frequency curve for the distribution of ascorbic acid content in 73 chili genotypes. (*Source:* Usha Rani, 1996.)

the greater than ogive curve, the cumulative frequencies greater than the lower limit are considered. For example, the grain yield/plant in grams in rice crop data of Palaniswamy (1990) collected in the IR 22 rice variety grown in an experiment at the International Rice Research Institute Experimental Farm (Los Boños, Laguna, Philippines) are used to illustrate the formation of cumulative frequency table (Table 1.4) and ogive curves (Figure 1.4).

Bar Diagrams

Vertical Bar Diagram

The vertical bar diagram consists of a number of rectangles called bars in which height varies according to magnitude of the measured factor. It is ex plained by taking equal widths for the rectangles. We use data relating to milk production and per capita availability in different years in India in this example.

TABLE 1.4. Table of cumulative frequencies.

Class limits	Class midvalue	Frequency	Cumulative frequency		Relative frequency	Relative frequency (%)
			Less than	Greater than		
2-6	4	2	2	200	0.010	1.0
6-10	8	6	8	198	0.030	3.0
10-14	12	12	20	192	0.060	6.0
14-18	16	12	32	180	0.060	6.0
18-22	20	21	53	168	0.105	10.5
22-26	24	27	80	147	0.135	13.5
26-30	28	35	115	120	0.175	17.5
30-34	32	29	144	85	0.145	14.5
34-38	36	26	170	56	0.130	13.0
38-42	40	18	188	30	0.090	9.0
42-46	44	5	193	12	0.025	2.5
46-50	48	4	197	7	0.020	2.0
50-54	52	0	197	3	0.000	0.0
54-58	56	3	200	3	0.015	1.5
Total		200			1.000	100.0

Note: The grain yield/plant (g) in rice crop data of Palaniswamy (1990) collected in the IR 22 rice variety grown in an experiment at the International Rice Research Institute Experimental Farm (Los Boños, Laguna, Philippines).

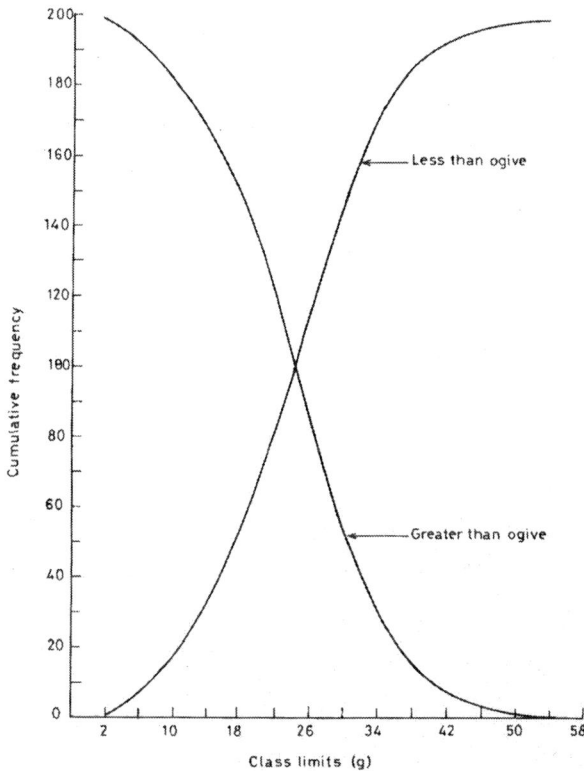

FIGURE 1.4. Diagram showing the less than and greater than ogive curves.

The milk production in different years in India is given in Table 1.5, and shown in a vertical bar diagram, Figure 1.5. We represent the data by using equal bar widths, yet the heights of the bars are proportional to the amount of production. A quick inference can be obtained by looking at Figure 1.5.

Here is another example: A company reports losses for four years as shown in the following list. The vertical bar diagram in Figure 1.6 is used to show these losses.

Year	Loss (percent)
1990	20
1991	16
1992	10
1993	5

TABLE 1.5. Milk production and per capita availability in different years in India.

Year	Milk production (million tons)	Per capita availability (g/d)
1950-1951	17.0	124
1960-1961	20.0	124
1970-1971	22.0	112
1980-1981	31.6	128
1990-1991	53.9	176
1996-1997	69.1	201
1997-1998	70.8	204
1998-1999	74.7	211
1990-2000	78.1	214

Note: The same data are illustrated as bar diagrams in Figure 1.5.

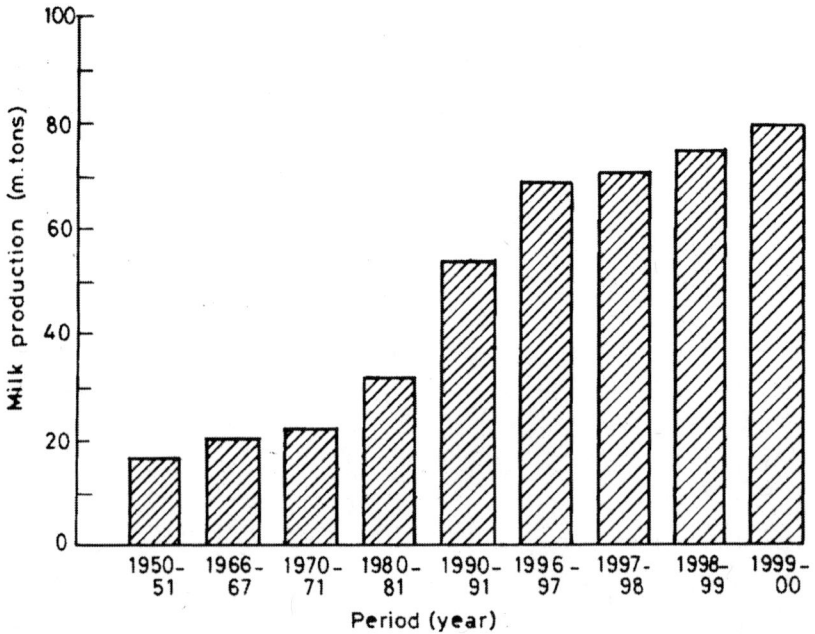

FIGURE 1.5. Vertical bar diagram showing production of milk in India.

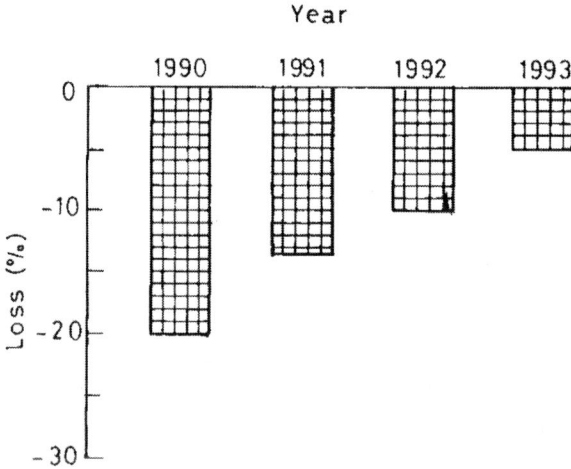

FIGURE 1.6. Figure showing a company's losses in different years.

Horizontal Bar Diagram

In this diagram, the bars (rectangles) are drawn horizontally. Data on causes of death in the United States during 1999 are given in Table 1.6 and also displayed as a horizontal bar diagram in Figure 1.7.

Subdivided Bar Diagram

In the subdivided bar diagram, the height or area of the vertical bar or rectangle is taken as unity or 100 percent. It is a tool to communicate the breakdown of the total into different components according to the magnitudes of the components. Such diagrams are useful when the relative differences in the sizes of the components are to be highlighted. Depending upon the nature of the data, we can use a single subdivided bar diagram or a multiple subdivided bar diagram.

Single subdivided bar diagram: Studies in chili (*Capsicum annuum* L.) fruit weight in 73 chili genotypes show that fruit weight is made up of three different component parts, namely (1) pedicel weight, (2) seed weight, and (3) skin weight (Usha Rani, 1997). The data recorded as follows are shown in a single subdivided bar diagram (see Figure 1.8).

TABLE 1.6. Causes of death in the United States during 1999.

Rank	Causes of deaths	Deaths	Deaths (in %)
1	Heart disease	724,621	36.3
2	Cancer	549,761	27.6
3	Stroke	167,261	8.4
4	Chronic lower respiratory disease	124.141	6.2
5	Unintentional injury	96,909	4.9
6	Hospital infections	90,000	4.5
7	Diabetes	68,394	3.4
8	Influenza and pneumonia	63,408	3.2
9	Alzheimer's disease	44,536	2.2
10	Nephritis	35,359	1.8
11	Septicemia	30,397	1.5
	Total	1,994,787	100.0

Source: Shnayerson, 2002, *Fortune* 146(6): 150.

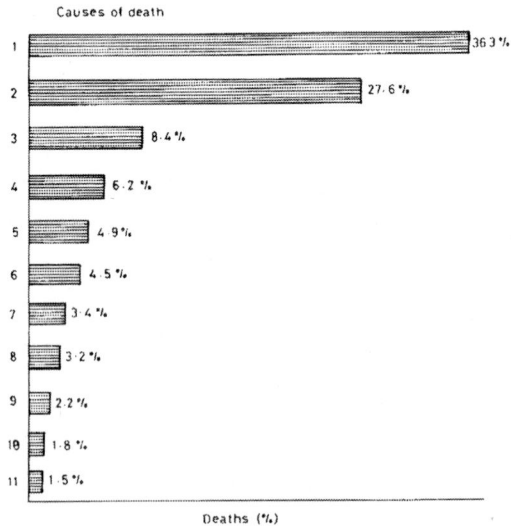

FIGURE 1.7. Horizontal bar diagram showing the causes of death in the United States in 1999. (*Source:* Shnayerson, 2002.)

(1. Heart disease; 2. Cancer; 3. Stroke; 4. Chronic lower respiratory disease; 5. Unintentional injury; 6. Hospital infections; 7. Diabetes; 8. Influenza and pneumonia; 9. Alzheimer's disease; 10. Nephritis; 11. Septicemia.)

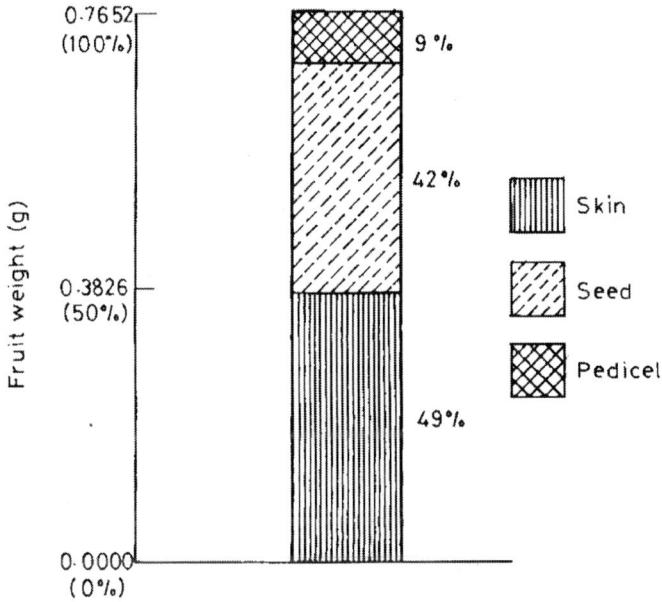

FIGURE 1.8. Single subdivided bar diagram showing relative weights of the different components of chili fruit. (*Source:* Usha Rani, 1997.)

Pedicel weight (g)	Seed weight (g)	Skin weight (g)	Total (g)
0.0671 (9%)	0.3179 (42%)	0.3802 (49%)	0.7652 (100%)

Multiple subdivided bar diagram: Subdivided bar diagrams can also be shown by taking the height of each bar as 100 percent, i.e., reducing to a common base for comparative purposes. In this case, the bars are of equal heights. The bars are then subdivided into different components according to relative proportions or percentages. The percentage figures discern to what extent the proportions differ. When we express in percentages, the original values are of secondary importance.

Example: The number of outpatients examined in different years during the five-year period at the Cancer Institute, Madras, India (Cancer Institute Cancer Registry, Madras, India, 1990) are shown in Table 1.7 and exhibited in multiple subdivided bar diagram, Figure 1.9.

Percentage bar diagram: The data in Table 1.7 are shown in the multiple subdivided bar diagram in original figures (Figure 1.9) and in percentages (Figure 1.10). When expressing the values in percentages, one can compare

TABLE 1.7. The number of outpatients examined in different years during the five-year period at the Cancer Institute, Madras, India.

	Year						
Sex	**1984**	**1985**	**1986**	**1987**	**1988**	**Total**	**Average**
Male	2,865	3,352	3,533	3,763	3,892	17,405	3,481
Female	3,723	4,319	4,354	4,750	4,867	22,013	4,403
Total	6,588	7,671	7,887	8,513	8,759	39,418	7,884

Source: Cancer Institute, 1990.

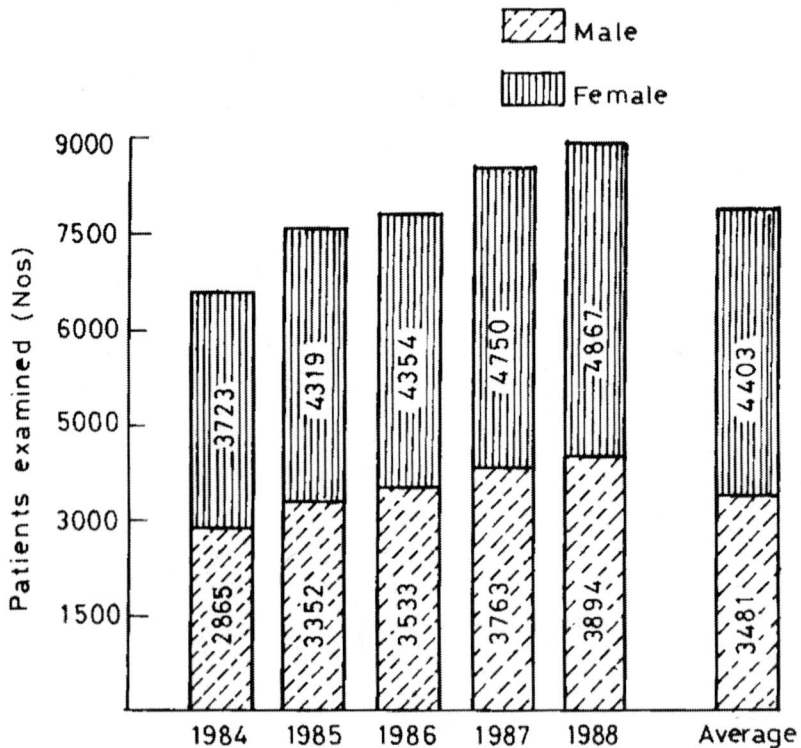

FIGURE 1.9. Multiple bar diagram showing the male and female patients examined in different years. (*Source:* Cancer Institute, Cancer Registry, Madras, India, 1990.)

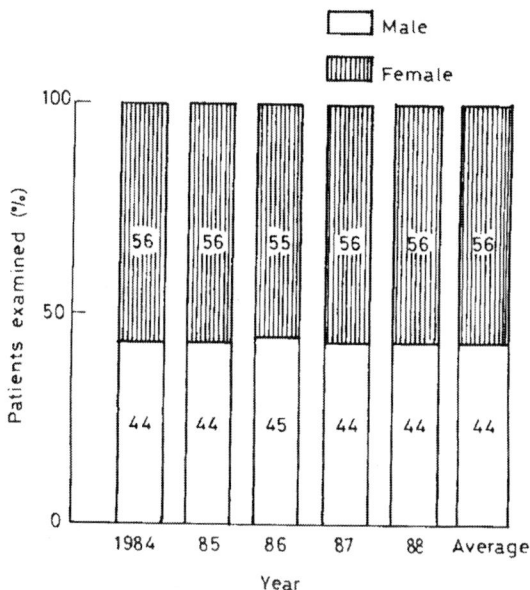

FIGURE 1.10. Multiple bar diagram showing the male and female patients examined in different years in percentages. (*Source:* Cancer Institute, Cancer Registry, Madras, India, 1990.)

the values in different years. The diagrams in percentages show more or less a constant ratio of male and female patients seeking treatment for cancer disease.

Multiple bar diagram: In this diagram, several variables can be depicted in situations in which comparative variations among the variables are of importance. Each character is represented in a vertical bar. The heights of the bars are proportional to the magnitudes of the characters. Here we use bars with equal widths. The different characters are represented in adjoining bars.

Example 1: Agricultural scientists are interested in knowing the variability that exists in different plant parts as they conduct research in their breeding programs. The coefficient of variations that exist in rice crop among different characters (Palaniswamy, 1990) is given in Table 1.8 and shown in Figure 1.11.

Example 2: Variation in grain weight/plant (g) and grain number/plant in the two different rice varieties and under three nitrogen levels (N1, N2, and N3; N1= 0 kg/ha, N2 = 60 kg/ha, N3= 120 kg/Ha) (Table 1.9) (Palaniswamy, 1990) (see Figure 1.12 and Figure 1.13).

TABLE 1.8. Coefficient of variations (percent) that exist in rice crops among different characters.

| Rank | Character | Variety | | |
		TKM 6	PTB 10	CO 29
1	Flowering duration (d)	4.91	4.32	4.56
2	Panicle length (in)	8.50	13.40	14.31
3	Plant height (in)	9.32	11.07	12.84
4	Boot leaf area (cm^2)	21.78	23.85	30.29

Source: Palaniswamy, 1990.

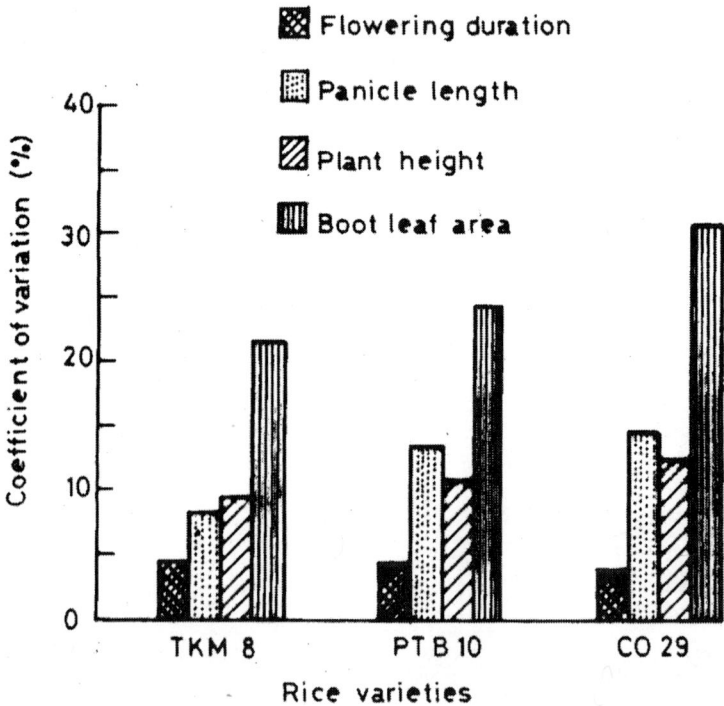

FIGURE 1.11. Multiple bar diagram showing coefficient of variation in different plant traits in three different rice varieties. (*Source:* Palaniswamy, 1990.)

Line Diagram

A line diagram is used to represent the frequency distribution of a discrete variable. Along the x-axis discrete variables are shown and on the y-axis frequencies are depicted. Straight lines are drawn parallel to the y-axis with heights proportional to the frequencies (Raghavarao, 1983).

For example, the number of outpatients examined/year at the Cancer Institute, Madras, India, on different days of the week are shown in the following list and in Figure 1.14. The figures are average for five years (1984-1988) (Cancer Institute, 1990).

Day of the Week	Mon.	Tues.	Wed.	Thur.	Fri.	Sat.	Total
No. of patients	1,911	1,177	1,249	1,164	1,331	981	7,813

Pie Diagram

Pie diagrams show different components by means of sectors of a circle. The total angle at any point is 360 degrees. The magnitudes of the components are expressed in terms of the corresponding angles.

For example, the total production of food grains in each category in India in the year 2000-2001 is given in Table 1.10 with the corresponding angles for drawing the pie diagram (Figure 1.15). The magnitudes are proportional to the angles. Since the total of the angles is 360 degrees, any component is converted in relation to 360 degrees.

Pictogram

The number of cars in production in three different years in a company can be shown by means of a pictogram (1 car pictogram = 1,000 cars).

TABLE 1.9. Variation in grain weight/plant and grain number/plant in two different rice varieties and under three nitrogen levels.

Variety	Grain weight per plant (g)			Gain number per plant		
	N1	N2	N3	N1	N2	N3
IR 22	17.2	22.6	30.0	733	943	1,266
IR 662	18.3	22.6	35.0	704	857	939

Source: Palaniswamy, 1990.
Note: N1 = 0 kg; ha^{-1}; N2 = 60 kg; ha^{-1}; N3 = 120 kg; ha^{-1}.

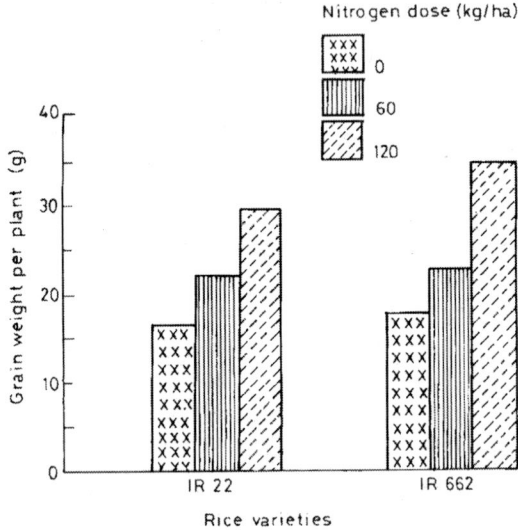

FIGURE 1.12. Multiple bar diagram showing differences in grain weight per plant in IR 22 and IR 662 rice varieties in different nitrogen fertilizer doses in respect of mean and range. (*Source:* Palaniswamy, 1990.)

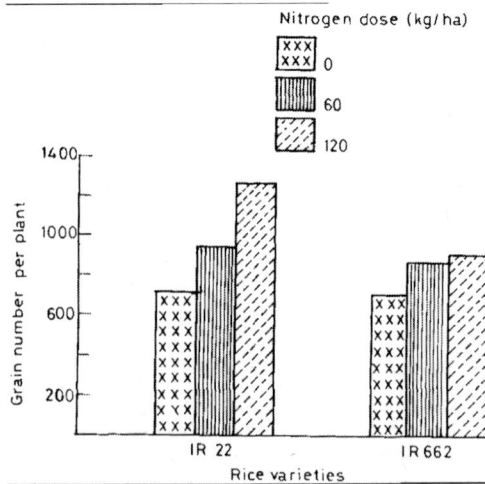

FIGURE 1.13. Multiple bar diagram showing differences in grain number per plant in IR 22 and IR 662 rice varieties in different nitrogen fertilizer doses in respect of mean and range. (*Source:* Palaniswamy, 1990.)

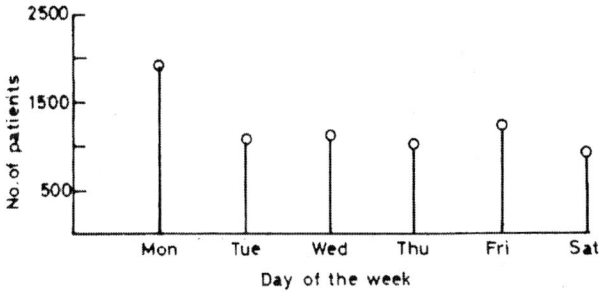

FIGURE 1.14. Line diagram showing the number of outpatients examined in different days in a week. (Data represent the average of a five-year period.) (*Source:* Cancer Institute, 1990.)

TABLE 1.10. The total production of food grains in each category in India in the year 2000-2001 and the corresponding angles for drawing the pie diagram.

Food grains	Yield (million tons)	Angle
Rice	86.8	157
Wheat	70.0	127
Coarse grains	29.9	54
Pulses	12.3	22
Total	199.0	360

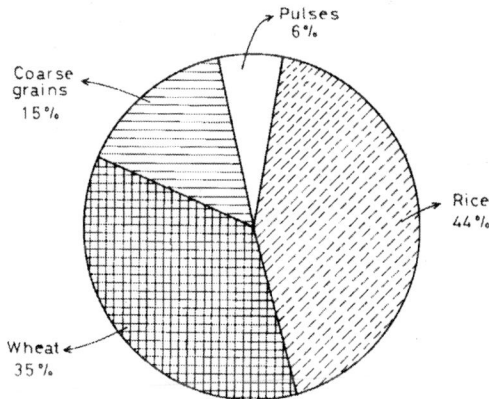

FIGURE 1.15. Pie diagram showing the different components of food grain production in India during 2000-2001. (*Source:* Anonymous, 2001.)

1990	2,000 cars	2 car symbols
1995	3,000 cars	3 car symbols
2000	7,000 cars	7 car symbols

The pictogram does not give a clear indication of the given relation. It is a representation of volume as in Figure 1.16. In a pictogram the subject matter is represented by figures, sketches, or diagrams.

Scatter Diagram

Researchers are interested in getting additional information from the data collected besides frequency distribution. For instance, a researcher may collect paired observations and wish to know the relationship between the two variables. A plot of paired values in a graph is called a scatter diagram.

For example, the grain number/plant and grain weight/plant in rice variety recorded in an experiment (Palaniswamy, 1990) are shown in Table 1.11 and exhibited in a scatter diagram (see Figure 1.17).

Zone Graph

Zone graphs are used to exhibit differences between minimum and maximum values in different years of a variable. For example, we can show average prices of a commodity over the years (see Table 1.12) as a graph (see Figure 1.18).

Scale: 1 car = 1,000 cars

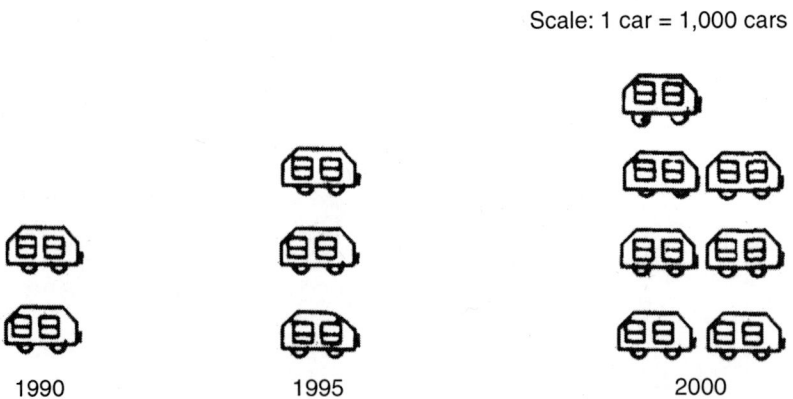

FIGURE 1.16. Pictogram showing the number of cars produced in different years in a company.

TABLE 1.11. The grain number/plant and grain weight/plant in rice variety recorded in an experiment.

Plant numbers	Character	
	Grain number/plant	Grain wt/plant (g)
1	972	22.7
2	429	10.5
3	765	18.3
4	530	12.8
5	973	22.2
6	1,091	25.9
7	917	22.3
8	272	6.5
9	1,274	29.5
10	454	10.7

Source: Palaniswamy, 1990.

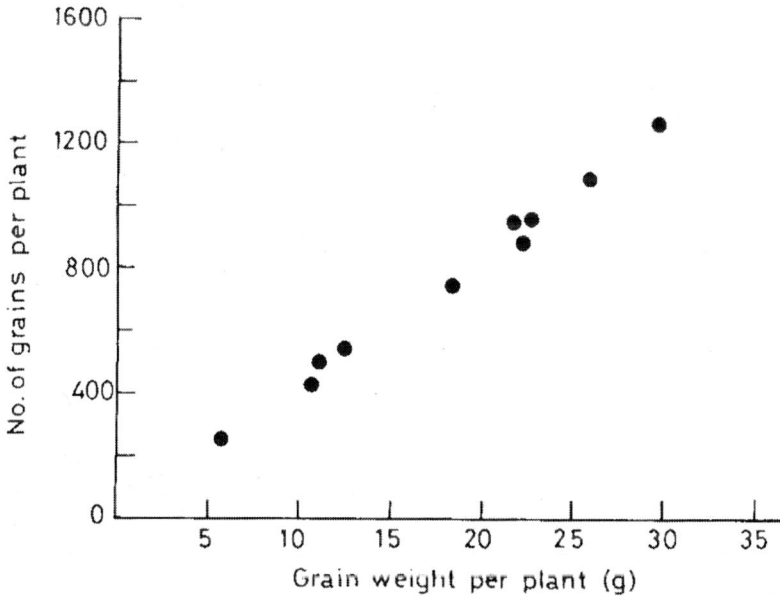

FIGURE 1.17. Scatter diagram showing the relationship between two variables (grain weight per plant and grain number per plant in rice crop).

Graphs of Time Series

One Variable

Values of a variable at different periods of time are shown in this type of graph. The time is shown along the x-axis. The points are plotted and joined by lines.

For example, the area and production of pulses in India (Misra and Gupta, 1999) are given in Table 1.13 and shown in a time series graph (see Figure 1.19).

Two Variables

A two-variable graph shows a time series with two variables having the same units. The crude incidence rate of stomach cancer in males and females in Madras City, India (Gajalakshmi and Palaniswamy, 1988), are presented in Table 1.14 and in Figure 1.20.

As another example, consider a graph of a time series with two variables having different units: wages in a company in countries A and B paid to the laborers/day as shown in Table 1.15 and in Figure 1.21.

More Than Two Variables Having the Same Units

Medical scientists are interested in knowing the crude incidence rate (CIR) for different diseases. For example, in cancer epidemiology studies, CIRs per 100,000 population are determined for different sites in males and

TABLE 1.12. Average prices of a commodity over the years 1980-1989.

Year	Price		
	Maximum	Minimum	Difference
1980	122	120	2
1981	123	118	5
1982	132	130	2
1983	134	128	6
1984	139	133	6
1985	135	125	10
1986	140	129	11
1987	143	133	10
1988	135	130	5
1989	139	129	10

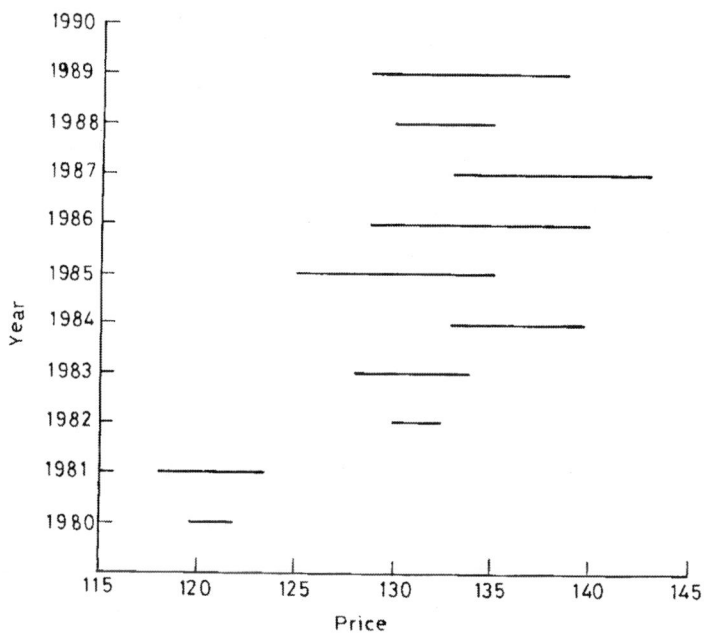

FIGURE 1.18. Zone graph exhibiting the differences between the maximum and minimum values. Average price of a commodity over the years 1980 to 1989.

TABLE 1.13. The area and production of pulses in India.

Year	Area (m ha)	Production (m tons)
1950-1951	19.1	8.4
1960-1961	23.6	12.7
1970-1971	12.5	11.8
1980-1981	22.5	10.6
1990-1991	24.7	14.3
1995-1996	23.8	24.1
1996-1997	15.0	15.5
1997-1998	24.4	14.5

Source: Misra and Gupta, 1999.

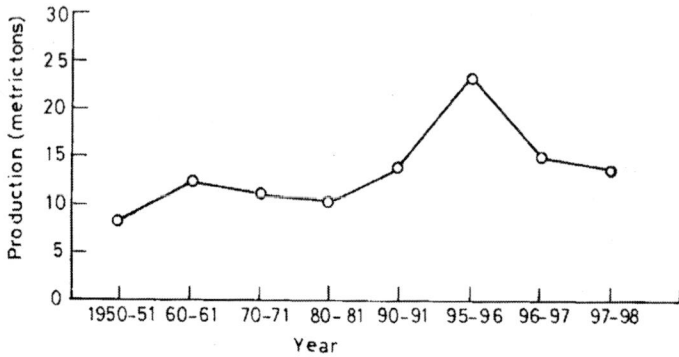

FIGURE 1.19. Graph showing production of pulses in different years in India.

TABLE 1.14. The crude incidence rate of stomach cancer in males and females in Madras City, India, per 100,000 people.

Year	Male	Female
1982	7.7	4.1
1983	9.0	4.4
1984	9.1	3.9
1985	9.8	4.2
1896	10.0	4.7

Source: Gajalakshmi and Palaniswamy, 1988.

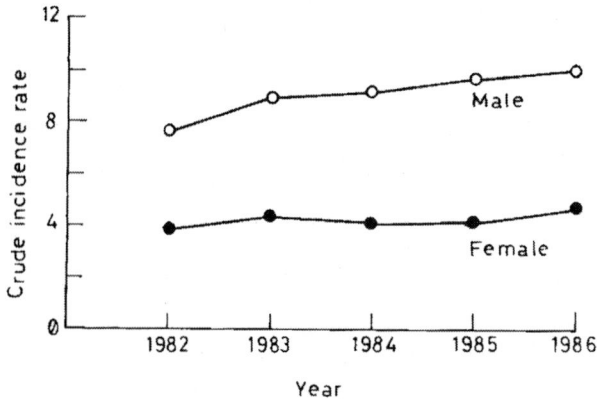

FIGURE 1.20. Graph of time series, two variables having the same units. Data represent the crude incidence rate of stomach cancer in males and females in Madras City, India. (*Source:* Gajalakshmi and Palaniswamy, 1988.)

females separately. The CIR in some sites in females are reported (Gajalakshmi and Palaniswamy, 1988) and data are given in Table 1.16 and shown in Figure 1.22.

Following is a comparison of tables and graphs:

Tables	**Graphs**
Tables give precise figures.	Only approximate figures can be read.
It is more difficult to interpret the data.	Data are easy to interpret.
More information can be shown.	Cannot display as much information in one graph.
They are not very impressive to view.	They are very impressive to view.
It takes time to formulate and verify figures.	They help to check computation.
It is not easy to determine the trend comparisons.	It is easy to determine the trend comparisons.
Suitable caption, sample size n, unit of measurement, standard deviation, standard error, etc., should be mentioned.	Same elements present as in tables.

TABLE 1.15. Wages in a company in two countries A and B paid to the laborers/day.

Country	1990	1991	1992	1993	1994	1995
A (in rupees)	120	130	150	180	190	200
B (in dollars)	60	65	68	70	80	90

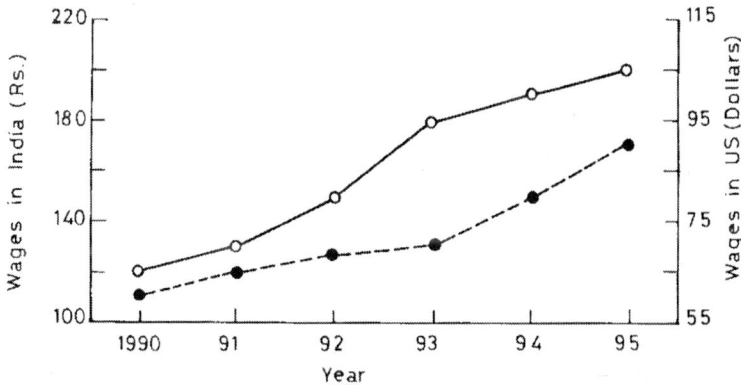

FIGURE 1.21. Graph of time series, two variables having different units. Data represent the daily wages in a company in two different countries (India and the United States).

TABLE 1.16. Crude incidence rate for leading cancer sites recorded 1982-1986 for females, per 100,000.

Site	1982	1983	1984	1985	1986	Average
Cervix	31.0	31.8	30.4	33.3	35.5	32.4
Breast	14.0	11.9	13.5	13.4	14.0	13.4
Oral cavity	6.6	6.4	5.0	5.7	6.2	6.0
Stomach	4.1	4.4	3.9	4.2	4.7	4.3
Leukemia	1.2	1.6	1.4	1.7	1.8	1.5

Source: Gajalakshmi and Palaniswamy, 1988.

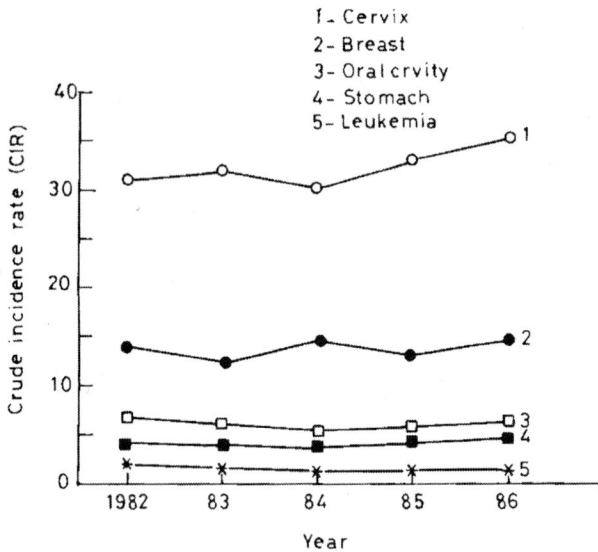

FIGURE 1.22. Crude incidence rate for cancer sites recorded in different years from 1982-1986 for females.

EXERCISES

1.1. The frequency distribution of the life (in hours) of a sample of 1,000 lightbulbs is as follows. Express the data graphically by a histogram, frequency polygon, and frequency curve.

Life (hrs)	700-749	750-799	800-849	850-899	900-949
Frequency	100	200	400	250	50

1.2. Illustrate graphically, from your specific field of study, a few examples for the study of relationships between two variables.
1.3. The incidence of measles (in percentages) was recorded for different age groups during their lifetimes in a village as shown in the following list.

Age (years)	Under 5	5-9	10-14	15-19	20-24	25-29
Percentage	15	45	62	75	84	87

  a. Draw a graph taking measles (%) along y-axis and age along x-axis.
  b. Draw a smooth curve through plotted points and interpret the frequency curve.
1.4. Graph of time series—two variables (having same units): Show the following data in a graph. Data relate to the average monthly rainfall and standard deviation in inches (average of ten years) on an experimental farm.

Month	Average rainfall (in)	Standard deviation (in)
July	0.02	0.09
August	0.01	0.02
September	0.23	0.58
October	0.79	1.02
November	1.78	1.86
December	3.48	2.90
January	3.76	2.82
February	3.10	2.21
March	2.47	1.84
April	1.47	1.84
May	0.70	0.81
June	0.14	0.26
Total	17.93	6.37

1.5. The yields (kg/ha) in the United States of a few crops in different years are given as follows. Show them in suitable diagrams.

	Yield (kg/ha)		
Crop	1992	1993	1994
Maize	8253	6321	8697
Rice	6429	6176	6719
Sorghum	4598	3760	4585
Barley	3382	3171	3026
Wheat	2635	2569	2526
Oats	2347	1950	2051
Pulses	1674	1636	1762

Source: FAO, 1994.

1.6. Food grain production in India in different years is shown in the following list. Exhibit these data in as many diagrams as possible.

Commodity	1995-1996	1996-1997	1997-1998	1998-1999	1999-2000	2000-2001
Food grains	180.4	199.4	192.3	203.5	208.9	199.0
Rice	77.0	81.7	82.5	86.0	89.5	86.8
Wheat	62.1	69.4	66.3	71.3	75.6	70.0
Coarse grains	29.0	34.1	30.4	31.2	30.5	29.9
Pulses	12.3	14.2	13.0	14.9	13.3	12.3
Oilseeds	22.1	24.4	21.3	24.7	20.9	18.6
Groundnut	7.6	8.6	7.4	9.0	5.3	6.2
Mustard	6.0	6.7	4.7	5.7	6.0	4.3
Soybean	5.1	5.4	6.5	7.1	6.8	5.2
Others	3.4	3.7	2.7	2.9	2.8	2.9

Source: Production of Food Grains and Oil Seeds 2000-2001, 2001.

1.7. Compare the use of tables and graphs listing their advantages and disadvantages.

1.8. How will you construct a pie diagram?

1.9. What are the advantages of a scatter diagram?

1.10. What is the difference between a horizontal bar diagram and a vertical bar diagram?

1.11. Illustrate with a suitable example how you will show two variables (different units).

1.12. What is (a) a line diagram? (b) a zone graph?

1.13. What is a subdivided bar diagram? Explain with a suitable example.
1.14. Express the following data about protein content of different pulse crops in a suitable graph: chickpea = 19.5 percent; pea = 24.9 percent; lentil = 24.8 percent; pigeon pea = 21.5 percent; mung = 25.3 percent; urd = 25.1 percent; and cowpea = 24.6 percent (Misra and Gupta, 1999).

REFERENCES

Cancer Institute (1990). Cancer registry. Annual report of Cancer Institute. Madras, India: Author.

Food and Agriculture Organization (1994). *FAO production year book.* Volume 48. pp. 68, 70, 75, 77, 80, 84, 97. Author.

Gajalakshmi, C.K. and K.M. Palaniswamy (1988). Incidence of cancer in Madras City. Cancer registry. *Abstract News Letter of the National Cancer Registry of India,* 3(2):7-8.

Misra, S.K. and D. Gupta (1999). Pathogen mediated disease in pulse crop—A major constraint and its solution. *India Grains,* 1(2):33-35.

Palaniswamy, K.M. (1990). Design of statistical field experiments and crop forecast in agriculture with special reference to rice. Doctoral thesis submitted to the University of Calicut.

Palaniswamy, U.R., J.D. Stuart, and C.A. Caporuscio (2002). Effect of storage temperature on the nutritional value of curry leaf. In J. Janick and A. Whipkey (Eds.), *Trends in new crops and new uses* (pp. 567-569). Alexandria, VA: ASHS Press.

Production of food grains and oil seeds 2000-2001 (2001). *India Grains,* 3(3):9.

Raghavarao, D. (1983). *Statistical techniques in agricultural and biological research.* New Delhi: Oxford IBM Publishing Co.

Shnayerson, M. (2002). The killer bug. *Fortune,* 146(6):150.

Usha Rani, P. (1996). Quality traits and their relation with other plant characters in chili (*Capsicum annuum* L.). *Mysore J. Agric. Sci.* 30:235-239.

Usha Rani, P. (1997). Seed recovery, seed production, seed prediction and variability studies in Chili. *Madras Agric. J.* 84(3):139-143.

Chapter 2

Review of Basic Mathematical Concepts Fundamental to Statistics

VARIABLES

Some basic concepts of mathematics are considered prerequisites to understand statistics. The subject of statistics deals with variables whose values vary from one to the other. Variation is studied using different statistical tools. For example, consider that an economist intends to study the income distribution in different families. It would be cumbersome to mention the characteristic frequently during the course of his or her investigation and hence it is imperative that we identify the variable by some algebraic notation. The variable under study (e.g., income) varies from one family to the other. Hence, income is called the variable, in statistical terminology. It is denoted by the letter x. If the incomes (in thousands of U.S. dollars) of the ten families are 4, 10, 5, 3, 20, 8, 7, 6, 7, and 10, then the incomes of the different families are referred to as follows:

$$
\begin{aligned}
x_1 &= 4 \text{ (income of the 1st family)} &\text{(read as sub 1)} \\
x_2 &= 10 \text{ (income of the 2nd family)} &\text{(read as sub 2)} \\
x_3 &= 5 \text{ (income of the 3rd family)} &\text{(read as sub 3)} \\
x_4 &= 3 \text{ (income of the 4th family)} &\text{(read as sub 4)} \\
x_5 &= 20 \text{ (income of the 5th family)} &\text{(read as sub 5)} \\
x_6 &= 8 \text{ (income of the 6th family)} &\text{(read as sub 6)} \\
x_7 &= 7 \text{ (income of the 7th family)} &\text{(read as sub 7)} \\
x_8 &= 6 \text{ (income of the 8th family)} &\text{(read as sub 8)} \\
x_9 &= 7 \text{ (income of the 9th family)} &\text{(read as sub 9)} \\
x_{10} &= 10 \text{ (income of the 10th family)} &\text{(read as sub 10)}
\end{aligned}
$$

In general the set of observations are denoted by $x_1, x_2 \ldots x_{10}$, and x_1 refers to the income of the first family, and so on. Generally, the size of the sample families is denoted by the letter n and the values by $x_1, x_2 \ldots x_i \ldots x_n$.

SUMMATIONS

If we are interested in finding the average income of a family, we add all the income values of the families and divide the total by the number of families:

$$\text{Mean} = (x_1 + x_2 + \ldots x_i \ldots x_{10})/10 \qquad (2.1)$$

The summation process is indicated by the notation Σ (sigma). The symbol Σx_i indicates the sum of all the values from 1 to 10; x_i indicates any value among the ten values. Symbol Σx_i indicates that the sum is from 1 to 10. Suppose we want to add the first six values. We would write Σx_i ($i = 1$ to 6) and for that add the values from 1 to 6, i.e., $4 + 10 + 5 + 3 + 20 + 8$. In

$$\sum_{i=1}^{10} x_i \qquad (2.2)$$

the i is called the index of summation and 1 to 10 are the limits of summation. Each family is termed an element. All ten elements constitute the sample.

If the economist wishes to study another variable, say expenditure, expenditure is y and the values of the elements $y_1, , \ldots y_n$. Similarly, data can be collected on another variable, e.g., number of members in the family, designated by the letter z as $z_1, z_2 \ldots z_n$. One can study the behavior of the data in pairs or in triplets. In this way, researchers generate their data from their experiments or surveys.

LOGARITHMS

The logarithm of a number to the base ten consists of two parts: (1) an integral part called the *characteristic,* and (2) a decimal part called the *mantissa.* For any number, the characteristic is written by inspection and the mantissa from the table of logarithms.

For example, take logarithms for three numbers: 47, 470, and 0.0047.

To find the logarithm, find the number 47 in the lefthand column of the log table. In that row and in that column headed by 0, we find the digit 6721. That is the mantissa for all the numbers whose significant digits are 47. After prefixing the characteristics, the log of the number will be

Log 47 = 1.6721
Log 470 = 2.6721
Log 0.0047 = −3.6721 (−3 + 0.6721)

Similarly, logarithms can be taken for any number.

Examples of Problems Using Logarithms

Find the Value for the Problem [(43.54 × 8.512)]/17.121

Let k = [(43.54 × 8.512)]/17.121
Log k = log [(43.54 × 8.512)/(17.121)]
= log 43.54 + log 8.512 − log 17.121
= 1.6387 + 0.9300 − 1.2335 = 1.3352
k = antilog 1.3352 = 21.64

The previous problem involves multiplication and division using logarithm.

Taking Root of a Number

Solve $5\sqrt{12.709}$
Let $x = 5\sqrt{12.709}$
Log x = Log $[5\sqrt{12.709}]$
= log $(12.709)^{1/5}$
1/5 log 12.709
= 1/5 × 1.1041 = 0.2208
x = antilog 0.2208
= 1.662

Solve $\dfrac{(33.52)^3 \times (6.52)^4}{(68.3)^5 \times (10.12)^{1/3}}$

Let $x = \dfrac{(33.52)^2 \times (6.52)^4}{(68.3)^5 \times (10.12)^{1/3}}$
= $3 \log 33.52 + 4 \log 6.52 − 5 \log 68.3 − 1/3 \log \times 10.12$
= $3 \times 1.5253 + 4 \times 0.8142 − 5 \times 1.8344 − 1/3 \times 1.0051$
= $4.5759 + 3.2568 − 9.1720 − 0.3350$
= $7.3827 − 9.5070$
= $−2.1243 = −3 + 0.8757 = 0.007511$

SQUARES AND SQUARE ROOTS

To find the square of a number, multiply the number by itself. For example,

Square of $16 = 16 \times 16 = 256$

Square Root of a Number

Example: Find the square root of 116964.

```
        3.4 2.
      _____
    3)116964
      9
      _____
    64)269
       256
       _____
       682)1364
            1364
```

Answer = 342

Find square root of 54002.

```
        232.38
      _____
    2)54002.00
      4
    43)140
       124
    462)1100
         924
    4643)17800
          13929
    46468)387100
           371744
            15356
```

Answer = 232.4 (corrected to first decimal)

Example: Find the square root of 0.5678.

```
        0.7535
    7)0.5678
        0.49
    145)778
        725
    1503)5300
        4509
    10565)79100
        75325
    150702)377500
           301404
            76096
```

Answer = 0.7535

Example: Find the square root of 0.00936.

```
        0.09674
    9)0.0093600
        81
    186)1260
        1116
    1926)14400
        13489
    19344)91100
           77376
           13724
```

Answer = 0.0967(corrected to fourth decimal)

QUANTITIES

Researchers collect observations from their experiments, and they constitute data. The data are classified into two groups: *quantitative* and *qualitative*. A quantity is one upon which the four mathematical operations, namely addition, subtraction, multiplication, and division, can be performed. Quantity is of two types: *constant quantity* and *variable quantity*.

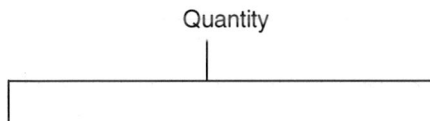

Quantity

```
        Constant                              Variable
            |                                    |
   ┌────────┴────────┐              ┌────────────┴────────────┐
Absolute    Arbitrary constant   Independent variable    Dependent
constant                                                  variable
```

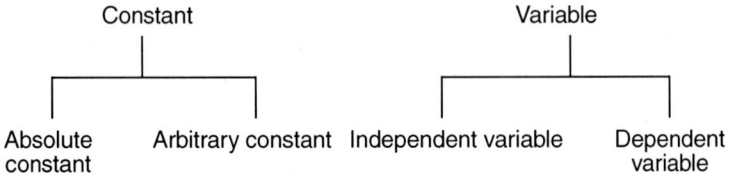

A *constant* is a quantity that possesses the same value or whose value remains unaltered throughout a mathematical operation. For example, 3, 9, 20, and 36 are constants.

An *absolute constant* has the same value in all kinds of mathematical investigations, e.g., π (or pi, having the value 3.1429).

An *arbitrary constant* takes one value in one investigation and another value in another investigation. Examples include $y = a + bx$ (a straight line), $y = a + bx + cx^2$ (quadratic curve).

A *variable* is a character which takes different values in a particular problem or which does not retain the same value during mathematical operations and takes different values. Examples include incomes of families, ages of persons in a family, number of patients in different hospitals, crop yields in different farms, and rainfall in a given place.

An *independent variable* is the variable whose value does not depend upon any other variable, e.g., fertilizer levels in an experiment, income of persons. It is usually determined by the researcher in an experiment.

A *dependent variable* is the variable whose value depends upon any other variable, which may be assigned any values, e.g., $A = l \times b$ (A = area; l = length; b = breadth; A depends on l and b); $y = bx$ (y = grain yield; x = leaf area of the crop; grain yield depends upon the leaf area of the crop).

PERMUTATION AND COMBINATIONS

Combinations

If we write 3!, it means "3 factorial" and it is equal to $3 \times 2 \times 1 = 6$. In general, if we write $n!$ it is read as "n factorial." Following are examples of the application of the principles of permutation and combination in statistical experiments.

Example 1

Suppose a cancer specialist has four patients: one, two, three, and four. The specialist wants to select three patients for a new treatment regime.

How can the specialist select three patients out of the four patients? The possible choices available to the specialist are four groups—Group1: Patients 1, 2, 3; Group 2: 1, 2, 4; Group 3: 2, 3, 4; and Group 4: 3, 4, 1. These selections are called "combinations of four patients taken three patients at a time." The number of combinations available for selection is four. It is denoted by the symbol $^4C_3$, meaning "selecting three out of four." Mathematically, it is written in a general way as ^nC_r, where n = total number of things available for selection and r = number of things selected out of n at a time.

Example 2

A plant pathologist wants to conduct an experiment in pots. In one treatment, the pathologist has six pots and needs to select two pots for the detailed study. The pathologist can select two pots in $^6C_2 = (6 \times 5) \div (2 \times 1) = 15$ ways.

Permutations

Permutations are particular arrangements of a given set of elements.

Example 1

Suppose we have $n = 4$, say 1, 2, 3 and 4, $r = 3$. We get $^4C_3 = (4 \times 3 \times 2)/(3 \times 2 \times 1) = 4$ combinations. They are 1, 2, 3; 1, 2, 4; 2, 3, 4; and 1, 3, 4. In this example, the first combination, namely 1, 2, 3, can be arranged as 1, 2, 3; 1, 3, 2; 2, 3, 1 ; 3, 1, 3; 3, 1, 2; and 3, 2, 1. Thus in each combination, we can have six different arrangements. Thus, for four combinations, there will be $4 \times 6 = 24$ arrangements.

Example 2

Suppose there are six treatments in an experiment and three replications. The experimenter can arrange the six treatments in the first replication in the 6! ("6 factorial") ways, i.e., $1 \times 2 \times 3 \times 4 \times 5 \times 6 = 720$ ways. Similarly, in II and III replications, we can arrange in 720 ways each. Thus, on the whole, the experimenter has $720 \times 720 \times 720 = 373,248,000$ possible arrangements of treatments in the conduct of the experiment. The experimenter finally selects one among the possible arrangements and conducts the experiment. This phenomenon is amazing and interesting, and emphasizes the importance and significance of the allotment of treatments in an experiment. Researchers have an enormous number of choices available to them in the selection of treatments for their experiments. Hence, random allotment of treatments will be the most efficient way to achieving this objective. Generally, if there are n objects, and if we take r of them at a time, then the number of arrangements or permutations formed can be arrived by using the formula $^nP_r = n(n-1) \ldots (n-r+1) = n!/(n-r)!$. If there are n objects, and if we take r of them at a time, then the number of combinations formed will be equal to $^nC_r = n!/r!(n-r)!$.

Example 3

In a hospital there are 20 patients. If the physician wishes to select two patients for an experiment, the physician can select any two patients in $^{20}C_2$ $= 20 \times 19/2 \times 1 = 190$ ways.

EXERCISES

2.1. Find the squares of the following numbers using logarithms: (a) 145, (b) 2555, (c) 0.52, (d) 0.00589, (e) 245.23.

2.2. Solve the following problems using log tables: (a) 123×234, (b) 3003×2502, (c) 55.05×55.05, (d) $(345 \times 567)/235$.

2.3. Find the square root of the given numbers: (a) 567, (b) 5986, (c) 0.5986, (d) 0.05986.

2.4. Define with examples the following terms: *variable, constant, independent variable, dependent variable*.

2.5. How will you classify quantity?

2.6. There are 100 patients in a hospital. You have to select five patients for a particular treatment. In how many ways can you select the patients?

2.7. Write down the combinatorial formula and explain.

Chapter 3

Nature of Statistical Data

RAW DATA

The data collected by researchers for analysis and drawing inferences and conclusions are called *raw data*. Examples follow.

1. Populations in various locations
2. Names and expenditures of different people
3. Grain yield/ha of different farms
4. Grade point average of the students in a school
5. Rainfall distribution in a place
6. Sales in a department store on different days of a month
7. Import and export of different commodities of a country
8. Daily maximum or minimum temperature in a place in different months
9. Number of patients admitted in a hospital during different days of a week/month/year
10. Height or weight of students
11. Number of students admitted to a college according to their nationalities
12. Farm size of different farms in a district
13. Milk yield of a cow on different days of a month in a dairy

Characteristics of the data in these examples all include observations in each category that exhibit *variation* or *variability*. The characteristics that show variation are hence called *variables*.

CLASSIFICATION OF DATA

Broadly, statistical data are classified into *quantitative* data or *qualitative* data.

1. Quantitative data examples: number of students in a college, number of patients in a hospital, GPA of students, income of people, heights and weights of students, rainfall distribution in a place
2. Qualitative data examples: students in a university according to their nationality, people described as honest or dishonest, taste of food described as good, better, sour, etc.

Data may also be classified based on the nature of observations or measurements as *continuous* or *discrete*.

1. Continuous data examples: income, expenditure, rainfall, farm output, height, weight, blood pressure
2. Discrete data examples: number of patients in a hospital, number of people in a family, number of students in different colleges or universities, number of diseased plants in an experimental plot

Besides these types, we also note other types of data in experiments such as the following:

1. *Chronological data:* This type of data shows the trend of observations in different times, e.g., production of food crops in a country in different years, import of a commodity in a country in different years, rainfall distribution in different years in a place.
2. *Geographical data:* When we collect data according to geographic regions, we get geographical data, e.g., population in different countries, forest area in different places.

EXERCISES

3.1. What is raw data? Give examples.
3.2. How will you classify the statistical data? Give some examples for each.
3.3. Define the terms *continuous data, discrete data, quantitative data,* and *qualitative data.*

Chapter 4

Measures of Central Tendency

Scientists working in various disciplines such as agriculture, horticulture, medicine, economics, industry, sociology, and other fields conduct various types of experiments or surveys and collect data following scientific principles and procedures to get reliable and valid information about the variables of their study. The data thus collected may relate to any character or variable such as income of individuals, number of members in a family, crop yield/ha, exchange rates, soil fertility levels, blood pressure, height, weight, age, weight of child at birth, number of students admitted in different years in a university. The observed variables exhibit variation or variability among themselves. Each set of observations generates a distribution, and we have to study the nature of such distributions in order to arrive at inferences. Often, observations cluster or concentrate around a particular value or a representative value or in the middle of the range. Values tend to approach a representative value. Such a phenomenon is called study of central tendency. Central tendency gives us the description of the average.

By central tendency, we infer the following:

1. We understand that a central value represents the center of the distribution.
2. The observations are spread around the central values and the spread or dispersion extends over a range of values.
3. The nature of the spread gives us some information about the shape of the curve.

The values that measure central tendency of a set of observations are called measures of central tendency. The three common measures of central tendency are mean, mode, and median.

MEAN

Mean is the representative value of the whole group of data. It is the most common measure of central tendency. The mean reveals the principle characteristic of the population. The population mean is denoted by the symbol μ (the Greek letter mu) and the sample mean by $\bar{x}$ (read as "x bar"). The population mean is a fixed quantity and is not subject to variation, whereas a sample mean is a variable quantity, since we can take different samples for a given size and each one gives different means.

The arithmetic mean is similar to the center of the moment, the balance point in a solid block. We seldom know the value of the population mean, μ. The sample mean, $\bar{x}$, becomes a better estimate of μ. As the sample size n increases, $\bar{x}$ approaches μ, i.e., $\bar{x} \rightarrow \mu$.

Computation of Population Mean

The mean of the observations is defined as the sum of the observed values divided by their number. Let X denote a variable of the population and its variate values $X_1, X_2, \ldots X_N$ where N indicates the size of the population. Thus

$$\mu = \frac{X_1 + X_2 + \ldots + X_N}{N} = 1/N \sum_{I=1}^{N} X_I \qquad (4.1)$$

where μ is the population mean, N the number of observations in the population size and $X_1, X_2, \ldots X_N$ are the variate values in the population and Σ (Greek letter sigma). The symbol Σ tells us to add all values of X ranging from $1, 2, \ldots i \ldots N$ for all N observations.

Computation of Sample Mean

Take a random sample from the population. Let n be the size of the sample and $x_1, x_2, \ldots x_n$ be the values. The sample mean is obtained by adding the values of the items and then dividing the sum by the number of items. Symbolically,

$$\bar{x}\ (\text{sample mean}) = \frac{x_1 + x_2 + \ldots + x_n}{n} = \frac{\Sigma x_i}{n} \qquad (4.2)$$

where Σ is the sum obtained after adding all the n values.

For example, a sample of five farms recorded 8, 8, 9, 11, and 14 t/ha yield of grain in rice crop. The mean grain yield/farm $(\bar{x})$ is $(8 + 8 + 9 + 11 + 14/5)$ = 50/5 = 10 t/ha. Here, the variable under study is the grain yield/ha; its mean is denoted by $\bar{x}$ and $\bar{x}$ is computed from a sample and is therefore referred to as an estimate. A statistic is an estimate of the population parameter.

Properties of the Mean

1. The sum of the deviations of the observations from the mean is zero. For example,

$$\Sigma(x_i - \bar{x}) = 0.$$
$$\begin{aligned}\Sigma(x_i - \bar{x}) &= (x_1 - \bar{x}) + (x_2 - \bar{x}) + \ldots + (x_n - \bar{x}) \\ &= x_1 + x_2 + \ldots + x_n - \bar{x} - \bar{x} \ldots \\ &= \Sigma x_i - \Sigma \bar{x} \\ &= \Sigma x_i - n\bar{x} \\ &= \Sigma x_i - \Sigma \bar{x} = 0 \text{ as } (\Sigma x_i / n = \bar{x} \text{ and } n\bar{x} = \Sigma x_i)\end{aligned}$$

Numerical Example

Consider the values mentioned, i.e, 8, 8, 9, 11, 14:

$$\begin{aligned}\bar{x} &= (8 + 8 + 9 + 11 + 14)/5 = 50/5 = 10 \\ \Sigma(x_i - \bar{x}) &= (8-10) + (8-10) + (9-10) + (11-10) + (14-10) \\ &= (-2) + (-2) + (-1) + (+1) + (+4) = (-5) + (+5) = 0\end{aligned}$$

The values $(x_1-\bar{x})$, $(x_2-\bar{x})$, ... i.e., $(8-10) = -2$, are called deviations of x values from the mean $\bar{x}$.

2. The sum of squares of deviations of the observations from the mean is minimum. It also means that the sum of the deviations from any other value is larger.

Example

Given the observations as 8, 8, 9, 11, and 14, with a mean 10, $\Sigma(x_i - \bar{x})^2$ is minimum.

Observed values	Deviation from the mean (d_i)	Squares of the Deviations (d_i^2)
8	8 − 10 = −2	4
8	8 − 10 = −2	4
9	9 − 10 = −1	1
11	11 − 10 = 1	1
14	14 − 10 = 4	16
Total	$\Sigma d_i = 0$	$\Sigma d_i^2 = 26$

Suppose we take deviations from any other value, say 5. Then we get the following:

Observed values	Deviation from the mean (d_i)	Squares of the deviations (d_i^2)
8	$8 - 5 = 3$	9
8	$8 - 5 = 3$	9
9	$9 - 5 = 4$	16
11	$11 - 5 = 6$	36
14	$14 - 5 = 9$	81
Total	$\Sigma\, d_i = 25$	$\Sigma d_i 2 = 151$

Thus $151 > 26$

Proof: $\Sigma\,(x_i - a)^2 = $ minimum

Let a be any point from which deviations are measured. Differentiating with respect to x and make it equal to 0, we get

$$-2\,\Sigma\,(x_i - a) - 1 = 0$$
$$-2\,\Sigma\,(x_i - a) = 0$$
$$\Sigma\,(x_i - a) = 0$$
$$\Sigma x_i - \Sigma a = 0$$
$$\Sigma x_i = \Sigma a \ (\text{because } \Sigma x_i = na \text{ and } a = \Sigma x_i/n = \bar{x})$$

3. Mean may be computed for any number of values.
4. Mean is affected by each and every value.

Example

Given the values 3, 4, 5, 7, 11, mean $= (3 + 4 + 5 + 7 + 11)/5 = 30/5 = 6$. Suppose we add 15 to the value 11 and find the mean. Now the given values are 3, 4, 5, 7, and 26. Then the mean $= (3 + 4 + 5 + 7 + 26)/5 = 45/5 = 9$. The new mean is now increased from 6 to 9. When the number 15 was added, it was distributed equally among all the values by an amount equal to the amount of its average, 3 ($15/5 = 3$). When 3 was added to each original value, we get 6, 7, 8, 10, and 14. The average of the numbers $(6 + 7 + 8 + 10 + 14)/5 = 45/5 = 9$, i.e., the new mean is increased by 3 from 6 to 9.

5. If each value of a distribution is multiplied by a constant, then the new mean is obtained by multiplying the old mean by that constant.

Example

Given 3, 4, 5, 7, 11, then $\bar{x}$ is 6. Multiply each value by a constant, say 2, then we get 6, 8, 10, 14, 22. The new mean is $(6 + 8 + 10 + 14 + 22)/5 = 60/5 = 12$ which is equal to the product of the original mean and the constant 2.

6. The variance of a population is defined as

$$\sigma^2 = \Sigma(x_i - \mu)^2 \ / \ N. \qquad (4.3)$$

If we know μ then the best estimate of σ from a sample is

$$s^2 = \Sigma(x_i - \mu)^2 \ / \ n$$

where n = number of variates in the sample. We do not know μ but we estimate it by a sample as $\bar{x}$. The sample mean is seldom equal to the μ. We know

$$\Sigma(x_i - \bar{x})^2$$

is lesser than the sum of squares of deviations from any value other than $\bar{x}$. Thus

$$\Sigma(x_i - \bar{x})^2$$

is lesser than

$$\Sigma(x_i - \mu)^2, \text{ meaning that } \Sigma(x_i - \bar{x})^2 \ / \ n$$

will give smaller estimate of σ^2. It is corrected by using $n-1$ in the denominator instead of n. On the average,

$$\Sigma(x_i - \bar{x})^2 \ / \ n - 1 = \Sigma(x_i - \mu)^2 \ / \ n = \sigma^2.$$

7. For a given set of observations there will be only one mean. Hence mean is a unique value in our observations.
8. The extreme value or values in a set of observations will affect the mean considerably. For example, if there are four observations recorded in an experiment (4, 5, 6, 5), then the mean is 5. If the four observations are recorded as 4, 5, 6, 25, then the mean is (4 + 5 + 6 + 25)/4 = 40/4 = 10. The mean 10 cannot be considered a representative value.
9. The arithmetic mean may take any value not observed in the original set of values.
10. The mean of the sample provides an unbiased estimate of the mean of the population from which the sample was drawn.
11. Mean is perhaps the most familiar measure of central tendency value.

The following sections show shortcuts to compute mean.

Assumed Mean Method

Sometimes, we find the variable takes values of more or less equal magnitude and under such situations, a shortcut method can be followed.

Example

Ginning percent in a cotton variety on 25 samples are observed as follows:

30	34	34	32	35
35	31	31	30	33
31	31	33	33	31
32	32	30	32	35
32	26	33	35	32

The mean = $\Sigma x_i / n$ = 803 / 25 = 32.12

Formula for Computing Mean

$$\text{Mean: } AM + \Sigma d_i/n \tag{4.4}$$

where AM = assumed mean and Σd_i = sum of the deviations of the values from the assumed mean and n = number of observations. Let AM = 30.
Then deviations are as follows:

0	4	4	2	5
5	1	1	0	3
1	1	3	3	1
2	2	0	2	5
2	−4	3	5	2

Σd_i = 53. Mean = 30 + (53/25) = 30 + 2.12 = 32.12

Calculation of Mean from Grouped Data

In a frequency table, we have class limits, midvalues, and class frequencies. We do not find the original observations. It assumes that the midvalue in each class occurs as many times as the number of frequencies in the class. Then the mean is given by

$$\bar{x} = \Sigma f_i x_i / \Sigma f_i \qquad (4.5)$$

Example

Let us recall the grain weight/plant data:

Class limits	Midvalue (x_i)	Frequencies (f_i)	$f_i x_i$
5-10	7.5	1	7.5
10-15	12.5	3	37.5
15-20	17.5	4	70.0
20-25	22.5	9	202.5
25-30	27.5	13	357.5
30-35	32.5	12	390.0
35-40	37.5	4	150.0
40-45	42.5	2	85.0
45-50	47.5	2	95.0
Total		50 (Σf_i)	1395.0 ($\Sigma f_i x_i$)

$\bar{x} = \Sigma f_i x_i / n = 1395.0/50 = 27.90$

Deviation Method

$$\bar{x} = AM + (\Sigma f_i d_i / \Sigma f_i) \times i \qquad (4.6)$$

where AM = assumed mean; f_i = frequency of the *i*th class; d_i = deviation in the *i*th class, i = class interval; $\Sigma f_i = n$

Midvalue	Frequency	Let $AM = 27.5$ $i = 5$	$f_i d_i$
7.5	1	-4	-4
12.5	3	-3	-9
17.5	4	-2	-8
22.5	9	-1	-9
27.5	13	0	0
32.5	12	1	12
37.5	4	2	8
42.5	2	3	6
47.5	2	4	8
Total	50		$\Sigma f_i d_i$ (-30) + (+ 34) = 4

Mean = $AM + (\Sigma f_i d_i / n) \times i = 27.5 + (4/50) \times 5 = 27.5 + 0.4 = 27.90$

Step Deviation Method

The following is the frequency table for a set of observations of $n = 220$. In this case, the step deviation method is explained ($I = 10$, $n = 220$).

class	f_i	Midvalue (x_i)	$d_i = (x_i-AM)/i$	f_id_i
0-10	42	5	-3	-126
10-20	44	15	-2	-88
20-30	58	25	-1	-58
30-40	35	35	0	0
40-50	26	45	1	26
50-60	15	55	2	30
Total	220			$(-272) + (56) = -216$

Assumed mean $= 35, \bar{x} = AM = 35 + (\Sigma f_id_i/\Sigma f_i)i/ = 35 + (-216/220) \times 10 = 35 - 9.8 = 25.2$

Some Characteristics of the Mean

All observations in the data are involved in the calculation of the mean. In some situations, the mean is affected by extreme values and under such circumstances mean will not be a representative value for the data. Mean may or may not be one among the observations. It is commonly computed in all investigations. It usually occurs the maximum number of times. The important feature is that it is amenable for further statistical analysis. It is centrally located.

Why Mean Alone Should Not Be Studied

Consider the following three sets of data, each with eight observations.

Set	Observations								Total	Mean
1	28	28	30	32	32	30	31	29	240	30
2	50	51	32	29	30	28	8	12	240	30
3	20	17	28	20	12	28	48	67	240	30

In all the three sets, the mean was 30. In Set 1, the range of values varied from 28 to 32 with a range of 4. In Set 2, the range $(51 - 8) = 43$; the mean is not a representative value. Likewise, the range in Set 3 is 55 $(67 - 12 = 55)$. Thus, the three sets vary in respect to variability or dispersion though all sets have an equal mean 30. Therefore, variability should also be studied in addition to mean.

Weighted Average

Example 1

Suppose a student reports the grades in different subjects:

Subject	Credit (w_i)	Grade (z_i)
A	1	4
B	2	3
C	2	4
D	3	2
Total	8	

The student's grade point average is

$$\bar{x} = \Sigma w_i z_i \,/\, \Sigma w_i, \qquad (4.7)$$

i.e., $[(1 \times 4) + (2 \times 3)+(2 \times 4)+(3 \times 2)]/(1 + 2 + 2 + 3)=24/8 = 3.00$. This is called weighted average. In this case, importance is given to the number of credits each subject carries. The number of credits is the weight. The student's ordinary average = total grades/number of subjects = $(4 + 3 + 4 + 2)/4$ = $13/4 = 3.25$.

Example 2

A shop sells three types of shirts with different prices. What is the average price of a shirt?

Shirt type	Price/shirt (x)	Number sold (w)	$w_i z_i$
A	$ 50	120	6,000
B	$100	50	5,000
C	$200	20	4,000
Total		190	15,000

We must compute the number of shirts sold in each type and the weights of each.

$$\text{Weighted mean} = \Sigma w_i z_i \,/\, \Sigma w_i = 15,000 \,/\, 190 = 78.9$$

Example 3

Suppose there are two factories, A and B. The monthly wages paid by them are $4,500 and $3,000, respectively. The average wages paid is ($4,500 + $3,000)/2 = $7,500/2 = $3,750. This average is correct if the two factories employ equal numbers of laborers. Instead, if factory A has 50 laborers and factory B has 20, then the average paid is computed by giving weights to the number of laborers working in the two factories. Thus, the weighted average is calculated as (50 × 4,500) + (20 × 3,000)/(50 + 20) = 280,000/70 = $4,071.40.

Geometric Mean

When we deal with variables that show geometric progression or exponential law, geometric mean (GM) is computed. A series of numbers, each of which is formed by multiplying the previous one by a constant number, is called geometric progression.

Example 1

For instance, let's consider observations (a) 1, 2, 4, 8, 16, 32, 64 (obtained by multiplying by a constant number), and (b) 1.20, 1.44, 1.728, 2.0736 (obtained by increasing by 20 percent). Geometric mean is used when we deal with growth rates, interest rates, price indices, etc. The GM is also obtained when dividing the previous number by a constant number instead of multiplying the previous number.

Example 2

Consider the following data obtained from an experiment: 32, 16, 8, 4, 2. In such a situation, the mean will be

$$\Sigma x_i \, / \, n = (32 + 16 + 8 + 4 + 2) \, / \, 5 = 62 \, / \, 5 = 12.4.$$

If we consider the arithmetic mean as our value of central tendency, it will be biased toward the largest number in the series. Diagrammatically, it is represented in Figure 4.1. Note the bend in the curve. Now plot the data in the log scale.

Original data	Log values
32	1.5051
16	1.2041
8	0.9031
4	0.6021
2	0.3010
62	4.5154

The results are arithmetic mean = 4.5134/5 = 0.9031, antilog = 8.000.

When we plot the data of GM in log scale, the curve becomes a straight line (see Figure 4.2). The GM is 4.5134/5 = 0.9031 and the antilog = 8.000. In general, the GM of n numbers is defined as the nth root of the product of n observations, i.e.,

$$GM = n\sqrt{(x_1, x_2, \ldots x_n)} \qquad (4.8)$$

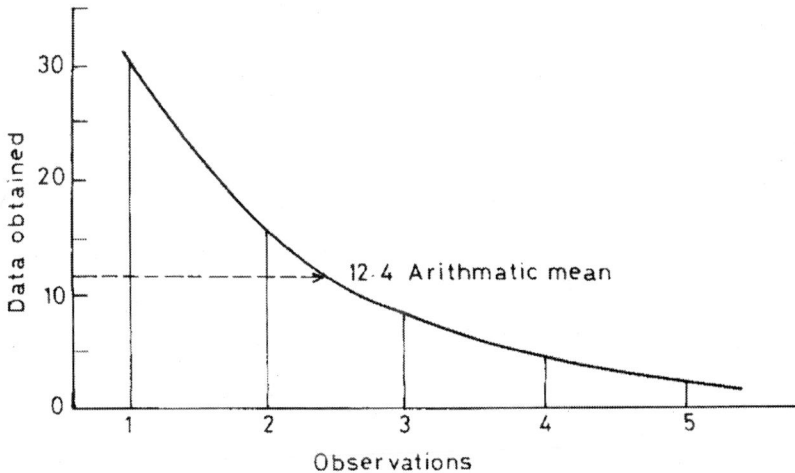

FIGURE 4.1. Position of arithmetic mean when data are in geometric progression.

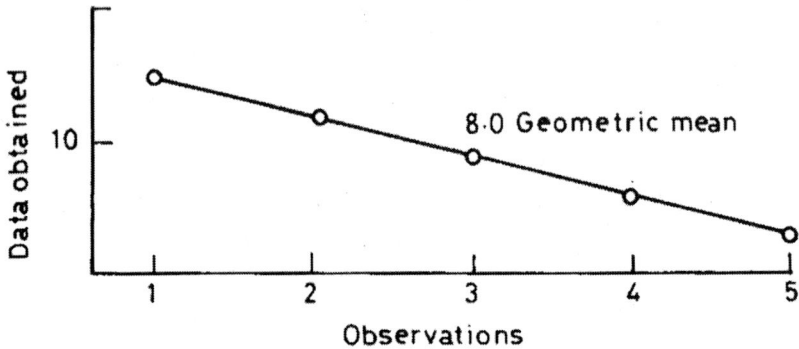

FIGURE 4.2. Position of geometric mean when data are in log.

Example 1

Consider the values 2, 4, 8, 16, 32. If we calculate AM for these data, it may not be a representative or central value as the figures are in geometric progression.

$$GM = \sqrt[5]{(2 \times 4 \times 8 \times 16 \times 32)} = (2 \times 4 \times 8 \times 16 \times 32)^{1/5}$$

$$\text{Log GM} = 1/n(\Sigma \log x_i)$$
$$= 1/n(\log 2 + \log 4 + \log 8 + \log 16 + \log 32)$$

$$= 1/5(0.3010 + 0.6021 + 0.9031 + 1.2041 + 1.5051)$$
$$= 1/5(4.5154) = 0.9031, \text{ the antilog } (0.9031) = 8.0$$

(geometric mean).We find that AM shows a bias toward the largest number of the series.

Example 2

The population in a place in 1960 was 250,000; in 1970 it was 490,000. What would be the population in 1965? In this situation, population increased proportionally and hence GM will be the appropriate central value. If the population increases with the same number, then the AM = 250,000 + 490,000/2 = 370,000. The GM $= 2\sqrt{250,000 \times 490,000} = 350,000$; the AM (370,000) is higher than the GM (350,000). In general, the AM is always higher than the GM.

Disadvantage of Geometric Mean

We cannot compute the GM for data that contain a zero value. Following is an example of GM for grouped data.

Class interval	f	x_i	$\log x_i$	$f \log x_i$
0-10	1	5	0.69897	0.69897
10-20	3	15	1.17609	3.52827
20-30	4	25	1.39794	5.59176
30-40	2	35	1.54407	3.08814
Total	10			12.90714

$$\text{Log GM} = 1/n\Sigma f \log x = 1/10 \times 12.90714 = 1.290714$$
$$\text{GM} = \text{antilog } 1.290714 = 15.53$$

Merits and Demerits of GM and Its Uses

1. It involves all observations.
2. It is difficult to compute.
3. If any item records a zero value, GM cannot be computed.
4. GM cannot be computed for negative values.
5. It is used when we give more weight to small numbers.
6. GM is less than AM.
7. It is useful when we deal with ratios or rates as in index numbers and values change proportionally.

Example

To calculate rate of increase in the population, we use the following formula:

$$Pn = Po\left(1 + \frac{r}{100}\right)^n \tag{4.9}$$

where Po = population in the beginning; Pn = population at the end of the period; r = rate of increase/100; and n = number of years in the period. Population at the end of n years is

$$Pn = Po\left(1 + \frac{r}{100}\right)^n$$

$$\left(1+\frac{r}{100}\right)=\left(\frac{Pn}{Po}\right)^{1/n}=n\sqrt{\frac{Pn}{Po}}$$

$$\text{arithmetic mean} = \left(\frac{Pn-Po}{n \times Po}\right) \times 100$$

Davies Test for Logarithmic Distribution

Under this test, the coefficient of skewness is calculated as

$$\frac{\log LQ + \log UQ - (2 \log MQ)}{\log UQ - \log LQ} \tag{4.10}$$

where LQ = value of the lower quartile; MQ = value of the middle quartile; UQ = value of the upper quartile. If the coefficient is more than 0.20, the data are symmetrical enough to use arithmetic mean. The Davies test is suitable when the data are asymmetrical in distribution; or when n is at least 50, because quartile values are unreliable when n is small.

MEDIAN

The median is also an important measure of central tendency. It is the middle most or most central value of a set of numbers. Median is defined as that value which divides the observations into two equal halves when arranged in either descending or ascending order. When the number of observations (N) is odd, then $(N + 1)/2$th term will represent the median value. When N is even, the average of the values of the terms $N/2$ and $(N + 2)/2$ will be the median.

Example

When N is Odd

The marks scored by seven students are 85, 91, 80, 79, 97, 89, 92. The values after ranking in ascending order are as follows:

Term	Marks (values)
1	79
2	80

3	85
4	89
5	91
6	92
7	97

The $(N+1)/2$th term is $(7+1)/2 = 4$th term. Therefore the median is 89. There are equal number of observations, i.e., 3 above and below the median. The median is not changed when the values 91, 92, and 97 are replaced by any values.

When N *Is Even*

The values are now 85, 91, 80, 79, 97, 89. Arrange in ascending order as before.

Term	Value
1	79
2	80
3	85
4	89
5	91
6	97

Steps: Find $N/2$ and $(N+2)/2$th terms, i.e., $6/2$th and $8/2$th terms, i.e., 3rd and 4th terms. The median is the mean of the observations, namely 85 and 89. Median is therefore $(85+89)/2=174/2=87$. When N becomes large, ordering of observations is time-consuming. When N is large, data are grouped and then median is computed.

Grouped Data

Example

Suppose the grouped data in respect to daily wages of 11 laborers in a factory are as follows:

Class limit	Midvalue (x_i)	f_i	Cumulative frequencies (cf)
90-92	91	3	3
92-94	93	2	5
94-96	95	3	8

96-98	97	0	8
98-100	99	3	11
Total		11	

The median in the grouped data is the value of that term whose cumulative frequency is N/2. This is determined using the following formula.

$$\text{Median} = 1 + \left(\frac{N/2 - C}{f} \right) x_i \qquad (4.11)$$

where l = lower limit of the median class (i.e., the class in which the median number N/2 lies; N = total number of observations; c = cumulative frequency of the class preceding median class; f = frequency of the median class; and i = class interval.

Median class = 11/2 = 5.5, i.e., the class 94-96; Median is given by

$$94 + \left(\frac{11/2 - 5}{3} \right) \times 2 = 94.00 + 0.33 = 94.33.$$

About the Median

Median is affected by the number of observations rather than the size. One extreme value will not affect the median where as mean is affected by the extreme value. Median is not amenable for further algebraic treatment. By adding or subtracting an extreme value to the distribution, the mean no longer represents the centrally located value. We have to depend on another value, namely median. Median is to be used as a measure of location to minimize the effect of one or more extreme values.

MODE

Mode is another common measure of central tendency. It is the observation that occurs with maximum frequencies in a set of observations. Examples follow.

1. The age of patients who visited a hospital in a place are 15, 10, 16, 10, 13, and 14. Here the frequency of 10 is maximum, i.e., 2 which occurs more than the frequencies of any other value and hence 10 is the mode. The data are said to have unimodal distribution.

2. The age of patients who visited a hospital in a place are 15, 16, 16, 10, 12, 18, 17, 15, 9, 17, 15, and 16. Here 15 and 16 occur equally 3 times more than any other values. Hence, 15 and 16 are the modes for the data, the distribution is bimodal. Usually, the traffic flow in a place shows bimodal distribution. In the mornings and in the evenings, the traffic will be high.

3. Consider the age distribution of 8 employees in a laboratory: 25, 29, 15, 22, 24, 20, 28, 30. Each value occurs once and no value occurs more than once and there is no mode in this distribution.

Calculation of Mode from Grouped Data

In a frequency table, the class in which the maximum frequency occurs is called the modal class. The midvalue of the class is considered as representative of that class. Therefore, the midvalue of the modal class may be considered approximately as the value of the mode. The following formula is used to calculate the mode.

$$\text{Mode} = 1 + \left[\frac{f_m - f_i}{(f_m - f_i) + (f_m - f_2)} \right] \times i \qquad (4.12)$$

where 1 = lower limit of the modal class; f_m = frequency of the modal class; f_1 = frequency of the preceding class to the modal class; f_2 = frequency of the following class to the modal class; and i = width of the modal class.

Example

The fasting blood glucose levels of 100 children in a hospital are given in the following frequency list.

Class limit	Frequency (f_i)
55-60	8
60-65	18
65-70	40
70-75	22
75-80	8
80-85	4
Total	100

$$\text{Mode} = 65 + \frac{(40-18)}{(40-18)+(40-22)} \times 5$$

$$= 65 + \left[\frac{22}{(22)+(18)}\right] \times 5 = 65 + (110/40) = 67.75$$

About Mode

Extreme values in the set of observations will not affect the mode. When heterogeneity in the observations occurs, we can get two or more modes. Not all the observations are involved in the computation of mode. If a histogram shows two peaks, we infer that there are two modes. The concept of mode is useful in describing some distributions. Mode is useful in business. The article, e.g., a cell phone, which has maximum sale, may be made available on a large scale. Meteorological forecasting is also done based on modal value.

In symmetrical distributions, the mean, mode, and median coincide. In skewed distributions, the difference between mean and mode is compared against dispersion (standard deviation) (see Table 4.1).

Skewness is measured by the equation

$$Sk = (\text{Mean–Mode})/(\text{Standard Deviation}) \qquad (4.13)$$

If the mean is greater than mode, positive skewness (Figures 4.3, 4.4) is seen and if the mean is less than mode, negative skewness (Figures 4.5, 4.6) is seen. If the distribution is symmetrical, the following relationship exists.

Mean – mode = 3 (Mean – median) or mode = mean minus 3 (mean – median). This is called Person's empirical formula to compute mode.

Measures of Skewness

$$
\begin{aligned}
Sk &= \frac{\text{Mean - mode}}{\text{Standard deviation}} \\[6pt]
&= \frac{\text{Mean - [Mean - 3(mean - median)]}}{\text{standard deviation}} \quad (\text{as Mode} = [\text{Mean - 3(mean - median)}]) \\[6pt]
&= \frac{\text{Mean - (mean - 3 mean + 3 median)}}{\text{standard deviation}} \\[6pt]
&= \frac{\text{Mean - mean + 3 mean + 3 median)}}{\text{standard deviation}} \\[6pt]
&= \frac{3(\text{mean - median})}{\text{standard deviation}}
\end{aligned}
$$

TABLE 4.1. Comparison of mean, median, and mode.

Measure	Advantage	Disadvantage
Mean	More reliable	Difficult to compute
	Located at the center of the distribution	Extreme values affect the mean
	Each and every value is involved in its computation	In skewed distributions it is not reliable
	Amenable for further analysis	
Median	Located at the center of the distribution	The observations need to be arranged in increasing or decreasing order
	Not affected by extreme values	A suitable measure of central tendency in skewed distributions.
Mode	Easy to compute	May not lie in the center
	Not affected by extreme values of the distribution	Occurs more number of times

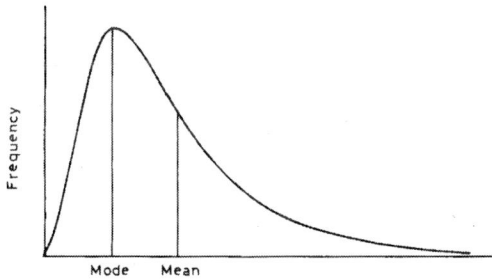

FIGURE 4.3. Positively skewed distribution.

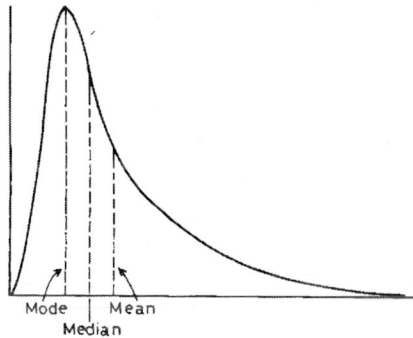

FIGURE 4.4. Positively skewed distribution (mean and median are to the right of the mode).

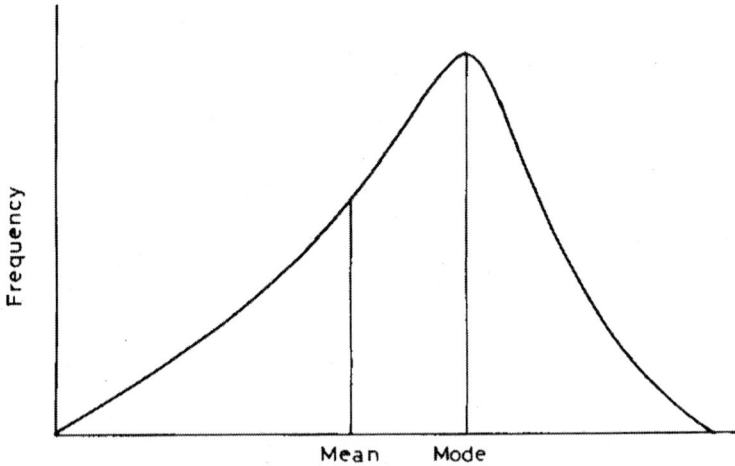

FIGURE 4.5. Negatively skewed distribution.

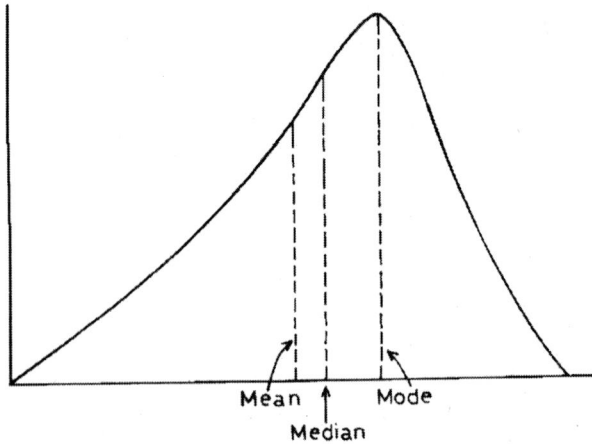

FIGURE 4.6. Negatively skewed distribution (mean and median are to the left of the mode).

Here the difference between the mean and median is divided by the standard deviation to make the description of the shape of the distribution independent of the units of measurements.

Example

In our grain yield data, we have = 27.90 g, median = 28.07 g.

Sk = 3(27.90–28.07)/8.53 = –0.06.

This value is negative and so small and that we can reasonably describe the grain yield/plant distribution as symmetrical.

$$Sk = 3(\bar{x} - Md) / SD,$$

i.e., the skewness of a distribution can be measured by taking 3 times the difference between the mean and median divided by the standard deviation. The values of skewness will generally range between –3 and +3. When a distribution is symmetrical, the skewness will be zero. Measures of skewness can be used to compare the skewness of the different distributions because the division by standard deviation has made the measures independent of the variability of the distribution.

Midrange

Midrange is the value that lies between the largest and the smallest values. It is easy to calculate.

Example

Find the midrange value of the data: 4, 5, 9, 1, 2.

$$\begin{aligned} \text{Midrange} &= (\text{Largest value} + \text{smallest value})/2 \\ &= (1 + 9) / 2 = 10/2 = 5 \end{aligned} \quad (4.14)$$

Quartiles, Deciles, and Percentiles

Quartiles are values of the variates that divide total frequency into four equal parts; deciles divide into ten equal parts, and percentiles divide into 100 equal parts.

Example

The age of the 20 patients who are admitted into a hospital are 97, 72, 87, 57, 39, 81, 70, 84, 93, 79, 84, 81, 65, 97, 75, 72, 84, 46, 94, and 77. To find the quartiles, arrange the values in increasing order as follows: 39 46 57 65 70 72 72 75 77 79 81 81 84 84 84 87 93 94 97 97

First quartile: $Q_1 = (n + 1)/4$th observation, i.e., $(20 + 1)/4 = 21/4 = 5.25$, i.e., average of 5th and 6th terms, i.e., $(70 + 72) / 2 = 71$.

Second quartile: $Q_2 = \text{median} = [2(20 + 1)]/4 = 10.5$, i.e., average of 10th and 11th values which is equal to $[(79 + 81)] / 2 = 80$.

Third quartile: $Q_3 = [3(n + 1)] / 4$th term $= [3(20 + 1)]/4 = 63/4 = 15.75$, i.e., the average of 15th and 16th values $= (84 + 87)/2 = 85.5$.

Inter Quartile Range

The inter quartile range is the difference between the third quartile (Q_3) and 1st quartile (Q_1) values, i.e.,

$$Q_3 - Q_1 = 85.5 - 71.0 = 14.5 \tag{4.15}$$

It contains 50 percent of the set with 25 percent falling above and 25 percent below the range. We have the central 50 percent of the set of observations, which are free from extreme values.

Quartile Deviation

Quartile deviation is the average of inter quartile range (IQR), i.e.,

$$(Q_3 - Q_i) / 2 = (85.5 - 71) / 2; 14.5 / 2 = 7.25.$$

The IQR is the difference between the 75th and 25th percentiles. A large IQR value indicates highly variable data. It is explained diagrammatically in the following representation and in Figure 4.7. Representation of partition values:

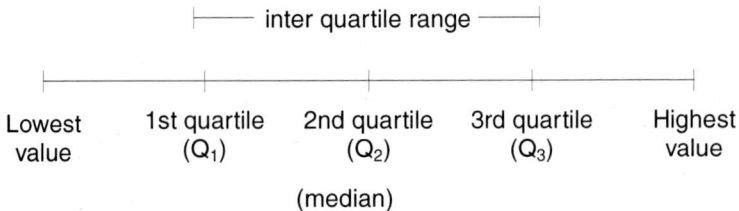

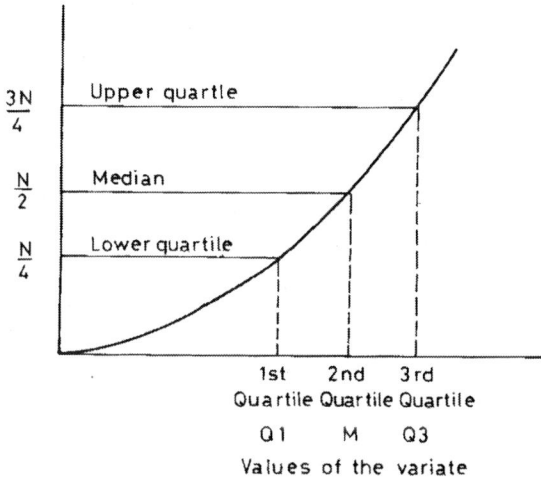

FIGURE 4.7. Graphical representation of the partition values.

Grouped Data

Usually, we collect many observations and it is advisable to calculate the quartiles from the grouped data instead of raw data. Formulas for computing the three quartile measures are:

$$Q_i = \left[1 + \frac{\left(\dfrac{iN}{4} - c\right)}{f}\right] \times h \qquad\qquad i = 1,2,3 \ldots$$

$$D_j = \left[1 + \frac{\left(\dfrac{jN}{10} - C\right)}{f}\right] \times h \qquad\qquad j = 1,2,3,4,5,6,7,8,9$$

$$P_k = \left[1 + \frac{\left(\dfrac{kN}{100} - C\right)}{f}\right] \times h \qquad\qquad k = 1,2,3, \ldots 99$$

where
  l = lower limit of the class in which the particular quartile lies,
  f = frequency of this class,
  h = width of the class interval,
  C = cumulative frequency preceding the class, and
  N = total frequency.

Example

The number of hours spent in a hospital for treatment by 70 patients are listed below in a frequency table.

Hours	No. of patients	Hours	No. of patients
1-3	6	9-11	21
3-5	53	11-13	16
5-7	85	13-15	4
7-9	56	15-17	4

A cumulative frequency table is formed as follows:

Class	f	c.f (less than)
1-3	6	6
3-5	53	59
5-7	85	144
7-9	56	200
9-11	21	221
11-13	16	237
13-15	4	241
15-17	4	245

Total = 245
  Using the formulas, we can find deciles and percentiles of our choice.

Skewness

See Table 4.2 for an example of positive skewness.

Harmonic Mean

Although arithmetic mean is the most common one in use, there are other measures of central tendency.

TABLE 4.2. Age group distribution of female population in Madras City, India, in 1981.

Age	Population	%	Age	Population	%
0-4	169,794	9.1	40-44	91,736	4.9
5-9	203,510	10.9	45-49	89,239	4.8
10-14	214,724	11.5	50-54	67,293	3.6
15-19	213,748	11.4	55-59	49,564	2.6
20-24	206,774	11.1	60-64	46,494	2.5
25-29	187,031	10.0	65-69	30,405	1.6
30-34	132,129	7.1	70-74	22,759	1.2
35-39	121,678	6.5	75-79	22,128	1.2
Total				1,869,006	100.0

Source: Shantha et al., 1980.

Example

The distance between two places, A and B, is 30 miles. A man travels by bicycle from A to B at 10 mph and then returns back to A at 30 mph. What is the average speed for the whole time?

Traveling time from A to B at 10 mph = 3 h
Traveling time from B to A at 30 mph = 1 h
Total distance traveled = 60 miles
Total time = 4 h
So average speed = 60/4 = 15 mph

This is an example of harmonic mean. Arithmetic mean as average speed for these data = (10 + 30)/2 = 20 mph; to cover 60 miles, it would have taken 3 h, which is not correct.

Skewness

Skewness is measured by the statistic

$$g_1 = /[(\sqrt{n})\Sigma(x - \bar{x})^3 / \{(x - \bar{x})^2\}^{3/2}].$$

If g_1 is close to zero, it implies that the distribution is approximately symmetrical.

Kurtosis

Kurtosis is a property of symmetric distribution. One symmetrical distribution is distinguished from another symmetrical distribution based on (1) length or thickness of their tails, and (2) the extent to which they are peaked when compared to normal frequency curve. Distributions that are thick in the tails and more peaked than the normal distribution are called leptokurtic; those that are flat rather than peaked (when compared to normal frequency curve) and short tailed are called platykurtic. The normal distribution itself is called mesokurtic.

Kurtosis is measured by the statistic

$$g_2 = [n\Sigma(x - \bar{x})^4] / [\Sigma(x - \bar{x})^2]^2.$$

For normal distribution, $g_2 = 0$. If g_2 is less than 0, it indicates platykurtic distribution and if greater than zero, it indicates leptokurtic distribution.

Example

Palaniswamy (1990) studied the grain weight/plant in three test conditions: 0 N kg/ha, 60 N kg/ha, and 120 N kg/ha in IR 22, and 0 N kg/ha, 60 N kg/ha, and 120 N kg/ha in IR 662 rice varieties and reported the g_1 and g_2 statistics as follows:

Sample (variety)	IR 22			IR 662		
N kg/ha	0	60	120	0	60	120
g_1	0.2555	0.0100	0.0265	0.0648	0.1963	0.2516
g_2	−0.7498	−0.4400	0.0548	−0.2203	−0.5814	−0.1893
n	145	200	200	200	200	200

The g_1 and g_2 coefficients were tested and found absent of skewness and kurtosis. Thus, the study of skewness and kurtosis is important to determine whether a character follows normal distribution.

EXERCISES

4.1. Define the term *central tendency*.
4.2. What are the different measures of central tendency?
4.3. Define *mean*. Explain the important properties of mean.
4.4. When will you recommend geometric mean?

4.5. Find the mean, median, and mode for these data sets:
   a. 1500, 50, 1500, 900, 900, 400, 400, 50, 800 (weight of tomatoes in g)
   b. 10, 4, 3, 6, 6, 9, 8, 3, 5, 6 (pedicel length in cm)
   c. 120, 100, 112, 130, 132, 105, 146, 113, 150, 143 (weight of students in lbs)

4.6. A scientist recorded the average value for 25 observations as 71.4. It was later found that one reading was misread as 79 instead of the correct value of 97. Find the correct average.

4.7. In an investigation, the data showed asymmetrical form. The mode and median are 40.9 and 47, respectively. Find the mean.

4.8. Serum calcium levels in 20 normal adults are given here: 9.5, 9.6, 9.8, 10.1, 10.3, 9.8, 10.6, 11.3, 13.0, 7.9, 12.0, 8.0, 8.8, 7.9, 9.0, 9.6, 12.7, 9.7, 9.8, 9.9. Form a frequency distribution. Find mean, median, and mode.

4.9. Two sets of data x_1, x_2, x_3, and x_4 and y_1, y_2, y_3, y_4, y_5, and y_6 gave the means as 1.5 $(\bar{x})$ and 2.2 $(\bar{y})$ and variances as 0.0456 (s_x^2) and (s_y^2) 0.0897, respectively. Assume that all ten values come from one set and compute the mean and variance.

4.10. The arithmetic mean and standard deviation in a set of data were found to be = 10 and standard deviation 3.1. The sample size was 500. One value 2.1 was wrongly written as 22.1. Compute the correct mean and standard deviation.

4.11. The income of 200 people are given as follows:

Income	No. of people
0-400	10
400-800	40
800-1200	70
1200-1600	40
1600-2000	20
2000-2400	10
2400-2800	7
2800-3200	3

Determine the correct mean and standard deviation. Is this distribution normal? Find the Davis coefficient and draw your conclusion.

4.12. The time spent watching TV in a week by 80 boys in a class is given in the following frequency table.

Time (hrs)	No. of boys
0-1	0
1-2	5

2-3	15
3-4	25
4-5	20
5-6	10
6-7	3
7-8	1
8-9	0
9-10	1

    a. Show these data in a suitable graph.

    b. Do you think the arithmetic mean is a better measure?

4.13. One set of 100 observations is found to have a $LQ = 2.37$, median $= 2.87$, and $UQ = 3.99$. Apply the Davies test and indicate which mean would be the best for these data.

4.14. a. State the assumption that you will make in finding the AM from grouped data.

    b. If the raw data tended to be concentrated around the lower limits of several classes, the computed mean could overestimate the true mean by a significant amount. Explain this.

4.15. State the locations of mean, median, and mode in a negatively skewed distribution and in a positively skewed distribution.

4.16. A sample of 10 plants gave the (yield/plant) mean of 79 g with a standard deviation of 10.5 g. A second sample of five plants gave a mean of 75 g and standard deviation of 10.2g. Find the mean and variance of the combined set of data.

4.17. In a hospital eight patients recorded blood sugar concentrations as follows: 103, 110, 120, 140, 150, 160, 115, 110. Find the mean, median, and standard deviation.

4.18. In an experiment, the number of ear-bearing tillers in rice crop was recorded as follows.

Class value	8.8	8.9	9.0	9.1	9.2	9.3	9.4	9.5	9.6	9.7
Frequency	1	2	17	20	43	62	56	3	7	2

Find mean, variance, and standard deviation. Find the median and quartiles. Find mode, 10th, and 90th percentiles of the data.

4.19. A set of a distribution 4, 1, 6, 2, 7 has mean of 4. What would the mean be if each value was multiplied by a constant of 3?

4.20. Describe the shape of the following distribution: 3, 3, 4, 4, 5, 5, 7, 8, 9, 10.

4.21. In the following table prepared by a researcher, one value is missing. The mean value is given as 35. Using this information, find the missing value. Also, calculate the median on the basis of the completed data.

Class interval	0-10	10-20	20-30	30-40	40-50	50-60	60-70
Frequency	10	20	missing	40	70	20	5

4.22. What are g_1 and g_2?
4.23. Why is the AM generally considered the best measure of central tendency?
4.24. Suppose each measurement in a distribution is multiplied by 2. What happens to the mean of the distribution? What happens to the variance of the distribution? What happens to the standard deviation? What happens to these three measures if 4 is added to each measurement?
4.25. Give an example of a situation in which the median might be a more appropriate descriptive measure than the mean.
4.26. Compare the values of mean, variance, and standard deviation for the following observations: 96, 100, 98, 100, 98.
4.27. Find mean, mode, and median for these data: 4, 9, 12, 16, 7, 8, 5, 4.
4.28. Three sets of data had the means 12, 20, and 30 based on 10, 30, and 50 observations, respectively. What is the mean of the three sets of data when combined?
4.29. Give an account of the relative merits and demerits of mean, mode, and median as measures of central tendency of a frequency distribution.
4.30. Show that the deviations of observations computed from their mean is zero.
4.31. Write a short definition of quartiles, deciles, and percentiles.
4.32. Show graphically how you will determine approximately the median for grouped data.
4.33. Write down the formulae for the computation of quartiles, deciles, and percentiles.
4.34. What is meant by the term *kurtosis?*
4.35. Given mean = 50, coefficient of variation = 40 percent, and coefficient of skewness = –0.4, determine standard deviation, mode, and median.
4.36. Sketch a histogram of two distributions having the same mean and total area such that one distribution has a larger variance and the other has a smaller variance.

REFERENCES

Palaniswamy, K.M. (1990). Design of statistical (field) experiments and crop forecast in agriculture with special reference to rice. Doctoral thesis submitted to the
University of Calicut, India.

Shantha, V., Palaniswamy, K.M., and Jeevamudha, M. (1980). Scenario of cancer
cervix. Paper presented at Annual Review Meeting of ICMR held at New Delhi,
November 16-18.

Chapter 5

Measures of Dispersion

Using measures of central tendency, we can measure the concentration of individual values around the central values, for example, mean, median, mode, geometric mean, and weighted mean. The distribution of the data is measured by these statistics. It is also equally important to know the magnitude of dispersion of observations from the central values for efficient research work. Study of dispersion, spread, scatter, or variability with reference to central values, particularly mean, is very important, and is considered as the major problem of statistical science. Variation among observations is universal and unavoidable. If we can account for variability, we can take steps or measures to minimize it, and consequently we can get precise estimates and inferences from the sample data.

VARIABLES

Scientists deal with variables such as crop yield, age, income, height, weight, number of patients being treated in a hospital, number of days in inpatient treatment, cost of treatment, number of members in the household, number of cigarettes used in a day by females in the age group 20 to 30 years, time spent viewing TV at home, temperature in a place, or rainfall in a city. Generally, these observations vary from case to case or time to time; hence they are called variables. A series of variates of a variable, such as body weights of persons, give what is called a *distribution*. The number of times a particular value occurs is referred to as the frequency of the value. A set of values with their corresponding frequencies is called a frequency distribution.

Central tendency and dispersion are the most important characteristics of a distribution. First, the basic terms, such as variance, standard deviation, and parameter are meaningfully defined. The following example illustrates the importance of central tendency and dispersion. Following are four sets of observed values of a particular character; each set contains seven observations:

			Values					Total	Mean
	1	2	3	4	5	6	7		
Set I	5	6	7	8	9	6	8	49	7
Set II	12	5	10	9	4	4	5	49	7
Set III	5	6	6	4	10	9	9	49	7
Set IV	7	7	7	7	7	7	7	49	7

Though the mean is the same, the observations in the first three sets show variation among themselves. In Set IV, all the observations are the same and hence in Set IV there is no variability. The mean does not show any difference in all the four sets. The variability is measured by quantities based on deviations from the mean.

MOST COMMON MEASURES OF DISPERSION

1. Range
2. Inter quartile range (IQR)
3. Quartile deviation
4. Median
5. Mean deviation
6. Variance (s^2)
7. Standard deviation (s)
8. Coefficient of variation (CV)
9. Standard error (SE)

Range

Ungrouped Data

Range is the difference between highest and lowest values in a set of observations. Thus, it is easy to compute. However, it involves only two extreme values and therefore it is a poor measure of dispersion.

Example: Six children receive treatment for hemolytic anemia. Their hemoglobin values (g/100 mL) are 8.9, 12.8, 6.9, 9.1, 11.6, and 8.6. The range in this example is:

highest value – lowest value = 12.8 – 6.9 = 5.9 (g/100 mL).   (5.1)

Grouped Data

In ungrouped data, we do not know the individual values in the frequency table. We assume that the values in each class were uniformly distributed. The range in the grouped data is determined as the difference between the upper boundary of the highest class and the lower class limit of the lower class.

Advantages
1. It is easy to calculate.
2. It is used in situations in which we are interested in the extreme values.
3. In quality-control science, range is commonly used.

Disadvantage. Only maximum and minimum values are involved in the calculation of range and thus it does not take into consideration the spread in the other values. It will give a very distorted picture of the true dispersion pattern of the distribution.

Inter Quartile Range (IQR)

Inter quartile range is the difference between the third quartile (75th percentile) and the first quartile (25th percentile) values. It tells us the range of variation among the middle 50 percent of the observations. It does not describe the variation in the first quartile (below 25 percent) of the observations and above 75 percent of the observations. The variability after eliminating the extreme values will sometimes be a good measure of variability. In educational fields, IQR is used frequently.

The semi-inter-quartile range is defined by the letter Q and is calculated as

$$Q = (Q_3 - Q_1)/2. \qquad (5.2)$$

It is superior to range. If two groups have the same quartile values, they likely possess similar patterns of variability.

Quartile Deviation (QD)

Quartile deviation is a measure of dispersion in terms of the distance between selected observation points. The interquartile range is the distance between Q_3 (third quartile) and Q_1 (first quartile). The QD is one-half the IQR. It is sometimes referred to as semi-inter-quartile range (SIQR). The formula for computing QD (SIQR) is as follows:

$$(Q_3 - Q_1) / 2 . \tag{5.3}$$

Smaller QD indicates the greater concentration of the observations in the middle half of the distribution.

Normal Distribution and QD

If distribution is normal in shape, 50 percent of the observations will include in the range: median + 1 QD because Q_1 and Q_3 will be at equal distance from the median. The Q_2 is the median. It can be explained as shown in Figure 5.1.

Example: Following is a frequency table for daily wages of 50 laborers in a factory:

Class limits	Frequency (f)	Cumulative frequency (cf)
80-90	2	2
90-100	6	8
100-110	10	18
110-120	14	32
120-130	9	41
130-140	7	48
140-150	2	50

$$Q_1 = LQ_1 + [(N/4 - cf)/fq_1)] \times i$$
$$= 100 + [(50/4-8)/10] \times 10 = 100 + 4.5 = 104.5$$
$$Q_3 = LQ_3 + [(3N/4-cf)/fq_3] \times 10$$
$$= 120 + [3 \times 50/4-32)/9] \times 10 = 120 + 6.11 = 126.11$$
$$IQR = Q_3 - Q_1 = 126.11 - 104.50 = 21.61, \text{ where}$$

$LQ_1 =$ lower limit of the first quartile value;
$LQ_3 =$ lower limit of the third quartile value;
$\ \ cf =$ cumulative frequency of all classes up to but not including the third quartile;
$\ \ Q_1 =$ first quartile value;

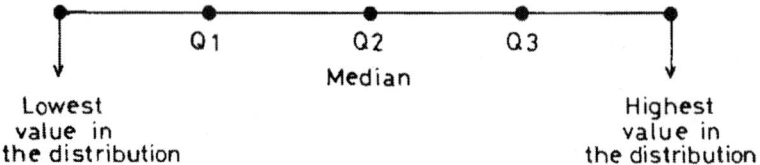

FIGURE 5.1. Diagram showing the position of median and other quartile values.

$Q_3 =$ third quartile value;
$fq_3 =$ frequency of the third quartile class; and
$i =$ size of the class interval.

$$IQR = 126.11 - 104.50 = 21.61$$

The IQR is therefore $Q_3 - Q_1 = 126.11 - 104.50 = 21.61$. If the laborers were arranged in order of the amount of wages they received, then the middle 50 percent of these people received wages between $104.50 and $126.11. This example shows the method of computation of Q_1 and Q_3, and also IQR. Compared to range, the virtue of IQR is that it excludes the extreme values.

$$QD = (Q_3 - Q_1)/2 = (126.11 - 104.50)/2 = 21.61/2 = 10.81 \qquad (5.4)$$

About Quartile Deviation

Quartile deviation is similar to range as it also considers two values, namely Q_1 and Q_3. It shows the spread in the middle 50 percent of the observations. Quartile deviation is easy to define and understand; it is used in skewed distribution. It is not affected by the extreme values so it is preferable to standard deviation when the distribution is a skewed one.

Median

Mediant is calculated using the formula

$$Q_2 = LQ_2 + [(2\ N/4 - cf)/fq_2] \times i \qquad (5.5)$$

$= 110 + [(\ 25 - 18\ )\ /\ 14] \times 10$
$= 110 + 70/14 = 110 + 5 = 115$

If the distribution is normal, 50 percent of the observations lie in between median $\pm$ 1 QD, i.e., $115 \pm 10.81 = 104.19$ to 125.81.

Mean Deviation

A given set of observations varies from one observation to another. This variation is calculated by measures of dispersion. One such measure is mean deviation. It is computed by calculating the absolute deviation, ignoring algebraic signs of each observation from the mean, and then taking the arithmetic mean of the sum of the deviations. The formula for computing the mean deviation is

$$\Sigma[|x_i - \bar{x}|] / n \qquad (5.6)$$

The symbols | | indicate modulus, i.e., ignore the signs. Example: Given the set of observations 5, 6, 10, 11 (t/ha) of four farms in paddy crop, find the mean:

$$\text{Mean} = (5 + 6 + 10 + 11)/4 = 32/4 = 8 \text{ t/ha}$$

Then take the deviations of observations from the mean and add ignoring the signs.

$$| (5 - 8) | + | (6 - 8) | + | (10 - 8) | + | (11 - 8) |$$
$$= 3 + 2 + 2 + 3 = 10$$

Finally, find the mean for the sum of the deviations: $10/4 = 2.5$ t/ha.

Mean deviation is therefore 2.5 t/ha. It indicates that on average each value deviates from the mean by 2.5 t/ha. The use of absolute values is necessary because the algebraic sum of the deviations from the mean is always zero. This absolute deviation must be used for computing the mean deviation.

Advantages
1. It gives equal weights to the deviation of every observation.
2. This is more sensitive than range as range is based only on two values.
3. It is an easy measure to calculate.
4. It is easy to understand.

Disadvantage. It is not useful for further calculations.

Uses of Mean Deviation

In normal distribution, a simple relation between the mean deviation (MD) and standard deviation exists, i.e.,

$$\text{standard deviation} = \sqrt{[(\Pi/2)]} \times (MD) = 1.25 \times$$
mean deviation, i.e., $\sigma \approx 1.25 \times MD$.

If we want the approximate standard deviation, we can use the previous formula and compute standard deviation from the mean deviation. The standard deviation is generally lesser than the mean deviation.

Grouped Data

Given the following grouped data, plant yields of 50 plants in rice crop (g/plant), find the mean.

Class limits	0-10	10-20	20-30	30-40	40-50
Class value	5	15	25	35	45
Frequency	5	8	15	16	6

$$\text{mean} = \Sigma f_i x_i \, / \, \Sigma f_i$$
$$= [(5\times5) + (8\times15) + (15\times25) + (16\times35) + (6\times45)] \, / \, 50$$
$$= 1350/50 = 27 \text{ g/plant}$$

Take the deviations of midvalues from the mean, multiply by the frequency, and then find the sum.

$$[(/5–27/) \, 5 + (/15–27/) \, 8 + (/25–27/) \, 15 + (/35–27/) \, 16 + (/45–27/) \, 6] = 472$$

Divide the sum by the total frequencies. 472/50 = 9.44 g/plant.

The result shows that the grain yield per plant varies on average to an extent of 9.44 g/plant.

About Mean Deviation

Mean deviation gives equal weights to the deviations of every observation. It is more sensitive than range or quartile deviation as they are based on only two values. Mean deviation is easy to calculate and easy to understand but its use is limited in statistics because it is not useful for further calculations.

Variance

Variance is a measure of dispersion of the observations in a given set. It is referred to as $Ó^2$ (sigma squared) for population and s^2 (s squared) for the sample. The variance of a sample is defined as the sum of squared deviations of the observations from the mean divided by the degrees of freedom (number of observations minus one). If n is the number of observations, then the degrees of freedom is $n - 1$.

If X is the variable and $X_1, X_2 \ldots X_n$ are the variate values, then the variance is defined by

$$s^2 = \{(x_i - \bar{x})^2 + (x_2 - \bar{x})^2 + \ldots + (x_n - \bar{x})^2 / (n-1)\}$$

$$\text{i.e., } s^2 = \Sigma(x - \bar{x})^2 / n - 1 \tag{5.7}$$

where $\Sigma(x - \bar{x})^2$ = sum of squares

$n - 1$ = degrees of freedom

n = number of observations

$s^2 = [(SS) / df]$ where SS = sum of squares and df = degrees of freedom

Note that the denominator is $n-1$ and not n as might be supposed, and s^2 is the observed or sample variance, whereas $\acute{o}^2$ denotes the true or population variance. This is expressed in units of squared deviations or squares of the units of the original numbers.

To calculate $\Sigma(x_i - \bar{x})^2$, we have to use the mean. The following formula can be used to find the sum of squares, which does not involve the mean. This is an alternative way of writing the previous formula or the working formula for calculating the variance.

$$SS = \Sigma(x_i - \bar{x})^2 = \Sigma xi^2 - (\Sigma x)^2 / n$$

Proof:

$$\Sigma(x_i - \bar{x})^2 = (x_1 - \bar{x})^2 + (x_2 - \bar{x})^2 + \ldots + (x_n - \bar{x})^2$$

$$= x_1^2 + \bar{x}^2 - 2x_1\bar{x} + \bar{x}_2^2 + x^2 - 2x_2\bar{x} + \ldots + x_n^2 + \bar{x}^2 - 2x_n\bar{x}$$

$$= x_1^2 + x_2^2 + \ldots + x_n^2 - 2\bar{x}(x_1 + x_2 + \ldots + x_n) + n\bar{x}^2$$

$$= \Sigma x_i^2 - 2\bar{x}\Sigma x_i + n[(\Sigma x_i)^2 / n^2] \text{ (as } \bar{x} = \Sigma x_i / n)$$

$$= \Sigma x_i^2 - 2(\Sigma x_i / n)(\Sigma x_i) + [(\Sigma x_i)^2 / n]$$

$$+ \Sigma x_i^2 - (\Sigma x_i)^2 / n$$

$\Sigma x_i^2 - (\Sigma x_i)^2 / n$ is called the computational method for finding the SS where Σx_i^2 = the uncorrected sum of squares or the crude sum of

squares,

$(\Sigma x_i)^2 / n$ = the correction factor (CF), and

$\Sigma x_i^2 - (\Sigma x)^2 / n$ is the corrected sum of squares or simply SS.

The important point to be remembered is the distinction between Σx_i^2 and $(\Sigma x)^2$.

Calculation of Variance (Ungrouped Data)

The heights of three women are given as 64, 66, and 74 inches. The variable under study is height and is denoted by X. Thus $X_1 = 64"$, $X_2 = 66"$, and $X_3 = 74"$.

Mean: $(64 + 66 + 74)/3 = 204/3 = 68"$

Value	Deviation from mean (d_i)	Squared deviations ($d_i{}^2$)
64	64-68 =–4	16
66	66-68 =–2	4
74	74-68 =+6	36
Total	0	56 ($\Sigma d_i{}^2$)

d_i = deviations of the observations from the mean

Note that the sum of the deviations is zero.

$\Sigma d_i{}^2$ = sum of the squared deviations

The number 56 reflects the variability of the heights. We divide it by $n-1$. The result is the variance. Thus,

$$\text{variance} = \Sigma(x_i - \bar{x})^2 / (n-1) = 56/2 = 28 \text{ inches}^2.$$

The variance is the mean of the sum of squares and hence the variance is also termed as mean square and denoted by s^2. The variance is expressed not in the original units but in the square of the original units. Variance is used extensively in statistics. Note that in the denominator $n-1$ is used and not n. To get an unbiased estimate of the population variance ($\acute{o}^2$), $n-1$ is used.
  Computation of variance using working formula:

$$SS = \Sigma x_i{}^2 - (\Sigma x)^2 / n = 64^2 + 66^2 + 74^2 - 204^2 / 3$$
$$= 13928 - 13872 - 56$$
$$\text{Variance} = SS/df = 56/2 = 28 \text{ (as before)}$$

Note that when s^2 is calculated, $n-1$ is used for the divisor.
  $\Sigma x_i{}^2$ = square each of the X values and then add them up
  $(\Sigma x)^2 / n$ = add all the X values first, and square the total and finally divide the squared total by the number of observations, n. In this case $n = 3$.

Note: In calculating variance, one must differentiate the terms $\Sigma\, x_i^2$ and $(\Sigma x)^2$.

$\Sigma\, X_i^2$ denotes sum of squares of each of the individual values.

$\Sigma\, (X)^2$ denotes square of the sum of the values.

Calculation of Variance (Grouped Data)

Method 1 (Assumed mean method): See the following frequency table.

Class interval	Mid value (x_i)	$(x_i{-}A)$	$(x{-}A)/$ $i{=}d_i$	f_i	f_id_i	$f_id_i^2$
1-3	2	−6	−3	3	−9	27
3-5	4	−4	−2	9	−18	36
5-7	6	−2	−1	25	−25	25
7-9	8	0	0	35	0	0
9-11	10	2	1	17	17	17
11-13	12	4	2	10	20	40
13-15	14	6	3	1	3	9
Total			0	$f_i = 100$	$\Sigma\, f_id_i = -12$	$\Sigma\, f_id_i^2 = 154$

The data relate frequency distribution of yield/plant in g.

A $\qquad$ = assumed mean = 8

Mean $\qquad$ = $A = [(\Sigma f_i d_i)/ N]\times i = 8+[-12/100]\times 2 = 8 - 0.24 = 7.76$

Variance (s²) $\;= [\Sigma f_i d_i^{\,2} -(\Sigma f_i d_i)^2\,/\,N]\times (i^2\,/\,N = [(154)-(-12)^2\,/100]$

$\qquad\qquad x4/100)$

$\qquad\qquad = \quad 4/100\,(154{-}1.44) = 6.10$ (gram square)

OR s^2 $\qquad = i^2\,/\,N^{\,2}\,[N\Sigma f_d^{\,2} -(\Sigma f_d)^2\,] = i^2\,/\,N[100\times 154 -(-12)^2\,]$

$\qquad\qquad = 6.10 g^2$

$\qquad\qquad = 4/10000\{100\times 154 - 144\} = 6.10$

Standard deviation: $= \sqrt{s^2} = s = \sqrt{6.10} = 2.47$ g

Shepperd's Correction

If the standard deviation is calculated from the grouped data, it is in error if the distribution is normal to the extent of $c^2/12$ (c = class interval). This error is due to the assumption that all observations are equal to the mid-value. In some cases, the midvalues are smaller and in some larger. These errors are positive in some cases and negative in others and they cancel each other while computing the mean. However, in computing variance, these errors, due to squaring, become large. To correct this grouping error, we have to apply the correction factor to the computed variance. This correction is called Shepperd's correction. Hence, the variance after Shepperd's correction in the previous example is $6.10 - 4/12 = 5.77$ g^2 and the standard deviation $\sqrt{5.77} = 2.40$ g.

Example: Calculation of standard deviation from the grouped data for grain yield data after Shepperd's correction.

Class interval	Midvalue (x_i)	$(x_i - A)$	$(x_i - A)/i$ $= d_i$	F_i	$f_i d_i$	$f_i d_i^2$
5-10	7.5	−20	−4	1	−4	16
10-15	12.5	−15	−3	3	−9	27
15-20	17.5	−10	−2	4	−8	16
20-25	22.5	−5	−1	9	−9	9
25-30	27.5	0	0	13	0	0
30-35	32.5	5	1	12	12	12
35-40	37.5	10	2	4	8	16
40-45	42.5	15	3	2	6	18
45-50	47.5	20	4	2	8	32
Total				50	$F_i d_i = +4$	$\Sigma f_i d_i^2 = 146$

$$\text{Variance } (s^2) = i^2 N[\Sigma f_i d_i^2 - (\Sigma f_i d_i)^2 / N]$$
$$= 25/50 \,(146 - 16/50) = 72.84 \text{ g}^2$$
$$\text{Standard deviation} = \sqrt{\text{variance}} = \sqrt{72.85} = 8.53 \text{ g}$$
$$\text{Or variance } (s^2) = i^2 / N^2 [N\Sigma f_d^2 - (\Sigma f_d)^2]$$
$$= 5^2 / 50 \times 50[50 \times 146 - 16] = 72.84 \text{g}^2$$
$$\text{Standard deviation} = \sqrt{72.84} = 80.53\text{g}$$

Variance after Shepperd's correction $= 72.84 - 25/12 = 72.84 - 2.08 = 70.76$ g^2 and standard deviation is 8.41 g.

Properties of Variance

Variance is computed invariably in many statistical problems such as the analysis of variance, comparison of treatments, component analysis, etc. It is an important statistic and hence it is better to know the essential properties of variance:

1. If a constant is added to the original observations, then the variance of the derived sample will not change, i.e., $V(x + a) = V(x)$.

Example: Consider the following as our original observations in an experiment and the derived observations after adding a constant to each.

Original observations = 5, 6, 7, 8, 9
Derived observations after adding a constant (e.g., 4) = 9, 10, 11, 12, 13
Variance of 5, 6, 7, 8, 9 = 2.5
Variance of 9, 10, 11, 12, 13 = 2.5
2. If the original observations are multiplied by a constant number, then the variance of the derived sample is equal to the variance of the original observations multiplied by the square of that constant number.
   Example: Original observations are 60, 61, 62, 61, 63
   Variance of the original observations is

$$= [(60^2 + 61^2 + 62^2 + 61^2 + 63^2)—(60+61+62+61+63)^2 /5]/ \, d_f \, (4)$$
$$= [18855 – (307)^2 / 5] / 4 = 5.2/4 = 1.3 \text{ unit}^2$$

Now derive the sample after multiplying each original observation by a constant, for example 2. Derived observations are 120, 122, 124, 122, 126

$$\begin{aligned} \text{Variance} = \ & [(120^2+ 122^2 + 124^2 + 122^2 + 126^2) \\ & -(120+122+124+122+126)^2 /5 \,]/ \, df \, (4) \\ = \ & [75420–614^2 / 5] / 4 = 20.8 / 4 \\ = \ & 5.2 \text{ which is } 1.3 \times 2^2 \end{aligned}$$

Standard Deviation (SD)

Standard deviation is a measure of dispersion of data. In any distribution, some observations are close to the mean and others are far from the mean. If the standard deviation is higher, there will be greater variability, i.e., data are spread widely and if the standard deviation is smaller, it is indicative of a smaller amount of variability among the observations. From the value of standard deviation, one can know the shape of the curve. If the standard deviation is zero, all values are identical to the mean and there is no curve. Standard deviation is always positive and it takes values from zero to infinity. It is expressed in original units. Standard deviation is used only when the deviations are measured from the mean and not from any other central value. The square root of variance is defined as standard deviation. The variance is a squared measure and it is difficult to interpret.
   Using symbols, standard deviation (SD) is shown as

$$s = \sqrt{\Sigma(x - \bar{x})^2 \, / \, (n - 1)} \qquad\qquad (5.7)$$

For example, if the original unit is in kilograms, the variance is expressed as squared kilograms. But the standard deviation would be in kilograms. It will be easy to interpret and understand. Suppose we have only one observation in our sample. Then we cannot get the standard deviation because $s^2 = [\Sigma (x - \bar{x})^2] / (n - 1) = 0/0$ as when $n = 1$, x, and $\bar{x}$ and are the same and degree of freedom is also zero.

Approximate Method for Estimation of Standard Deviation

If the distribution is normal, one-sixth of the range can be used as a rough estimate of the standard deviation.

Characteristics and Uses of Standard Deviation

1. It measures the amount of variability in the distribution. If the standard deviation is small, it indicates lesser variability, and if the standard deviation is larger, it shows wider variability. It takes any value from zero to infinity.

2. All the observations are involved in the computation of standard deviation. Therefore, it is a good measure of dispersion.

3. If two distributions have same means and are in the same units, we can compare them by using standard deviations. The distribution having the smallest standard deviation should be considered as the most representative value, since there is consistency among the individual values.

4. Variance is expressed in square units whereas standard deviation is expressed in the original units.

5. Relationship between mean and standard deviation: In large populations, about 68 percent of observations will lie within $\bar{x} \pm 1$ SD, about 95 percent within $\bar{x} \pm 2$ SD and about 100 percent within $\bar{x} \pm 3$ SD. If we are interested in knowing the number of observations that lie in between $\bar{x} \pm 1.96$ SD and $\bar{x} \pm 2.58$ SD, it is seen that 95 percent of the values lie in between $\bar{x} \pm 1.96$ SD and 99 percent in between $\bar{x} \pm 2.58$ SD, where 1.96 and 2.58 are the values of t from the t table for 0.95 and 0.99 probability levels, respectively.

Example: In an experiment in rice crop (variety CO 29), the plant heights (cm) of 20 observations were recorded as follows:

94	87	63	83	97	95	79	67	89	108
96	73	78	85	117	93	69	82	103	125

Mean: $(\bar{x}) = 89.15$ cm; SD $= 16.17$; $n = 20$

$\bar{x} \pm 1$ SD $= 89.15 \pm 1 \times 16.17 = 105.32 - 72.98$ (14 observations, i.e., 70 percent lie in this range) (see Figure 5.2).

$\bar{x} \pm 2$ SD $= 89.15 \pm 2 \times 16.17 = 121.49-56.81$ (19 observations, i.e., 90 percent lie in this range) (see Figure 5.2).

$\bar{x} \pm 3$ SD $= 89.15 \pm 3 \times 16.17 = 137.66-40.64$ (all the observations, i.e., 100 percent lie in this range). Graphically, it is represented in Figure 5.2.

$\bar{x} = 89.15$ cm

6. Approximate estimation of SD: SD can be estimated approximately as mentioned before.

7. If each value in the observations is multiplied by a constant number, for example k, then the SD of the derived sample is given by $k \times$ SD.

Example: Given the observations as 60, 61, 62, 61, 63; its variance $= 1.43$; and SD $= \sqrt{1.43} = 1.14017$.

Derived sample observations after multiplying the original observations by a constant number, i.e., 2 ($k = 2$), results in 120, 122, 124, 122, 126. Its variance $= 5.2$ and SD $= \sqrt{5.2} = 2.28035$ which is $k \times$ SD $= 2 \times 1.14017 = 2.28035$. This property is used in many areas.

For example:

a. In agricultural experiments, plot yields are recorded after harvest of the crop. The plot sizes often vary from experiment to experiment. For comparison purposes, the plot yields are converted to the standard unit, i.e., kg/ha. Suppose in an experiment with plot size 4.00 m $\times$ 5.00 m, i.e., 20 m^2 is used, and the SD of 0.5 kg/plot is reported. The factor 500 can be used to convert the plot size into hectares to express the yield/ha. The SD can also be expressed in kg/ha by multiplying 0.5 kg/plot with the factor 500. Thus, the SD in the experiment is 250 kg/ha.

b. In an experiment, an entomologist recorded the weights of five insects as 0.008, 0.009, 0.008, 0.011, and 0.009g.

Mean: $(0.008 + 0.009 + 0.008 + 0.011 + 0.009)/5 = 0.045/5 = 0.009$ g

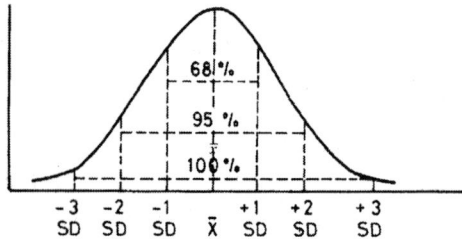

FIGURE 5.2. Graphic representation of the number of observations that lie between mean and different standard deviations ($\bar{x} = 89.65$ cm).

Its variance = 0.0000015 g^2; SD = 0.00122 g

Now to get rid of the decimals, we multiply each value by 1,000 and get the following derived observations as 8, 9, 8, 11, 9.

Its mean = 45/5 = 9 g. Original mean = 9/1000 = 0.009 g (because we multiplied by 1,000, we are now dividing by the same number by 1,000 to get the original value).

Variance = 1.5 SD = 1.22

To get the SD for the original observations, we have to divide SD obtained from the derived observations by 1,000, i.e., 1.22/1000 = 0.00122 g^2.

8. Standard deviation and variability.

These two terms convey different meanings in statistics. The standard deviation is one measure of variability whereas variability in general refers to the variation among the observations.

9. Standard deviation and standardized normal curve.

The standardized normal curve is distributed with mean 0 and SD 1. The point of inflection occurs at 1 SD on either side of the curve. This information is useful when a normal curve is drawn. The mean μ and SD σ are the parameters of the normal distribution.

How to Find Approximate Standard Deviation

1. If N is large and the distribution is normal, the SD can be estimated approximately from the range of the data. Rough estimate of SD = range/6, where six SDs (three on either side of the curve) encompass virtually all the data.
2. The range and SD are related to each other. If there are n observations in the sample and $\overline{R}$ is the mean value of range, SD can be obtained from the relationship SD $= \overline{R} \times d$. This equation is applicable when the distribution is normally distributed. The value of d can be obtained from Table 5.1. For n between 3 and 12 d varies approximately as $1\sqrt{n}$ (Neville and Kennedy, 1934).

Standard Deviation and Mean

Consider two distributions that give histograms in which the population mean $\mu = 20$ with different standard deviations (see Figures 5.3 and 5.4).

Suppose we have a value $x = 22.85$. It will have different meanings in the two distributions. In one of the distributions (Figure 5.3), it refers to one of

TABLE 5.1. Table value for *d*.

No. of observations	Value of *d*	No. of observations	Value of *d*
2	0.8862	13	0.2998
3	0.5908	14	0.2935
4	0.4857	15	0.2880
5	0.4299	16	0.2831
6	0.3945	17	0.2787
7	0.3698	18	0.2747
8	0.3512	19	0.2711
9	0.3367	20	0.2677
10	0.3249	24	0.2567
11	0.3152	50	0.2223
12	0.3069	100	0.1994
		1,000	0.1543

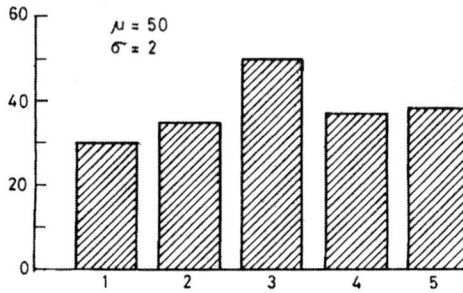

FIGURE 5.3. Values are similar and hence lesser variability.

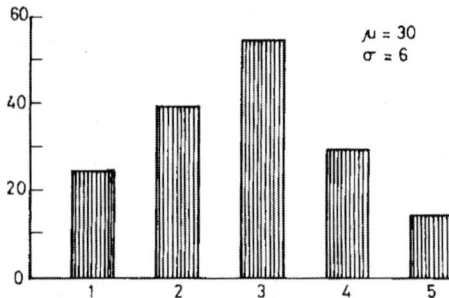

FIGURE 5.4. Here the differences between observations are higher and hence higher variability.

the highest values in the low variability group. In the second distribution (see Figure 5.4), it is the value in the higher variability group.

Standard Deviation and Researchers

If the distribution has μ=20 and low variability or low standard deviation, and if the researcher wants to select an observation, he or she can be confident of obtaining a value close to μ=20. On the other hand, if if the researcher wants to select a value from a distribution having higher variability (i.e., having high SD), the chance of getting a high value is better.

Standard deviation is also one of the critical components of inferential statistics. In inferential statistics, we use sample data as the basis for drawing conclusions about the population. A sample will not give an accurate representative value of its population. Thus, the sample will give error. SD provides a measure of how big this error will be. Suppose we have $x = 40$. Will this predict μ? It depends upon the SD. Suppose SD = 5. It means the distance between the value and μ is 5. In general, the smaller the SD, the more accurately a sample represents its population.

Standard Deviation and Standard Scores

In a school there are 100 students, and thus, 100 score values. Their mean is 18.75 and SD 2.60. Suppose a student scored 20. The relative position among the 100 is not immediately known. We can transfer the score in terms of SD as $(x-18.75) / 2.60$ where x is the score of the student. Thus $(20-18.75) / 2.60 = 0.48$ SD units. Thus, the score is 0.48 SD above the mean. Let the transformed value be denoted by z.

Then, consider $z_i = (x_i-) / SDx$. For each student, we can determine the z values. Then $z_i = 0$ and SD of $Z = 1$. These z values are called standard scores. The higher the standard score, the higher the efficiency.

Proof (Neville and Kennedy, 1934):

$$z_i = (x_i - \bar{x}) / SDx$$
$$\bar{Z} = \Sigma(\bar{x} - i) / (SDx) \times 1 / n$$
$$= 1 / n \times SDx \, \Sigma(X_i - \bar{x}) = 1 / n \, SDx \, (0) = 0$$
$$s^2_z = [\Sigma(z_i - \bar{Z})^2] / n - 1 = \Sigma z_i^2 / n - 1 \, (as \, \bar{Z} = 0)$$
$$= [\Sigma(x_i - \bar{x})^2 / s^2 x] / (n - 1).$$
$$= 1 / s^2 x [\Sigma(\bar{x} - x_i)^2 / (n - 1)] = (1 / s^2 x) \times (s^2 x) = 1.$$

Coefficient of Variation (CV)

Coefficient of variation is another measure of variability. It is a dimensionless index of variability (Cox, 1987). When we want to compare two distributions, the standard deviations may differ in their units, e.g., when we want to compare the distributions of heights, expressed in cm with that of weight, expressed in kg. Since the units differ, we cannot compare these two distributions based on their standard deviations.

For comparative purposes a relative measure of variation is therefore required. Such a relative measure is called coefficient of variation (CV), which is obtained by expressing the SD as percentage of the mean. Symbolically,

$$CV = (s / \bar{x}) \times 100 \qquad (5.9)$$

This formula was developed by Karl Pearson. Variability not revealed by standard deviation is measured by CV. The CV depends upon the nature of the experiment. For example, the CV in a rice varietal trial is lower than the CV obtained in a fertilizer trial (Bose et al., 1995).

Uses of Coefficient of Variation

Example 1: In an experiment, two variables, plant height and grain weight/plant, along with their coefficient of variations are as follows (Palaniswamy, 1990).

	Character	
Statistic	Plant height (cm)	Grain wt/plant (gm)
Mean	90.00	20.78
Standard deviation	3.60	7.7
CV (percent)	4.00	37.1

From this information, we can infer that based on SD, we cannot evaluate the extent of variability among the different characters. Based on CV, we find that grain weight/plant is more variable than plant height.

Example 2: Usha Rani (1996) observed in chili crop (*Capsicum annuum* L.) the means and CVs for various characters as follows:

Character	Mean	CV (%)
Fruit weight (g)	0.765	28.1
Fruit length (cm)	8.193	35.5

Fruit diameter (cm)	1.221	23.1
Fruit seed weight (g)	0.070	33.2
No. of seeds/fruit	64.9	32.8
1000 seed weight (g)	4.9	17.4
Capsanthin content (%)	0.245	22.7
Ascorbic acid content (mg/100 g)	130.0	24.7
Capsaicin content (%)	10.7	59.0

This information shows that different characters in chili crop exhibit different CVs, which will be useful for chili breeders. These values can be considered as standard for comparative purposes.

Suppose the same experiment is conducted by two different researchers, and sample statistics recorded are as follows:

Researcher	Mean	SD	CV (%)
1	30.5	1.18	3.9
2	20.5	1.18	5.8

A comparison indicates that there is more variation in experiment 2 though the standard deviations are the same, and hence the results of experiment 1 are more reliable.

Sample Size (n) *and Coefficient of Variation* (CV)

When planning sample surveys, we require the information on the variability of the variable under survey. If we know the CV of the character and CV of the mean of that character, we can estimate the sample size required for the survey by using the formula: $n = (cv_x / cv\ \bar{x})^2$.

Coefficient of variation varies from character to character. The information on the CVs already established by the researchers can be used as guideline values in their experiments. For example, in rice crop, the CVs for different characters namely grain yield, plant height, and number of ear-bearing tillers/plant are 10, 5, and 30 percent, respectively.

In experimental design, CV expresses the experimental error as percentage of mean. The higher the CV, the lower the reliability of the experiment. CV indicates the degree of precision with which treatments are compared (Gomez and Gomez, 1984).

CV is used to measure yield variability of crops (Singh, 1989; Kandaswamy, 1988).

Use of CV in Finding the Sample Size

The difference between two treatments' means is declared significant if it exceeds the least significant difference (LSD).

The difference is σ percent of the general mean.

It is equal to $\bar{y} \times \sigma / 100$ percent

i.e., $\bar{y} \times \sigma / (100) > LSD = t \times \sqrt{2s_e^2 / r}$

i.e., $(\bar{y}^2 \times \delta^2) / (100)^2 >= t^2 \times (2s_e^2 / r)$

$r(\bar{y} \times \delta^2) > (t^2 \times 2s_e^2 \times 100^2)$

$r > (t^2 \times 2s_e^2 \times 100^2) / (\delta^2 \times \bar{y}^2)$

$r > [(2t^2) / \delta^2] \times (CV)^2$

We know that LSD $= t \times \sqrt{2s^2 / r} = t \times \sqrt{2}\sqrt{s^2 / r}$

$= 2.2 \times 1.414 \times SE = 3 \times SE$

$r > 2 \times 2.18^2 \times [(CV)^2 / (\delta^2)]$ i.e., $r > 10 \times (CV)^2 / \delta^2$

The relation shows that as CV increases, the number of replications also increases.

Standard Error (SE)

Standard error is quite different from SD. The purpose of statistics is to draw inferences about the population from the samples. Specifically, we are interested in estimating the true mean of a population. The extent to which the sample mean varies from the population mean is measured by the standard deviation among the means. This measure is called standard error of the mean or simply standard error (SEm).

The mean of a sample is a more reliable estimate of the population mean (μ) than a single observation. There is a relationship between standard deviation of the observations and the sample size. By using this relationship, we compute the SEm. It is obtained by using the following formula:

$$SE = \sigma / \sqrt{n}, \text{ where} \tag{5.9}$$

σ = standard deviation of the population parameter, is estimated from a sample taken from the population, and is denoted by s. Thus, standard error of a sample mean $= s / \sqrt{n}$. SE is inversely proportional to the square root of the sample size, n.

Standard Deviation and Standard Error

SD and SE are quite different. We are interested in estimating μ of a population for which we have taken a sample, say 25 cases. One sample may give $y_1 = 102$, another sample $y_2 = 99$, and so on. Suppose we continue this process repeatedly. We get a large number of means. These sample means form a normal distribution. $\bar{y}$ is an unbiased estimate of the μ and its variance is

$$\sigma^2 / n$$

and its SD (or standard error of the mean) is

$$\sigma / \sqrt{n}.$$

$\bar{y}$ is an unbiased estimate of μ:

$$
\begin{aligned}
&= E[1/n(y_1 + y_2 + y_3 + \dots y_n)] \\
&= 1/n[E(y_1 + y_2 + y_3 + \dots + y_n)] \\
E(\bar{y}) \quad &= 1/n[Ey_1 + Ey_2 + \dots + Ey_n] \\
&= 1/n(\mu_1 + \mu_2 + \dots + \mu_n) \\
&= n\mu/(n) = \mu
\end{aligned}
$$

Variance of $\bar{y} = \sigma^2 / n$

Proof:

$$
\begin{aligned}
(\sigma^2 \bar{y}) &= E(\bar{y} - \mu)^2 \\
&= E[(y_1 + y_2 + \dots + y_n)/n - \mu]^2 \\
&= E \frac{(y_1 + y_2 + \dots + y_n - n\mu)^2}{n} \\
&= E[(y_1 - \mu)/n + (y_2 - \mu)/n + \dots + (y_n - \mu)]^2 \\
&= 1/n^2 E[(y_1 - \mu) + (y_2 - \mu) + \dots + (y_n - \mu)]^2 \\
&= 1/n^2 [E(y_1 - \mu)^2 + E(y_2 - \mu)^2 + \dots + E(y_n - \mu)^2 + \\
&\quad 2E(y_1 - \mu)(y_2 - \mu) + \dots + 2(y_n - 1 - \mu)(y_n - \mu)] \\
&= 1/n^2 [E(y_1 - \mu)^2 + E(y_2 - \mu)^2 + \dots + E(y_n - \mu)^2 + \\
&\quad 2E(y_1 - \mu)(y_2 - \mu) + \dots + 2E(y_n - 1)(y_n - \mu)] \\
&= 1/n^2 [\sigma^2 + \sigma^2 + \dots + 0] \\
&= n\sigma^2 / n^2 = \sigma^2 / n
\end{aligned}
$$

Standard error of the mean is calculated as follows: Let there be N observations. $x_1, x_2 \ldots x_N$ are drawn at random from the population having variance, σ^2. We have,

$$\bar{x} = 1/N(x_1 + x_2 + x_3 + \ldots + n_N)$$
$$V(\bar{x}) = V[1/N(x_1 + x_2 + \ldots + x_N)]$$
$$= [1/N^2 V(x_1 + x_2 + \ldots + x_N)]$$
$$= 1/N^2 [V(x_1) + V(x_2) + \ldots + V(X_N)]$$

$$\text{Since } V(x_1) = V(x_2) = \ldots = V(x_N) = \sigma^2$$
$$V(\bar{x}) = N\sigma^2 / N^2 = \sigma^2 / N$$
$$SE(\bar{x}) = \sigma / \sqrt{N}$$

Since σ is not known and it is estimated by a sample standard deviation (s), and the standard error $[SE(\bar{x})] = s/\sqrt{n}$.

EXERCISES

5.1. What do you understand by dispersion? Discuss the relative merits of various measures of dispersion.

5.2. What is standard deviation? How is it superior to other measures of dispersion?

5.3. From the following frequency table, find the mean and standard deviation.

Class limits	f
290-294	17
295-299	86
300-304	240
305-309	104
310-314	43
315-319	10

5.4. a. Explain what you understand by skewness of a frequency distribution. Give different measures of skewness.

b. Skewness of a frequency distribution is 0.3. The difference between mean and mode is 1.2. Find the standard deviation for the distribution.

c. Why is it necessary to measure dispersion in a set of data?

5.5. When computing the standard deviation, we ignore the algebraic signs. Why?

5.6. How does the method of computation of standard deviation in raw data differ from the computation of mean deviation?

5.7. Define variance. Show its computation from a set of observations consisting of four values.

5.8. Discuss the relationship that exists between (a) mean deviation and standard deviation, and (b) median and quartile deviation.

5.9. List the importance of range, mean deviation, standard deviation, and quartile deviation.

5.10. Calculate the sample standard deviation for the following data:
   a. 4, 4, 8, 2, 5, 7, 3, 1, 6, 3
   b. 29, 30, 25, 30, 28, 27, 28, 30, 26, 30, 24, 22, 24, 27, 26, 28, 27, 26, 25, 28

5.11. Given two sets of data, compare the two means and their standard deviations and give your comments.

	1	2	3	4	5	Total	Mean	Variance	SD
Set I	45	30	42	45	52	214	42.8	64.7	8.0
Set II	42	45	42	38	45	212	42.4	8.3	2.9

5.12. A sample of 25 observations has a variance of 100. Find (a) standard deviation, and (b) sum of squares (SS).

5.13. A population consists of four observations. Data were expressed as (i) original values, (ii) modified values after adding a constant to each value, and (iii) after multiplying by a constant. A standard deviation in a set of observations is zero. Does it mean that there is variability? Verify this statement with an example.

5.14. The heights (cm) of seedlings in a rice plot experiment were as follows:

14	20	14	26	16	17	20	23	19	19
23	15	19	13	22	17	19	20	11	21
24	12	27	19	17	20	20	24	18	21
15	22	22	19	24	20	22	19	13	23
29	24	23	16	11	26	28	27	21	12

Find the mean, range, variance, standard deviation, and coefficient of variation. Form a frequency table and find the measures from the grouped data. Do you find any difference between grouped data and ungrouped data in respect to these measures?

5.15. List the advantages and disadvantages of range. What is the common reason for its use?

5.16. What is semi-inter-quartile range? What are the advantages of this measure? How does it differ from range and standard deviation?

5.17. State the method of calculation of standard deviation.

5.18. The yields of sugarcane (t/ha) in different farms are tabulated in the following table.

Yield	No. of farms
7-9	5
9-11	10
11-13	13
13-15	22
15-17	26
17-19	35
19-21	23
21-23	21
Total	150

a. Compute range within which 50 percent of the farms fall.
b. Compute standard deviation.

5.19. State the relationship between mean, median, and mode in (a) a normal distribution, (b) positively skewed distribution, and (c) negatively skewed distribution.

5.20. The coefficient of skewness is zero for a normal distribution. Why?

5.21. Do you think that the following two distributions have the same skewness?

Distribution A: Mean = 100, Median = 90, SD = 10
Distribution B: Mean = 90, Median = 80, SD = 10

5.22. The mean height of males is 64" and standard deviation is 2.5"; in females the mean height is 60" and standard deviation is 2.5". Compare the variability.

5.23. The marks scored by 50 students in a statistics class follow:

75	31	54	77	26	34	43	57	18	15
74	50	36	18	30	48	62	16	27	58
79	6	46	52	8	98	88	19	32	60
43	44	52	69	30	86	38	94	24	32
68	99	60	63	45	19	23	97	75	74

a. Calculate mean, median, range, and semi-inter-quartile range.
b. Which measure is more suitable to these data?

5.24. Two researchers conducted the same experiment and reported the mean and standard deviation as 30 and 25, and 26 and 16, respectively. Who's data are more reliable?

5.25. Two researchers estimated the protein content (in percent) in rice as follows. Whose results are more reliable?

     Experimenter A: 12.5 11.9 12.8 12.0 12.8 12.4 12.1 11.9 12.2 12.3
     Experimenter B: 12.1 12.2 12.5 12.0 12.2 12.3 12.5 12.4

5.26. Calculate various measures of skewness on the data that follow:

     Mean = 75
     Median = 72
     Mode = 67
     LQ = 62
     UQ = 84
     SD = 13

5.27. Find the sum of squares, variance, and standard deviation in the following observations (data relate to yield/plot (kg) in corn crop): 76.1, 62.0, 73.5, 78.5, 66.9.

5.28. True or false: For most of the data, the range will be from three to five times the standard deviation. Explain your answer.

5.29. Given the following data, verify the statements that when a constant is added to each value, the sum of squares, variance, and standard deviation do not change and when each value is multiplied by a constant, the standard deviation is also multiplied by the constant.

	Original value	After adding a constant (3)	After multiplying by a constant (2)
	2	5	4
	0	3	0
	8	11	16
	2	5	4
Total	12	24	24
Mean	3	6	6

5.30. Define standard score and state its important uses.

REFERENCES

Bose, G.K., D.N.R. Paul, and K. Rahim (1995). Distribution pattern of coefficient of variation of some biometric characters of rice. *Ann. Bangladesh Agric.,* 5(4):23-26.

Cox, C.P. (1987). *A handbook of introductory statistical methods.* New York: John Wiley and Sons.

Gomez, K.A. and A.A. Gomez (1984). *Statistical procedures for agricultural research,* Second edition. New York: John Wiley and Sons.

Kandaswamy, A. (1988). Commercial crops in India. *Indian J. Agric. Econ.,* 43:444-445.

Neville, A.M. and J. B. Kennedy (1934). *Basic statistical methods for engineers and scientists.* Scranton, PA: Intl. Text Book Co.

Palaniswamy, K.M. (1990). Design of statistical (field) experiments and crop forecast in agriculture with special reference to rice. Doctoral thesis submitted to the University of Calicut, India.

Singh, I.J. (1989). Agricultural instability and farm poverty in India. *Indian J. Agrl. Econ.,* 44:1-16.

Usha Rani, P. (1996). Studies on fruit weight and its related characters in Chili (*Capsicum annuum* L.). *Inter. Trop. Agric.,* 14(1-4):123-130.

Chapter 6

Normal Distribution

Observations collected by the researchers are studied critically by applying statistical techniques such as the formation of frequency distributions, histograms, frequency polygon, frequency curve, and other methods. Different characters give rise to similar or dissimilar distributions. Among the distributions, normal, binomial, and Poisson distributions have more practical importance and hence we study these distributions in more detail. Once we confirm that a particular character follows a specified mathematical distribution, we can study more about the phenomenon by employing the mathematical properties of the specific distribution. Mohr (1990) defines normal distribution as a curve given by a certain formula; many statistics have a sampling distribution that is normal. Numerous statistical techniques concerned with sampling involve the use of normal distribution.

Normal distribution is the most important theoretical distribution in statistics as it has a wide range of practical applications (Campbell, 1974). It describes continuous data. Many characteristics such as crop yields, blood pressure, height of persons in a particular age group, and test scores are known to follow normal distribution. The concept of normal distribution has universal appeal. The formula follows:

$$f = \frac{N}{\sigma\sqrt{2\pi}} e^{-\frac{(x-\mu)^2}{2\sigma^2}} \tag{6.1}$$

where f = frequency of occurrence of any given variate (it is a dependent variable),
μ = population mean (population parameter, μ),
N = number of observations in the population,
σ = standard deviation (population parameter),
e = base of the natural system of logarithms (e = 2.7183; it is a constant),
x = independent variable (any measurement made on the variable, which can take any value from $-\infty$ to $+\infty$, i.e., $-\infty < x < +\infty$), and

$\pi = 3.1429$ (a constant, lower-case Greek letter).

The normal curve will look like the graph in Figure 6.1.

$$Y = \frac{1}{\sigma\sqrt{2\pi}} e^{-(x-\mu)^2 / 2\sigma^2}$$

For any given σ, an infinity of normal curves can be obtained with a different μ. Likewise, for any given μ an infinity of normal distributions will result, each with a different value of ó. One need not remember the complicated formula.

PROPERTIES OF NORMAL DISTRIBUTION

1. The curve is symmetrical about the mean μ and asymptotic, i.e., it extends from the mean in the right and left directions toward infinity. The curve extends from $-\infty$ to $+\infty$, because the number of observations needed to form a normal curve is infinite.
2. When depicted graphically, the curve is bell-shaped and unimodal.
3. The mean is at its maximum height. The frequency of observations is greatest at the mean and declines as the distance from the mean increases.
4. Mean, median, and mode are the same.
5. If two ordinates are erected at $\bar{x} = +\sigma$ and $\bar{x} = -\sigma$, the area under the curve between these two ordinates is 68 percent of the total area. Of

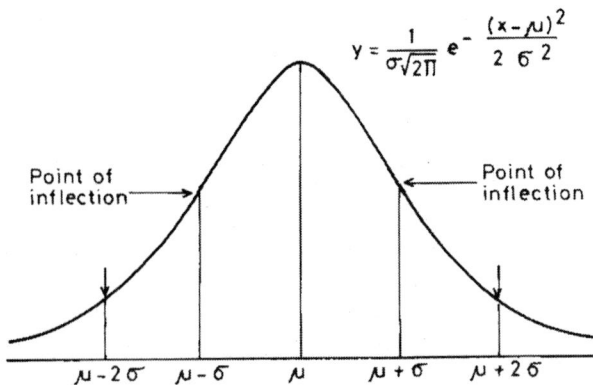

$$y = \frac{1}{\sigma\sqrt{2\pi}} e^{-\frac{(x-\mu)^2}{2\sigma^2}}$$

Point of inflection

Point of inflection

$\mu - 2\sigma \quad \mu - \sigma \quad \mu \quad \mu + \sigma \quad \mu + 2\sigma$

FIGURE 6.1. Normal curve is symmetrical about its mean. Tails do not touch the x-axis.

the remaining 32 percent, 16 percent lie on one side and 16 percent on the other side. If two ordinates are erected at $\bar{x} \pm 2$ SD, the area under the curve between these two ordinates is 95 percent of the total area under the curve. The remaining 5 percent (2.5 percent on each side) lies on both sides. The values in the normal population usually do not exceed $\bar{x} \pm 2$ SD. The proportion of the cases within $\bar{x} \pm 3$ SD of the normal curve is 99 percent of the total area under the curve. It is therefore common practice to illustrate only those cases contained between the arbitrary limits $\mu \pm 3$ SD (see Figures 6.2 through 6.4).

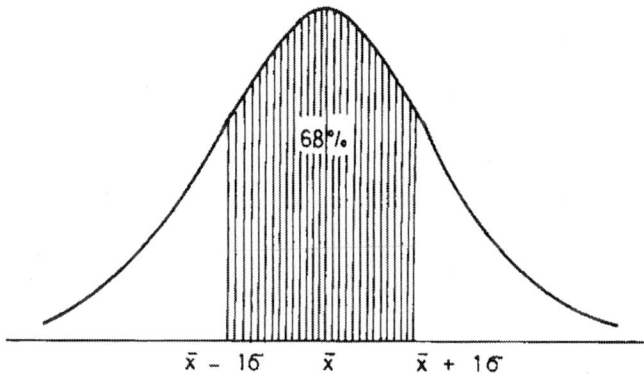

FIGURE 6.2. Graphic representation of the number of observations that lie between the mean and one standard deviation.

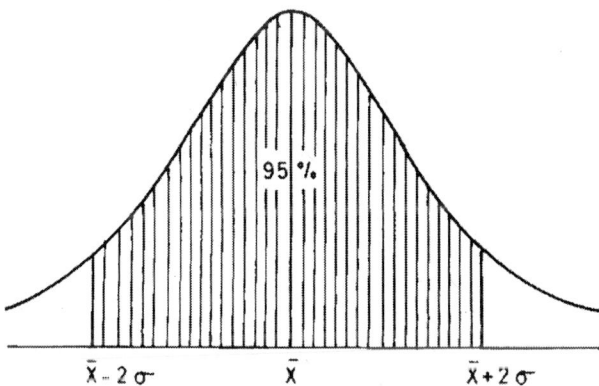

FIGURE 6.3. Graphic representation of the number of observations that lie between the mean and two standard deviations.

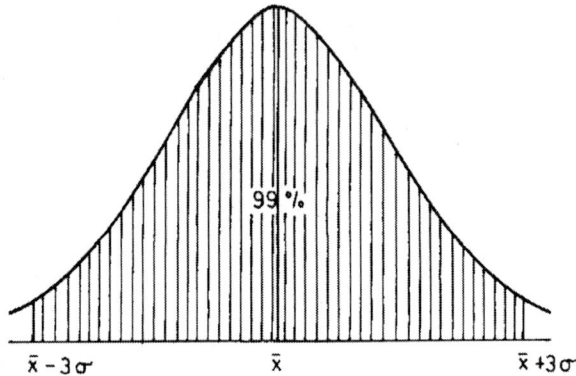

FIGURE 6.4. Graphic representation of the number of observations that lie between the mean and three standard deviations.

6. Standard normal distribution: If the variable x follows normal distribution with mean μ and standard deviation ó, then the transformed variable is called z, and $z = (x - \mu) / \sigma$ is called the standardized normal variable. The distribution of z is called the standardized normal distribution. The total area of this distribution is 1. The mean and variance of z is 0 and 1, respectively. By definition,

$$z = (x - \mu) / \sigma$$
$$E(z) = E[(x - \mu) / \sigma] = 1 / \sigma[E(x - \mu)] = 1 / \sigma[E(x) - E(\mu)]$$
$$= 1 / \sigma(\mu - \mu) = 0$$

OR

$$z = (x - \bar{x}) / s$$

where x = is the observed value; $\bar{x}$ = mean; s = standard deviation

$$\bar{z} = \Sigma z / n = \Sigma[(x - \bar{x}) / s] / n \text{ as } [z = (x - \bar{x}) / s]$$
$$= 1 / n[\Sigma(x - \bar{x}) / s] = 0 \text{ [as } \Sigma(x - \bar{x}) = 0]$$
Variance of z: $s^2 = \Sigma(z - \bar{z})^2 / n = \Sigma z^2 / n \text{(as } \bar{z} = 0)$
$$= \Sigma[(x - \bar{x}) / s]^2 / n = \Sigma(x - \bar{x})^2 / ns^2$$
$$= \Sigma(x - \bar{x})^2 / \Sigma(x - \bar{x})^2 \text{ as } ns^2 = \Sigma(x - \bar{x})^2 = 1$$

OR

$$\text{variance of } z \text{:} \ \sigma^2 z = E^5[(z - E^5)z]^2 = E^5(z^2) - (0)^2$$
$$= E^5[(x-\mu)/\sigma]^2 = 1/\sigma^2[E^5(x-\mu)^2] = 1$$

The standard normal distribution is shown graphically in Figure 6.5.

The area above the interval from $z = 0$ to $z = 1$ SD is 0.3413. Since the distribution is symmetrical, the area above the interval from $z = -1$ SD to $z = +1$ SD is $2 \times 0.3413 = 0.6826$ of the total area. The area above the interval from $z = 0$ to $z = 2$ SD is 0.4772 and hence the area from z–2 SD to $z + 2$ SD $= 0.9544$ and the area from $z = 0$ to $z = 3$ is 0.4987. These important results are summarized as follows:

Distance from the mean	Area (%)
±1 SD	68.26
±2 SD	95.44
±3 SD	99.74

These results will apply to any normal distribution.

Area under the normal curve for some selected values of z:

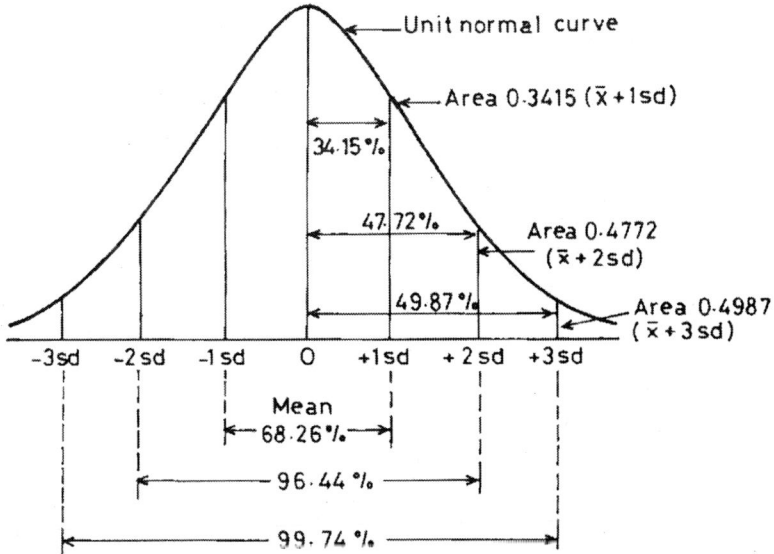

FIGURE 6.5. Standard normal distribution curve with mean 0 and standard deviation 1.

z	Area	z	Area
0.00	0.0000	1.96	0.4750
0.20	0.0793	2.00	0.4772
−0.50	0.1915	2.50	0.4938
1.10	0.3643	2.58	0.4950
−1.10	0.3643	3.00	0.4987

Any normally distributed random variable can be transformed into standardized normal form by standardizing x to z. The derived z values follow standard normal distribution with mean 0 and standard deviation 1.

Since the normal curve is asymptotic with reference to the x-axis, we can divide the abscissa into equal parts indefinitely. The proportion of cases beyond ± 3 SD from the mean of the normal curve is so small that they are generally ignored. Therefore, it is common practice to illustrate only those cases contained between the arbitrary limits ± 3 SD.

Example

If $\bar{x} = 100$ and $s = 15$, find the corresponding z value for $x = 121$. See Figures 6.6 and 6.7

$$zi = (121 - 100)/15 = 1.4$$

By converting the values to z values, we can obtain the probability or proportion of cases within any two z_i values by making use of the standard normal distribution table. In this table, z_i values (whose mean is zero and

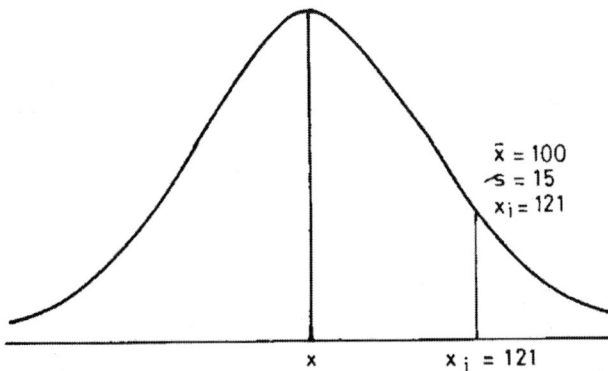

FIGURE 6.6. Before transformation of the value.

variance 1), proportion of the area (or probability) between $\bar{x}$ and $\pm$ zi and the proportion of area (or probability) beyond $\pm$ zi are given.

PROPERTIES OF STANDARD NORMAL DISTRIBUTION

1. A unit normal curve (see Figure 6.8) has a mean of 0 and a standard deviation of 1. The area above the interval from $z = 0$ to $z = 1$ (see Figure 6.8) is 0.3413. Since the distribution is symmetric, the area above the interval from

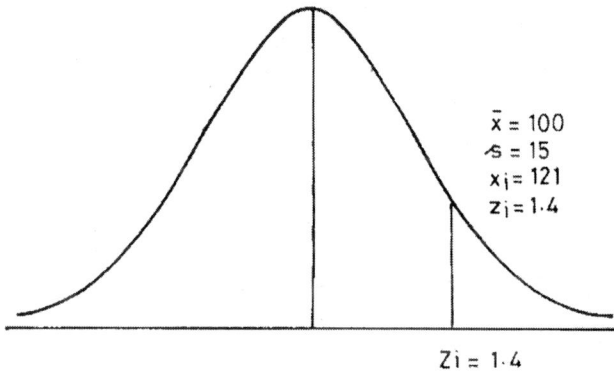

$$\bar{x} = 100$$
$$s = 15$$
$$x_i = 121$$
$$z_i = 1.4$$

$$Z_i = 1.4$$

FIGURE 6.7. After *z* transformation.

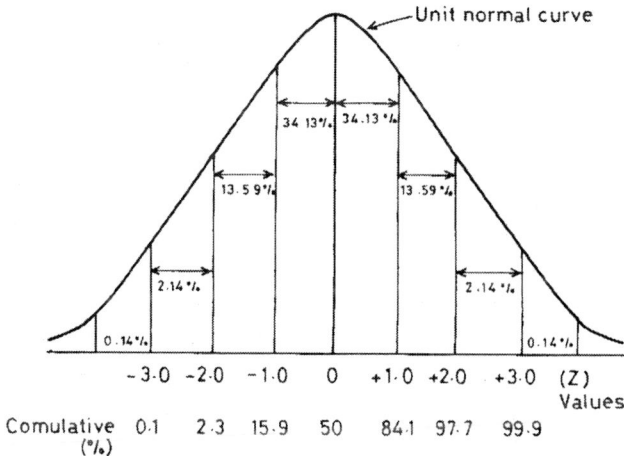

34.13% 34.13%

13.59% 13.59%

2.14% 2.14%

0.14% 0.14%

| | −3.0 | −2.0 | −1.0 | 0 | +1.0 | +2.0 | +3.0 | (Z) Values |

| Comulative (%) | 0.1 | 2.3 | 15.9 | 50 | 84.1 | 97.7 | 99.9 | |

FIGURE 6.8. Areas between selected *z* values under the standard normal distribution and cumulative percentage.

$z = -1$ to $z = +1$ is $2 \times 0.3413 = 0.6826$ or 68.26 percent of the total area. The area above the interval from $z = 0$ to $z = 2$ is 0.4772, and the area from $z = 0$ to $z = 3$ is 0.4987. These results apply to any normal distribution.
The area under the normal curve for some selected values of z are given next:

z	$F(z)$	$2 \times F(z)$
0.00	0.0000	0.0000
0.25	0.0987	0.1974
0.50	0.1915	0.3830
0.67	0.2486	0.4972
0.6745	0.2500	0.5000
0.68	0.2518	0.5035
1.00	0.3413	0.6826
1.96	0.4750	0.9500
2.00	0.4772	0.9544
2.50	0.4938	0.9876
2.576	0.4950	0.9900
3.00	0.4987	0.9974

2. The sum of z values is equal to zero.
3. The sum of squares of z values is equal to the number of observations in the distribution is $\Sigma z_i^2 = N$.

Proof:
$$
\begin{aligned}
zi &= (x_i - \mu) / \sigma = \Sigma z_i^2 = \Sigma[(x_i - \mu)/\sigma]^2 = 1/\sigma^2 \Sigma(x_i - \mu)^2 \\
&= 1/[\Sigma(x_i - \mu)^2 / N] \text{ since } \sigma^2 = \Sigma(x_i - \mu)^2 / N \\
&= [(\Sigma x_i - \mu)^2 / \Sigma(x_i - \mu)^2] \times N = N
\end{aligned}
$$

4. The total area under the standard normal curve is 1. The area between any two z values is the proportion or probability of the observations in the experiment. We can find the probability between any two z_i values using the tabulated values.

Example
1. The plant height in a rice crop is normally distributed with a mean of 75 cm and $s = 5$. A selected plant is 85 cm. The proportion of plants that lie between 75 and 85 cm is

$$
\begin{aligned}
z_i &= (x - \bar{x})/s \\
&= (85 - 75)/5 = 10/5 = 2.00, \text{ i.e., } P(0 \le z \le 2) = 0.4772
\end{aligned}
$$
(from tables)

i.e., about 47.72 percent of the plants have heights between 75 and 85 cm.

2. Suppose we record the mean plant yield in rice crop as 16.71 g and standard deviation as 7.0608 g. What proportion of the plant population gives 10 percent higher grain yields than the mean? Now we have $\bar{x}$ = 16.71 g and x = 16.71 + (16.71) × (10/100) = 18.38 g. Then z = (18.38 – 16.71)/7.0608 = 0.2365 (see Figure 6.9).

The area up to z = 0.5000 + 0.0935 = 0.5935. The area beyond z is 1 – 0.5935 = 0.4065 or 40.65 percent. Therefore, the probability of plants recording 10 percent more than the mean grain weight/plant is 0.4065 percent or 40.65 percent. In a similar way, the researcher can estimate the probabilities of plants giving 11 percent, 12 percent more than the mean (Palaniswamy, 1990).

A normal distribution is completely characterized by its two parameters—namely mean and variance. Because of many possible values for the mean and standard deviation, there are many possible distributions. Even then, one single table is sufficient to calculate probabilities associated with the observations. The parameter, μ the center of the distribution and the shape of the curve depends upon σ^2. Therefore, the normal distribution is described by its two parameters μ and σ^2.

Sample Mean and Normal Distribution

Even if the distribution under study is not normal, the sample means taken from the population follow approximately normal distribution. The curve obtained from the means of samples shows that it is narrower than the original observations. The narrowness is very important as it shows im-

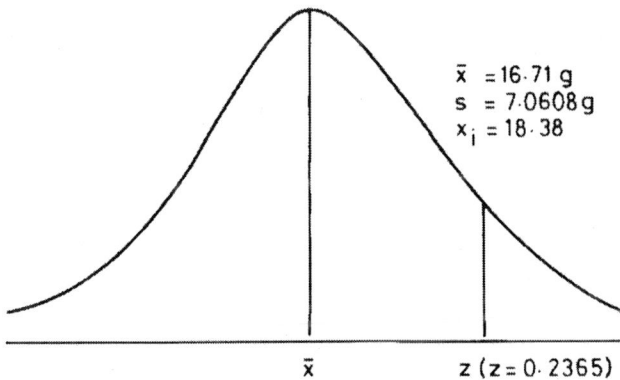

FIGURE 6.9. The position of z value.

provement in precision. The sample means cluster more closely around the population mean than do the single observations (Figure 6.10). The reduced variability is represented by the quantity of the standard deviation of the distribution of the sample means. Its value is $SEm = \sigma / \sqrt{n}$, where n = number of observations in the sample and σ = standard deviation.

Effect of Sample Size on Mean

The curve for $n = 1$ represents normal distribution. When n is increased to 2, we get a narrower curve than the previous one. When n is increased to 3, we get still a narrower and steeper curve than the previous two (see Figure 6.11). Note that when we increase sample sizes, the means remain unchanged.

Case 1 (see Figures 6.12 and 6.13)

The shape of the curve depends upon mean and standard deviation.

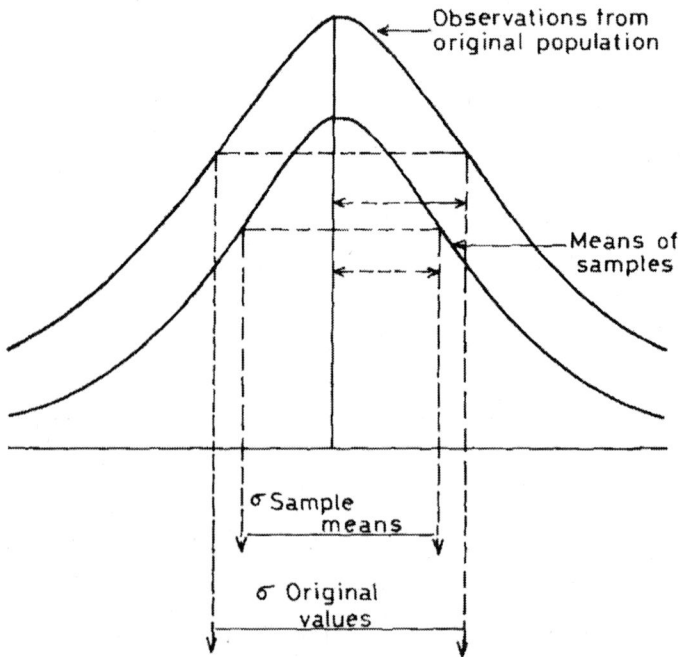

FIGURE 6.10. The shape of the curves when individual observations and sample means are involved.

__ Example: Suppose there are three variables, *X, Y,* and *Z.* Their means are $\bar{X} = \bar{Y} = \bar{Z}$; their standard deviations are 8 *(S1),* 16 *(S2),* and 24 *(S3),* respectively. The result is three different normal curves. The greater the SD the more spread out the normal curves. They are of the same pattern but located at different points along the horizontal axis. In normal distribution, the probability of a continuous distribution is expressed in terms of the center of the curve between an interval.

The shaded area under the curve between the two values *a* and *b* (see Figure 6.14) gives the probability that a continuous variable will take on the interval from *a* to *b.* The probability is designated as $P(a \le y \le b)$. The area under the normal curve is 1.

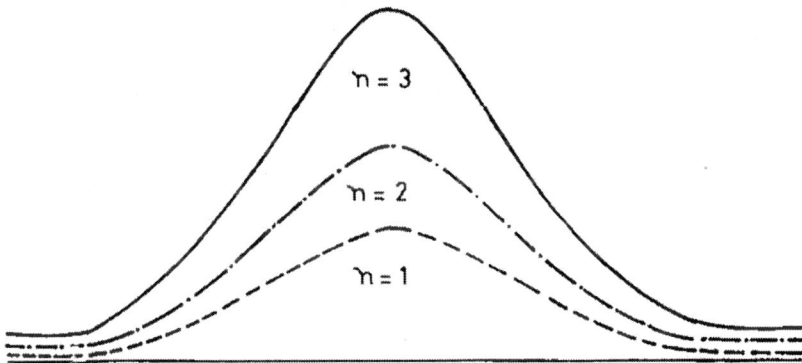

FIGURE 6.11. The shapes of the curves when $n = 1$, $n = 2$, and $n = 3$.

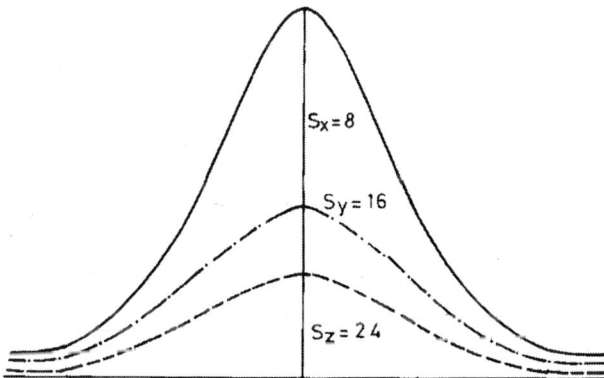

FIGURE 6.12. The shapes of the curves under different standard deviations.

Sx = Sy = S

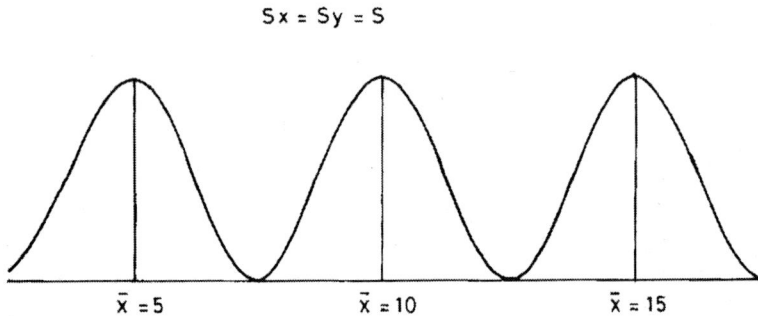

$\bar{x} = 5$ $\bar{x} = 10$ $\bar{x} = 15$

FIGURE 6.13. The shapes of the curves when means and standard deviations are equal.

The area between the interval from a to b as a continuous variable is the proportion of the outcomes that assume the values from a to b. Thus the probability between an interval cannot be negative, and the total probability under the normal curve is always 1.

PROBABILITY INTEGRAL

We defined the probability of an event as the number of cases favorable to the event divided by the total number of equally likely cases. In a frequency curve, the probability of an individual to have a value less than x_1, greater than x_2, or between x_1 and x_2 is the ratio between the shaded area and the total area under the curve as shown in Figures 6.15 through 6.17.

The shaded area represents the number of favorable cases and the total area represents the total number of cases, which are all equally likely. The ratio between the shaded area and the total area is the proportion of the shaded area. We are interested in the proportion of the individuals having characters below x_1 or above x_2 and between x_1 and x_2. This is the proportion of the area lying under the frequency curve between the appropriate limits. The proportion of the area lying under the curve below a given value of the variate is called probability integral. The probability integrals are given as fractions of the whole. The actual frequency is obtained by multiplying the total number of observations (N) by the probability integral (P), i.e.,

$$f = NP,$$

where f = frequency between x_1 and x_2,

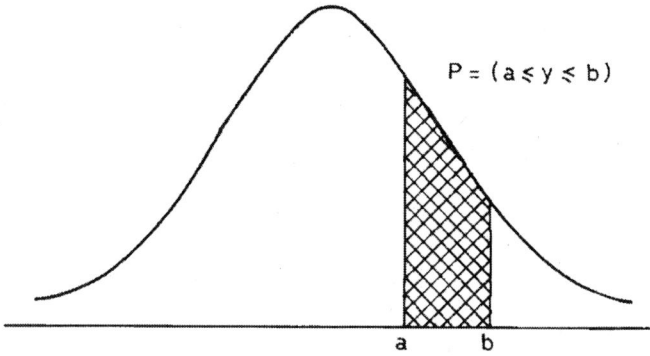

FIGURE 6.14. Area between two values *a* and *b*.

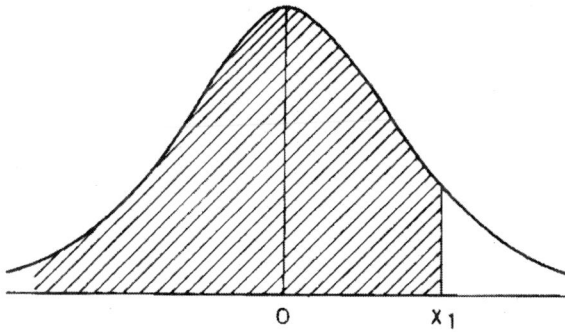

FIGURE 6.15. Probability of an individual value less than X_1.

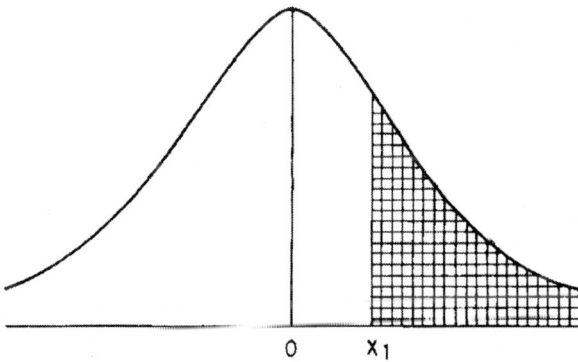

FIGURE 6.16. Probability of an individual having value greater than X_1.

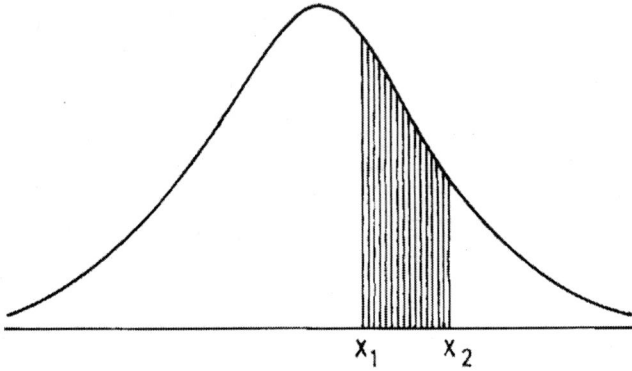

FIGURE 6.17. Probability of an individual having a value between X_1 and X_2.

P = proportion of the area lying between x_1 and x_2, and
N = total number of observations.

Normal Probability Integral Table

In this table, the proportion of the areas under the curve to the left of the ordinate at x_1 are given. This is generally represented by $1/2(1 + \alpha)$. The variate x_1 is the normal deviate $(x - \mu)/\sigma$.

Area to the right $= 1 - [1/2(1 + \alpha)]$
$$= 1 - (\text{area to the left, i.e., } 1/2(1 + \alpha) = 1 - 1/2(1 + \sigma)).$$

The normal deviate is represented by z. Thus

$$z = (x - \mu)/\sigma.$$

(See Table 6.1.)
By interpolation, we can calculate normal deviate value between any two values.
Example: Given $\bar{x} = 10.78$, and $s = 1.25$, find the probability of occurrence of the variate value (1) 13.72 or more, (2) 8.23 or less, (3) 8.23 and more, and (4) ± 2.05.
1. $(x - \mu)/\acute{o} = (13.72 - 10.78)/1.25 = 2.352$
  From tables:
       $2.3 = 0.98928$
       $2.4 = 0.99183$

For 0.1 difference we have 0.00255 for 0.052
$$= (0.00255/0.1) \times 0.052 = 0.00133$$
Area for 2.352 = 0.98928 + 0.00133 = 0.99061
Area to the left of 13.72 = 1 − 0.99061 = 0.00939
Thus P of a value equal to or greater than 13.32 is 0.00939 or 1 percent.
2. $(x − \mu)/\sigma = (8.23 − 10.78)/1.25 = −2.04$
From tables:
$$2.0 = 0.97725 \text{ and } 2.1 = 0.98214;$$
$$2.04 = 0.97723 + 0.00196 = 0.97921$$
And $1/2 (1−\alpha) = 1 − 1/2 (1 + \alpha) = 1 −0.97921 = 0.02079$
Thus P of a value equal to or less than 8.23 = 0.02079 = 2 percent.
3. Between 8.23 and 13.72

$$z_1 = (8.23 − 10.78)/1.25 = −2.04$$
$$z_2 = (13.72 − 10.78)/1.25 = 2.352$$
$$z_1 \text{ and } z_2 = 0.99059 − 0.02079 = 0.96980$$

Probability of occurrence of a variate value between 8.23 and 13.72 is 0.96980 or 97 percent.
4. Here $(x − \mu)/\sigma = \pm (2.05/1.25) = \pm1.64$
Normal deviate value for 1.6 = 0.94520 and 1.70 = 0.95543. By interpolating we get the value for 1.64 = 0.94929.

$$1/2 (1- \alpha) = 1 −1/2 (1 + \alpha) = 1 − 0.94929 = 0.05071$$
The area on both sides = $2 \times 1/2 (1 − \alpha) = 2 \times 0.05071 = 0.10142$

TABLE 6.1. Normal probability integral table showing areas under the normal curve for different z values.

$z (x−)/$ (Normal deviate)	Area to the left of the ordinate ½	z	Area	z	Area
0.0	0.50000	1.1	0.86433	2.1	0.98214
0.1	0.53983	1.2	0.88493	2.2	0.98610
0.2	0.57926	1.3	0.90320	2.3	0.98928
0.3	0.61791	1.4	0.91924	2.4	0.99180
0.4	0.65542	1.5	0.93319	2.5	0.99379
0.5	0.69146	1.6	0.94520	2.6	0.99534
0.6	0.72575	1.7	0.95543	2.7	0.99653
0.7	0.75804	1.8	0.96407	2.8	0.99744
0.8	0.78014	1.9	0.97128	2.9	0.99813
0.9	0.81594	2.0	0.97725	3.0	0.99865
1.0	0.84134				

Thus the probability of a variate deviating from the mean ± 2.05 is 0.10142, or 10 percent.

PRACTICAL APPLICATION EXAMPLES

Palaniswamy (1990) reported the distribution of the flowering duration of the ear-bearing tillers (panicles) in three different rice varieties (populations). They followed normal distribution (Figure 6.18). From this study, the rice researchers can estimate the number of days required for 50 percent, 75 percent, etc., of the tillers for flower in rice crop (Figure 6.19). Based on this information, one can know the flowering duration of the crop in different treatments in an experiment. The cumulative frequency table (Table 6.2.) is also given for easy reference and to show the utility of the normal distribution.

Palaniswamy (1990) studied the distribution of grain yield per plant in IR22 and IR 662 rice varieties tested under three nitrogen levels (0, 60, and 120 kg/ha). The expected number of plants yielding greater than the mean values for different percentages are presented (Figures 6.20 and 6.21).

EXERCISES

6.1. What is a normal distribution? How many normal curves are there?

6.2. How will you interpret the area under a section of a normal curve?

6.3. How can you check to see whether a sample is approximately normally distributed?

6.4. Explain the difference between the standard deviation of the sampling distribution of single observations and the standard deviation of the sampling distribution of the means. Why is the standard deviation of the means especially useful?

6.5. Given mean = 145.8, $s = 51.8$, approximately what proportion of the observations are in the interval mean ± 2 SD?

6.6. State the important properties of normal distribution.

6.7. State the use of normal distribution in some practical problems.

6.8. What is standard normal distribution? How will it be useful?

6.9. What percentage values in the normal curve relate to when the z values are 1.96 and 2.58, respectively?

6.10. Name the three important population distributions to be studied.

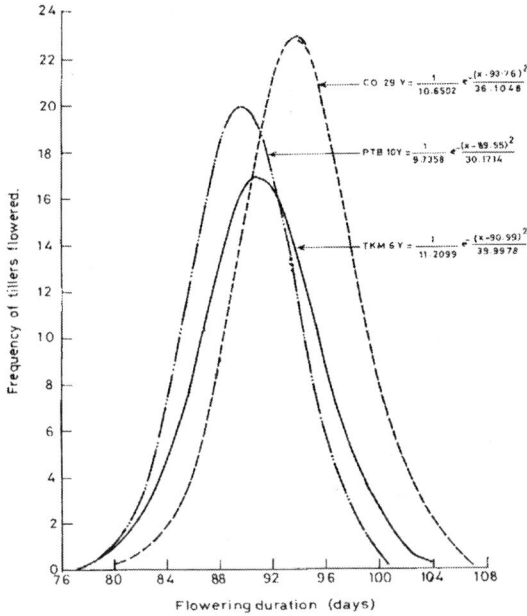

FIGURE 6.18. Normal curves for the flowering duration of different tillers of three rice varieties.

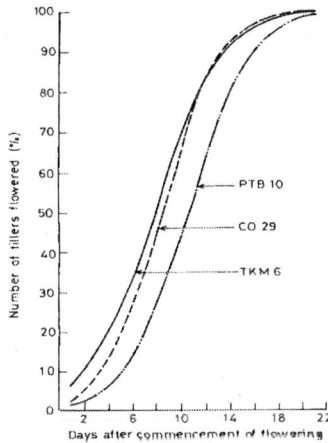

FIGURE 6.19. Cumulative distribution of number of tillers flowered at different days after commencement of flowering in three different rice varieties.

TABLE 6.2. Number of tillers flowered (percent) in different days and cumulative number of tillers flowered after commencement of flowering with corresponding flowering duration in three rice varieties.

						Variety					
	TKX 6				CO 29				PTB 10		
1	2	3	4	1	2	3	4	1	2	3	4
1	84	5.9	5.9	1	82	1.4	1.4	1	84	1.1	1.1
2	85	3.1	9.0	2	83	3.2	4.6	2	85	0.9	2.0
3	86	4.1	13.1	3	84	3.0	7.6	3	86	1.4	3.4
4	87	5.6	18.7	4	85	4.5	12.1	4	87	2.2	5.6
5	88	6.4	25.1	5	86	6.0	18.1	5	88	3.1	8.7
6	89	7.5	32.6	6	87	7.4	25.5	6	89	4.4	13.1
7	90	8.7	41.3	7	88	9.0	34.5	7	90	5.8	18.9
8	91	8.7	50.0	8	89	9.9	44.4	8	91	6.9	25.8
9	92	8.1	58.1	9	90	10.4	54.8	9	92	8.3	34.1
10	93	9.3	67.4	10	91	9.6	64.4	10	93	8.8	42.9
11	94	7.5	74.9	11	92	9.2	73.6	11	94	8.7	51.6
12	95	6.6	81.5	12	93	7.7	81.3	12	95	9.8	61.4
13	96	5.4	86.9	13	94	6.2	87.5	13	96	8.8	70.2
14	97	4.1	91.0	14	95	4.4	91.9	14	97	7.4	77.6
15	98	3.2	94.2	15	96	3.3	95.2	15	98	6.5	84.1
16	99	2.1	96.3	16	97	2.1	97.3	16	99	5.0	89.1
17	100	1.5	97.8	17	98	1.3	98.6	17	100	3.8	92.9
18	101	1.9	92.0	18	99	0.7	99.3	18	101	2.6	95.5
19	102	0.5	99.5	19	100	0.4	99.7	19	102	1.9	97.4
20	103	0.3	99.8	20	101	0.3	100	20	103	1.1	98.5
21	104	0.2	100					21	104	0.8	99.3
								22	105	0.7	100

Col 1 = Number of days after commencement of flowering
Col 2 = Flowering duration (in days)
Col 3 = Number of tillers flowered (%)
Col 4 = Cumulative number of tillers flowered (%)

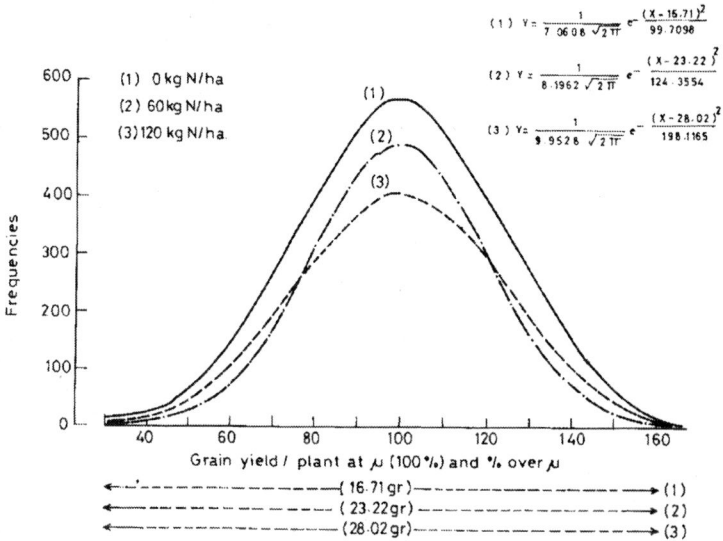

$$(1)\ Y = \frac{1}{7.0608\ \sqrt{2\pi}}\ e^{-\frac{(X-15.71)^2}{99.7098}}$$

$$(2)\ Y = \frac{1}{8.1962\ \sqrt{2\pi}}\ e^{-\frac{(X-23.22)^2}{124.3554}}$$

$$(3)\ Y = \frac{1}{9.9528\ \sqrt{2\pi}}\ e^{-\frac{(X-28.02)^2}{198.1165}}$$

FIGURE 6.20. Normal probability curve for the grain yield/plant data of IR 22 at three nitrogen rates having different μ and σ.

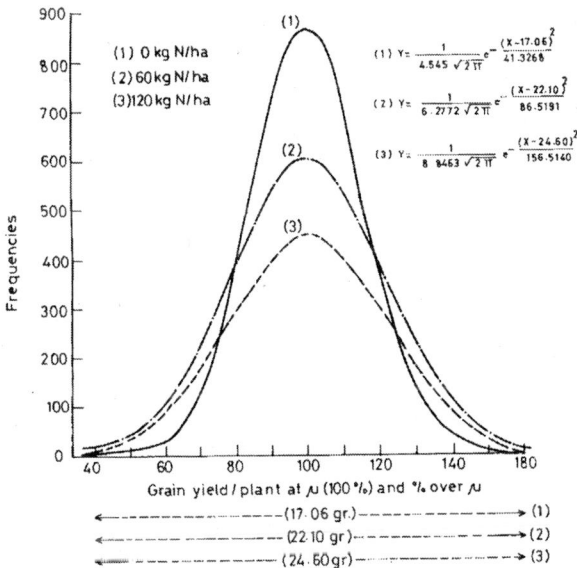

$$(1)\ Y = \frac{1}{4.545\ \sqrt{2\pi}}\ e^{-\frac{(X-17.06)^2}{41.3268}}$$

$$(2)\ Y = \frac{1}{6.2772\ \sqrt{2\pi}}\ e^{-\frac{(X-22.10)^2}{86.5191}}$$

$$(3)\ Y = \frac{1}{8.8463\ \sqrt{2\pi}}\ e^{-\frac{(X-24.60)^2}{156.5140}}$$

FIGURE 6.21. Normal probability curve for the grain yield/plant data of IR 662 at three nitrogen rates having different μ and σ.

REFERENCES

Campbell, R.C. (1974). *Statistics for biologists.* second edition. Cambridge: Cambridge University Press.

Mohr, L.B. (1990). *Understanding significance testing.* Newbury Park, London: Sage Publications.

Palaniswamy, K.M. (1990). Design of statistical (field) experiments and crop production in agriculture with special reference to rice. Doctoral thesis submitted to the University of Calicut, India.

Chapter 7

Probability

EXPERIMENTS

Scientists conduct experiments to test hypotheses, and to verify hypotheses to arrive at new findings or to acquire knowledge. As a result they collect data from their experiments, analyze appropriately, and draw inferences. In order to arrive at meaningful inferences, one must have some knowledge of probability. Simple definitions of the topics dealt with in this book are explained in this chapter. In statistical language, an experiment is defined as performance of some act or process, which can produce observations on a variable.

Examples:

1. Observing the market impact of a newly introduced consumer item
2. Throwing a coin and observing the outcome as either heads or tails
3. Conducting a germination test and observing the percentage of germination
4. Applying herbicide and observing its effect on weeds
5. Administering several drugs to several patients and observing their effects
6. Observing the effect of a new drug after administering to a patient
7. Testing the effect of a teaching method in a class
8. Comparing two teaching methods in a class

Experiments are classified into two types: determinate and nondeterminate.

Determinate Experiments

In determinate experiments, the observed result is not subject to change, i.e., if we repeat the experiment several times under the same conditions, we get the same results. The results do not change when the experiments are conducted repeatedly, e.g., (1) Ohm's law, (2) the angle of reflection of a ray of light equals its angle of incidence when focused on a certain surface, and (3) the relationship between the area of a circle and the equation: area $= \pi r^2$. The outcomes of these experiments are determined by the experimental conditions under which the experiments are conducted.

Nondeterminate Experiments

Contrary to determinate experiments, experimental conditions do not completely determine the outcomes in nondeterminate experiments. Some chance factors affect the results and consequently we do not get identical or similar results when the experiments are repeated. For example, (1) the effect of a drug administered in the same dose on different patients produces different results, (2) a fertilizer applied in the same quantity to a particular crop in the same place does not result in the same production when repeated, or (3) the same coin tossed several times does not give identical outcomes, i.e., heads and tails.

These results are uncertain because the experimenter is not able to either identify or control all the experimental conditions. In all such cases, there is an element of uncertainty of outcomes or results. A scale or mechanism is necessary to measure the extent of uncertainty in quantitative terms. The extent of uncertainty is measured by probability. There is no single commonly accepted definition of probability. The term *probability* is defined in two different ways.

1. *Definition for priori or classical probability:* The definition of this type of probability is explained with an example. A coin is tossed. A coin has two sides, a head and a tail. When we toss a coin, it will show either the head or the tail. The possibility of getting either the head or the tail is 1/2. If we calculate the probability this way, it is known as *priori* probability. It means that we can find the probability without actually performing the experiment of tossing a coin, i.e., probability is obtained *prior to* the event. Examples follow.
   a. A die has 6 faces showing 1, 2, 3, 4, 5, and 6. The probability of an event, say (i) rolling a 1, is 1/6, (ii) rolling an even number is 3/6.
   b. Getting a diamond card from a standard pack of playing cards is 13/52.
   c. Getting a success in an experiment is 1/2.
2. *Posteriori or frequency or empirical probability:* In many situations, we cannot enumerate all possible ways in which an event can occur and hence we use the relative frequency definition. It is based on

events that have already occurred. It is called a *posteriori* definition of probability. Examples follow.

a. Suppose 800 people (male) die of a heart attack in a city of 400,000 people. Then the probability of a person (male) dying in that city with a heart attack is 800/400,000. To arrive at a probability, we must specify the information on total probability.

b. In an agricultural experiment in rice crop, the frequency distributions of plant heights are recorded as follows:

Range (cm)	Frequency	Relative frequency	Proportion or probability
Below 85	27	27/200	0.135
85-90	78	78/200	0.390
90-95	75	75/200	0.375
95-100	20	20/200	0.100
Total	200	1.000	1.000

The relative frequencies and the corresponding probabilities are the same. They would become similar as sample size increases. The set of probabilities shows the probability distribution for the character. The sum of the probabilities for the four classes in the sample is one.

BASIC CONCEPTS OF PROBABILITY

Some of the terms frequently used are defined as follows.

definition for probability: If an event can happen in a ways and fail to happen in b ways and all these are equally likely, then the probability of the happening of an event is $a/(a + b)$ and the probability of the event not happening is $b/(a + b)$.

Let $p = a/a + b$ and $q = b/a + b$, $p + q = (a/a + b) + (b/a + b) = (a + b)/(a + b) = 1$

Where p = favorable cases out of the total and q is the unfavorable cases out of the total.

equally likely: All cases are said to be equally likely when there is no reason to expect one to occur more number of times than the other.

Examples: (1) Consider a pack of cards. There are 52 cards in the pack. Draw a card. Any card may appear in the draw. The 52 cases are said to be equally likely. (2) Consider that tossing a coin is an experiment. It yields one of the observations namely head or tail. There is no reason to expect a head to appear more often than a tail. We regard each of these two possible outcomes as equally likely. We expect head and tail in equal proportions. Suppose x relates to the number of heads obtained in n tosses, the proportion x/n tends to the value 1/2. We call this value the probability of observing a head when tossing a coin. Here the two outcomes, head and tail, are said to be equally likely.

event: An event is a set of possible outcomes. If an event has only one outcome, the outcome is the event itself. An event may consist of more than one outcome. The event may be a simple event or a joint event. Examples of a simple event: obtaining a head from a coin toss; example of a joint event: among several causes for death, death due to cancer.

exhaustive: This is the total number of all possible outcomes of an experiment. For example, when we throw a die, only one of the six faces (1, 2, 3, 4, 5, 6) may turn up, and there are only six possible outcomes. Hence, there are only six exhaustive cases or events.

laws of probability: The two important laws of probability are (1) the addition law, and (2) the multiplication law.

1. *Addition law:* Let $A, A_2, \ldots A_N$ be N mutually exclusive events, then the probability of occurrence of any one of the events is the sum of the probabilities of the separate events, i.e., $P(A_1 \text{ or } A_2 \text{ or } \ldots A_N) = P(A_1) + P(A_2) + \ldots + P(A_N)$.

Example 1: What is the probability of throwing a number greater than 4 when a die is thrown? The event can happen in two ways: either it can be 5 or 6. The probability of throwing 5 with a single die is 1/6 (p_1) and the probability of throwing 6 (p_2) is 1/6. The required probability is $1/6 + 1/6 = 2/6 = 1/3$.

Example 2: What is the probability of drawing either a diamond or heart from a pack of cards? A pack contains 52 cards. There are four suits and in each suit 13 cards. The probability of drawing a diamond is 13/52, and similarly the probability of drawing a heart is also 13/52. Probability of drawing either a diamond or a heart is $13/52 + 13/52 = 26/52 = 1/2$.

Example 3: There are ten students in two classes, A and B. In class A there are six students, and in class B there are four. What is the probability of selecting one student for these classes? The probability of selecting a student from class A is 6/10 and in class b is 4/10. Since A and B are mutually exclusive, the probability of AB = 0. Thus $P(\text{A or B}) = P(A) + P(B) = 6/10 + 4/10 = 1$.

Example 4: Let there be two circles *A* and *B*. There are 8 points in *A* and 6 points in *B*. The total points is 12, as shown in the following illustration:

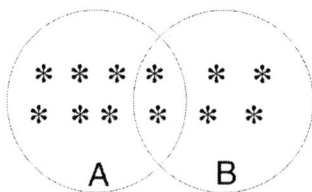

The $P(A) = 8/12$, $P(B) = 6/12$. It is not correct to find the probability as $P(A) + P(B) = 8/12 + 6/12$, which has a value greater than 1. The probability will not exceed 1. Two points lie in both A and B. They are no longer mutually exclusive, and $P(AB) = 2/12$. This probability is counted twice. Therefore, the correct probability is $P(A)$ or $P(B)$ is $P(A) + P(B) - P(AB) = 8/12 + 6/12 - 2/12 = 12/12 = 1$.

  2. *Multiplication law:* Let A occur in *a* ways, and another event B occur in *b* ways independently. The probability of A occurring is 1/2 and the probability of B occurring is 1/2. The probability of the two events occurring is $1/2 \times 1/2 = 1/4$.

mutually exclusive: Outcomes of an experiment are said to be mutually exclusive if it is impossible for any two or more of them to occur simultaneously. In general, all possible outcomes are mutually exclusive and the sum of those mutually exclusive outcomes is unity.

odds: The odds against an event are defined as the ratio of the number of unfavorable outcomes to the number of favorable outcomes. All outcomes are equally likely. Following is a table that shows an example of determining odds.

	Total events	No. of events Favorable to A	No. of events not favorable to A (B)	P	Odds against the event
Obtaining a diamond king from a pack of 52 cards (A)	52	1	51	1/52	51 to1
Obtaining a 9 (B)	52	4	48	4/52	48 to 4 or 12 to 1

properties of probability: The probability of the occurrence of an event is indicated by a number ranging from 0 to 1. If the probability is 1, it means that the event is certain to occur and if the probability is 0, it means that the event is certain not to occur. A probability of 0.99 refers to an event that may be expected to occur 99 times in 100, while probability of 0.05 refers to an event that may be expected to happen 5 times in 100. In general, the larger the probability, the more likely the event is to happen. When we say that the probability of an experiment has a success of 0.95, we mean that 95 percent of the experiments will give successful results. The probability of an event to occur definitely is 1, and the probability of an impossible event is 0, thus $0 \le p(A) \le 1$.

Example: Suppose in a birth register in a place, it appears that 52 out of 100 births are males. The probability of a child being a male is then 0.52. In calculating probability, the number of observations must be large. For example, among ten births, seven might be boys, but among 1,000 or 10,000, the ratio 0.52 would certainly appear.

random experiment or trial: Experiments conducted under a given set of conditions do not always give the same outcomes but rather different outcomes which follow a sort of statistical regularity. The outcomes are subject to chance. The outcome is affected by several factors. Such experiments are called random experiments.

sample space: A set of all possible outcomes from an experiment is called a sample space.

EXERCISES

7.1. What are two different types of experiments?
7.2. What is probability?
7.3. Define the *terms exhaustive, equally likely, frequency,* and *relative frequency.*
7.4. State the addition law of probability and explain with a suitable example.
7.5. State the multiplication law of probability and explain with a suitable example.

Chapter 8

Set Theory

SAMPLE SPACES

Set theory is a basis for the development of probability theory and hence it is discussed briefly in this section. A set is defined as the collection of all possible outcomes of an experiment or test. These outcomes are considered as sample points. All the sample points are in a space called sample space denoted by S.

Example: We are interested in studying some characteristics of a population, for example, average age of all students in a college. The set, which consists of the ages of all students in our population, is called the universal set or sample space.

If S contains only a finite number of points, then we can associate probability to each point such that the sum of all probabilities corresponding to all points equal 1. Thus, the sample space is also called a probability space. Sample spaces may be either discrete or continuous. A sample space whose elements are finite or infinite but countable is called a discrete sample space. A sample space whose elements are infinite or uncountable is called a continuous sample space. The sample points in the sample space vary depending upon the nature of the experiment.

Examples:

1. When we toss a coin, we get either heads or tails. Let heads be denoted by 1 and tails by 0. Thus we get two points, 0 and 1. These points are called sample points and the collection of the sample points is called sample space.
2. When we toss two coins, we get 1 1, 1 0, 0 1, and 0 0, and these four sample points are included in the sample space.
3. When we toss a single die, we get one of the six numbers, 1, 2, 3, 4, 5, or 6. Thus $S = \{1, 2, 3, 4, 5, 6\}$.
4. Life of a vacuum tube $S = \{t = 0 \text{ to } \infty\}$.

VENN DIAGRAMS

In a Venn diagram the sample space is represented by a rectangle and the sample points inside the rectangle. The sample space is represented by a rectangle. Subsets A and B are shown inside the rectangle. A and B are subsets and have no elements in common (see Figure 8.1).

Equality of two sets: A = (3, 4) and B = (4, 3). Here A and B are equal as both the sets contain the same elements and order of the elements may not be the same.

Event: An event is a subset of outcomes contained within S. It is designated by a capital letter other than the letter S. In the Venn diagram, A and B are two events. Since an event (A) is a subset of probabilities of space, S, the probability of an event is greater than or equal to zero (≥ 0) or less than or equal to one (≤ 1), i.e., $0 \leq P(A) \leq 1$.

Examples:

1. Tossing a coin is an event. If A is an event equal to obtaining a head, then A = (H), which is a subset of S (H, T).
2. Tossing a die gives several events. Suppose A = (1, 2, 3), which is a subset of S = (1, 2, 3, 4, 5, 6) or A may be A = (1, 4, 6) or A = (5).

Mutually exclusive events: In this sample space, the event A, B, or AB could occur. To find one of the events A or B, it is necessary to include the portion of the probability space covered by B. As A and B overlap AB, adding area A with area B would result in double counting of AB. To avoid double counting, we subtract AB from the sum of the areas A and B. Thus A or

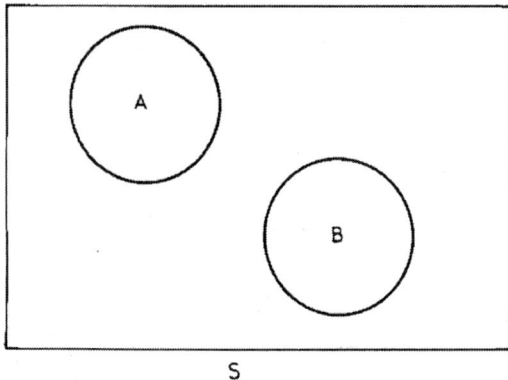

S

FIGURE 8.1. Venn diagram.

B = A + B–AB. In terms of probability, $P(A \text{ or } B) = P(A) + P(B) - P(AB)$ (see Figure 8.2).

When A and B are independent or mutually exclusive, as in Figure 8.3, we see AB = Φ, then $P(A \text{ or } B) = P(A) + P(B)$ as $P(AB) = 0$. The outcome AB is a null event and A and B are mutually exclusive, i.e., A and B have no points in common. Thus if AB = Φ = 0 (which means that the corresponding events are mutually exclusive).

Compound event: When an event is decomposable into a number of simple events, then it is called a compound event. Example: When rolling 2 dice, we get 7 from two dice. It is a compound event.

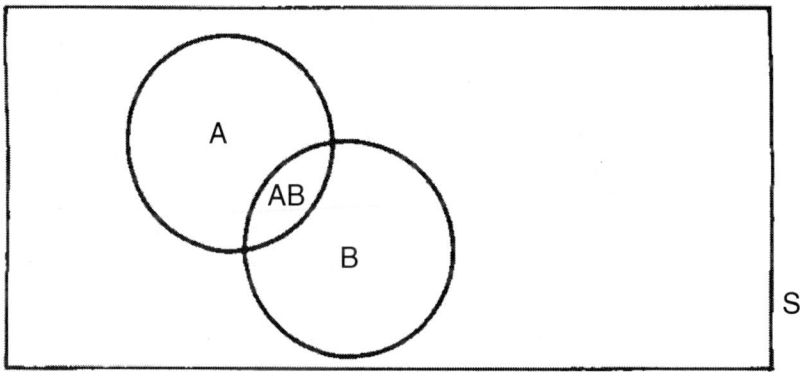

FIGURE 8.2. Figure explaining not mutually exclusive events.

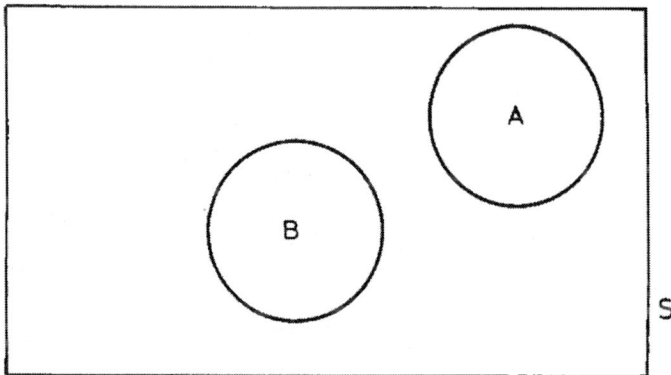

FIGURE 8.3. Figure explaining mutually exclusive events.

Simple problems:

1. Let $S = 1, 2, 3, 4, 5, 6, A = \{1, 2, 3\}, B = \{5, 6\}$. Then A^c (A complement) $= \{4, 5, 6\}$, i.e., S–A. B^c (B complement) $= \{1, 2, 3, 4\}$, i.e., S – B.
2. Compound events.
   a. Both A and $B = AB = \Phi$ and B^c
   b. Both A^c and $B^c = (4)$
   c. Both A and $B^c = AB^c$ $\{1, 2, 3\}$
   d. Both A^c and $B = A^cB = \{5, 6\}$
3. $A = \{2, 7, 8, 5\}, B = \{2, 7, 8, 5, 11\}$. Then $AB = \{2, 7, 8, 5\}$.
4. Given $S = \{1, 2, 3, 4, 5, 6, 7\}$, $A = \{2,3,4\}$, and $B = \{3, 4, 5, 6\}$.
      Then, $A \cup B = \{2, 3, 4, 5, 6\}$ (AUB reads as "A union B")
      $A \cap B = \{3, 4\}$ (A∩B reads as "A intersection B")
      $A' = \{1, 5, 6, 7\}$ and $B' = \{1,2,7\}$
5. Given: $S = a, b, c, d, e, f$ and $A = \{a,b\}, B = \{c, d, e, f\}$, then $(AUB) = \{a, b, c, d, e, f\}$, and $(AUB)' = 0$.
6. Given: $A = x$, x is an integer, and $10 < x < 14$, and y, y is an integer and $12 < y < 16$, then $A \cap B = 13$ as $A = 11, 12, 13$ and $B = 13, 14, 15$.
7. Given: $A = x$ and x is an integer $2 \le x \le 5$, $B = y$, and y is an integer greater than 3.
      Then, $A \cap B = \{4, 5\}$ as $A = 2, 3, 4, 5$ and $B = 4, 5, 6, 7, \ldots \infty$.
8. $S = \{x, y, z\}$. List all subsets of S. We get $A = x, B = y, C = z, D = x, y\ E = x, z\ F = y, z\ G = \{x, y, z\}\ H = \Phi$.

EXPECTED VALUE

Mathematical expectation: If we tell many 50-year-old men that they can expect to live 30 more years, it does not mean that they expect to live until their 80th birthdays and then die the next day. Some will live 10 years more; some 19 years; others 30 years. The life expectancy of 30 more years must be interpreted as an average or, as we call it, a mathematical expectation. Any numerical quantity whose value is determined by the outcome of an experiment is called a random variable. The random variable possesses the expected value. The expected value of a random variable X is denoted as $E(X)$ and is read "expected value of X." The expected value is also called mean value.

Example: Let X be a random variable taking the values $x_1, x_2, \ldots x_n$ with probabilities $p_1, p_2, p_3 \ldots p_n$ respectively, then

$$E(X) = x_1 p_1 + x_2 p_2 + \ldots + x_n p_n = \Sigma x_i p_i.$$

The expected value of a random variable is obtained by considering the various values that the variable can take, multiplying these by their corresponding probabilities, and then summing these products. The expected value of x in the binomial distribution, i.e., $E(X) = np$, i.e., the expected number of successes in n trials is defined as np.

EXERCISES

8.1. What is the expected value of the number of points that will be obtained in a single throw with an ordinary die?

8.2. In a game, a person can win a prize of $25 and its probability is 0.2, and win a prize of $10 and its probability is 0.4. What is the average prize?

8.3. A bar expects a profit of $300 per day if it rains. If it does not rain, it will lose $60 per day. The probability of rain for a specific day is 0.3. What is the expectation for the day?

8.4. One purchases a raffle ticket with probabilities 0.001 and 0.003 for winning the first prize of $5,000 or second prize of $2,000, respectively. What should be the fair price to pay for the ticket?

8.5. Given these X values and their probabilities, find $E(X)$, $E(X^2)$, and $E(X-\bar{X})^2$

X	8	12	16	20	24
$P(X)$	1/8	1/6	3/8	1/4	1/12

Chapter 9

Special Techniques in Descriptive Statistics

FREQUENCY DISTRIBUTION

Researchers collect data from their experiments in a systematic, scientific way in the order in which they occur. The data they collect are primary data, and usually bulky in nature as they consist of many observations. These data are called *raw data* or ungrouped data or unclassified data. These data are only a sample from a population of observations. The population may be of any size. Using the sample, we estimate the population values. An important characteristic of the data is variation. Variation is common in a myriad of characteristics. Some characteristics that show variation are: age, sex, height, weight, crop yields, blood pressure, blood sugar, temperature, sale of commodities, rainfall, wages of agricultural laborers, number of employees in supermarkets, soil fertility, amount of sunshine, etc. One cannot make any inference or conclusion as the values of the mass of data and the degree of variation show variation from one datum to the other, and the degree of variation varies from character to character.

The process of condensation of data is done by classification. Classification is done by grouping the observations of similar nature into homogeneous groups to obtain more clarity of the observations. There are two types of data: qualitative and quantitative. In the qualitative type of data, observations are not characterized by numerical numbers for example, sex, hair color, blood group, nationality, religion, etc. These data are counted and grouped based on the natural difference present in the individuals. For example, in plant genetics, scientists classify plants based on color such as pigmented or nonpigmented, or on height such as tall or dwarf. Sometimes, observations are classified based on attributes, e.g., in a college the students are classified as those who are graduate students and those who are doctoral students, grouped into male and female. They may be expressed in proportions or percentages.

Quantitative data may be classified into two categories: discrete and continuous. In the discrete type, the values are in integers (whole numbers) and in the continuous type, the values are in numerical measurements.

When the observations, discrete or continuous, occur in large numbers, it is useful to form groups by taking similar values, and counting their frequencies in each group. When grouping, we have to consider three important features, i.e., the number of class intervals, size of the class interval, and class limits.

Need For Grouping Data

Consider the following data on grain weight/plant (g) in 20 plants selected at random from 50 plants of IR 22 rice variety in an experiment conducted at the International Rice Research Institute, Philippines (Palaniswamy, 1990).

	26.9	34.3	23.5	21.4	26.1
	33.2	25.7	28.9	26.6	48.2
	20.3	40.3	40.2	25.2	28.2
	29.4	21.7	16.5	19.4	15.5
	37.5	30.8	35.0	27.3	11.8
	12.6	33.4	30.2	35.5	20.9
	35.4	29.1	26.4	16.2	12.1
	33.5	49.5	21.5	20.1	22.4
	32.7	32.8	5.9	24.6	33.6
	27.9	33.3	35.5	33.1	29.2
Column total:	289.4	330.9	263.6	249.4	248.0
					Total 1381.3

The data relate to ungrouped or unorganized observations. From the previous lists, one cannot describe the characteristics of the data and draw any conclusion or inference. The data can be studied in different ways to derive useful information.

Array

In the first stage, the data may be arranged in descending or ascending order. The data thus obtained are called *array* or the data are said to be arrayed. After arranging in descending order, the grain yield/plant (g) data are shown as follows.

49.5	33.6	29.4	26.1	20.3
48.2	33.5	29.2	25.7	20.1
40.3	33.4	29.1	25.2	19.4

40.2	33.3	28.9	24.6	16.5
37.5	33.2	28.2	23.5	16.2
35.5	33.1	27.9	22.4	15.5
35.5	32.8	27.3	21.7	12.6
35.4	32.7	26.9	21.5	12.1
35.0	30.8	26.6	21.4	11.8
34.3	30.2	26.4	20.9	5.9

The following information can be obtained from the arrayed data.
Advantages:

1. The maximum and the minimum values are easily located. Range of observations can be computed. Range is given by maximum minus minimum. Thus the range in the data is $49.5 - 5.9 = 43.6$ g.
2. The number of observations between two values can be calculated. For example: the values between two numbers, 33.6 and 29.4, is 20.
3. Some information on extreme values can be found. This will help the researchers to check whether there are extreme values. If so, they can be omitted and analysis performed.
4. A rough idea about the distribution can be obtained.
5. Since the data are arranged, we can find the median. For example, the median in this example is 1/2 (n/2th + 1th term) = +n/2 1/2 (25th, 26th term) = 1/2 (28.2 + 27.9) = 28.5.
6. If the distribution is normal, one can observe smaller differences in the middle values and larger differences among the values in the extremes.

Disadvantages of arraying the data: When the number of observations are numerically large, for example 100, it is difficult to array the data and hence we have to resort to other methods.

Grouping of the Data

When grouping data, values of similar magnitude are grouped and analyzed. The following steps will facilitate grouping in an efficient way. Grouping of data is very important and is similar to the process of sorting letters in a post office. The grouped distribution of the grain weight/plant is given in Table 9.1. The steps involved in this process follow.

TABLE 9.1. Distribution of grain weight/height of rice crop in grams (class interval = 5).

Class limits or class intervals	Midvalue (x_i)	Tallies	Frequency (f_i)	$x_i f_i$
5-10	7.5	/	1	7.5
10-15	12.5	///	3	37.5
15-20	17.5	////	4	70.0
20-25	22.5	//// ////	9	202.5
25-30	27.5	//// //// ///	13	357.5
30-35	32.5	//// //// //	12	390.0
35-40	37.5	////	4	150.0
40-45	42.5	//	2	85.0
45-50	47.5	//	2	95.0
Total			50	1395.0

Step 1. Find range. Find the highest and lowest values in the set of observations and compute range by subtracting the lowest value from the highest. In the example the range is 49.5 – 5.9 = 43.6.

Step 2. Selection of number of classes. Suppose we want the number of classes or groups to be 10 (grouping the individuals together usually means that their identities are lost). The class width or class interval is worked out on range/numbers of classes, i.e., 43.6/10 = 4.4 approximately. The value is taken as 5 for easy mathematical operations. The number of groups to be arrived at depends upon the magnitude of dispersion or scatter of observations in the data. Generally, the number of groups should not be smaller than 6 or 8 and not greater than 16. When variability is high, more groups are recommended and vice versa. Two methods are suggested.

- *Method 1:* Sturges' formula: Sturges (1926) rule states as follows:

$$k = 1 + 3.322 (\log_{10} n)$$

where k = approximate number of groups,
n = the total number of observations in the sample
$\log n$ = common log value to n.
For example: if $n = 80$, then
$$k = 1 + 3.322 \log 80$$
$$k = 1 + 3.322 \times 1.90 = 7.31$$
We can have seven classes.

- *Method 2:* Yule (1929) recommended the following formula to arrive at approximate number of classes:
$$\text{Number of classes} = 2.5 \times (\sqrt[4]{n}).$$
Where n is the number of observations in the given data. If $n = 50$, then the number of classes will be $2.54 \times 4\sqrt{50} = 6.6$ or 7.
- *Method 3:* Doane (1976) proposed the method for calculating the number of classes for an approximately symmetrical distribution as $k = 1.44 \log_e (n) = 3.3 \times \log_{10} (n)$.

Doane (1976) further suggests increasing the value of k by 2 or 3 when skewness is present. The value of k may be taken nearest integer to k. Equal class intervals are preferable. For the class intervals, if we use multiples of 5, 10, 100 it will be easy to work on the analysis. Table 9.1 shows formation of frequency.

The groups are called classes; the boundary defines class limits: the lower limit and the upper limit. The difference between the upper limit and the lower limit is called *class range* or *class width* or *class interval.* The average of the upper limit and lower limit is called *midvalue* or *class value* or *central value.* All of the values included in a particular class are assumed equal to the midvalue of that class. If the observed values are close to the midvalues, the bias in the estimation of the mean will be minimized and hence the midvalue in a class should be the representative value for that particular group.

Selection of Lower Limit in the First Class and Upper Limit in the Last Class

In the first group, the value of the lower limit may be fixed slightly below the minimum value (5.9 g is the minimum value and the lower limit is fixed as 5). Each class should begin with a multiple of the class interval. The first class begins with 5, the second class 10, the third 15, and so on. The class interval with midvalue 7.5 has the limits 7.5 ± 2.5. The class limits are lower limit $7.5 - 2.5 = 5$ and upper limit $7.5 + 2.5 = 10$ g/plant.

Formation of Frequency Table—Continuous Data

To summarize the data, a frequency table is formed. The tally method is followed in Table 9.1. A tally mark is made for each and every observation that falls within a particular class. This method in classification is employed invariably when we analyze continuous variables. The lower limit of any class is the same as the upper limit of the previous class. The values that are equivalent to the upper limit are included in that class and excluded in the next class whose lower limit is the same as the upper limit of the previous class. This method of grouping is called *exclusive method*. For example, consider the classes 30 to 35 and 35 to 40. The lower limit in the class 30 to 35 is 30 and the upper limit 35. If 35 is an observation, it is included in the class 30 to 35 and excluded in the next class, 35 to 40. The midvalue of the class 30 to 35 is the average of the lower and upper limits, i.e., (30 + 35)/2 = 32.5. The frequencies in the class 30 to 35, i.e., 12, are called class frequency. The class frequency is arrived at by the number of tally marks in that class. The difference between the upper and lower limits (35 – 30 = 5) is the class interval. Class intervals are maintained equally. Sometimes unequal class intervals may also be found.

DISCRETE VARIABLES

In cases of discrete variables, a gap will exist between the upper limit of the class and the lower limit of the following class.

Example: If the number of children in 15 randomly selected families are 5, 0, 4, 3, 8, 9, 7, 7, 4, 2, 3, 1, 3, 4, and 1, then the frequency table for this data by taking 2 as the class interval will be as in Table 9.2.

Sometimes this method is called the *inclusive method*.

Check: The sum of the frequencies should be equal to the total number of observations.

Selection of classes: By adopting Yule's formula, the approximate number of classes may be arrived at. If too many classes are fixed, some classes will have no frequencies and the distribution may not depict the true behavior of the distribution. If too few classes are fixed, many frequencies would appear in a single class or in a few classes and valuable information would be lost. After knowing the range and the number of classes, the class interval is fixed. The choice of intervals and the study of frequency distribution is arbitrary.

TABLE 9.2. Frequency table for discrete data.

Class limits	Tallies	Frequency
0-1	///	3
2-3	////	4
4-5	////	4
6-7	//	2
8-9	//	2

Equal class intervals facilitate other mathematical operations. Sometimes unequal class intervals are also recommended depending upon the circumstances. In some situations, e.g., when the range is less, or in smaller class intervals and more variation, greater class intervals are preferred. The information on the amount of variability, number of classes, and class intervals will help in the efficient summarization of the data.

Variability, Numbers of Classes, and Class Intervals

A rough idea of the class interval may be obtained when we know the variability, i.e., range in the set of observations. The class interval should not be more than 1/4 (one-fourth) of the standard deviation. It is difficult to obtain the standard deviation from the observations. The relationship between range, standard deviation, and size of sample are given in Table 9.3 for guidance.

In the grain weight/plant data, $n = 50$, range $= 43.6$. The multiplier for $n = 50$ is 0.222. Thus, the estimated standard deviation is $43.6 \times 0.222 = 9.6792$. Thus, the approximate class interval to be fixed is $1/4 \times 9.6792 = 2.42$.

The statistical measures computable from grouped data or frequency distribution should be close to the analogous values calculated from the ungrouped raw data. If the inference drawn from the frequency data differs considerably from the inference obtained from raw data, distortion of the characteristics of the data occurs. Hence, proper procedure is to be followed in the computation of frequency distribution, e.g., the arithmetic mean obtained from the ungrouped data to the grain weight/plant data ($\Sigma x_i/n$, i.e., $1381.3/50 = 27.63$ g) will be compared with the frequency distribution data ($\Sigma f_i x_i/\Sigma f_i$, i.e., $1395/50 = 27.90$ g). The difference $(27.90 - 27.63)$ of 0.27 g is due to grouping and hence, it is termed as *grouping error* (it is about 0.98 percent).

TABLE 9.3. Relationship between range, standard deviation, and size of sample.

N	Range SD	N	Range SD	N	Range SD_1	N	Range SD
2	0.886	8	0.351	18	0.275	150	0.189
3	0.591	9	0.337	20	0.268	200	0.182
4	0.486	10	0.325	30	0.245	300	0.172
5	0.430	12	0.307	40	0.231	400	0.169
6	0.395	14	0.304	50	0.222	500	0.164
7	0.370	16	0.283	75	0.208	700	0.159
				100	0.200	1000	0.154

Frequency Distribution with Tables and Graphs

Often, frequency distribution of a variable is shown in figures, graphs, and tables. The most important utility of statistics is analysis of numerical data by suitable statistical techniques. The frequency distribution, if depicted graphically, reveals the pattern of class frequencies more clearly than with numbers per se. Three methods of graphic representations are used: histogram, frequency polygon, frequency curve, and cumulative frequency curves.

Histogram

The frequency distribution on grain weight/plant (Table 9.4) is represented graphically in a coordinate system. This type of figure is called a histogram.

A histogram is a bar graph in which frequencies are shown along the y-axis (vertical axis) and class limits are shown along the x-axis (horizontal axis) (see Figure 9.1).

In Figure 9.1, the frequency of the class is directly proportional to the ordinate on the vertical axis. In the second vertical axis (on the right side) the frequencies are expressed as percentages. Thus, researchers have the choice of interpreting frequencies either in numbers or in percentage values. The area of the bar (rectangle) represents the frequency in the interval. When class intervals are equal, the heights of the rectangles as well as the areas are proportional to the concerned class frequencies represented. The class in-

TABLE 9.4. Frequency distribution of grain weight/plant data and computation of mean.

Grain weight/plant (g)	Numbers of Plants (f_i)	Midpoint (x_i)	$f_i x_i$
5-10	1	7.5	7.5
10-15	3	12.5	37.5
15-20	4	17.5	70.0
20-25	9	22.5	202.5
25-30	13	27.5	357.5
30-35	12	32.5	390.0
35-40	4	37.5	150.0
40-45	2	42.5	85.0
45-50	2	47.5	95.0
Total	50		1395.0

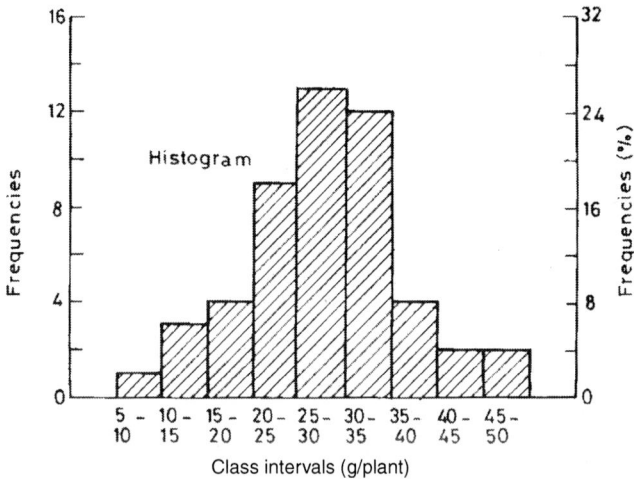

FIGURE 9.1. Histogram for the frequency distribution of grain weight per plant (g) data on 50 plants.

tervals should be chosen is such a way that the characteristic features of the distribution are emphasized.

Nine classes were derived from the data and correspondingly there are nine vertical bars or rectangles. These bars, erected at each class interval, constitute the histogram. The whole area of the histogram is equal to the total number of frequencies or observations. In general, no less than six or more than fifteen classes are recommended (Freund, 1971).

Relative frequency. The frequency in each class can be expressed as a fraction of the total frequency. This is called the relative frequency of the class to the total frequency. Thus, the relative frequency is the proportion of times a value occurs. The total of the relative frequencies would always add to one or 100 percent of the cases. The total area occupied by the rectangles (histogram) is also one. In the figure, the area of the rectangle is defined as height of the rectangle (class frequency) times base (class interval).

Uses of histogram. Histograms roughly show the center of the distribution, the spread, and shape. The shape gives an idea as to how the entire population should look. For example, a shape that is symmetrical about the center with more observations in the central region might reflect normal population. In our grain weight/plant data, the sample size is 50 and the sample mean is 27.63 g/plant. The histogram shows that the mean lies in the class 25-30, which has the highest frequency.

Frequency Polygon

In the histogram (Figure 9.1), there are several bars or rectangles. Plot the midpoints of the rectangles at different class intervals. Join the successive midpoints with straight lines. Figure 9.2 is a frequency polygon formed by a connecting series of ordinates whose heights are proportional to the various frequencies.

Suppose there are two classes at the endpoints, e.g., 0-5 and 50-55, then the frequencies in these classes are 0. The polygon may be completed by joining these two points at the x-axis. This is done to make the area under the polygon equal to the area under the corresponding histogram. It will also show two different distributions in the same graph, e.g., distributions of breast and cervical cancer cases compared in one frequency polygon (see Figure 9.3). The data are shown in Tables 9.5 and 9.6.

Frequency Curve

To explain a frequency curve, instead of $n = 50$ in our grain weight/plant data, consider that n is increased considerably and the class intervals of the frequency distribution are very small. In this case, we will get a larger number of rectangles with closer and closer midvalues. When the midvalues lie closer and when all the class intervals tend to zero and at the same time the number of intervals approach to infinity, then the outline of the histogram

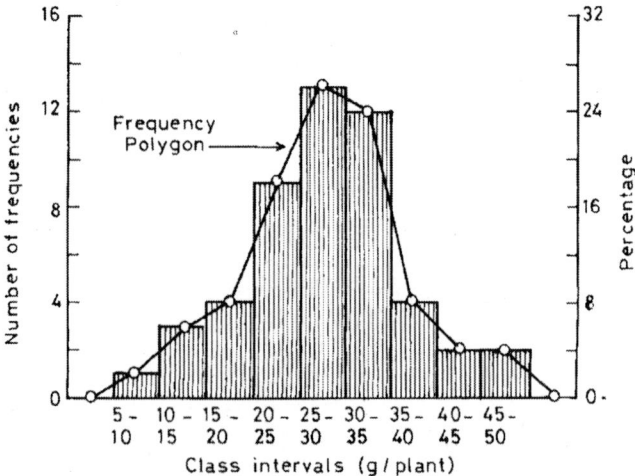

FIGURE 9.2. Frequency polygon for the grain weight (g) per plant data on 50 plants.

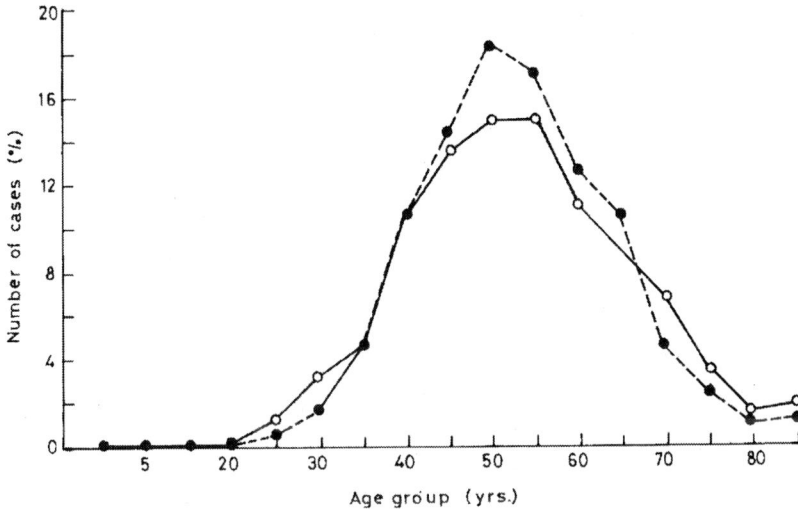

FIGURE 9.3. Age-group distribution of incidence of cancer cases in cervix and breast in females for the years 1982-1987. Data represent average of 1982-1987. (*Source:* Cancer Institute, 1982.)

and frequency polygon would approach to a smooth curve. Such a limit in a histogram or polygon is called a frequency curve. The frequency curve takes the shape of the parent population. The area under the curve is 1 (Figure 9.4).

Cumulative Frequencies

Sometimes we want to know the information on the number of observations (frequencies) less than, greater than, or equal to specified value(s). This information can be obtained from cumulative frequency distribution. Cumulative frequency up to any class is defined as the sum of the frequencies of all classes up to that class. The last cumulative frequency is equal to the total frequencies.

The following list explains the method of forming cumulative frequencies (see Figure 9.4).

TABLE 9.5. Age-group distribution of incidence of cervical cancer in females for the years 1982 to 1987.

Age group	1982	1983	1984	1985	1986	1987	Mean	%
0-4	0	0	0	0	0	0	0.0	0.0
5-9	0	0	0	0	0	0	0.0	0.0
10-14	0	0	0	0	0	0	0.0	0.0
15-19	0	0	0	0	0	0	0.0	0.0
20-24	2	5	2	4	2	1	2.7	0.5
25-29	6	13	9	13	10	10	10.2	1.8
30-34	25	17	22	32	26	30	25.3	4.5
35-39	78	71	50	51	61	54	60.8	10.8
40-44	61	79	83	83	83	92	80.2	14.2
45-49	87	86	99	112	132	105	103.5	18.3
50-54	81	93	84	100	122	101	96.8	17.1
55-59	63	80	60	74	79	76	72.0	12.7
60-64	62	53	59	58	71	58	60.2	10.6
65-69	18	25	26	29	35	30	27.2	4.8
70-74	7	10	19	21	15	15	14.5	2.6
75-79	6	1	5	9	6	8	5.8	1.0
80+	12	7	8	5	5	1	6.3	1.1
Total	508	540	526	591	647	581	565.0	100.0

Source: Cancer Institute, 1982.

1. On x-axis, mark the class intervals of the data.
2. Mark cumulative frequencies on the y-axis.
3. At the upper boundary of each interval, place a dot on the height of the cumulative frequency for that interval.
4. The lower boundary of first interval is zero. There are no frequencies less than this point.
5. Connect adjacent dots with straight lines and you have the cumulative frequency polygon.

Two kinds of cumulative frequencies (ogives) (see Figure 9.5) are discussed as follows:

1. *Less than cumulative frequency curve:* This relates to the total numbers of frequencies *less than* the upper limit. It is also called less than ogive curve. The cumulative frequencies are plotted against the upper

limits, i.e., the points in the upper limits are joined. The curve takes S shape.

2. *Greater than cumulative curve:* In this curve, the cumulative frequencies are formed considering greater than the lower limit criterion. The graph looks like a reversed S-shaped figure (sloping downward from the upper left to the lower right on the graph). The cumulative frequency is plotted against the lower limit and the cumulative frequencies diminish.

Example: We have reviewed frequency, relative frequency, frequency in percent, cumulative frequency less than upper limit, cumulative frequency less than upper limit expressed in percent, cumulative frequency greater than lower limit, and cumulative frequency greater than lower limit expressed in percent for the grain weight/plant data (see Table 9.7). To convert a frequency distribution (or cumulative distribution) into a corresponding percentage distribution, we have to divide each class frequency (or each cu-

TABLE 9.6. Age-group distribution of incidence of breast cancer cases in females for the years 1982 to 1987.

Age group	1982	1983	1984	1985	1986	1987	Mean	%
0-4	0	0	0	0	0	0	0.0	0.0
5-9	0	0	0	0	0	0	0.0	0.0
10-14	1	1	0	0	0	0	0.3	0.1
15-19	0	0	1	0	0	0	0.2	0.1
20-24	2	5	2	2	2	3	2.7	1.1
25-29	8	4	9	6	8	12	7.8	3.3
30-34	12	10	12	12	10	10	11.0	4.6
35-39	28	15	24	25	29	34	25.8	10.8
40-44	34	32	36	33	33	30	33.0	13.8
45-49	41	30	36	38	36	34	35.8	15.0
50-54	28	35	33	29	46	44	35.8	15.0
55-59	17	27	25	38	22	30	26.5	11.1
60-64	22	29	24	29	29	23	26.0	10.9
65-69	16	17	15	8	23	21	16.7	7.0
70-74	9	6	6	13	9	10	8.8	3.7
75-79	1	3	7	2	6	5	4.0	1.6
80+	11	3	4	3	2	4	4.6	1.9
Total	230	217	234	238	255	260	239.0	100.0

Source: Cancer Institute, 1982.

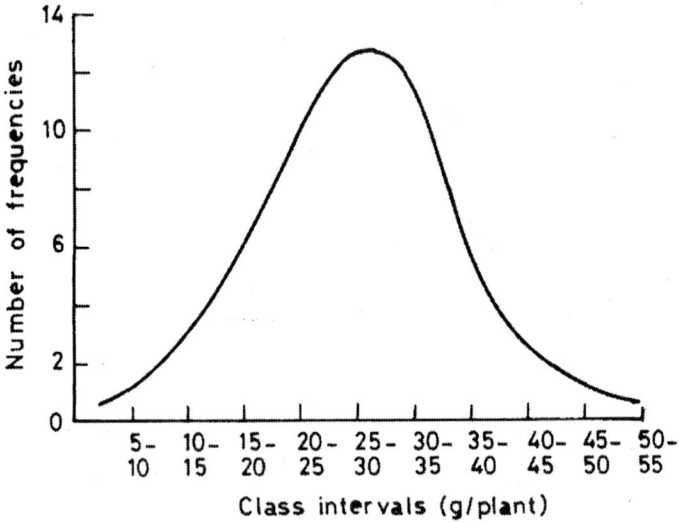

FIGURE 9.4. Frequency curve for the distribution of grain weight per plant data given in Table 9.4.

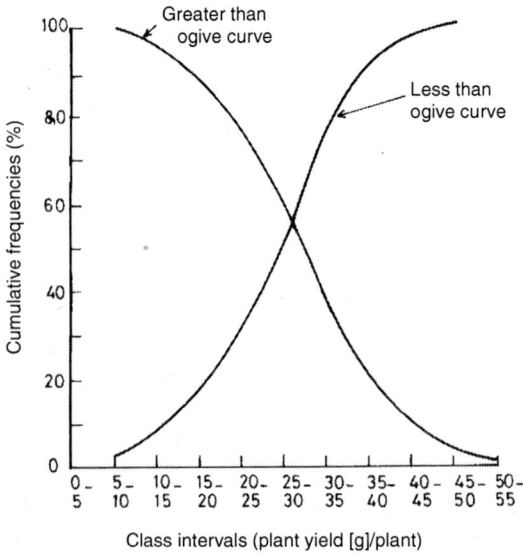

FIGURE 9.5. Ogive curves for the distribution of grain weight per plant data.

TABLE 9.7. Cumulative frequencies for grain weight/plant data.

Class limit	Midvalue	Frequency	Relative freq.	Freq. in %	Cum. freq. less than upper limit	Cum. freq. less than upper limit in %	Cum. freq. greater than lower limit	Cum. freq. greater than lower limit in %
5-10	7.5	1	0.02	2	1	2	50	100
10-15	12.5	3	0.06	6	4	8	49	98
15-20	17.5	4	0.08	8	8	16	46	92
20-25	22.5	9	0.18	18	17	34	42	84
25-30	27.5	13	0.26	26	30	60	33	66
30-35	32.5	12	0.24	24	42	84	20	40
35-40	37.5	4	0.08	8	46	92	8	16
40-45	42.5	2	0.04	4	48	96	4	8
45-50	47.5	2	0.04	4	50	100	2	4
Total		50	1.00	100				

145

mulative frequency) by the total number of items grouped and multiply by 100. Percentage distributions are particularly useful when comparing two or more sets of data.

Uses of Ogive Curves

1. Suppose we want to know the number of plants producing grain weight/plant, e.g., 20 g or less than 20 g. Draw an ordinate at 20 g in the x-axis so that it intersects the curve at point A. Then join A and the corresponding point in the y-axis, which give the frequencies. If the value is 8, then we can infer that 8 plants (16 percent) gives yield less than or equal to 20 g.
2. If the number of plants yielding 40 g or more is required, a similar estimate may be made from greater than ogive curve.
3. Cumulative frequencies can be expressed in percentages. Percentage cumulative frequency curves can be used to compare different distributions.
4. We can determine the number of plants with grain yields within any given interval.

Dot Diagram

To determine the distribution of the data, we can also use a dot diagram (see Figure 9.6). From the dot diagram we can discern the following:

1. The general location of the observations
2. The spread of the observations

Following is a comparison between a histogram and a frequency polygon.

Histogram	*Frequency polygon*
Frequency distribution is represented by a bar diagram called a histogram.	Frequency distribution is represented by a line chart called a frequency polygon.
Frequency of occurrence in a class is represented by a rectangle.	Frequency of occurrence is represented by a single point.
If the objective is to emphasize the differences between the frequency of occurrences of classes, it is preferred.	If the objective is to visually depict the pattern of the data, it is preferred.
For discrete data, a gap may be left between rectangles.	No gaps.
In the same graph two histograms are not possible.	Two or more distributions can be plotted in the same diagram.

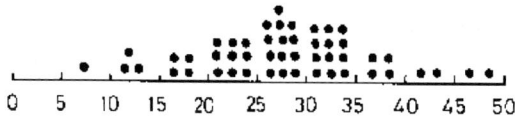

FIGURE 9.6. Dot diagram showing the distribution of grain yield (g/plant) data.

EXERCISES

9.1. The annual rainfall (inches) recorded in an agricultural research station is given as follows:

18.5	9.5	14.9	9.9	16.4	12.6	11.2	9.3
10.6	18.4	10.4	8.8	15.7	17.8	8.9	11.7
8.3	16.8	11.6	9.3	16.0	10.1	11.6	16.0
13.1	11.7	11.9	16.9	14.2	10.5		

    a. Arrange the data in ascending order.

    b. Prepare a frequency table with a class width of 1.0 inch and start 8.0 inches as the lower limit of the first class.

    c. Draw a histogram, a frequency polygon, and a frequency curve diagram.

    d. In how many years did the precipitation exceed 13.5 inches and in how many years was it lower than 10 inches?

9.2. What information can be obtained from a frequency distribution?

9.3. What are the advantages and disadvantages in fixing larger numbers of intervals?

9.4. Define the terms *histogram, raw data, discrete variable, continuous variable, array, grouping of data, relative frequency,* and *frequency curve.*

9.5. The following figures give the weights in pounds of 60 pigs used for a feeding trial.

202	195	191	179	191	227	228	200	201	180
224	203	201	204	219	192	194	188	184	195
209	204	186	199	178	248	227	190	167	200
150	209	218	197	150	171	199	149	180	191
229	220	200	174	219	173	224	234	200	173
140	195	217	185	181	215	207	169	204	204

Form a frequency table using eight groups.

9.6. What is grouping error?

9.7. The moisture content of 40 samples of clay was measured as follows:

10.4	11.3	10.5	12.2	9.6	11.0	11.4	9.1	12.1	10.4
8.9	11.4	8.3	9.3	9.2	9.8	11.4	10.3	11.7	10.2
9.7	10.2	8.7	11.4	10.6	13.1	9.8	10.3	11.4	10.4
11.6	12.2	11.7	7.8	10.7	8.7	10.1	10.2	8.3	9.1

a. Compute grouped frequency distribution for these results.
b. Draw a histogram.
c. Draw a frequency polygon and a cumulative frequency diagram.

9.8. In an industry, the numbers of rejected articles in 40 batches are given as follows:

B. No	No. Rej	B. No	No. Rej	B. No	No. Rej	B. No	No. Rej
1	8	11	6	21	2	31	0
2	5	12	2	22	3	32	1
3	0	13	5	23	1	33	0
4	1	14	4	24	7	34	2
5	2	15	1	25	0	35	4
6	3	16	3	26	5	36	7
7	0	17	3	27	3	37	6
8	1	18	2	28	2	38	2
9	2	19	6	29	3	39	0
10	3	20	2	30	3	40	1

Draw a frequency distribution and a histogram for these data.

9.9. A pediatrician began testing the cholesterol levels of young patients and was alarmed to find that a large number had cholesterol levels over 200. List of readings of 40 high-level patients are:

210	219	215	208	210	212	218	202
209	211	219	219	210	212	208	195
212	219	207	208	199	215	221	200
208	221	210	213	213	221	221	214
205	206	203	208	219	213	221	206

Group the data using a class width of 5.

9.10. A statistician interested in obtaining information about the weights (lbs) of newborn babies gathered the following data from hospital records:

9.8	6.5	9.5	5.1	4.5	8.8	6.5	9.5	7.7	6.9	6.6	6.0 7.9	7.8
7.7	6.9	6.6	5.8	7.1	6.8	8.4	6.9	5.8	7.1	6.8	8.5 9.8	5.0
5.7	3.8	7.4	7.3	8.6	9.8	3.8	10.3	7.4	5.7	4.6	7.7 10.3	7.4
5.7	8.0	7.8	8.9	5.8	8.6	7.8	8.5	5.6	7.4	6.7	4.5 6.7	6.0
7.8	6.6	7.4	5.7	8.8	9.4	6.0	5.9	7.4	8.8	9.4	6.0 7.2	7.0
9.4	7.4	8.9	7.2	10.5	8.4	10.4	7.8	5.0	5.6	8.0	10.4 3.0	7.8

Form a frequency table using 1.0 lb as class width and starting at 3.0 lbs.

9.11. A social scientist conducted a survey on households and collected the following data on ages of the family members in 20 households. The data are given as follows:

(1) 40, 40, 17	(2) 27, 3, 5	(3) 32, 28, 6, 5	(4) 66, 62
(5) 28, 29, 28	(6) 36, 11, 14	(7) 5 8 , 27, 24, 1	(8) 35, 12, 8
(9) 35, 33, 11, 8	(10) 29, 11, 10	(11) 50, 46	(12) 47, 42, 16, 15
(13) 29, 27	(14) 39, 35, 13, 10	(15) 32, 30, 3	(16) 49, 43, 17, 15, 14
(17) 39, 39	(18) 76, 44, 38, 17, 15	(19) 43, 38, 18	(20) 53, 24

    a. Draw a histogram for the ages with 10 as class width. Start with 0-9.

    b. Find the frequencies and relative frequencies in the class width (i) 0-9 and (ii) 0-4.

    c. Compare the two groupings of the data and decide which one is more useful.

9.12. Fill in the gaps in the following sentences.

    a. In a histogram, frequencies are represented by _____.

    b. The following frequency data are given:

Class limits	Frequency
30-39	nil
40-49	6
50-59	24
60-69	64
70-79	50
80-89	29
90-99	14
100-109	6
110-119	3
120-129	1

The total frequency is _____.

    c. The interval width is _____ and the interval midpoints
       are _____.

    d. When drawing a histogram for this distribution, the bases of the
       rectangles would have their endpoints _____

       _____.

    e. When the ogive is drawn, the vertical scale will represent _____
       _____ and the horizontal scale _____.

9.13. Why should the number of classes in a frequency distribution be nei-
ther too small nor too large?

9.14. Give some examples for qualitative data and quantitative data.

9.15. How will you decide approximately the number of classes to be
fixed while forming a frequency table for a set of data?

9.16. What are the differences between a histogram and a frequency polygon?

9.17. Consider the following data relating to the weight of apples (g).

112	76	80	89	102	130	126	84	116	90
108	89	119	188	100	116	80	202	132	100
118	99	112	80	83	138	127	128	111	74
60	128	125	89	70	78	108	102	96	89

Draw a frequency table, a histogram, a frequency polygon, and a frequency curve.

9.18. How many classes would be appropriate if you want to construct a
frequency distribution for 60, 250, and 2,000 observations?

9.19. What is grouping error?

9.20. Given 200 numbers such that the smallest is 14.5 and the largest
65.3. Set up a table that would be suitable for grouped data. Give the
class marks as well as boundaries.

9.21. If the class limits in a frequency distribution are 15-18, 19-22, 23-26,
and 27-30, what are the boundaries of the classes? Class marks?

9.22. Draw a frequency distribution to represent the following test scores:
0, 2, 1, 2, 3, 2, 2, 3, 3, 3, 5.
Draw a histogram and a frequency polygon to illustrate the frequency distribution.

9.23. Explain the concept of frequency curve.

9.24. How do you find class width?

9.25. What is a dot diagram? What information can be obtained from it?

9.26. What is array? State its uses.

REFERENCES

Cancer Institute (1982). Cancer registry annual reports for the years from 1982 to 1987. Adyar, Madras, India: Author.

Doane, D.P. (1976). Aesthetic frequency classifications. *Amer. Stat.,* 30(4).

Freund, J.E. (1967). *Modern elementary statistics.* Englewood Cliffs, NJ: Prentice Hall.

Palaniswamy, K.M. (1990). Design of statistical (field) experiments and crop forecast in agriculture with special reference to rice. Doctoral thesis submitted to the University of Calicut, India.

Strurges, H.A. (1926). The choice of class intervals. *J Amer. Statistical Assoc.* 21:65-66.

Yule. (1929). An introduction to the theory of statistics. Ninth edition. In C. Griffin (Ed.), pp. 211-213.

Chapter 10

Population, Sample, and Statistical Inference

POPULATION

Population is also sometimes called universe. Population is the totality of elements that have one or more characters in common. In statistical investigations, researchers are interested in the study of the characteristics of their population. Therefore, the population concerned is to be specified carefully. Suppose we are interested in a particular character in the population, and we observe that this character takes different values, and these values constitute a distribution to that particular character. Our goal is to study the character of our interest and make inferences. Examples of populations are provided here:

1. Number of farmers in a village
2. Number of people in a city
3. Number of trees in a forest
4. Children studying in a school
5. Machines produced in a factory
6. Heights of people in a city or in a country
7. People in a particular age group in a place
8. Population of trees infested with a particular disease
9. Population of patients affected with a disease, e.g., cancer
10. Number of books stocked in a library
11. Suppose we consider that people in a city constitute our population. If we take the blood pressure (systolic and diastolic) of each individual in the population, then we get two types of population values—one relating to systolic and the other to diastolic.
12. Population of pigs in a given place
13. Number of patients admitted to hospitals in a city

Population may be finite or infinite. If the elements in the population are countable, then the population is finite. Finite populations may vary in size from small to very large containing a number of elements.

In an infinite population, the elements in the population are uncountable. Populations may be real or hypothetical. Scientists propose a hypothesis, and then test its validity based on the data collected from the population. When the population size is large, it is impossible to collect data from each and every member (element) of the population. Hence, an explicit definition of the population will help accurate data collection and in drawing inferences or conclusions about the population. Second, proper definition of the population will help in the selection of a representative sample after identifying each and every element in the population.

SAMPLE

Scientists cannot study whole populations. They do not have resources such as money, expertise, labor, and time. So, scientists are compelled to study only a part of the population called a sample, and relate inference to the population from which the sample was drawn. Scientists expect the sample to represent the population in all respects. Thus, the nature of the population is inferred from the known sample. This process is related to what is called statistical inference. Sometimes, the sample does not represent the population and it will lead to erroneous conclusions. Therefore, it is imperative to fully understand the population, sample, and sampling techniques and procedures in drawing samples. Like the population, a sample also varies in size. In short, only a properly selected sample can yield precise results.

PARAMETERS AND STATISTICS

From the population as well as the samples, we can determine important statistics such as mean, mode, median, variance, standard deviation, standard error, correlation, etc. A quantity calculated from a population is called a *parameter*. A quantity calculated from a sample is called an *estimate* (or *statistic*). The symbols used in denoting some of the parameters and estimates are shown here:

Character	Parameter	Estimate
Mean	μ (mu)	$\bar{x}$ ("x bar")
Variance	σ^2 (sigma square)	s^2 ("s squared")
Standard deviation	σ (sigma)	("s")
Standard error	$\sigma / \sqrt{N}$	"$s/\sqrt{n}$"
Correlation coefficient	P (rho)	"r" (small letter)
Regression coefficient	β (beta)	"b" (small letter)

The population parameters are fixed and hence there are no distributions whereas sample estimates differ from sample to sample and do possess a distribution. The sample we select from the parent population should represent the population so that the characteristics of the population can be estimated. For example, a representative sample of 100 is generally preferable to an unrepresentative sample of a large number, e.g., 1 million.

Statistical Inference

Any conclusion relating to a population made on the basis of a sample of observations is called statistical inference.

Some examples of sampling: A rice breeder identifies the quality of a rice variety based on the examination of a few grains. A soil chemist declares the pH value of the field (larger area) after determining the pH value from a small soil sample. A farmer determines the percentage germination of the seed lot of a crop by determining the germination percentage from a sample of seeds. A person tastes a grapefruit before buying a large quantity. The quality of oranges in a crate or lot is tested by tasting a few before purchase. Patients in a clinical trial serve as a sample as they provide information for a larger group of patients. The protein content of a crop is determined from a few grains selected at random. Thus, sample and sampling are important tools in making decisions in our everyday life.

Sample and Sampling

Sample and sampling refer to the techniques and procedures for selecting a sample from the population in order to make inferences about the entire population. A properly selected sample can give precise estimates sufficient for almost any reasonable purpose. The aim of sampling is to estimate population parameters or to indicate the limits within which the parameters are expected to lie with a certain degree of confidence.

Random sample: A sample is random if every element of the population has an equal chance of being included in the sample. A random sample will give close estimates of the population and eliminate bias.

Sampling Error

The mean of the sample may or may not be equal to the population parameter. The difference between the estimate obtained from the sample (statistic) and the parameter value of the corresponding population is called sampling error. Suppose we take k samples each of size n from a population.

Each sample will give an estimate and we will have k estimates. The mean is represented by μ in the population and $\bar{x}$ in the sample. Similarly, the variance in the population is represented by σ^2 and in the sample by s^2. The standard deviation of the means is called standard error of the mean (SEm). The standard error tells us the precision of the sample mean. The measurements of a character will have a distribution characterized by its mean and variance. These two statistics are obtained from different samples drawn from the same population. Thus mean and variance vary from sample to sample whereas the mean and variance of the population do not vary and they are constants for a particular population.

Sampling Distribution of Means

Suppose we have a population whose mean and variance are μ and σ^2, respectively. We select k samples each of size n. For each sample, we get mean, median, mode, variance, standard deviation, etc. Each one is a statistic. The distribution of a statistic over k samples of size n from any well-defined population is called sampling distribution of the statistic. The idea of sampling distribution is fundamental for understanding statistical inference. This idea is difficult to grasp in one reading and hence, students will have to study this topic very carefully to understand and familiarize themselves with this concept. We are mainly concerned with sampling distribution of the mean and the variance. We find that the k sample means are more close to the population mean, μ, than the single observations. If we study the standard deviation of the means, the resulting value is called the standard error of the mean (SEm) (Mohr, 1990). It is shown by the formula $s/\sqrt{n}$, i.e., standard deviation of the individual observations divided by the square root of the sample size. The standard deviation of any statistic has a special name called standard error. The distribution of the means will give rise to normal distribution whether the sample means are drawn from skewed, binomial, rectangular, normal distribution, etc. Since the means are normally independently distributed (NID), it follows that about 95 percent of the means could be within 2SEs of the population mean. The experimenter has only one sample mean and is interested in getting an idea of how far away μ is likely to be from the particular sample mean; 95 percent of the sample means will fall within 2 SEs of μ. The value $t \times$ SEm will show, for the given probability and sample size, the interval in which μ lies.

SAMPLE SIZE AND SAMPLING DISTRIBUTION
OF MEAN AND VARIANCE

When n = 2

A population consists of four elements ($n = 4$) whose values are 4, 5, 6, 9. Now we want to take a sample of size 2 ($n = 2$). Then we can draw 4C2 samples, i.e., 6. The 6 samples with their mean, variance, standard deviation, and standard error are given in the following table. The population mean (μ) is $(4 + 5 + 6 + 9)/4 = 24/4 = 6$ and population variance (σ^2) is 14/3 (note that the divisor is 4–1= 3) and standard deviation (σ) is 2.1602.

Sample no.	Elements in the sample	Mean $(\bar{x})$	Variance (s^2)	Standard deviation (s)	Standard Error (SEm) $(s/\sqrt{n})$
S1	4,5	4.5	0.5	0.7071	0.5001
S2	4,6	5.0	2.0	3.5355	2.5000
S3	4,9	6.5	12.5	3.5355	2.5000
S4	5,6	5.5	0.5	0.7071	0.5000
S5	5,9	7.0	8.0	2.8284	2.0000
S6	6,9	7.5	4.5	2.1213	1.5000
Total		36.0	28.0	11.3136	8.0000
Mean		6.0	28/6=14/3	1.8887	1.3333

Standard error of the mean is the standard deviation of the sample means namely 4.5, 5.0, 6.5, 5.5, 7.0, and 7.5 is calculated as

$$SEm = \sqrt{[(\Sigma x_i^2) - (\Sigma x)^2 \ / \ n] / (n-1)}$$
$$= \sqrt{[(4.5^2 + ... + 7.5^2) - (36)^2 \ / \ 6] / 5} = \sqrt{7/5} = \sqrt{1.4} = 1.18$$

The standard error for each sample mean (SEm) is calculated using the mathematical relationship $s/\sqrt{n}$. The standard errors for different samples are given in the last column of the table. Sample 1 and sample 4 show the least standard error of 0.50.

Properties of Mean and Variance

The average of sample means (6.0) and sample variance (14/3) are exactly equal to the population mean and variance, respectively. Hence, the sample mean and sample variance are called unbiased estimates of population mean and population variance, respectively. Note that the average sample standard deviation (s) of the samples is not equal to the population standard deviation (σ). Standard error indicates that the sample mean deviates from μ to the extent of the value of standard error 0.5.

When n = 3

From the previous population, consider samples of size 3, i.e., $n = 3$. We can take $^4C_3 = 4$ samples. As before, we form the following table to study the effect of sample size.

S. no.	Elements in the sample	Mean ($\bar{x}$)	Variance (s^2)	Standard deviation (s)	Standard error ($s/\sqrt{n}$)
S1	4,5,6	15/3 = 5.0	6/6	1.0000	0.5774
S2	4,6,9	19/3 = 6.3	38/6	2.5166	1.4530
S3	5,6,9	20/3 = 6.7	26/6	2.0816	1.2019
S4	4,5,9	18/3 = 6.0	42/6	2.6457	1.5276
Total		72/3 = 24	112/6	8.2439	4.7599
Mean		24/4 = 6	[(112/6)/4]= 14/3	2.0610	1.1900

Now find the standard deviation for the means (15/3, 19/3, 20/3, and 18/3). As in the previous case, we get the standard error as 0.72. In this case also, we find that the sample mean and sample variance are exactly the same as the population mean and population variance, respectively. Hence, the sample mean and sample variance are unbiased estimates of the population mean and population variance, respectively. Note the standard deviation among the sample means where $n = 2$ and $n = 3$ are 1.18 and 0.72. This indicates that as the sample size increases the deviation among the sample means reduces.

Following is a summary of the calculations:

Sample size	Number of samples	Range in the sample mean	Standard deviation among the means
1	4	9–4 = 5	2.16
2	6	7.5–4.5 = 3	1.18
3	4	20/–15/3 = 1.7	0.72

These data show that as the sample size increases, the deviation among the sample means decreases. When $n = 2$, the range in the sample means is 3.0 and when $n = 3$, it is 1.7. As the sample size increases, the sample mean approaches the population mean. When $n = 4$, it is the size of the population and hence, no sampling is involved.

SOME POINTS

1. If $n = 1$, then x and $\bar{x}$ are equal. The sampling distribution of $\bar{x}$ and the parent population are the same. When $n = 1$, the mean of a sample of size 1 is the same as the single observation.
2. As n increases, the variability of $\bar{x}$ decreases.
3. The sample means obtained from different samples tend to follow normal distribution. This is true when the sample size n increases. This result is called *central limit theorem* (CLT). Let X be the variable in which we are interested. The frequency distribution of X may give normal or skewed distribution. If the distribution is normal, the sample means also follow normal distribution. If the distribution of X is not normal, the sample means may not be normal but tend to normal distribution as the sample size increases. Snedecor and Cochran (1967) observe that the central limit theorem is the most important theorem in statistics from the theoretical and applied points of view. According to Mood and Graybill (1963), the central limit theorem is one of the most remarkable theorems in the world of mathematics, even if the parent population is not normal, if $n \geq 25$, the sampling distribution of the means will be approximately normal.

Diagrammatic representation of central limit theorem (CLT): The population is normal, $\mu = 50$ and $\sigma = 8$ (see Figures 10.1 through 10.4).

Example: A population consists of the heights (in inches) of 5 persons: 67, 73, 75, 76, 84; population mean, $\mu = 75$; Variance $(\sigma^2) = 37.5$; SD $(\sigma) = 6.12$. The distribution of sample means when $n = 2$, $n = 3$, and $n = 4$ is shown here:

S. No	$n = 2$	$n = 3$	$n = 4$
1	70.0	71.7	72.75
2	71.0	72.0	74.75
3	71.5	74.7	75.00
4	75.5	72.7	75.50
5	74.0	75.3	77.00
6	74.5	75.7	
7	78.5	74.7	
8	75.5	77.3	
9	79.5	77.7	
10	80.0	78.3	

These data are illustrated in the dot diagrams of Figures 10.5 to 10.7.

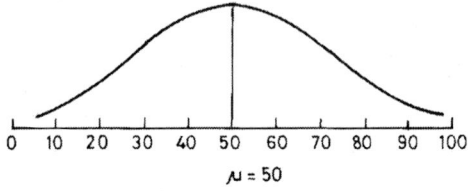

FIGURE 10.1. Normal distribution of original values ($\mu = 50$, σ - 8).

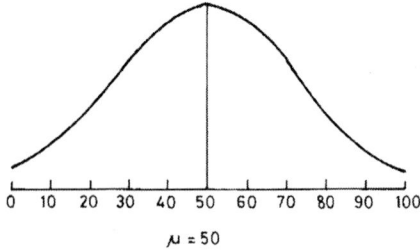

FIGURE 10.2. Normal distribution of means with $n = 1$ ($\bar{x} = 49.5$, SEm = 7.8).

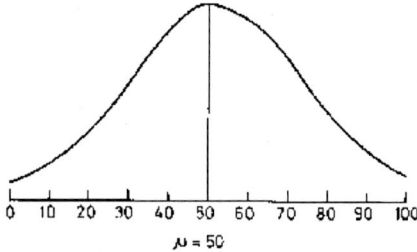

FIGURE 10.3. Normal distribution of means with $n = 2$ ($\bar{x} = 50.2$, SEm = 5.66).

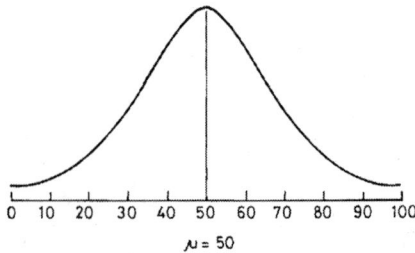

FIGURE 10.4. Normal distribution of means with $n = 5$ ($\bar{x} = 50.07$, SEm = 3.57).

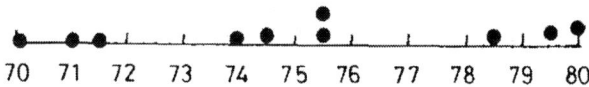

FIGURE 10.5. Dot diagram showing the distribution of sample means with the sample size $n = 2$ for the data given in the example on height of persons.

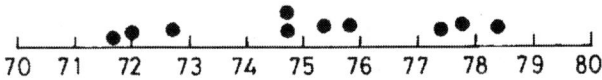

FIGURE 10.6. Dot diagram showing the distribution sample means with the sample size $n = 3$.

FIGURE 10.7. Dot diagram showing the distribution of sample means with the sample size $n = 4$.

The positions of values tend to cluster around the true mean μ as the sample size increases. The sample mean is equal to μ regardless of sample size.

Relation between number of observations (n) and degrees of freedom (n-1)

When calculating the variance from a sample, we use $n - 1$ as the divisor. When n is large, the divisor n or $n - 1$ is not very important; it is more important for small values of n. When $n = 1$, there is a single observation and the variance is $SS/df = 0/0$, an indeterminate quantity. This is so because we cannot find variance from a single observation, i.e., it cannot vary from itself. Again, a single observation taken from a population cannot give any information about the variability in the population.

Estimation of Population Mean (M)

A population is described by its mean and variance and hence, they should be computed. If the population is small, we can compute population mean and population variance exactly. In many cases, the size of the population is large and therefore it will be impractical, impossible, and extremely expensive to compute population mean and population variance. Moreover, we can obtain sufficiently accurate information about the parameters by just taking a sample from the population. As we frequently deal with the sample statistics and population parameters, the symbols, and formulae relating to sample estimates and parameters are given here:

Character	Calculation of parameter	Calculation of estimate
Mean	$(\Sigma Y/N) = \mu$; N = population size	$\Sigma x/n = \bar{x}$ n = sample size
Variance	$\sigma^2 = [\Sigma(x-\mu)]^2 / N$	$s^2 = [\Sigma(x-\bar{x})^2]/(n-1)$
Standard deviation	$\sigma = \sqrt{[\Sigma(x-\mu)]^2} / N$	$s = \sqrt{[\Sigma(x-\bar{x})]}/(n-1)$
Variance of the mean	$\sigma\bar{x}^2 = \sigma^2 / n$	$s^2\bar{x} = s^2 / n$
Standard error of the mean	$\sigma\bar{x} = \sigma / \sqrt{n}$	$s\bar{x} = s / \sqrt{n}$

Law of Large Numbers

Consider a sample from the population and estimate population mean (μ). Population mean depends upon the sample selected. Sample means vary for a given sample size. If the sample size is increased, it is likely that the sample mean approaches the population mean (μ). This is called the law of large numbers.

Example: Suppose one is interested in the estimation of height (cm) of students in a class. The population of students gives a mean value, μ, which is unknown. If one takes a sample of 10 students and their measurements are given as 62, 65, 64, 62, 60, 58, 60, 61, 63, and 60 the mean is 615/10 = 61.5 inches. The μ is about 61.5 inches. If another sample gives a mean value of 60.7 inches the μ is about 60.7 inches. The value of μ therefore depends upon the sample selected. Although $\bar{x}$ varies from sample to sample, if the sample size is increased, it is likely that $\bar{x}$ approaches μ, i.e, $\bar{x} \rightarrow \mu$, i.e., the larger the sample size, the closer is our estimate of μ.

STANDARD ERROR

Researchers collect observations from their experiments with respect to the character(s) of their interest. The mean is considered as an index of central tendency of the data. To know more about the variability of the data, we also calculate standard deviation. The standard deviation is used to describe the dispersion of individual observations whereas the standard error is used to draw inferences about the population mean (μ) from which the sample was taken. The standard error of the sample mean is $s/\sqrt{n}$. It therefore depends upon sample size, n and shows that we can get precise results from a larger sample than from a smaller one. If all the observations in the population have equal values, then any sample will estimate the population mean (μ) exactly and our estimate is a precise one. The purpose of statistics is to draw inferences from samples of data to the population from which these samples are derived from.

Uses of Standard Error

1. To compare the precision obtained in one experiment with that obtained from another experiment
2. To estimate the size of the sample
3. To construct the confidence limits to the unknown population mean
4. To compute $CV(\bar{y})$ by using SE; $CV(\bar{y}) = (SE\bar{y}/\bar{y}) \times 100$
5. The variance of the mean is given by $\sigma\bar{x}^2$, which is in square units. Its square root will have the same units as the original measurement.
6. The standard error of the mean (SEm) is $s/\sqrt{n}$. It therefore depends upon the sample size. Thus, by increasing the sample size, we can get more precise estimates of the population mean.

Standard Error of Difference Between Two Means

The standard error of difference between two means is the standard deviation of a theoretical distribution of differences between means.

Procedure: Let there be two treatments. Find their means $\bar{x}_1$ and $\bar{x}_2$. Obtain the difference between the means $(\bar{x}_1 - \bar{x}_2)$. If we do this infinite number of times, we would get a distribution of differences between sample means of two groups. These differences would be normally distributed. Their mean would be the true average differences between the populations of two treatments, and the standard deviation of this distribution is called standard error of difference between two means. It is calculated as

$$\sigma\bar{x} - \bar{y} = \sqrt{(\sigma^2{}_x / n_x + \sigma^2{}_y / n_y)}$$

where $\sigma^2{}_x$ = variance of population x;
$\sigma^2{}_y$ = variance of population y,
n_x = sample size of x,
and n_y = sample size of y.

For samples having the common variance then,

$$SE\bar{x} - \bar{y} = \sqrt{[s^2{}_p(1/n_x + 1/n_y)]}$$

where $s^2{}_p$ = pooled variance from sample data

$$= [(n_x - 1)s_x{}^2 x + (n_y - 1)s_y{}^3 y]/(n_x + n_y - 2),$$

i.e., $s^2{}_p = [(\Sigma(x_i - \bar{x})^2) + (\Sigma(y_i - \bar{y})^2)]/(n_x + n_y - 2)$

Estimation

Scientists study a population by taking samples. Population values (parameters) are completely unknown. Therefore, they are estimated to draw inferences only from samples drawn from the population. Three kinds of inferences are commonly drawn about the parameters. They are (1) point estimate, (2) confidence interval, and (3) significance tests.

Point Estimate

Suppose one is interested in the estimation of protein content in the rice seeds. The population of seeds will give rise to a mean, which is unknown. A sample of seeds on estimation gives the value 11.3 percent. Now we can say that μ will be about 11.3 percent. An estimation of a population mean based on the sample mean is called point estimate. If we take a second sample, which may give the mean value of 12.8 percent, then 12.8 percent is the point estimate. The value of μ depends upon the sample selected.

Comparison of two estimates: Let μ be the population mean, and A and B be the two sample means that estimate the parameter. The efficiency of A relative B is computed as

[(Var B)/ (Var A)] × 100.

Example: Let Var B = 180 and Var A = 115. Then efficiency of A related to B is (180/115) × 100 = 565 percent.

Interval Estimate. In an interval estimate, we construct a range of values (interval) rather than a single value in such a way that we have confidence that the interval does include the unknown population parameter. These interval estimates are called confidence intervals.

Confidence Interval for the Mean

More information can be gained from calculating a confidence interval for the mean than from making a hypothesis test. We obtain the range of possible values of μ, which, with a probability of 0.95, could have yielded the samples of data that have been observed. Using observed sample values of $\bar{x}$ and s^2, the 95 percent confidence interval for the true mean μ is calculated from:

$$\bar{x} - [t \times (s / \sqrt{n})] \text{ to } \bar{x} + [t \times (s / \sqrt{n})]$$

where t stands for the critical value of t_{n-1} at 5 percent (two tail) point. With 95 percent, this interval contains the true value of μ.

Thus the interval depends upon the values t, s, n, and degree of confidence. The width of the confidence interval depends on confidence probability. If we have no confidence in our interval estimate, we find $t = 0$, then we get $\bar{x} = μ$, i.e., the interval estimate reduces to point estimate.

Some values of t are given as follows:

Confidence coefficient (%)	4	8	12	20	30	infinity
95	2.776	2.306	2.179	2.086	2.042	1.960
99	4.604	3.355	3.055	2.815	2.750	2.576

The value of t increases for small sizes (n). For n greater than 25, 95 percent confidence interval is $\bar{x} \pm t \, x$ (standard error of the sample mean).

Suppose the degrees of freedom is 20 and the confidence coefficient is 95 percent, the t value is 2.086 and for 99 percent the t value is 2.815 (see previous data). When the confidence coefficient is reduced, the t value becomes higher.

Examples:

A sample of 25 observations is taken from a distribution with unknown mean μ. The sample mean is 16 and the sample standard deviation is 6. Then SEm is given by $s/\sqrt{n} = 6 / \sqrt{25} = 6 / 5 = 1.2$. The t value for the 95 percent confidence interval is 2.06. Then confidence interval for the true mean μ is $\bar{x}$ ± 2.06 × s $/\sqrt{n}$ = 16 ± 2.06 × 1.2 = 16 ± 2.47 = 18.47-13.53. We now state that we are 95 percent confident that the true mean is between 13.53 and 18.47. The confidence interval is 13.53 to 18.47 when the value of t for 0.95 probability level and 25 degrees of freedom is 2.06. It can be concluded with a confidence coefficient 0.95, 95 percent of the intervals obtained from large numbers of samples will include population mean μ. This 95 percent confidence interval of 13.53 to 18.47 does not necessarily include the population mean, but there is a chance that μ does not lie within the limits. When many 95 percent confidence intervals are made, one can expect to be correct 95 percent of the time (i.e., 19 out of 20) and wrong 5 percent of the time

(1 out of 20). In this example, 13.53 is called the lower limit of confidence interval and 18.47 the upper limit of confidence interval.

Similarly, we can set up confidence intervals for the unknown μ with different confidence coefficients. The quantity $1-\alpha = 1-0.05 = 0.95$ is called the confidence coefficient.

The smaller the SE, the smaller will be the confidence interval meaning that we have estimated μ more precisely. A large n will result in smaller SE. Thus a parameter estimated from a large sample is more precise than an estimate of the same parameter from a small sample.

The 99 percent confidence interval will be $\bar{x} \pm t \times SE$, where $t_{0.01,\ 24} = 2.80$. Thus, the confidence interval is $16.00 \pm 2.80 \times 1.2 = 12.64–19.36$. Thus, if we increase our confidence, we increase the width of the confidence interval. If our confidence coefficient is 100 percent, then our confidence interval would be $-\infty$ to $+\infty$. We utilize the two-tailed test for computation of confidence interval as we are getting limits on both sides of μ.

Procedure

1. Find the sample size, n.
2. Find sample mean $\bar{x}$ and sample standard deviation, s.
3. Use the formula $\bar{x} \pm 1.96 \times s/\sqrt{n}$.
   Example: Given $\bar{x} = 5$, $n = 100$, $s = 3$.
   a. 90 percent confidence interval for μ: Find $1.65\ s/\sqrt{n} = 1.65\ (3/\sqrt{100}) = 0.495$. 90 percent confidence interval for μ is 5 ± 0.495, i.e., 4.505 to 5.495. We are confident that μ is somewhere between 4.505 and 5.495.
   b. 99 percent confidence interval for μ is $\bar{x} \pm 2.58\ (s/\sqrt{n}) = \bar{x} \pm 2.58\ (3/\sqrt{100}) = 0.774$. CI $= 5 \pm 0.774$; i.e., 4.226 to 5.776, i.e., we are 99 percent confident that μ is somewhere between 4.226 and 5.774.

Confidence Interval for Proportions

Suppose we have a population in a country and we are interested in knowing the proportion of males among the population. This number is called a population proportion (P). It is estimated by sample, and it is called by the letter p. The central limit theorem for sample proportions tells us that if the sample size n is large, the random quantity, p, will have normal distribution with mean, p, and standard deviation $\sqrt{\{[p(1-p)]/n\}}$. Using the central limit theorem, we can find p and confidence intervals for P for different confidence coefficients.

Using the following formula, we can compute confidence intervals for different confidence coefficients.

Confidence coefficient (%)	Confidence intervals
90	$p \pm 1.65\sqrt{p(1-p)/n}$
95	$p \pm 1.96\sqrt{p(1-p)/n}$
99	$p \pm 2.58\sqrt{p(1-p)/n}$

Numerical example: In a city, in a random sample of 100 people, 65 percent are males. Find 90 percent confidence interval for the population proportion p of males in the city.

Given $p = 65$ percent, i.e., $p = 0.65$ $n = 100$,

$$90 \text{ percent CI} = p \pm t \times \sqrt{[p(1-p)/n]}$$
$$= 0.65 \pm 1.65\sqrt{[0.65(1-0.65)]/100} = 0.08$$

The 90 percent confidence interval for the proportion of males is 0.65 ± 0.08, i.e., 0.57 to 0.73.

EXERCISES

10.1. Given a set of numbers: 2, 3, 5, 6, 8, and 12, at random, draw samples of two from the data and find the sampling distribution of sample means. Show that the mean of the sampling distribution is the same as the population.

10.2. Select three groups of five numbers from any single digit column in a random number table. Compute s^2, s, and $\bar{x}$ for each group and compare.

10.3. Find the standard error of the mean to systolic blood pressure (mm Hg) of 12 people: 125, 128, 134, 136, 138, 139, 141, 144, 145, 151, and 150.

10.4. Collect three samples of 6, and one sample of 10 from a random table. Compute s^2, s, and $\bar{x}$, and compare.

10.5. Define the terms *parameter* and *statistic*.

10.6. What is meant by sampling distribution of a statistic, for example, mean.

10.7. Define *population* and *sample*. Give some examples for the population.

10.8. What is sampling error?

10.9. Explain the law of large numbers.

10.10. Write a short essay on various uses of standard error. What is standard error of difference?

10.11. Distinguish between parameter and statistic. Are these constants?

10.12. Provide two uses of statistics.

10.13. Why are we not measuring parameter?

10.14. How many random samples of size 3 could be taken from a population of 50?

10.15. What information do you get from the sampling distribution of any statistic?

10.16. What do you understand by confidence interval? Show how you will construct confidence intervals for population mean for a given confidence coefficient.

10.17. Find the standard error of the mean (SEm) to the systolic blood pressure (mm Hg) of the following 12 people: 121, 125, 128, 134, 136, 149, 138, 139, 141, 144, 145, and 151.

10.18. A population has six values: 2, 3, 5, 6, 8, 12. Draw samples of size 2 from these data and find the sampling distribution of the sample means. Show that the mean of the sampling distribution is the same as the population mean. Also show that the sampling variance is equal to the population variance.

10.19. A population consists of four observations: $x_1 = 2, x_2 = 4, x_3 = 8, x_4 = 10$. How many samples of size 3 could be drawn? From these samples, compute mean and median for each one. Using data from the sampling distribution of the mean and the sampling distribution of the median for samples of size 3, which of these distributions has the smallest variance? If you had a single sample of three observations, which measurement, the mean or the median, would be more likely to be close to the population mean?

REFERENCES

Mohr, L.B. (1990). *Understanding significance testing.* London: Sage Publications.

Mood, A.M. and F.A. Graybill (1963). *Introduction to the theory of statistics,* Second edition. New York: McGraw Hill.

Snedecor, G.W. and W.G.Cochran (1967). *Statistical methods,* Sixth edition. Ames: The Iowa State University Press.

Chapter 11

Hypothesis and Test of Significance

The general principle in scientific experiments is to formulate a meaningful hypothesis (i.e., statement) and then test it or verify it. A statistical hypothesis is a statement either about the values of one or more of the parameters of a given distribution or about the form of the distribution itself (Afifi and Azen, 1972). For this purpose, investigators should collect sufficient data from well-designed and well-executed experiments. The purpose of the design is to provide the experimenter sufficient data for testing of the hypothesis. We are therefore generally concerned with testing the validity of the hypothesis.

NULL HYPOTHESIS

A hypothesis of no difference that we formulate and test statistically is called a null hypothesis. Examples follow:

1. One brand of toothpaste is as good as others in preventing tooth decay.
2. One corn variety gives similar production as others.
3. One piece of laboratory equipment is more efficient than others.
4. Girls are as intelligent as boys.
5. One insecticide is as good as another in the control of pests in a crop.
6. No relationship can be shown between smoking and blood pressure.
7. No difference exists in the yielding capacity of two sugarcane varieties.
8. No difference is found in the living standards of people in two different countries.
9. No relationship exists between outdoor temperature and ice cream sales.

Null hypothesis applies to the population and because population cannot be studied, we take a representative sample, collect the relevant data, and test it statistically. The hypothesis to be tested is called null hypothesis and is generally designated as H_o. The researcher draws the inference or conclusion based on the results of the test of null hypothesis. This is called testing of statistical hypothesis. Hypothesis in statistics is always set up with a certain probability level. Testing of hypothesis is the procedure for deciding whether to accept or reject the hypothesis under the desired probability levels.

Example: H_o: sample mean $(\bar{x})$ is equal to the population parameter value, μ. Procedure: Take a random sample and perform the experiment. Depending upon the outcomes, we will either accept or reject the hypothesis. It is the sample that determines the action to be taken on the hypothesis. It is illustrated in Figure 11.1.

Figure 11.1 indicates the sample space. It contains two parts, A and R. If the sample corresponds to a point in A, we accept the null hypothesis. If the sample corresponds to a point in R, we reject the null hypothesis. The region of rejection is called the critical region, i.e., the set of values of the test criterion that leads to the rejection of the hypothesis. On the other hand, the set of values that leads to the acceptance of the null hypothesis forms the acceptance region.

The rejection of the null hypothesis indicates significant difference. In statistical problems, the experimenter's aim is to establish statistical significance of difference. Rejection of the null hypothesis indicates success in statistical problems. Testing null hypothesis is a rule indicating when to accept the null hypothesis and when to reject the null hypothesis.

For example: we want to test the effects of two kinds of fertilizers, A and B, on crop yields.

H_o: A and B give the same yields, i.e., there is no difference between A and B in their effect on crop yields. If the data give evidence to reject the

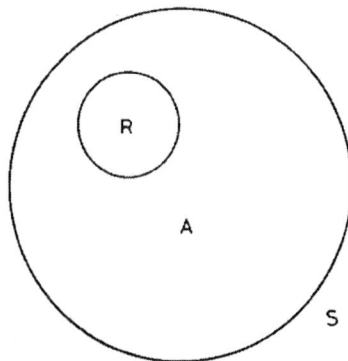

FIGURE 11.1. Diagram showing the acceptance and rejection regions.

null hypothesis, we could say that the yields obtained under A and B differed significantly.

To test the null hypothesis, we collect data and observe the difference between two treatment groups. If the observed difference is sufficiently greater than zero, we reject the null hypothesis, and if we reject the null hypothesis we accept the alternate hypothesis. The reason for setting up the null hypothesis that there is no difference between the two groups is to be stated.

Let there be two treatments, X and Y. The difference between $\overline{X}$ and $\overline{Y}$ may be 20 percent, 15 percent, 10 percent, 8 percent, 5 percent, etc. To test whether 20 percent, 15 percent, 10 percent, 8 percent, or 5 percent are significantly higher, we have to carry out several tests. But if we set up the null hypothesis so that there is no difference, i.e., $\overline{X} - \overline{Y} = 0$ percent, then we have to test only 0 percent. If the observations show no difference, we cannot state that one group is better than the other.

When we test a hypothesis, we commit two types of errors:

1. Rejection of hypothesis when it is true
2. Acceptance of hypothesis when it is false

These errors occur with certain probabilities. We cannot eliminate errors but we must take steps to reduce the probabilities that we will make such errors. When we reject the null hypothesis, we infer that the results are statistically significant. Significance means that the sample results are not likely to be due to chance, i.e., the sample results are not attributed to the fluctuations associated with the random sampling procedure. One should note that the term *statistically significant* does not mean that the results are important or interesting. This is to be determined within the context of the research.

TEST OF HYPOTHESIS

A test that is used to decide whether to accept or reject the null hypothesis is called test of significance. A statistic is computed and is tested for rejection under the assumption that it is true. If the computed value of the test statistic falls in the rejection region, the test is said to be significant. The hypothesis can be rejected and cannot be proved. The test criterion is the statistic that is used.

LEVEL OF SIGNIFICANCE

The probability level that is used as the criterion for rejecting the null hypothesis is called the level of significance. It is represented by the letter α. Usually, we use two levels of α, i.e., 0.05 and 0.01. In many areas of research, these two levels are accepted in the test of hypothesis. However, the choice of the level of significance may be decided based on the nature of the problem. Suppose we choose $\alpha = 0.05$. We say that the result is significant at the 5 percent level. It is equivalent to saying that the event will occur 5 times or less in 100.

When the null hypothesis is not rejected, one should not say that the null hypothesis is accepted. We should say that the null hypothesis is not rejected. The aim of hypothesis testing is to permit the researcher to make decisions. The function of a testing of significance is to determine whether the apparent treatment effect could reasonably be attributed to chance alone. Significant tests can effectively demonstrate the true real effect.

Testing a Single Population Mean (Small Sample) When Population Variance is Known

Rarely do we know the population variance of a variable of interest.

Example: Consider a population of weights of eggs. Can we say that the mean egg weight in the population is different from 62.0 g? The null hypothesis is: $H_o : \mu = 62.0$ g.

Now a sample of 12 eggs are randomly selected and their weights recorded, which are as follows:

60.2	53.8	67.2	56.9	58.6	60.0
66.3	50.7	56.0	63.3	58.2	64.1

$H_o: \mu = 62.0$; $H_A\mu/62.0$

The test statistic t is defined as

$$(\bar{x} - \mu) / [(\sigma / \sqrt{n})]$$

where μ = population mean, $\bar{x}$ = sample mean, σ = standard deviation (in the population), n = sample size, and $\sigma / \sqrt{n}$ = standard error of the mean.

Procedure

1. Compute the mean: $\bar{x} = \Sigma x_i / n = 715.3/12 = 59.61$ g.

2. Variance: $(s^2) = (42911.01 - 42637.84) / 11 = 273.17/11 = 25$ g^2.

Assumption: The sample mean comes from normal population and σ^2 is known. Compute t.

$$t = \frac{(59.1 - 62.0)}{\sqrt{(25/12)}} = -2.9 / 1.44 = -2.01$$

Our test statistic is distributed as Student's t with $n-1$ degrees of freedom if the H_o is true. If the H_o is true and $\mu = 62.0$ g, then $\bar{x}$ will be near 59.1 and t will be near 0.

Let $\alpha = 0.05$, then t for 0.05, and 11 df is 2.201 from the table of t (see Appendix Table A.5).

Decision: The acceptance and rejection regions are shown in Figure 11.2. If computed, t is either greater than or equal to 2.201 or less than equal to -2.201, we reject H_o. The computed $t(-2.01)$ falls in the acceptance region. We conclude that the mean of the population from which the sample came is 62.0 g.

How to Read t *value from the* t *Table*

There are many t distributions, one for each number of degrees of freedom (see Appendix Table A.5). In the first column of the t table, the number of degrees of freedom for the variance estimate involved is given and along the corresponding row, values of t for different probabilities are given. For

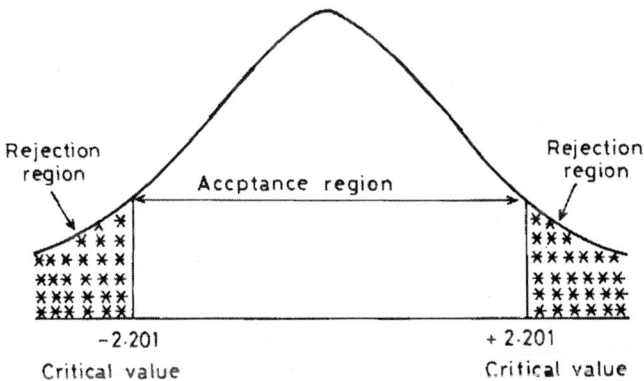

FIGURE 11.2. Diagram showing the acceptance and rejection regions while testing null hypothesis.

example, if s^2 is calculated for 12 observations, the $df = 11$. In the row $t_{0.05} = 2.201$, i.e., for probability $= 0.05$ so that the probability is 0.05 that a random t variate with 11 df will exceed 2.201 (see Figure 11.3).

INTERPOLATION

It is not possible to provide t values for an infinite number of degrees of freedom. For example, the t value for 6 df at $P = 0.05$ is 2.45, and for 9 df 2.26. To find t value for 8 df, we have to take the proportional value. The proportional value for 8 df would be $2.264 + (2.45 - 2.26) \times (1/3) = 2.32$ (correct to two decimal places) (Mack, 1967).

TESTING OF A POPULATION MEAN

When population variance is unknown:

Case 1

Consider the observations recorded in TKM 6 (rice variety) in the character, number of panicles/plant in a sample of ten plants. Observations are 6, 6, 7, 6, 7, 6, 6, 6, 6, 7.

The population mean is 6. Test the null hypothesis: H_o: $\mu = 6$, at 5 percent level of significance or test whether the sample mean was taken from the population whose mean is 6. H_o: $\mu = 6$; $H_A \neq 6$.

Solution: Compute sample mean, $\bar{x}$.

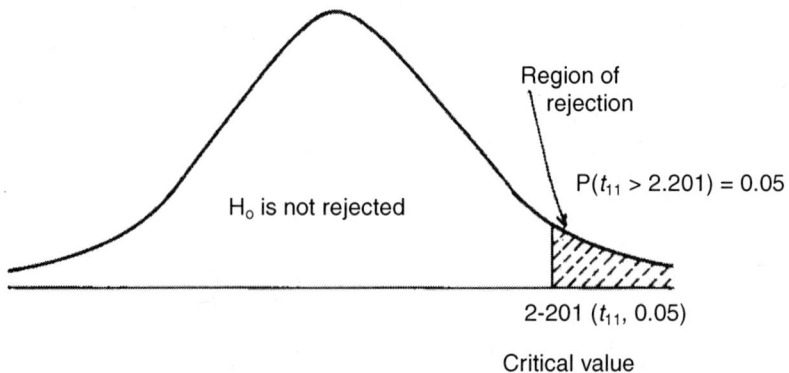

Region of rejection

$P(t_{11} > 2.201) = 0.05$

H_o is not rejected

2-201 (t_{11}, 0.05)

Critical value

FIGURE 11.3. Student's t test distribution $\alpha = 0.05$ with region of rejection, two-tailed test.

$$\bar{x} = \Sigma x / n = 63 / 10 = 6.3.$$

Since the population variance is unknown, the test statistic

$$t = \frac{\bar{x} - \mu}{s / \sqrt{n}} \text{ where}$$

sample variance $s^2 = \dfrac{\Sigma x_i^{\,2} - (\Sigma x)^2 / n}{n - 1} = 0.23$ and $s = \sqrt{0.23} = 0.479$.

Standard error of the mean (SEm) $= s/\sqrt{n} = 0.479 /\sqrt{10} = 0.479/3.16 = 0.152$.

The value of t at $P = 0.05$ and $n{-}1$ $(10 - 1)$ df is 2.262. As we did not estimate μ, but estimated variance, hence we lose 1 df. From the sample,

$$t = \frac{\bar{x} - \mu}{\text{SEm}} = (6.3 - 6.0) / 0.152 = 0.3 / 0.152 = 1.97$$

Since the calculated t (1.97) is lesser than the table value of t (2.262), we accept H_o. That means that the sample comes from the population whose mean is 6. The difference of 0.3 is attributed to sampling. This process is explained with a flow chart (see Figure 11.4). The observed value of the test statistic does not fall in the critical region.

Conclusion: The sample mean 6.3 does not differ significantly from the population mean, 6.0, and $t = 1.97$ is not significant.

Example: The protein content in rice grain is estimated. Nine different samples give the results as 10.7, 11.6, 11.1, 12.0, 11.3, 11.4, 10.9, 11.7, and 12.2.

Is the mean of these determinations significantly different from 12.1 percent?

Procedure: $\bar{x} = 11.43$, $s^2 = 0.24$, $s = 0.49$.

Test statistic $t = (\bar{x} - \mu) / (s / \sqrt{n})$

where s is the standard deviation calculated from the sample.

$$t = (11.43 - 12.1) / (0.49 / \sqrt{9}) = -4.1$$

If the null hypothesis is true, the test statistic should follow a t distribution with 8 $(n{-}1)$ df and $t_{0.05,\,8} = 2.306$. As the calculated t (-4.1) is greater than the table value of t, the difference is significant at 5 percent level ($t_{0.01\,8}$

= 3.355) and hence, the result is significant at $P = 0.01$. It shows that the percentage protein content in the sample is not 12.1 percent.

Case 2

Independent t test (or nonpaired t test or t test for comparison of two independent means).

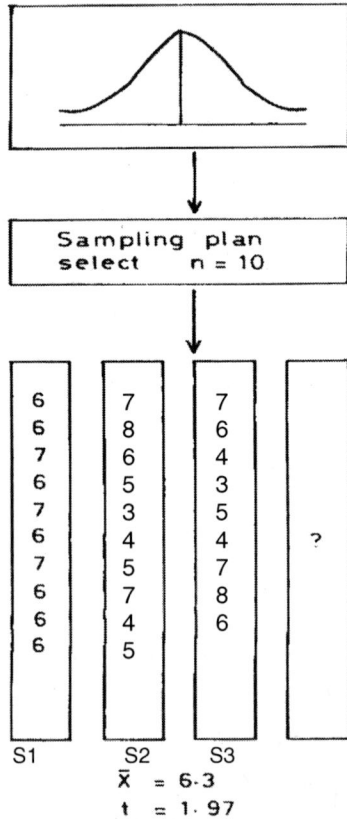

FIGURE 11.4. Flow chart for generating samples of size = 10. Observations for three samples (S1, S2, S3) are shown.

Population Variances Are Unknown

To test the difference between two population means, we draw samples on the basis of which we will make inferences about the populations from which they came. The test statistic for the test of two independent means is

$$t = \frac{\bar{x}_1 - \bar{x}_2}{s_p \sqrt{(1/n_1 + 1/n_2)}} \qquad H_o:\mu_1 - \mu_2 = 0 \text{ or}$$

$$H_o:\mu_1 - \mu_2 \geq 0 \text{ or}$$
$$H_o:\mu_1 - \mu_2 \leq 0$$

where $\bar{x}_1$ and $\bar{x}_2$ are sample means, n_1 and n_2 are sample sizes for samples 1 and 2, respectively, and s_p pooled standard deviation, which is calculated as

$$s_p = \sqrt{[(n_1 - 1) \times s_1^2 + (n_2 - 1) \times s_2^2 / (n_1 + n_2 - 2)]}$$

where s_1^2 and s_2^2 are the sample variances calculated from the two samples separately. The statistic t is distributed as Student's t distribution with $n_1 + n_2 - 2$ degrees of freedom.

Assumptions

1. Samples are drawn independently from the two populations.
2. The two populations are normally distributed with equal standard deviations.

Example: Two fertilizers F1 and F2 were evaluated on grain yields of wheat crop. There are 16 plots available for the experiment. F1 was applied in 8 plots and F2 in 8 plots. Random procedure was adopted in the application of fertilizers. Plot yields (kg/plot) were recorded. We have to compare the two means. The layout of the field and the yields recorded in each plot are shown in the Figure 11.5.

The data in Figure 11.5 are arranged as follows for statistical analysis purpose:

Fertilizer	1	2	3	4	5	6	7	8	Total	Mean
F1	12	15	12	15	10	14	14	13	105	13.1
F2	9	13	13	12	7	10	12	13	89	11.1

F 1	F 2	F 2	F 1	F 1	F 2	F 1	F 1
12	9	13	15	12	13	15	10
F 2	F 2	F 1	F 2	F 1	F 2	F 1	F 2
12	7	14	10	14	12	13	13

FIGURE 11.5. Field plan for the fertilizer trial. Figures indicate the plot yields in kg.

Computation:

$$\Sigma x_1 / n = \bar{x}_1 = 105/8 = 13.1$$
$$\Sigma x_2 / n = \bar{x}_2 = 89/8 = 11.1$$

Variance of F1 : $s_1^2 = 2.982$

Variance of F2 : $s_2^2 = 4.982$

Pooled variance $s^2{}_p = [(n_1 - 1)s_1^2 + (n_2 - 1)s_2^2] / (n_1 + n_2 - 2)$
$$= [(8-1)(2.982) + (8-1)(4.982)] / (8+8-2)$$
$$= 55.748/14 = 3.982 \text{ and } s = 1.995$$

$$t = \frac{\bar{x}_1 - \bar{x}_2}{S_p\sqrt{(1/n_1 + 1/n_2)}}$$

$$= \frac{13.1 - 11.1}{1.995\sqrt{(1/8 + 1/8)}} = 2.0000/0.998 = 2.004$$

The table value of t for $n_1 + n_2 - 2 = 14$ degrees of freedom at 5 percent and 1 percent level of probability is 2.145 and 2.977, respectively. When the sample sizes are equal, we can calculate the pooled variance by taking the average of the two variances as

$$s^2{}_p = (s_1^2 + s_2^2)/2$$
$$= [(3.982 + 4.982)]/2 = 3.982 \text{ and}$$
$$s_p = \sqrt{s^2{}_p} = \sqrt{3.982} = 1.995.$$

According to H_o, the two samples are random samples drawn from the same population, and $\bar{x}_1$ and $\bar{x}_2$ both estimate the same population mean, μ.

$(H_o: \mu_1 = \mu_2)$.

In order to reject the null hypothesis, the computed t value must be larger than the critical value, i.e., 2.145 ($t_{0.05,14}$) for 0.05 level of significance.

In this experiment, the H_o is not rejected. The t curve is shown in Figure 11.6. If calculated t is less than -2.145 or greater than $+2.145$, we have to reject H_o. But t calculated (2.004) lies to the left of the upper critical value (2.145), our decision is to accept H_o, i.e., the calculated t (2.004) is less than $t_{0.05,14} = 2.145$. Hence, H_o is not rejected.

Conclusion: Both fertilizers give the same yield.

Alternate Method of Analysis Using F Statistic

The data can also be analyzed using the analysis of variance (ANOVA) technique following the completely randomized design (CRD). The ANOVA data are given in Table 11.1.

In medicine, it is more common to use a two-tailed test of significance since we often do not know in which direction the difference may turn out. It is important to report whether we are using a one-tailed or two-tailed test.

Testing two means using t test when the number of observations is different, i.e., $n_1 \neq n_2$.

Sometimes, researchers will record different numbers of observations in two samples either due to lack of materials for the experiment or due to some other reasons.

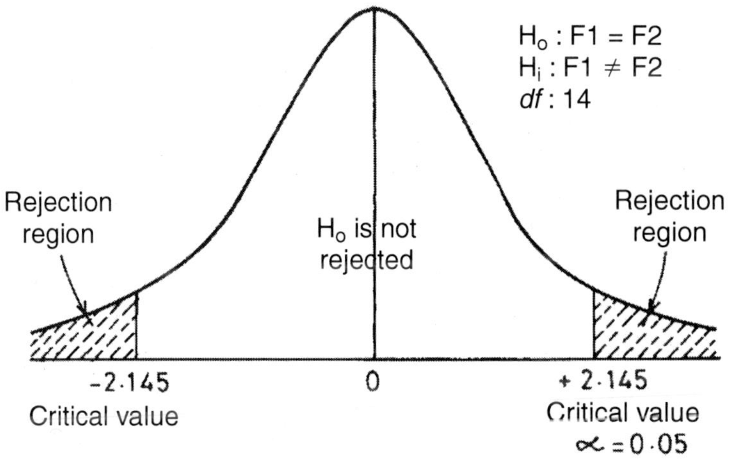

FIGURE 11.6. Student's t distribution with 14 degrees of freedom.

TABLE 11.1. ANOVA data.

Source of variation	df	SS	MS	F	Table value (5 %)	Table value (1 %)
Treatments	1	16.00	16.00	4.02ns	4.60	8.86
Error	14	55.75	3.98			
Total	15	71.75				

ns = not significant
Note the relationship between t and F values, i.e., $F = t^2$, i.e., 4.02 = 2.004².

Example: A and B are two treatments. They give the following results:

Replications	A	B
1	28	24
2	30	25
3	27	20
4	32	18
5	34	21
6	36	
7	30	
Total	217	108
Mean	31	21.6

Sample size	7 (n_1)	5 (n_2)	$n_1 + n_2 = 7 + 5 = 12$
Sum of squares	6787-CF	2366-CF	
	= 6789–6727	= 2366–2332.8	
	= 62	= 33.2	
Variance	10.33	8.3	
Degrees of freedom	6	4	

$$t = (\bar{A} - \bar{B}) / s_p \sqrt{[(1/n_1 + 1/n_2)]}$$

Before pooling the two variances, s_1^2 and s_2^2, we have to test for their homogeneity. A homogeneity test of two variances is done using F test. It is done as follows. Calculate the F statistic as F = greater mean square/smaller mean square = 10.33/8.3 = 1.24. Since F computed (1.24) is smaller than the $F_{0.05,6,4}$ = 6.16 (from the F table), the two variances are homogeneous.

$$s^2_p = [(n_1 - 1)s_1^2 + (n_2 - 1)s_2^2]/[n_1 + n_2 - 2]$$
$$= [(7-1)10.33 + (5-1)8.3]/[7+5-2] = 98.18/10 = 9.51$$

and hence $s = 3.08$, $SE = 3.08\sqrt{(1/7 + 1/5)} = 1.803$.

$$t = (9.4)/[3.08\sqrt{(1/7+1/5)}] = 9.4/1.839 = 5.21$$

The table value of t for 10 df for 0.05, 0.01, and 0.001 probability levels are 2.228, 3.169, and 4.587, respectively. It means we would get a value of t as high as this only once out of 1,000 by chance, if H_o is true. So we reject H_o.

Alternate Method Using F Test

The data may be considered a result of conducting CRD with two treatments, A and B with 7 and 5 replications, respectively, i.e., $n_1 \neq n_2$. The results follow:

Tr	1	2	3	4	5	6	7	Total	Mean	n
A	28	30	27	32	34	36	30	217	31	7
B	24	25	20	18	21			108	21.6	5

Replication

ANOVA

Source of variation	df	SS	MS	F	0.05	0.01	0.001
Treatments	1	257.7	257.7	27.069***	4.96	10.04	17.14
Error	10	95.2	91.52				
Total	11	352.9					

*** = significant at P = 0.001

F is significant at $P= 0.001$ as before. Verify the relationship between F and t as $F = t^2$ when the number of treatments are two in the experiment.

Confidence limits (CL) for a mean:

$x_1 \pm tx$ SEm where SEm is the standard error of the mean which is equal to $s/\sqrt{n}$

CL (for $\overline{A}$) $= 31.00 \pm 3.14/\sqrt{7} = 31.0 + 3.14/2.65 = 31 + 1.18 = 32.18 - 29.82$

CL (for $\overline{B}$) $= 21.60 \pm 3.14/\sqrt{5} = 21.6 + 3.14/2.24 = 21.6 + 1.40 = 23.00 - 20.2$

Example: Consider the fertilizer experiment with the following data where $n_1 \neq n_2$. The data are:

Fertilizer	1	2	3	4	5	6	7	8	Total	Mean
F1	12	15	12	15	10	14	14	13	105 (ΣF1)	13.13
F2	9	13	13	12	7	12	10		76 (ΣF2)	10.86

Replication

SS F1: = 20.875   *SS* F2: = 30.86
Variance = 2.98   Variance = 5.14
Before pooling the two variances, s_1^2 and s_2^2, we have to test for their homogeneity. A homogeneity test of two variances is done using the *F* test.

$$F = \text{greater variance/smaller variance} = 5.14/2.98 = 1.72^{ns}$$

Since it is lesser than the table value of *F* which is 3.87 for $P_{0.05, 6, 7}$ degrees of freedom, we conclude that the two variances are homogeneous.

$$
\begin{aligned}
\text{Pooled variance} &= [(n_1 - 1) \times s_1^2 + (n_2 - 1) \times s_2^2] / [n_1 + n_2 - 2] \\
&= [(7 \times 2.98) + (6 \times 5.14)] / 13(8 + 7 - 2) \\
&= (20.86 + 30.84) / 13 = 51.70 / 13 = 3.98
\end{aligned}
$$

Now we apply *t* test as follows:

$$t = (13.13 - 10.86) / [1.99\sqrt{(1/8 + 1/7)}] = 2.27 / 1.03 = 2.203$$

The table value of *t* for 13 *df* at $P = 0.05$ and $P = 0.01$ are 2.160 and 3.012, respectively. Since *t* computed (2.203) is greater than *t* from the tables (2.160), there is significant difference between the two means. Therefore H_o is rejected.

The t Test for Difference Between Two Means (One-Tailed t Test)

In an experimental research station, a new fertilizer (N) is to be evaluated to replace the existing fertilizer, (S). The experiment was laid out in rice crop in a CRD design with five replications each. The plot yields are given as follows:

Fertilizer	Replication					Total	Mean
	1	2	3	4	5		
N	22.4	27.7	23.5	29.1	25.8	128.5	25.7
S	21.4	23.6	24.8	22.4	26.3	118.5	23.7

The null hypotheses here is $H_o: \overline{N} > \overline{S}$, i.e., 25.7 > 23.7.
Use *t* test. Also set up confidence limits for the difference at $P = 0.99$. The standard error of difference between two means is (SE- $\overline{N}$-$\overline{S}$).

$$= \sqrt{[s^2 / n_1 + s^2 / n_2]}$$

$$s^2 = \frac{[\Sigma x_i^2 - (\Sigma x_1)^2 / n_1] + [(\Sigma x_2^2 - \Sigma x_2)^2 / n_2]}{n_1 + n_2 - 2}$$

$$= [31.30 + 14.96] / 8 = 5.7875$$

$$\text{SEd} = \sqrt{[s^2 / n_1 + s^2 / n_2]} = \sqrt{5.7875 / 5 + 5.7875 / 5} = 1.52$$

$$t = (25.7 - 23.7) / 1.52 = 2 / 152 = 1.315$$

Statistical inference: Since the computed t value is lesser than table value of t (2.9), the difference is not significant. N does not give higher yield than S.

Confidence limits: We estimate the population value by means of sample. It is desirable to know not only the estimate but also how precise the estimate is. This precision is expressed by stating the limits within which the true value lies with some probability. We state that the true value does not exceed a certain limit, is not less than a certain value or does not lie beyond certain limits. The degree of confidence may be set up at desired level.

For example, confidence limit for a mean at $P = 0.99$ is $\bar{x} \pm 3 s \sqrt{n}$.

Confidence limit for a difference between two means is (at $P = 0.99$) is

$$\bar{x}_1 - \bar{x}_2 \pm 3s \sqrt{[(1 / n_1 + 1 / n_2)]}$$

Confidence Limit for Population Mean

Suppose we want to estimate the population mean of a character, which has a normal distribution. We take a random sample of size n and compute mean $\bar{x}$. The mean $(\bar{x})$ is a point estimate. Although $\bar{x}$ is an unbiased estimate of population mean, it will not be equal to μ. We therefore want to estimate μ in some interval.

We can estimate μ of the distribution as in Figure 11.7. In a normal distribution, $\bar{x} \pm (t \times \text{SEm})$ provides us the confidence limits for the sampling distribution.

Example: We are interested in finding the systolic blood pressure of females in the age group 40 to 60 years in a population. We select a random sample of six persons. We record the BP as 115, 130, 125, 144, 140, and 150. This sample gives the sample mean of 134 (804/6) and sample variance of 170 ($SS/df = 850/5$). The sample mean (134) and sample variance (170) are unbiased estimates of population mean and population variance, respectively.

We can now specify the confidence limits for μ (μ is unknown) by using the formula

$$\bar{x} \pm t \times \text{SEm}$$

where $\bar{x}$ is the sample mean, SEm is the standard error of the mean, and t is the value from the t table for the desired probability. $\text{SEm} = s / \sqrt{n} = \sqrt{170 / 6} = 13.04/2.45 = 5.32$. Now we set up the confidence limits for 95 percent probability level. The t value for 5 df and 95 percent probability is 2.57. The confidence interval for μ is given by

$$\bar{x} \pm t \times \text{SEm} = \bar{x} \pm t_{5,0.05} \times \text{SEm}$$
$$= 134 \pm (2.57)(5.32) = 134 \pm 13.67 = 147.67 - 120.33$$
(see Figure 11.8).

For each sample, we can construct confidence limits and compute an infinite number of confidence intervals, out of which 95 percent intervals would include μ, i.e., we are confident with probability 0.95 that μ lies within the range 120.33–147.67. The upper confidence limit is $134 + 13.67 = 147.67$. The lower confidence limit is $134 - 12.67 = 120.33$. The interval between these limits is called the confidence interval. If we take a large number of samples from the population and calculate confidence limits for each of them, 95 percent of the intervals would contain the μ and 5 percent would not contain it. But we take one sample only. We can say that we are 95 percent confident that the interval does contain μ. We can also calculate confidence interval for 99 percent probability as $\bar{x} \pm t_{5,0.01} \times \text{SEm}$, i.e., $\bar{x} \pm 3s / \sqrt{n}$. This interval is explained in Figure 11.8.

Thus the confidence interval for μ is a range of values that includes the true population mean with a specific probability for μ. A 95 percent confidence interval for the population mean indicates that 95 percent of the intervals constructed in this manner will include the unknown population mean.

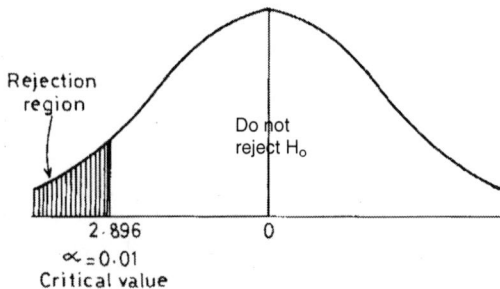

FIGURE 11.7. Sampling distribution of the statistic t; one-tailed test, 0.01 level of significance.

μ = Unknown
$\bar{x}$ = 134.00
SEm = 5.32
t = 2.57

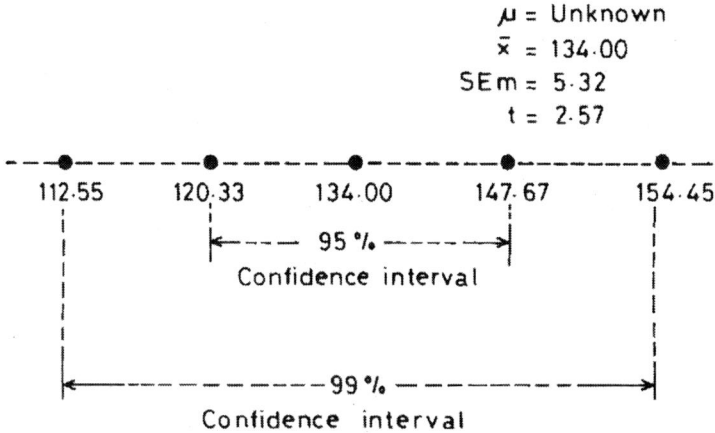

FIGURE 11.8. Figure showing the confidence intervals for the population mean μ at 95 and 99 percent levels.

Example: In a previous exercise on egg weights, we had the sample observations on randomly selected 12 eggs as 64.1, 60.2, 53.8, 67.2, 56.9, 58.6, 60.0, 66.3, 50.7, 56.0, 63.3, and 58.2 g. The sample mean is = 59.61 g, sample variance = 24.8336, and SE = 1.44. We can now construct 95 percent confidence interval for the population mean (μ = 62.0) as follows:

$$\bar{x} \pm t \times \text{SEm} = 59.61 \pm 2.201 \times \sqrt{24.8336/12} = 59.61 \pm 2.201 \times 1.4306$$
$$= 59.61 \pm 3.166 = 62.78 - 56.44 \text{ g.}$$

The value μ (62) is included in the interval and is therefore an acceptable value for μ. No value between 56.41 and 62.78 would be declared significant at $P = 0.05$. Thus we get more information about the population from the confidence interval analysis.

Now suppose we take a second sample of the same size and the sample data are as follows: 62.8, 67.2, 68.0, 68.4, 64.4, 60.9, 63.2, 61.6, 63.5, 74.3, 71.2, and 58.7

Say that x_2 gives the mean, $(\bar{x}) = 65.35$ and variance, and $(s^2) = 20.6918$.

Since s_1^2 and s_2^2 are equal, we can pool the two variances and estimate the pooled variance as

$$s_p^2 = \{(n_1 - 1)s_1^2 + (n_2 - 1)s_2^2\}/\{(n_1 + n_2 - 2)\}$$
$$= \left[(11 \times 24.8336) + (11 \times 20.4918)\right]/22 + 22.7627 \text{ and } s = 4.77$$

$$t = \left(\bar{x}_1 - \bar{x}_2\right)/s\sqrt{\left(1/n_1 + 1/n_2\right)} = \left(59.61 - 65.34\right)/\sqrt{2/12(22.7627)}$$
$$= -5.73/1.948 = -2.94 \text{ which is significant at } P = 0.01.$$

Hence, we reject the H_o of no difference.

Confidence Interval for the Difference Between Two Means

From a population having mean μ, a sample is drawn. Thus μ lies in between $\bar{x} \pm t \times \text{SEm}$. $\bar{x} + (t \times \text{SEm})$ gives the upper limit and $\bar{x} - (t \times \text{SEm})$ the lower limit. In a similar way, the limits of an interval for a difference between two means are:

Estimated difference $\pm$ ($t \times$ standard error of difference (SEd))
where $\text{SEd} = \sqrt{(s^2/n_1 + s^2/n_2)}$

Numerical example: Given two sample means with $n_1 = n_2 = 12$,
$\bar{x}_1 = 59.61$, $\bar{x}_2 = 65.35$, pooled variance (s^2) = 22.7627
The 95 percent confidence interval for the true difference ($\mu_1 - \mu_2$) is

$$\text{CL} = \text{difference between two means} \pm (t \times \text{SEd})$$
Observed difference = $65.35 - 59.61 = 5.74$
$$\text{SEd} = \sqrt{2s_2/r} = \sqrt{(2 \times 22.7627)/12} = 1.948$$
The value of $t_{22,\,0.05} = 2.074$.

$$\text{CL} = 5.74 \pm 2.074 \times 1.948 = 5.74 \pm 4.04 = 1.70 - 9.78.$$

The difference is significant as zero is not included in the interval. The wide interval indicates greater variability in the egg weights.

Testing the Difference Between Two Means

Situations in which paired t test is efficient: We know that no two experimental units are alike and that they always exhibit variation. If the variation is more, the real effects are masked.

Examples in which paired t test is recommended:

1. To find out the effect of fungicide on leaves, the experiment can be conducted on selected leaves. The leaves in plants usually vary in size, shape, position, age, smoothness of surface, etc. Thus, for efficient estimate of treatment effect, the experimenter should eliminate

the leaf-to-leaf variation. Leaves appear in pairs. A pair of leaves can be selected; one leaf is to be selected at random and the treatment applied, and another leaf may be kept as a control. Another method of selection is that one leaf may be divided into two halves, and one half selected at random and the treatment applied. The other half may be kept for control treatment. The treatment differences may be evaluated among the pairs of leaves or half leaves. We are interested in differences. These differences are assumed to follow normal distribution, and, consequently, paired *t* test is applied to find out the significance of the effect of the treatment.

2. In medical research, pairing can be done on the basis of age, sex, and physical conditions.

3. In animal experiments, pairs can be formed on littermates.

4. Any data collected before the start of the experiment and after the completion of the experiment are automatically paired. For example, in a clinical trial, a blood-glucose-reducing drug is administered to each patient, and the glucose level before and after the treatment is evaluated.

5. In agricultural field trials, two treatments are compared in blocks of two plots that lie side by side so that comparisons are made within blocks. When the experiment is carried out in pairs, it is the difference between each pair of measurements that is of interest.

t test: When paired observations are involved, the sample sizes for the two groups are equal. The formula for the *t* statistic is

$$t = \bar{d} / \text{SE } \bar{d} \text{ where}$$
$$\bar{d} = \text{mean difference between differences of pairs}$$
$$\text{SE } \bar{d} = \text{SD}_d / \sqrt{n} \text{ where SD}_d = \text{standard deviation}$$
$$\text{of the differences and}$$
$$n = \text{number of differences, i.e., the number of pairs}$$
$$\text{of observations.}$$
$$\text{SD}_d = \sqrt{[\Sigma d_i^2 - (\Sigma d_i)^2 / n] / (n-1)}.$$

The test statistic (calculated) has a *t* distribution when the null hypotheses is true. The calculated *t* value is compared with the table value of *t* at n–1 *df* and α level, and a decision is made on the difference between the two treatments.

Example: Usha Rani (1996) estimated capsaicin content in 29 chili (*Capsicum annuum* L.) genotypes from ground chili samples at the time of harvest and after one year of storage with a goal of finding whether the capsaicin content deteriorated significantly due to storage. Usha Rani (1996) applied a paired t test to find whether the two means (mean at harvest and the mean after one year of storage) differ significantly.

The H_o: $\mu_1 = \mu_2$ H_1: $\mu_1 \neq \mu_2$.

The data are shown in Table 11.2.

$$t = \bar{d} / SE\bar{d}$$

$$S_d^2 = \frac{[(\Sigma d_i^2) - (\Sigma d_i)^2 / n]}{n - 1}$$

$$s^2 = (8.558005 - 5.8806027) / 28 = 2.677402 / 28 = 0.095622$$

$$s = 0.309227$$

$$t = \bar{d} / (s / \sqrt{n}) = \frac{0.450}{0.3092227 / 5.385165} = \frac{0.450}{0.057422} = 7.84 **$$

$** = $ significance $P = 0.01$

The t value for $n - 1$ ($29 - 1 = 28$) df for $P = 0.05$ and 0.01 from t tables are 2.0484 and 2.7633. Since t calculated (7.84) is greater than the value from the table (2.7633), the difference is significant at $P = 0.01$. We reject the null hypotheses that there is no deterioration of capsaicin content in storage, i.e., there is significant deterioration during storage in the capsaicin content.

6. If the correlation is given between the observations at the beginning of the experiment and at the end of the experiment and also the means and variances for the two treatments, we can use the formula:

$$S_D^2 = s_1^2 - 2r_{12}s_1s_2 + s_2^2 \text{ (Walker and Lev, 1965; Klugh, 1974)}$$

where

S_D^2 = variance of the differences;

r_{12} = correlation coefficient between the pairs of the observations;

s_1^2 = variance of the observations in the treatment 1;

s_2^2 = variance of the observations in the treatment 2;

s_1 = standard deviation of the observations in treatment 1; and

s_2 = standard deviation of the observations in treatment 2.

If the correlation is high, the standard error of $\bar{d}$ is reduced and we get a significant test.

Example: With reference to the previous example, we have

For treatment 1, $\bar{x}_1 = 0.770$ $s_1 = 0.4351$ $r_{12} = 0.720$

For treatment 2 $\bar{x}_2 = 0.320$ $s_2 = 0.24524$

$$D = 0.770 - 0.320 = 0.450$$

TABLE 11.2. Capsaicin content comparison.

S. No.	Genotype	Capsaicin content at		Difference (d_i)
		Harvest	One year after storage	
1	Ducale	1.810	0.475	1.335
2	G3	0.575	0.125	0.450
3	IHR 289	1.626	1.295	0.331
4	Shankeshwar	0.862	0.280	0.582
5	Byadagi	0.502	0.120	0.382
6	Ronyal local	0.627	0.300	0.327
7	IHR-310-2	0.532	0.350	0.182
8	Gowribidnur	1.275	0.515	0.760
9	IHR-324-16	0.708	0.285	0.424
10	IHR-318-25	1.023	0.405	0.618
11	IHR-332-10	0.398	0.040	0.358
12	IHR 344	1.046	0.195	0.851
13	IHR 347-14	0.115	0.055	0.060
14	IHR 348-4	0.830	0.360	0.470
15	Dabigai	3.643	0.180	0.463
16	JCA232	0.135	0.035	0.100
17	IHR 360	0.990	0.410	0.580
18	IHR-525	0.980	0.470	0.510
19	G4	0.547	0.360	0.187
20	IHR 304-3	0.531	0.155	0.376
21	IHR-306-1	0.234	0.195	0.039
22	IHR-307-4	0.647	0.290	0.357
23	IHR-309-4	0.670	0.637	0.033
24	IHR-315	1.707	0.550	1.157
25	IHR-327	0.840	0.390	0.450
26	IHR-358	0.978	0.250	0.728
27	IHR-531	0.785	0.250	0.535
28	JCA-154	0.105	0.060	0.045
29	Jwala	0.627	0.260	0.367
	Mean	0.770	0.320	13.059/29 = 0.450

$$S_D^2 = s_1{}^2 - 2\, r_{12}\, s_1\, s_2 + s_2{}^2$$
$$= 0.1892841 - 2 \times 0.720 \times 0.4351 \times 0.24524 +$$
$$0.0601413 = 0.0957718$$
$$S_d^2 = 0.0957718/29 = 0.0033624$$
$$t = 0.450/\sqrt{0.0033024} = 0.450/0.0574665 = 7.83$$
$$\text{** significant at P} = 0.01$$

Significance of Correlation Among the Pairs of Observations:

Consider the two sets of data given in the examples that follow.

| | Set 1 | | | Set 2 | | |
| | Paired values (x_1-x_2) | | | Independent values (uncorrelated or independent) $x_1 - x_2$ | | |
	x_1	x_2	d_i	x_1	x_2	d_i
	2	3	−1	2	5	−3
	5	4	1	5	3	2
	15	10	5	15	4	11
	10	8	2	10	8	2
	8	5	3	8	10	−2
Total	40	30	10	40	30	10
Mean	8.0	6.0	2.0	8.0	6.0	2.0

Correlation: $x_1 x_2 = 0.97$**   Correlation $x_1 x_2 = 0.09$
$SD_d = 2.236$ $SD\ d = 5.52$
$SE_d = 2.236/\sqrt{5} = 1.00$ $SEd = 2.46$
$t = 2.0/1.0 = 2.00$ $t = 2.0/2.46 = 0.81$ ns
(Table value of t,0.05,4 = 2.78)   (Table value of $t_{0.05,4} = 2.78$)
** = significant at P = 0.01   ns = not significant
When there is correlation betwteen   When there is no correlation between
pairs, we find hight value close to the   pairs of observations, we do not get
level of significance.   significant difference between the two
                                           treatments.

The data for correlated and uncorrelated sets are the same. The data in the uncorrelated set are obtained from random assignment but in the correlated table, the data are paired. The group means ($\bar{x}_1$ and $\bar{x}_2$) and mean difference (2.0) are the same. The variability in the differences is not the same. The standard deviation (SD) in the correlated data is 2.236 whereas in the uncorrelated data it is 5.52. Since the standard deviation in the correlated data is less, the standard error of the differences is also less (1.0). The reduction in the standard error of differences will yield a larger t value for the correlated group even though $\bar{D}$ is the same in both cases. The high correlation decreased $S\bar{d}$ and increased the power of the t test.

Paired Test

Example: In an experiment on growth studies in maize crop, the plant heights recorded at weekly intervals and the plant heights predicted during the growth curve were compared. The difference between actual mean height and the predicted mean height is compared.

$$H_o : \mu_1 - \mu_2 = 0 \text{ or } \bar{\mu d} = 0$$
$$H_1 : \mu_1 \neq \mu_2 \text{ or } \mu_d \neq 0$$

The data are given as follows:

Plant height (cm)					Age (days)					
	7	14	21	28	35	42	49	58	63	70
Actual	13.70	27.70	42.90	74.70	99.80	128.40	141.00	149.70	153.40	153.50
Predicted	11.58	23.53	45.00	74.20	104.17	126.86	140.40	147.42	150.82	152.27
Difference	2.12	3.87	-2.1	0.50	-4.37	1.54	0.60	2.28	2.58	1.23

Total difference = 8.25

	Actual difference	Predicted difference
Total	984.80	976.55
Mean	98.480	97.655

Mean difference = 98.480 – 97.655 = 0.825

Variance of differences: $s^2 d = [\Sigma d_i^2 - (\Sigma d_i)^2] / n - 1$
$$= n-1 = 40.814 - 8.25^2/10$$
$$= 40.814 - 6.806/9 = 3.779$$
$$Sd/ \sqrt{3.779} = 1.942$$
$$t = d/SEd = 0.825/1.942/\sqrt{10} = 0.825/0.615$$
$$= 1.341$$

The table value of t for 9 df and at $P = 0.05$ is 2.269. Since the t test is not significant, the actual data do not differ from the predicted data. In this experiment, the observations in actual data are correlated with the predicted data. The data are treated as in pairs. As calculated "t" value is less than "t" table value. The H_o is accepted. Because 1.341 lies between –2.62 to +2.62 we accept null hypothesis H_o (see Figure 11.9).

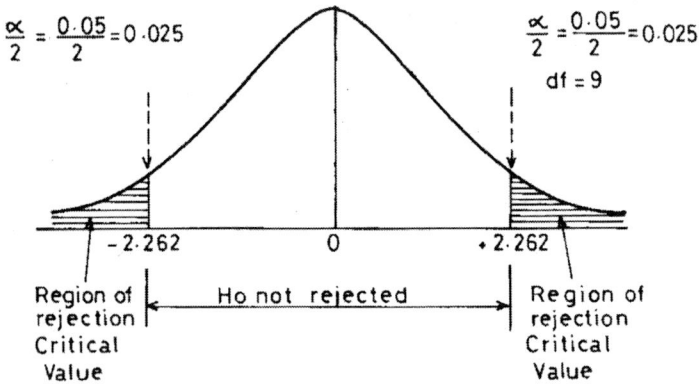

$$\frac{\alpha}{2} = \frac{0.05}{2} = 0.025 \qquad \qquad \frac{\alpha}{2} = \frac{0.05}{2} = 0.025$$

$$df = 9$$

FIGURE 11.9. Sampling distribution showing the rejection regions at 0.05 significance level.

Example: An experiment was conducted to evaluate two fertilizers F1 and F2 in wheat crop in a paired plot design. The layout is shown in Figure 11.10. There are 8 blocks (replications) and in each block there are two plots of equal size for the treatments (F1 and F2), which were allotted at random in each block.

Analysis: The yield data are arranged in the table as follows:

Tr/block	B1	B2	B3	B4	B5	B6	B7	B8	Total	Mean
F1	9	13	13	12	7	10	12	13	89	11.1
F2	12	15	12	15	10	14	14	13	105	13.1
Total	21	28	25	27	17	24	26	26	194	

Except for F1 and F2 application, the crop was grown under identical conditions. The yields of each plot in a block constitute a set of paired observations. The significant difference, if any, between the means is due to the effect of treatments, and also due to experimental error. In such a design, the difference between the observations in each pair is to be considered as the variable for the test. The test statistic for the t test of paired design is $t = \overline{d} / sd\sqrt{n}$ where n = number of pairs and sd = is the standard deviation of the differences between the paired observations. Difference (d_i) is calculated for each pair. The test statistic t follows a t distribution when the null hypotheses is true.

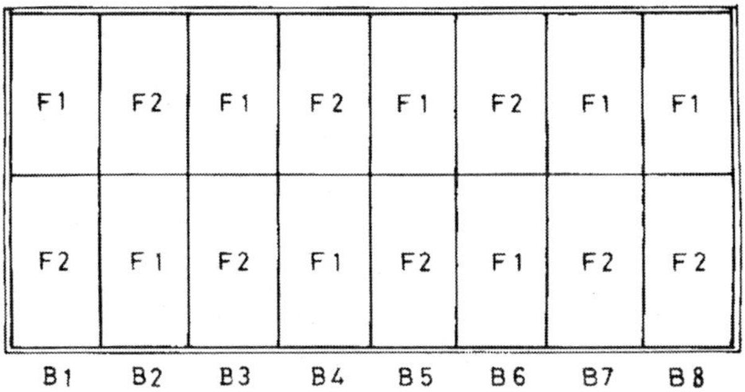

F 1	F 2	F 1	F 2	F 1	F 2	F 1	F 1
F 2	F 1	F 2	F 1	F 2	F 1	F 2	F 2
B 1	B 2	B 3	B 4	B 5	B 6	B 7	B 8

FIGURE 11.10. Layout plan for the experiment: Paired plot design with eight blocks and two treatments in each block.

Analysis: Form the following table:

Block	F1	F2	Difference (d_i)
1	12	9	+3
2	15	13	+2
3	12	13	−1
4	15	12	+3
5	10	7	+3
6	14	10	+4
7	14	12	+2
8	13	13	0
Total	105	89	16 (Σd_i)
Mean	13.1	11.1	+2 ($\bar{d}$)

$$S^2_d = \{\sqrt{[\Sigma d_i^2 - (\Sigma d_i)^2 / n] / (n-1)}\}$$
$$\text{(variance for the differences)}$$
$$= [52 - (16 \times 16) / 8] / 7 = 20 / 7 = 2.857$$
$$S_d = \sqrt{2.857} = 1.690$$
$$SEd = 1.690 / \sqrt{8} = 1.690 / 2.8284 = 0.5975$$
$$t = (\bar{x}_1 - \bar{x}_2) / s\bar{d} / \sqrt{n} = (13.1 - 11.1) / 0.5975 - 3.347$$

The table value of t for 7 df at $P = 0.05$ and $P = 0.01$ are 2.365 and 3.499, respectively. Our null hypotheses is that there is no difference between F1 and F2. Since the calculated t (3.347) is greater than the tabular t, the difference is significant at $P = 0.05$ level and we conclude that the difference is not due to chance or chance factors. We therefore reject the H_o since the risk involved in such a decision is less than 5 percent.

SIGNIFICANCE OF ASTERISKS

The sample value of t is marked with a single asterisk if the difference is significant at $P = 0.05$ (in which case we say *significant*), and with double asterisks if the difference is significant at $P = 0.01$ (in which case we say *highly significant*).

Alternate method using F statistic: The paired t test data can also be analyzed using the analysis of variance technique. It is similar to the design called randomized complete block design (RCBD).

Calculations:

SS due to blocks:	$(21^2 + \ldots + 26^2) / 2 - 194^2/ 16$ (CF)
	$= 2398 - 2352.25 = 45.75$
SS due to treatments:	$(F1^2 + F2^2)/r - CF = (105 + 89)/8 - CF$
	$= 2368.25 - 2352.25 = 16$
Total SS:	$(12^2 + 15^2 + \ldots + 13^2) - CF = 2924 -$
	$2352.25 = 71.75$

(see Table 11.3)

If the H_o is not true, i.e., the population means are not the same, then s^2 (T) (16.00) will be relatively large compared with 5^2, and F ratio will tend to be much greater than 1. If H_o is true, the two population means, and consequently the two populations themselves are identical.

In this example, verify the relationship that $t^2 = F$ ($3.347^2 = 11.20$).

Comparison of Paired t Test and Nonpaired t Test

1. The number of degrees of freedom is greater in the independent t test than the paired t test. A design with a comparitively more number of degrees of freedom is usually preferred as it gives lesser error mean square. However, the gain achieved by pairing is usually greater than the slight loss due to decreased degrees of freedom (later in this chapter, an example is given to this effect).
2. In a paired t test, we do not assume the equality of variances and normality as in a nonpaired t test. We assume that the differences follow from a normally distributed population of differences.
3. If there is pair-wise correlation of the data, the paired t test is more powerful.

Example: To show that paired t test is superior to nonpaired t test.

TABLE 11.3. ANOVA data.

SV	Df	SS	MS	F	F from tables (5 %)	1 %
Blocks	7	45.75	6.53	4.57		
Treatments	1	16.00	16.00	11.20*	5.59	12.25
Error	7	10.00	1.43			
Total	15	71.75				

* = significant at $P = 0.05$

Consider an experiment in which $n_1 = n_2 = 10$ and the x_1 and x_2 (treatments) are recorded as follows:

Tr/Rep	1	2	3	4	5	6	7	8	9	10
X_1	142	140	144	144	142	146	149	150	142	148
X_2	138	136	147	139	143	141	143	145	136	146
Diff(d)	4	4	-3	5	-1	5	6	5	6	2

$$\overline{X}_1 = 144.7 \; \overline{d} = 144.7 - 141.4 = 3.3 \; SE_{\overline{d}} = 0.97 \; \Sigma_d / 3.3$$
$$\overline{X}_2 = 141.4$$
$t = \overline{d}/ SE\overline{d} = 3.3/0.97 = 3.402*$ (the table value of t for 9 df and $P = 0.05$ is 2.262)

Analysis in nonpaired design (independent t test analysis)
$$s_1^2 = 104.10 \qquad\qquad s_2^2 = 146.40$$
$$n_1 = 10 \qquad\qquad n_2 = 10$$
$$df = 9 \qquad\qquad df = 9$$

$$s^2p = (104.10 + 146.40) / 9+9=18 = 13.917$$
$$t = \overline{x}_1 - \overline{x}_2 / \sqrt{s_p^2 (1/n_1 +1/n_2)} = 3.3/1.67 = 1.976 \text{ ns}$$

(as t value from table for 18 df and at $P = 0.05 = 2.101$)

The results show that we do not reject the H_o as calculated t (1.976) is lesser than the table value of t (2.101). This analysis reveals that the paired design is more powerful than the unpaired design.

Efficiency of paired design over nonpaired design: The efficiency of the two designs can be seen by comparing the confidence intervals for the mean difference (3.3).

Confidence interval for $\mu_1 - \mu_2$ is $\pm t \times s\bar{d}/\sqrt{n}$

$$S^2_d = [\Sigma d_i^2 - \frac{(\Sigma d_i)^2}{n}]/n - 1$$

$$= (193 - \frac{33^2}{10})/9$$

$$= 9.344$$

$$S = \sqrt{9.344} = 3.056 \quad SE_d = \frac{s}{\sqrt{n}} = \frac{3.056}{\sqrt{10}}$$

$$= 0.97$$

$$95\% \text{ CI} = 3.3 \pm 2.262 \times 0.97 = 1.11 - 5.49$$

CI for unpaired data:
We estimate pooled variance (s^2) as $(104.10 + 146.40)/18 = 139.17$, $s = \sqrt{139.17} = 3.73$

$$S^2 d = \left\{ \sqrt{\left[\Sigma d_i^2 - (\Sigma d_i)^2/n\right]/(n-1)} \right\}$$

$$95\% \text{ CI} = \bar{A} - \bar{B} \pm t \times s\sqrt{(1/n_1 + 1/n_2)}$$

$$= 3.3 \pm 2.101 \times 3.73\sqrt{1/10 + 1/10} = 3.3 \pm 3.50$$

$$= -0.2 - 6.80$$

Now compare the CI for the two tests:

Paired test : CI = 3.3 ± 1.46 (1.84 – 4.76)
Nonpaired test: CI = 3.3 ± 3.50 (–0.2 – 6.80)

In the paired analysis, we get 2.4 times (3.50/1.46) less CI than the other design. Since the standard deviation of the difference ($\bar{A} - \bar{B}$) is inversely proportional to the square root of the sample size, it would require approximately $(2.4)^2 = 6$ times as many observations in the unpaired design to reduce the width of the unpaired CI to that obtained from the paired design.

Example (paired design): In wetland rice cultivation, manuring in seedbed is useful in producing vigorous seedlings and higher grain yield. The results of an experiment conducted by Palaniswamy (1969) at Rice Research Station, Tamil Nadu Agricultural University, India, in paired design are given as follows. The two treatments are: (1) no manuring the seedbed (NM), and (2) manuring the seedbed at 100 kg N/ha + 110 kg P$_2$O$_5$/ha + 140 kg K$_2$O/ha (M). The plot size used in the experiment was: 3.5 ft by 15.5 ft. The figures are given in kg/ha.

Replication	NM	M	d_i
1	2630	2865	−235
2	2070	2127	−57
3	1950	2577	−627
4	2085	2697	−612
Total	8735	10266	−1531
Mean	2183.75	2566.5	−382.75

Variance of $d_i = \{\Sigma d_i^2 - (\Sigma d)^2 / n\} / (n-1) = (82617 - 585990)/3 =$ 240156.75/3 = 80052.25

$$s = 282.94 \; SE = s / \sqrt{n} = 282.94 / 2 = 141.46$$
$$t = \bar{d} / SE\bar{d} = -382.75 / 141.46 = 2.71 \text{ ns}$$

The table value of t for 3 df and at $P = 0.05 = 3.18$.

Conclusion: Manuring has no effect on the seedbed in rice crop.

One-Sided t Test

Sometimes, researchers conduct a one-sided test, and in this test the t tables are generally modified slightly to allow for this option. For example, a sugarcane breeder evolved a new variety and wishes to replace the existing old variety. The breeder claims that the new variety gives higher production than the standard one. In this instance the type of critical region is called a one-sided critical region or a one-tail critical region. Here H_o: $\mu_1 > \mu_2$; H_1: $\mu_1 < \mu_2$.

In this case, the researcher is interested in detecting a difference in one direction only. There is only one critical value for t (Figure 11.11).

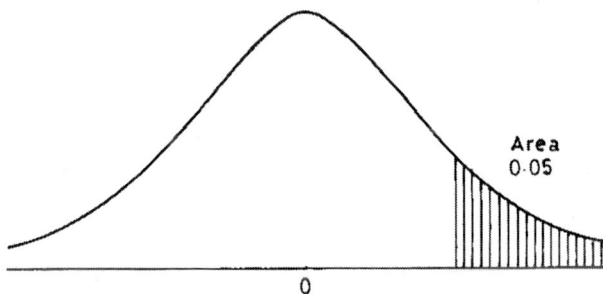

FIGURE 11.11. Student's t test distribution with n−1 degrees of freedom for one-sided t test.

The t is defined as $t=(\bar{x}_1 - \bar{x}_2)/S(\bar{x}_1 - \bar{x}_2)=(\bar{x}_1 - \bar{x}_2)/(\sqrt{s^2/n_i}+s_2 p/n_2)$.

Assumptions: Both samples come from normal populations with equal variances. Many studies showed that t test is robust enough to stand considerable departures from its theoretical assumptions especially when sample sizes are equal or nearly equal (Box, 1953). Of the two assumptions, the equality of variances is more important. If severe deviation from normality and equal variances is observed, nonparametric tests may be done.

Example for one-sided t test: A new variety of rice crop (Y) and a standard variety (X) were compared in an experiment in paired design and the plot yields (kg/plot) recorded are given in the following table for analysis.

Tr./Bl	1	2	3	4	5	6	7	8	9	10	11	12	13	14	15
Y	25	20	26	26	20	23	14	18	18	20	22	27	19	27	21
X	18	19	16	21	16	20	20	14	11	22	19	29	16	27	15
d_i	7	1	10	5	4	3	-6	4	7	-2	3	-2	3	0	6
d_i^2	49	1	100	25	16	9	36	16	49	4	9	4	9	0	36

$\Sigma Y_i = 326$; $\Sigma d_i = 43$; $\bar{Y} = 21.7$; $H_o:\mu_d =0$; $H_1:\mu_1 >0$; $\Sigma X_i = 283$; $\bar{d}$ (Mean of d_is) $= 2.9$; $\bar{X} = 18.8$.

Since the sample size is 15, the distribution of t is $=d/[SEd/15]$. Follows t curve with 14 df (Figure 11.12).

$$t = 2.9/(4.1/\sqrt{15}) = 2.74$$

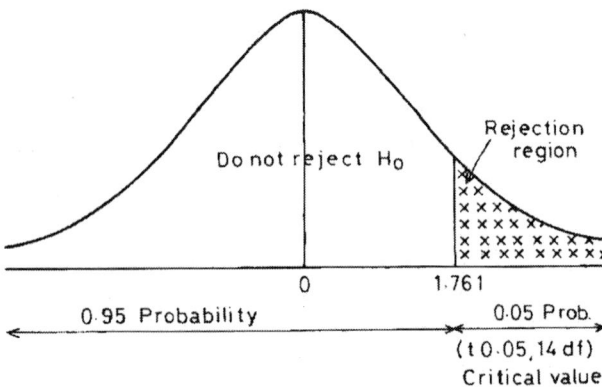

FIGURE 11.12. Distribution of t statistic, one-tailed test, 0.05 level of significance.

Since $2.74 > 1.76$ at $P = 0.05$, we should reject H_o and conclude that Y (new variety) records higher production than X (standard). If $P = 0.05$, the critical value for a two-tailed test is 2.14 and for a one-sided test the critical value is 1.76. Thus with a smaller departure, the sample difference can be declared significant at a given level for one-sided procedure. In case of doubt, it is better to perform a two-tailed test procedure (Remington and Schork, 1970).

Estimation of Proportion

The relative frequency of an event is the probability of occurrence of that event. If there are n cases, of which x are favorable to an event, then x/n is the sample proportion, which estimates the corresponding population or true proportion. This is denoted by the letter p. Then, mean $= \mu = np$, and $\sigma = \sqrt{np\,(1-p)}$, $SEp = \sqrt{pq\,/\,n}$.

Example: In a sample population of 400 females to whom a new drug was given, only 136 expressed relief. Find the sample proportion and confidence interval (CI) for the true population proportion.

Solution: $p = 136/400 = 0.34$ $SEp = \sqrt{pq\,/\,n} = \sqrt{0.34} \times 0.66/400 = 0.0236$.

The p estimates the population proportion P. This is not known. But we can say that P lies within a certain interval called CI.

The 95 percent confidence interval for P:

$$= p \pm t \times SEp$$
$$= 0.34 \pm 1.96 \times \sqrt{(0.34 \times 0.66)\,/\,400}$$
$$= 0.34 \pm 1.96 \times \sqrt{0.2241\,/\,400}$$
$$= 0.34 \pm 1.96 \times \sqrt{0.002244\,/\,4}$$
$$= 0.34 \pm 1.96 \times \sqrt{0.0005588}$$
$$= 0.34 \pm 0.046 = 0.34 \pm 0.05 = 0.29 \text{ to } 0.39.$$

The CI 0.29 to 0.39 includes the population proportion.

Similarly, we can compute 99 percent confidence interval for P:

$$CI = 0.34 \pm 2.58 \times 0.0236 = 0.34 \pm 0.06 = 0.28 \text{ to } 0.40.$$

Testing of Sample Proportion

Given $P = 0.80$ in a population, in a random sample of 320 ($n = 320$), 267 are in favor of an event. $H_o: p = 0.80$; $H_1 : p \neq 0.80$; $\sigma = 0.05$.

The statistic $Z = (z-np) / \sqrt{np(1-p)} = 267-(320 \times 0.80)\sqrt{320(0.80)(0.20)}$ = 1.47. Since the calculated Z is less than 1.96, we cannot reject the null hypothesis.

Testing Goodness of Fit

Example: 120 patients did not receive physical therapy and 120 patients did receive physical therapy. After six months, their conditions were examined and classified under three categories, namely, deteriorated, unchanged, and improved. Test whether there is an effect on physical therapy.

Condition	No therapy	Therapy	Total
Deteriorated	6	11	17
Unchanged	65	31	96
Improved	49	78	127
Total	120	120	240

The chi-square value calculated is 20.13. As it is significant, the therapy is rejected.

Large Sample Tests of Proportions

Let P1 and P2 be the proportions of the same character in the population and p_1 and p_2 are the two estimates, respectively, in the samples. Let the sample sizes be n_1 and n_2. To test the hypothesis that the proportions P1 and P2 are the same, the test of significance is given by

$$Z = \frac{p_1 - p_2}{\sqrt{P(1-P)(1/n_1 + 1/n_2)}}$$

where $P = (n_1 p_1 + n_2 p_2) / (n_1 + n_2)$ and $Q = 1-P$.

In large samples, Z is distributed with mean 0 and variance 1. Therefore, if the observed Z is greater than 1.96 (the 5 percent value of the two-tail test for normal variate), we conclude that the two population proportions are different. If $Z < 1.96$, we conclude that the proportions P1 and P2 do not indicate significant difference at 5 percent level of significance.

Example: In a certain poultry farm, out of 1,000 birds, 700 contracted a disease during a particular season. During the same season, in another farm, it was observed that out of 1,100 birds, 810 contracted the same disease. Test whether the proportions of the disease in both places are different.

Solution:

$p_1 =$ proportion attacked in farm 1 $= 700/1000 = 0.7$
$P_2 =$ proportion attacked in farm 2 $= 810/1100 = 0.7364$
$P = (n_1 p_1 + n_2 P_2) / (n_1 + n_2) = [(0.7 \times 1000) + (0.7364 \times 1100)] /$
$1000 + 1100 = 1510 / 2100 = 0.7190$ or $(700 + 810) / (1000 + 1100) /$
$1510/2100 - 0.7190$
$Q = 1 - P = 1 - 0.7190 = 0.2810$

$$Z = \frac{P_1 - P_2}{\sqrt{PQ(1/n_1 + 1/n_2)}} = \frac{0.700 - 0.7364}{\sqrt{(0.7190 \times 0.2810)(1/1000 + 1/1000)}}$$
$= -0.0364 / 0.0196 = -1.86$

In large samples, Z is distributed normally with mean 0 and variance 1. Calculated Z is less than 1.96, so we conclude that there is no difference between the two proportions.

Testing Proportions

Testing differences between two proportions. In many problems, we have to decide whether two sample proportions or percentages are significant or may be attributed to chance.

Example 1: In an engineering process, one firm produces 28 defective articles in a sample size of 400, and in another firm the process produces 15 defective articles in a sample size of 300. We want to find whether $28/400 = 0.07$ and $15/300 = 0.05$ are the same. The sampling distribution of the difference between two samples proportions $(p_1 - p_2)$ follows normal distribution with mean $p_1 - p_2$ and standard deviation $= \sqrt{(P_1 Q_1 / n_1 + P_2 Q_2 / n_2)}$.

The statistic Z is defined as

$$Z = \frac{p_1 - p_2}{\sqrt{p_1 q_1 (1/n_1 + 1/n_2)}}$$

where $p =$ pooled proportion value
$H_o : p_1' = p_2'$ $H :: p_1' \neq p_2'$
$P = (28+15) / (400+300) = 0.061$, $Q/1-0.001=0.939$

Therefore $$Z = \frac{(28/400) - (15/300)}{\sqrt{[(0.061)(0.039)(1/400 + 1/300)]}} = 1.10$$

Z at $P = 0.05$ from table is 1.96. As calculated Z is less, H_o cannot be rejected. We cannot conclude that one process is better than the other.

Example 2: Shantha et al. (1989) reported that the total number of female cancer cases in 1984 was 2,166, out of which 1,171 cases were cervical cancer cases. During 1987, it was observed that out of 2,708 cases, 1,414 were cervical cancer cases. Is the difference in proportions the same in the incidence of the disease in both the years?

It is assumed that the population proportions are p_1 and p_2 in 1984 and in 1987, respectively. The sample sizes n_1 (2,166) and n_2 (2,708) are taken and the respective proportions p_1 and p_2 calculated. The test statistic is

$$Z = \frac{p_1 - p_2}{\sqrt{P(1 - P)(1 / n_1 + 1 / n_2)}}$$

where $P = (p_1 n_1 + p_2 n_2)/(n_1 + n_2)$
Calculations: $p_1 = 1171/2166 = 0.541$ and $p_2 = 1414/2708 = 0.522$

$$P = [(0.541 \times 2166 + 0.522 \times 2708)]/(2166 + 2708) = 2585.4/4874 = 0.530$$

$$Z = \frac{0.541 - 0.522}{\sqrt{[0.530 \times 0.470(1 / 2166 + 1 / 2708)]}} = 0.019 / 0.0144 = 1.319$$

Under the null hypothesis, Z is approximately distributed as a standard normal deviate. Because the calculated Z value is less than 1.96, we accept the null hypothesis that the proportions of cervical cancer cases out of the total female cancer cases are the same in both the years.

Testing many proportions. Example: Consider the following contingency table which shows the effect of a drug on disease in 100 patients. Is there a relation between cure and drug effect?

	High cure	Moderate cure	Mild cure	Total
Drug	116 (98)	141(147)	37 (49)	294
No drug	84 (102)	159 (153)	63 (51)	306
Total	200	300	100	600

The figures in parentheses relate to the expected frequencies. Now calculate the chi-square value as chi-square $= (116 - 98)^2 /98 + \ldots + (63 - 51)^2/51 = 12.73$.

The table value of chi-square for $(3 - 1)(2 - 1)$ degrees of freedom at $P = 0.05$ is 5.991. The null hypothesis of no relation is rejected.

t *and* F *Tests*

We propose t and F tests of significance when (1) we test the hypothesis that the sample mean against the population mean has some specific value, and (2) we test the hypothesis that the two sample means are not significantly different. When we compare more than two samples' means, comparison of each pair of means using t test is difficult. For example, we have four treatments T1, T2, T3, and T4. If we use t test, we can compare two means at a time. The possible pairs are T1 and T2, T1 and T3, T1 and T4, T2 and T3, T2 and T4, and T3 and T1.

In other words, if we have four treatments' means, then there will be 4C2 $= 4 \times 3/2 \times 1 = 6$ pairs of t tests. If we have seven means, then three will be 21 separate pair t tests. When we use $P = 0.05$, we take the risk of being wrong as often as 5 percent of the time in our rejection of the H_o. The precision of the experiment is affected. In this particular situation, the F test becomes more powerful. When there are more than two treatments (sample means), we might consider using F test to test the significance of the differences among all the means at once.

EXERCISES

11.1. Explain with a suitable example the following terms: *statistical hypotheses, test of significance, level of significance, critical region, error of the first kind, error of the second.*

11.2. The random variable x is normally distributed with mean μ (unknown) and variance σ^2. How you would test for the hypothesis H_o: $\mu = \mu_o$? Discuss both the cases when σ^2 is known and unknown.

11.3. A sample of rural families and urban families was taken to study the differences in television-viewing habits in the evenings. The data are given for the number of hours/week. Assume the variances are equal. Do you think the difference in viewing time is real among the two classes of the people? Construct 95 percent confidence limit for the difference given. Rural $n_1 = 15$; mean of (rural) $= 27$; variance 13.5; urban: $n_2 = 20$; mean $= 17$, and variance 80.

11.4. An experimenter observed the new variety (N) of sugarcane crop grown in six plots and the standard variety (S) grown in ten plots recording the yields as follows:

New variety: 25, 21, 24, 20, 24, 23
Standard: 22, 19, 18, 21, 21, 17, 23, 20, 17, 22

Do you think that the observed difference is a significant difference? Also construct 95 percent confidence limits for the true difference.

11.5. What is type I error? Why should it occur?

11.6. A newly produced corn variety was tested in two different climatic zones, Z1 and Z2 in a state by a corn breeder. He wants to test the suitability of the variety for these two zones. The yields (kg/plot) recorded in the two zones in different plots are given as follows:

Zone 1: 26, 27, 26, 23, 34, 27, 21, 20, 20, 23

Zone 2: 25, 30, 25, 28, 30, 32, 30, 28, 25, 32

Test the significance of the difference in the grain yield in the two different zones.

11.7. In a fertilizer trial in corn crop, two different amounts of nitrogen (N = 60 and N = 120) were compared. The researcher collected 150 ear heads from the low-fertilizer plot and recorded the mean grains ear head as 110 with a standard deviation of 15. The researcher collected 100 ear heads from the high-fertilizer plot and recorded the mean number of grains/ear head as 153 with a standard deviation of 27. Do you conclude that the highly fertilized plot has more grains/ear head at 0.05 level of significance?

11.8. Palaniswamy and Palaniswamy (1973) investigated the hybrid vigor in rice crop by growing F1 seeds and one parent (ASD 5 variety) in a paired plot design. The plant yields (g) recorded by them are given as follows:

Hybrid: 380, 133, 75, 204, 73, 102, 134, 105, 48, 180, 107, 207, 85, 50, 159, 190

Parent: 135, 60, 35, 60, 60, 67, 75, 92, 85, 82, 87, 87, 77, 50, 65, 110

Analyze the data and find out whether the hybrid vigor is present or not.

Suppose the experiment had been conducted following independent *t* test procedure, what would have been the consequence? Also analyze the data in randomized block design and find the relation between *F* and *t* values.

11.9. What formula will you use to set up confidence interval for the difference between two means for a paired design? An unpaired design?

11.10. Two types of trees, X and Y, were planted in 10 plots each by a forest breeder. The recorded height measurements (m) after five years follow:

X: 3.1, 2.8, 3.2, 1.9, 2.9, 1.9, 2.9, 3.0, 3.2, 3.3

Y: 2.2, 1.9, 2.4, 2.0, 2.6, 2.0, 1.9, 2.5, 2.1, 1.0

Set 95 percent confidence interval for the difference in means.

11.11. The *t* test is suggested to compare two treatment means. This procedure cannot be used in all situations. Why?

11.12. The population mean in a set of data is 20. A sample of size 12 was taken and the data are as follows: 20, 21, 23, 18, 21, 30, 29, 23, 28, 30, 28, 19.
Calculate that the sample came from the population whose mean is 20.

11.13. A farm manager wants to find out which variety is better between two varieties, A and B. He selected eight plots of land in his farm and divided each plot into two halves and allotted A and B at random in each half. The yields are as follows.
    A: 80, 79, 110, 89, 105, 89, 92, 112
    B: 78, 72, 80, 112, 120, 95, 99, 120
Which variety do you think is better? Set up the confidence interval for each.

11.14. State whether there is any difference between a two-tailed test and a one-tailed test.

11.15. Under what circumstances is the paired test more efficient than the nonpaired t test?

11.16. A sample of size 20 has mean = 14 kg and variance = 5.9822 kg^2. Calculate the confidence interval for population mean.

11.17. Differentiate between *assumption* and *hypothesis*.

11.18. What are the assumptions in t test of significance?

11.19. In the one-sided t test, the rejection region is only on one side. Why?

11.20. The differences in weights before and after administering a drug in several animals are 0.2, –0.4, –1.3, –1.6, –0.7, 0.4, –0.1, 0.0, –0.6, –1.1, –0.8. Apply one-tailed t test and state whether you accept or reject the null hypothesis.

11.21. Plant heights in a sample of plants varied from standard-fertilizer(s) and new-fertilizer (N) plots. Is the variability in height due to the new fertilizer?
    S: 68.2, 54.6, 58.3, 46.8, 51.4, 52.0, 55.2, 49.1, 49.9, 52.6
    N: 52.3, 52.4, 55.6, 53.2, 61.3, 58.0, 58.3, 52.8, 59.8, 54.8

11.22. The yield per plant of a rice variety follows normal distribution (mean = 25 g/plant and standard deviation 4). A new variety was tried and the yields of 10 plants taken randomly were recorded as follows: 18, 21, 24, 18, 19, 16, 13, 17, 20, 21. Can you state that the new variety gives greater yield?

11.23. A pharmaceutical company claims that the medicine produced by it is effective in relieving pain in 90 percent of people. Among a sample of 250 people who are using the medicine, 150 reported that they got relief. Determine whether the claim is justified.

11.24. One breed of rat shows a mean gain in weight of 60 g during the first two months of life. Six rats were fed with a specified diet from birth and after two months the weight gains were 63, 68, 67, 56, 60, and 70. Do you think that the diet caused increase in weight?

11.25. Two laboratories carry out independent estimates of protein content in rice. Eight estimates were made by each laboratory. At each time, one sample was taken; half the quantity was sent to Lab 1 and the other half to Lab 2. The results follow.

Sample No.	1	2	3	4	5	6	7	8
Lab 1	8	10	8	10	10	9	11	9
Lab 2	9	11	9	11	10	8	10	8

Do the two laboratories report the same results?

11.26. When you will compute pooled variance?

REFERENCES

Afifi, A.A. and S.P. Azen (1972). *Statistical analysis: A computer oriented approach.* New York: Academic Press.

Box, G.E.P. (1953). Non-normality tests on variances. *Biometrika* 40: 318-335.

Klugh, H.E. (1974). *Statistics: The essentials for research.* New York: John Wiley and Sons, Inc.

Mack, N. (1967). *Essentials of statistics for scientists and technologists.* New York: Plenum Press.

Palaniswamy, K.M. (1969). A note on nursery manuring in rice. *Oryza* 6(2):111-113.

Palaniswamy, K.M. (1974). Prediction of plant height in corn. *Madrid Agricultural J.* 61(9): 751-752.

Palaniswamy, K.M. and S. Palaniswamy (1973). Heterosis, growth, and variability in rice. *Madras Agric. J.* 60(9):31-32.

Remington, R.D. and M.A. Schork (1970). *Statistics with applications to biological and health sciences.* Englewood Cliffs, NJ: Prentice Hall.

Shanta, V., K.M. Palaniswamy, and M. Jeevamudha (1989). Scenario of cancer cases. Paper presented at special symposium on scenario of cancer of the cervix in India, held at Annual Review Meting of ICMR, New Delhi, November 16-18, 1989.

Usha Rani, P. (1996). Evaluation of chili (*Capsicum annuum* L.) germplasm for capsanthin and capsaicin contents and effect of storage on ground chili. *Madras Agric. J.* 83(5): 288-291.

Walker, H.M. and J. Lev (1965). *Statistical inference.* Calcutta: Oxford and IBM Publishing Co.

Chapter 12

Correlation

Many research studies in different fields involve simultaneous study of two or more variables. The purpose of such studies is to investigate the relationships among the variables. In relationship studies, observations are recorded either in pairs, triples, or more as they exist naturally in the environment. Although recording observations, no attempt is made either to control or manipulate the variables. The most important factor in correlation studies is covariation. Covariation refers to change in one variable accompanied by a corresponding change in another variable, i.e., how the two variables simultaneously covary. When our study involves only two variables, it is called *simple correlation* or *simple linear correlation*. If our study relates to more than two variables, it is called a *multiple correlation study*. In this chapter, we concentrate on the relationship between two variables. In correlation studies, our interest is on the strength of the relationship between the variables, i.e., how well the variables are correlated. Correlation is therefore a statistical technique that is used to measure and describe the relationship between two variables. It plays an important role in agriculture, medicine, industry, education, biology, social sciences, and many other areas. A few examples will illustrate the importance and usefulness of correlation studies in several areas:

1. Air pressure and altitude (A decrease in air pressure occurs as the altitude increases, and they show a negative relationship.)
2. Rainfall and rise of water levels (When rainfall increases, the water level in dams and rivers increases. Increase in one variable [rainfall] is relative to the increase in another variable [water level]. These two variables are said to be positively correlated.)
3. Grades and class attendance of the students in a class
4. Time and population
5. Age and blood pressure
6. Ages of husbands and wives
7. Rainfall and agricultural production
8. Age and insurance premium paid

9. Amount of fertilizer used and amount of crop production
10. Tree trunk diameter and height
11. Diet and body weight
12. Temperature and sale of cool drinks
13. Age of car and number of miles driven
14. The number of miles the car is driven and the resale value
15. Sale proceeds and amount spent on advertising
16. Incidence of cancer and smoking habits
17. Smoking and incidence of heart attack
18. Blood pressure and weight loss
19. Age and pulse rate
20. Amount of fluoride concentration in drinking water and prevalence of tooth decay among children

Some examples where no correlation is found:

1. Number of human births and fertilizer sales
2. Number of books in libraries and car accidents
3. Grade point averages of students and the number of patients in hospitals
4. Number of births in India and rainfall in the United States

METHODS OF STUDYING CORRELATION

The most important methods of displaying and studying correlation are described as follows:

1. Scatter diagram
2. Coefficient of correlation
3. Coefficient of rank correlation
4. Regression line
5. Intraclass correlation

Scatter Diagram

In a scatter diagram the data take the form of paired values of two variables such as y and x or x_1 and x_2, etc. Each pair is represented in a graph by a dot (x, y). Thus the data plotted in the graph give the scatter diagram.

Uses of Scatter Diagram

On a scatter diagram, one can infer, approximately, the relationship between two variables. Some examples of different types of scatter diagrams showing different relationships are illustrated in Figures 12.1 through 12.9.

If $r = 0$, the variables are said to be uncorrelated or independent. This implies Sxy (a quantity measuring covariation) is zero, and it occurs when x and y are independent. The data points are scattered around the horizontal line and hence the data points do not indicate a positive or negative line or relationship. From the scatter diagrams, one can guess how the two variables move in their relationship. A positively sloped line indicates the positive correlation, and a negatively sloped line the negative correlation. When

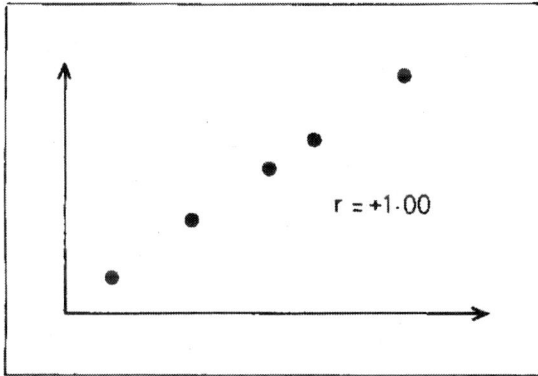

FIGURE 12.1. Scatter diagram showing perfect positive correlation.

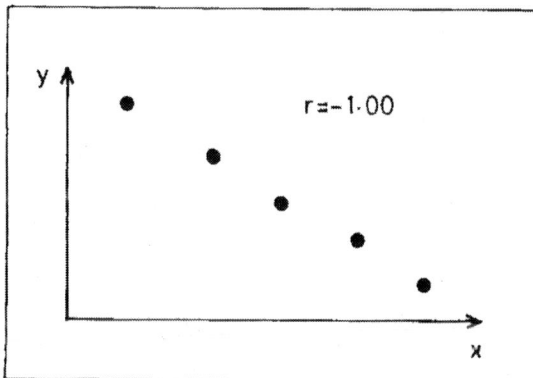

FIGURE 12.2. Scatter diagram showing perfect negative correlation.

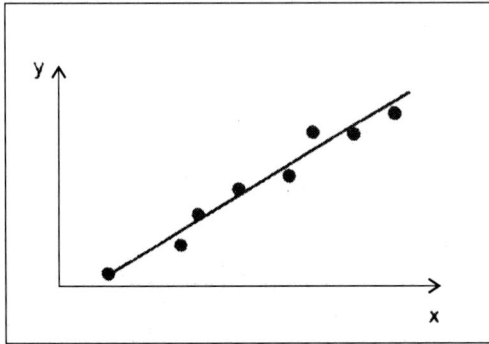

FIGURE 12.3. Scatter diagram depicting high positive correlation between *y* and *x*.

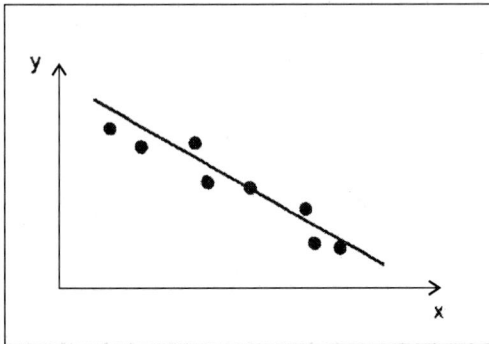

FIGURE 12.4. Scatter diagram showing high negative correlation.

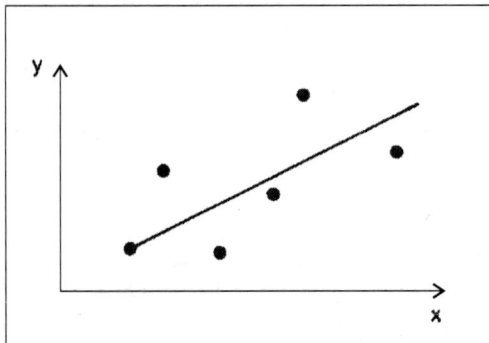

FIGURE 12.5. Scatter diagram showing low positive correlation.

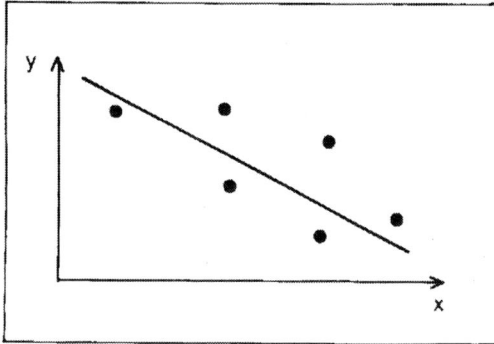

FIGURE 12.6. Scatter diagram showing low negative correlation.

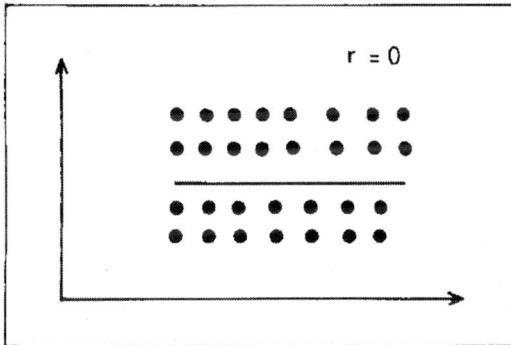

FIGURE 12.7. Scatter diagram showing zero correlation (*x* and *y* are not linearly correlated).

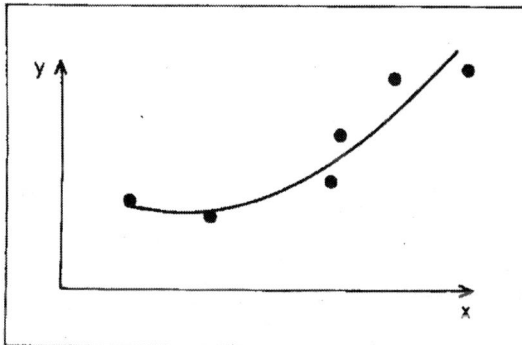

FIGURE 12.8. Direct curvilinear correlation.

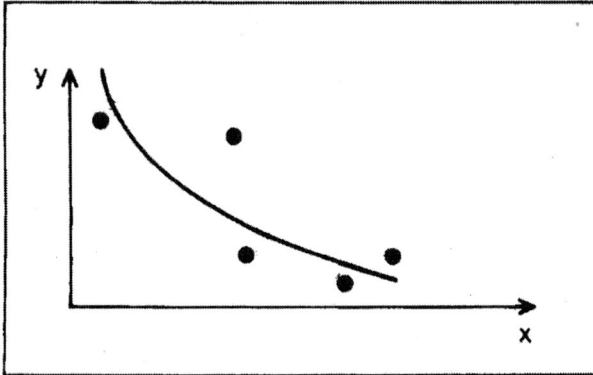

FIGURE 12.9. Inverse curvilinear correlation.

the correlation coefficient is less about 0.05 to 0.10, it is difficult to infer the presence of correlation. Sometimes, when low correlation is present, it is difficult to know the direction of the correlation. The exact relationship cannot be determined from scatter diagrams and hence, we must resort to coefficient of correlation method.

Coefficient of Correlation Method

Since we want to quantify the exact relationship, it must be measured and expressed by a numerical value called the correlation coefficient. The correlation coefficient determined from sample data is denoted by the lower-case letter r and the correlation coefficient determined from a population is referred to as P (rho). It is computed from the sample data by using the formula:

$$r = \frac{\text{Covariance}(x, y)}{\sqrt{\text{Variance } x \text{ Variance } y}}$$

$$= \frac{\{[\Sigma(x - \bar{x})(y - \bar{y})] / (n - 1)\}}{\sqrt{[\Sigma(x - \bar{x})^2 / (n - 1)]}\sqrt{[\Sigma(y - \bar{y})^2 / (n - 1)]}}$$

$$= \frac{\Sigma(x - \bar{x})(y - \bar{y})}{\sqrt{\Sigma(x - \bar{x})^2 \Sigma(y - \bar{y})^2}} = \frac{\Sigma xy - (\Sigma x)(\Sigma y) / n}{\sqrt{[\Sigma x_i^2 - (\Sigma x)^2 / n]}\sqrt{[\Sigma y_i^2 - (\Sigma y)^2 / n]}}$$

$$= \frac{\text{corrected sum of products}}{\sqrt{(\text{corrected sum of squares for } x)}\sqrt{(\text{corrected sum of squares for } y)}}$$

$$r = \frac{Sxy}{\sqrt{Sx^2\, Sy^2}}$$

where

Sxy = Corrected sum of products
Sx^2 = Corrected sum of squares of x
Sy^2 = Corrected sum of squares of y

Example: In a study, the following pairs of values are obtained:

$x = 3\ 5\ 4\ 6\ 7$
$y = 2\ 3\ 2\ 4\ 4$

Method 1

Computation of correlation coefficient using the definition formula from the following table.

	y	x	$(y-\bar{y})$	$(x-\bar{x})$	$(y-\bar{y})^2$	$(x-\bar{x})^2$	$(x-\bar{x})(y-\bar{y})$
	3	2	−2	−1	4	1	2
	5	3	0	0	0	0	0
	4	2	−1	−1	1	1	1
	6	4	1	1	1	1	1
	7	4	2	1	4	1	2
Total	25	15	0	0	10	4	6
Mean	5	3			Sy^2	Sx^2	Sxy

$$r = \frac{\Sigma(x - \bar{x})(y - \bar{y})}{\Sigma(x - \bar{x})^2\, \Sigma(y - \bar{y})^2} = \frac{Sxy}{\sqrt{Sx^2\, Sy^2}} = \frac{6}{\sqrt{4 \times 10}} = 0.95$$

Steps involved in the computation of r:

1. Compute the totals of x and y and compute the mean of x ($\bar{x}$) and the mean of y ($\bar{y}$), i.e., Σx, Σy, $\bar{x}$, $\bar{y}$.
2. Compute the deviations of each case from the mean of x, i.e., $x-\bar{x}$.

3. Compute the deviations of each case from the mean of y, i.e., $y-\bar{y}$.
4. Compute the product of $(x-\bar{x})(y-\bar{y})$ and get xy (x deviation times the y deviation).
5. Sum these xy products and get Sxy. This is the value of covariance, the numerator in the formula. This is the most important factor in the correlation. It indicates the change in one variable accompanied by a change in a corresponding variable. The sign of covariance (Sxy) indicates the direction of the relationship between two variables.
6. Compute the squares of each x deviation, x^2.
7. Compute the squares of each y deviation, y^2.
8. Sum x^2 deviations and get Sx^2.
9. Sum y^2 deviations and get Sy^2.
10. Apply the following formula to solve for the r value.

$$r = \frac{Sxy}{\sqrt{Sx^2\,Sy^2}}$$

Method 2

In Method 1, we use the mean and get the deviations. In this method, we calculate Sx^2, Sy^2, and Sxy directly from the original observations without involving the mean values as follows.
Compute:

$$n = 5, \; \Sigma x = 15, \; \Sigma y = 25, \; \bar{x} = 3, \; \bar{y} = 5$$

Calculate:

1. $Sx^2 = \Sigma x^2 - CF = \Sigma x^2 - (\Sigma x)^2/n = 2^2 + 3^2 + 2^2 + 4^2 + 4^2 - 15^2/5$
   $= 49{-}45 = 4$
2. $Sy^2 = \Sigma y^2 - CF = \Sigma y^2 - (\Sigma y)^2/n = 3^2 + 5^2 + 4^2 + 6^2 + 7^2 - 25^2/5$
   $= 135{-}125 = 10$
3. $Sxy = \Sigma xy - \dfrac{(\Sigma x)(\Sigma y)}{n} = (2 \times 3) + (3 \times 5) + (2 \times 4) + (4 \times 6) + (4 \times 7) -$
   $\dfrac{15 \times 25}{5} = 81 - 75 = 6$
4. Calculate $r = \dfrac{Sxy}{\sqrt{(Sx^2)(Sy^2)}} = \dfrac{6}{\sqrt{4 \times 10}} = 6/6.325 = 0.95$ (as before)

Note the similarity in the formula used in the computation of sums of squares (SS) and sum of products (SP), i.e., Sxy.

1. Sum of squares of $x = Sx^2 = \Sigma(x-\bar{x})^2$ (deviations squares in x are added) $= \Sigma x^2 - (\Sigma x)^2/n$ (computational formula; where $(\Sigma x)^2/n$ is the correction factor for the mean).
2. Similarily for y: Sum of squares of $y = Sy^2 = \Sigma(y-\bar{y})^2$ (deviations squares in y are added). $= \Sigma y^2 - (\Sigma x)^2/n$ (computational formula; where $(\Sigma y)^2/n$ is the correction factor for the mean).
3. Sum of products $= Sxy = \Sigma(x-\bar{x})(y-\bar{y})$ (deviations of products in x and y are added). $= \Sigma xy - (\Sigma x)(\Sigma y)/n$ (computational formula) $= \Sigma xy - (\Sigma x)(\Sigma y)/n$ (computational formula).

Thus $r = \dfrac{\text{Corrected sum of products}}{\sqrt{(\text{corrected sum of squares of } x)(\text{corrected sum of squares of } y)}}$

$= \dfrac{Sxy}{\sqrt{SxSy^2}}$ (computational formula)

We find $Sxy = \Sigma(x-\bar{x})(y-\bar{y})$, i.e.,

$$\Sigma(x-\bar{x})(y-\bar{y}) = \Sigma(x_i y_i - x_i \bar{y}_i - \bar{x}y_i + \bar{x}\bar{y})$$

$$= \Sigma x_i y_i - \bar{y}\Sigma x_i - \bar{x}\Sigma y_i + n\bar{x}\bar{y}$$

$$= \Sigma x_i y_i - (\frac{\Sigma y_i}{n})(\Sigma x_i) - (\frac{\Sigma x_i}{n})\Sigma y_i + n(\frac{\Sigma x_i}{n})(\frac{\Sigma y_i}{n})$$

$$= \Sigma x_i y_i - [(\Sigma x_i)(\Sigma y_i)]/n - [(\Sigma x_i)(\Sigma y_i)]/n + [(\Sigma x_i)(\Sigma y_i 0]/n$$

$$= \Sigma x_i y_i - [(\Sigma x_i)(\Sigma y_i)]/n$$

Example: Usha Rani (2001) studied the correlation between fruit surface area (cm^2) and fruit length (cm) in 15 chili varieties (*Capsicum annuum* L.). See Table 12.1.

Method of Calculation

Draw a scatter diagram to show the trend of the data. You will get some idea of the trend of the collected data (see Figure 12.10).
Calculations

$\Sigma y = 295.70$	$\Sigma x = 155.26$	
$\Sigma y^2 = 5869.20$	$\Sigma x^2 = 1673.1704$	$\Sigma xy = 3092.8021$
$CF = 5829.2326$	$CF = 1607.0444$	$CF = 3060.6921$
$Sy^2 = 39.9674$	$Sx^2 = 66.1260$	$Sxy = 32.11$

TABLE 12.1. Fruit surface area and fruit length of 15 chili varieties.

Variety	Fruit surface area (cm², y)	Fruit length (cm, x)
Belgaum local	22.70	14.60
Pusa Juwala	22.12	12.25
Kortical	17.58	13.18
Kalyanpur Red	17.79	10.09
Byadegi	22.06	13.64
Sardana	20.77	9.52
CA 960	17.83	8.41
Ronyal	18.21	7.59
Examba	20.72	11.39
Meerut	20.57	10.20
IHR 525	20.01	8.81
Kalyanpur	19.06	9.90
Arun	18.96	8.21
Arun (L)	18.93	8.67
Mulata	18.39	8.80

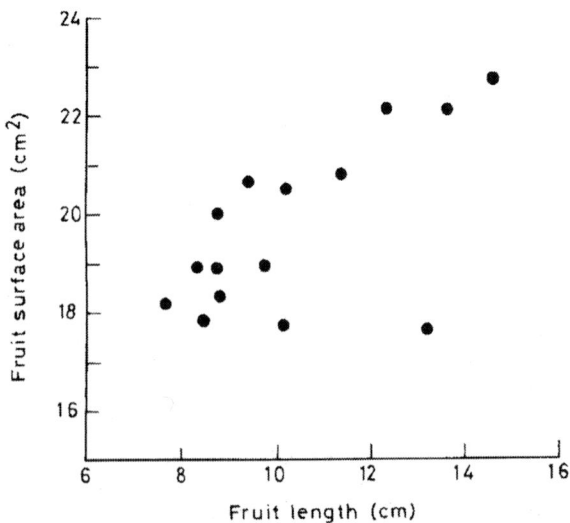

FIGURE 12.10. Scatter diagram depicting the correlation between fruit length and fruit surface area in chili (*Capsicum annuum* L.).

$$r = \frac{Sxy}{\sqrt{Sx^2 \, Sy^2}} = \frac{32.11}{\sqrt{66.1260 \times 39.9674}} = \frac{32.11}{51.41} = 0.6246$$

Test of significance for *r*:

The test of significance for the *r* value is to be done because the correlation coefficient was computed from a sample of data taken from a population. The correlation coefficient in the population is ρ (rho), which is unknown. The *r* obtained from the sample is only an estimated value of ρ. The sample *r* is therefore subject to error. The estimated correlation coefficient value may be either due to chance or may show the real relationship. Furthermore, based merely on the magnitude of the *r* value, one should not conclude the significance. Sometimes, small values of *r* may be significant and sometimes, high values may not be significant. Significance of correlation coefficient should be carried out using an appropriate test of significance. The *r* test of significance is described as follows.

Test of significance for *r* value and null hypothesis:

Researchers make the null hypothesis as H_o: ρ = 0 (i.e., no linear relationship).

H_1: ρ≠0 (i.e., there is positive or negative relationship between the two variables).

When there is a relationship, we say that there is a significant relationship between the two variables. It is tested using the *t* statistic. The statistic *t* is given by the following formula:

$$t = \frac{r}{\sqrt{(1 - r^2)}} \sqrt{(n - 2)} \text{ with } n-2 \text{ degrees of freedom}$$

With reference to our example,

$$t = \frac{0.6246}{\sqrt{1 - (0.6246)^2}} \sqrt{(15 - 2)} = \frac{0.6246}{0.7809} \times 3.6056 = 2.88$$

If the calculated *t* value is greater than the table value of *t*, we conclude that the sample *r* value is significant, i.e., differs significantly from zero. On the other hand, if the calculated *t* value is lesser than the table value of *t*, we infer that the sample is consistent with the hypothesis of no correlation between the variables under investigation. In our example,

Reject				Reject
Rho=0				Rho=0
−1				+1
	−2.160	0	+2.160	
	C		C	

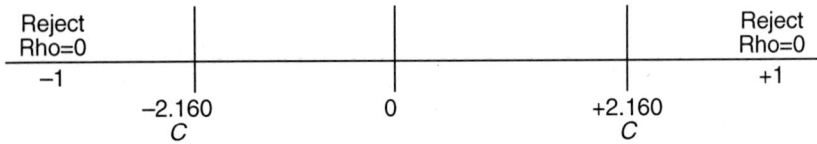

The t value (2.88) is greater than C; we reject the null hypothesis. The value of t from the table for $n - 2$ degrees of freedom, i.e., 13 degrees of freedom at $P = 0.05$ and $P = 0.01$ are 2.160 and 3.012, respectively. Since the calculated t value at $P = 0.05$ is greater than t value, we say that the correlation is significant between fruit surface area and fruit length. This infers that as the fruit length increases, there is a corresponding increase in fruit surface area in chili (*Capsicum annuum* L.).

Some Properties of Correlation Coefficient

The following information about the correlation coefficient will be useful to the researchers while performing correlation analyses and in the interpretation of the results.

1. The correlation coefficient r ranges from +1 (perfect positive linear relationship) through zero (no relationship) to −1 (perfect negative relationship).
2. The sign + or − indicates the direction of the relationship. When r is positive, all the points lie on or near the straight line with positive slope (Figures 12.1, 12.3, 12.5). When r is negative, all the points lie on or near the straight line with negative slope (Figures 12.2, 12.4, 12.6). When there is no correlation, the points lie near the line, which is parallel to the x-axis (Figure 12.7).

Example: In an experimental station, a germination test was conducted in corn crop in eight different locations having different soil temperatures. The data on number of days for germination (Y) under different soil temperatures (X) (°F) are given as follows. Find the correlation between number of days for germination and soil temperature.

First, draw a scatter diagram to show the approximate relationship between the two variables as shown in Figure 12.11.

				Locations				
	1	2	3	4	5	6	7	8
X	58	55	45	42	42	36	38	40
Y	20	20	25	30	26	40	41	37

$\Sigma X =$ 356	$\Sigma Y =$ 239	$\overline{Y} =$ 29.875	$\overline{X} =$ 44.5
$\Sigma X^2 =$ 16282	$\Sigma Y^2 =$ 7651.000	$\Sigma XY =$ 10215.0	
$CF =$ 15842	$CF =$ 7140.125	$CF =$ 10635.5	
$Sx^2 =$ 440	$Sy^2 =$ 510.875	$Sxy =$ −420.5	

$$r = \frac{Sxy}{\sqrt{Sx^2\,Sy^2}} = \frac{-4205}{\sqrt{440 \times 510.875}} = \frac{N\,4205}{474.115} = -0.887$$

Test of Significance for r

$$t = \frac{r}{\sqrt{1-r^2}}\sqrt{n-2} = \frac{-0.887}{\sqrt{1-(-0.887)^2}} \times \sqrt{8-2} = -\frac{0.887}{0.4617} \times 2.449$$

$= -4.705$ (the values of t from tables are 1.943 and 3.143, at $P =$ 0.05 and $P = 0.01$ for six degrees of freedom, respectively)

Conclusion: Since the calculated t value is greater than the table value of t, the correlation is significant. The null hypothesis is therefore rejected. A significant negative correlation exists between soil temperature and the number of days required for germination.

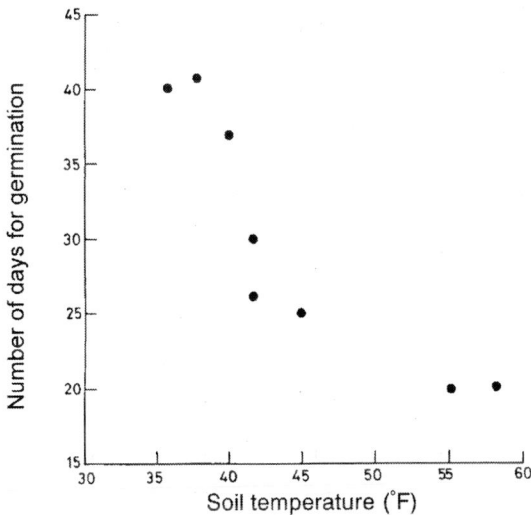

FIGURE 12.11. Scatter diagram showing the relation between soil temperature and days for germination.

3. For interpretation purposes, the following guidelines may be followed roughly:

0.00-0.25 correlation is doubtful

0.26-0.50 fair correlation

0.51-0.75 good correlation

0.76-1.00 high correlation

If $r = 0$, the variables are said to be uncorrelated or independent.

4. In some areas of research, the amount of correlation may be considered high for a particular value and the same correlation coefficient may be considered low in some other research area.

5. The strength of the relationship is based on the numerical value and not the sign. For example, an r value of -0.69 is higher than an r value of $+0.57$ for the same variables.

6. Suppose one wants to compare the two r values $r_1 = 0.80$ and $r_2 = 0.40$. One cannot infer that r_1 is twice that of r_2. One can say that r_1 (0.80) is a much higher degree of correlation than r_1 (0.40).

7. Cause and effect relationships do not exist in correlation studies. That is, a high positive or negative correlation does not mean that one variable is responsible for a change in another variable. For example:

a. Cigarette smoking and lung cancer

b. Age and high blood pressure

c. Fatty food consumption and heart attack

d. Fertilizer application and yield

One cannot say that cigarette smoking causes lung cancer, age causes high blood pressure, fatty food consumption causes heart attack, and fertilizer application produces yield.

8. Comparison of increase or decrease in r values in two different studies: Suppose one gets an increase in correlation from 0.52 to 0.82 in a study and an increase from 0.21 to 0.51 in another study. We cannot say that equal amounts of increase in correlation occurred in both the studies.

9. Sometimes a low amount of correlation may give better results in prediction.

10. Prediction of a dependent variable from an independent variable depends upon the degree of relationship between two variables. When correlation is high, one can expect a precise prediction, i.e., lesser error in prediction.

11. Based on numerical value of r, one should not interpret the correlation.

Extreme pairs of values affect the r value, e.g., consider the two sets of pairs of values given:

	Set 1			Set 2
x	*y*		*x*	*y*
1	3		1	3
3	5		3	5
6	4		6	4
4	3		4	3
5	2		5	2
14	12			
	$r = 0.88$			$r = 0.06$

In Set 1, $n = 6$, the estimated r value is 0.88. In Set 2, $n = 5$, the estimated r value is 0.06. In set 1, the extreme pair (14, 12) is present and this caused high correlation. When the extreme pair is deleted as in Set 2, the correlation is only 0.06. The interpretation of r is changed. Hence, in correlation analysis, one has to check for extreme pairs of observations before the analysis is undertaken. It is advisable to discard extreme pairs after careful checking.

13. To some extent, the degree of correlation depends upon the accuracy with which one can measure the variables. For example, in biological experiments, one can measure the variables accurately and expect higher r values. However, in areas such as psychology, the variable cannot be measured accurately and under such circumstances, higher r values cannot be obtained.

14. The correlation between two variables depends upon the nature of the data. When the data are heterogeneous, one must compute the correlation with caution.

15. Sample size and correlation coefficient: The correlation in the population is denoted by ρ, the population correlation coefficient, which is unknown, and is to be estimated from sample of data. Estimates of r from small sample sizes are unreliable. As the sample size increases, the difference between actual rho and estimated r decreases. An accurate value of r could be obtained when the sample size is 25 or more.

16. r is a pure number and independent of units of measurements of X and Y.

17. The correlation coefficient does not change if every value in either or both the distributions is increased or decreased or multiplied or divided by a constant. Data transformation of this type may affect the regression coefficient but not the correlation coefficient.

18. Measurement errors and correlation coefficient: Measurement errors in x and y affect the correlation coefficient. Errors greatly re-

duce the r values. The greater the errors, the lower the r values. Hence, when errors are present, the real relationship cannot be assessed.

COEFFICIENT OF DETERMINATION

The square of the correlation coefficient, i.e., r^2 value is defined as the coefficient of determination. An important relationship exists between correlation and regression coefficients. More information about regression is discussed in Chapter 13.

We know that

$$r = \frac{Sxy}{\sqrt{Sx^2 \, Sy^2}} = \frac{Cov(x, y)}{\sqrt{(\text{var } x) \times (\text{var } y)}}$$

Regression of y on x is defined as $byx = Sxy/Sx^2$
Regression of x on y is defined as $bxy = Sxy/Sy^2$

$$byx \times bxy = \frac{Sxy}{Sx^2} \times \frac{Sxy}{Sy^2} = \frac{Sxy \; Sxy}{Sx \; Sx \; Sy \; Sy}$$

$$= \frac{Sxy}{\sqrt{Sx^2 \, Sy^2}} \frac{Sxy}{\sqrt{Sx^2 \, Sy^2}} = r \times r = r^2$$

If $r = 1$, the sum of squares of deviations from either line of regression is zero. Consequently, each deviation is zero and all the points lie on both lines of regression. As r^2 comes to 1, Sx^2 and Sy^2 approach to zero so that all the data points are closer to the lines of regression. This departure of the values of r^2 from unity is a measure of departure of the relationship between the two variables from linearity.

Relation Between Correlation and Regression

1. For any data set, the null hypothesis either accepts or rejects it in both the tests of correlation and regression coefficients. If one is doing regression analysis and already tested for $\beta = 0$, one need not test for ρ (rho) $= 0$.
2. From the scatter diagram of correlation studies, we can infer the following information about regression:

a. If $r = -1$, all the points lie on the regression line with negative slope.
b. If $r = +1$, all the points lie on the regression line with positive slope.
c. When r is negative, the data points are scattered around the line with negative slope. If the points are farther from -1, the points are scattered farther from the line. If r is negative, the line will have a negative slope.
d. If r is close to $+1$, the data points lie near the line. If the value of r is far from $+1$, the points also lie far from the regression line. When r is positive, the regression line has a positive slope.

Uses of Correlation

1. The estimated sample correlation coefficient indicates the real relationship that exists in the population between the variables.
2. Correlation analysis can be used for prediction purposes. The dependent variable Y can be predicted from the independent variable X. The higher the r value, the higher the precision in estimation. When r value is low, the error in prediction is high.

   The estimated sample correlation coefficient indicates the real relationship that exists in the population. If the relationship is established from the sample data, the information can be utilized by researchers. Usha Rani (1996) reported highly significant positive correlation ($r = 0.834$**) between capsaicin content and amount of reduction in capsaicin content after storage for a period of one year in chili (*Capsicum annuum* L). The highly significant positive correlation ($r = 0.3618$ **) reported by Usha Rani (1995) between capsanthin content and ascorbic acid content in chili can be utilized in breeding work by plant scientists.
3. Amount of prediction of variability in the dependent variable: When r value is $+0.90$ or -0.90, the r^2 value is 0.81 meaning that 81 percent of variability is accounted for by the independent variable X, i.e., r^2 indicates the proportion of $\Sigma(y-)^2$, which is linearly related to X.
4. Prediction of dependent variable (Y) using the r^2 values varies depending upon the specific field of application.
5. Palaniswamy (1990) reported high significant positive correlation between grain yield per plant and other plant traits namely the tiller height ($r = 0.9677$), panicle length (0.9684), number of grains per panicle (0.9712), and grain weight per panicle (0.9702), and reported that these plant traits can be used for prediction of grain yield in rice crop.

ASSUMPTIONS IN CORRELATION ANALYSIS

Certain assumptions exist in the correlation model. Violation of these assumptions leads to wrong inferences:

1. The relationship between Y and X is assumed linear. This holds for both normal and nonnormal distributions.
2. Both Y and X distributions should be homogeneous.
3. The Y and X should follow bivariate normal distribution.
4. Observations are recorded in pairs. The pairs in the sample represent a random sample taken from the population.

Correlations Between Different Variables

In research, two or more variables are recorded simultaneously and researchers are often interested in knowing the relationship between pairs of variables. Correlation analysis may reveal the relationship between variables.

Example: In an experiment, the maximum temperature (°F) ($X1$), the minimum temperature (°F) ($X2$), and relative humidity ($X3$) are recorded. We can find r_{12}, r_{13}, and r_{23}. For example, data are given here for weekly totals:

Weekly S. No.	X1	X2	X3
1	635	528	649
2	660	527	641
3	672	538	574
4	675	532	622
5	646	514	655
6	662	528	626
7	675	525	582
8	672	520	574
9	631	520	618
10	637	520	637
11	641	492	611
12	626	510	645
13	619	483	594
14	620	494	575
15	600	507	653
16	571	505	666
17	601	514	650

| 18 | 434 | 359 | 467 |
| Total | 11277 | 9116 | 11039 (n = 18) |

Calculations:

$\Sigma x_1^2 = 7118650$	$\Sigma x_2^2 = 4643426$	$\Sigma X_3^2 = 6808597$
$CF = 7065040.5$	$CF = 4616748$	$CF = 6769973$
$Sx_1^2 = 53609.5$	$Sx_2^2 = 26678$	$Sx_3^2 = 38624$
$\Sigma X_1 X_2 = 5745656$	$\Sigma X_1 X_3 = 6937621$	$\Sigma X_2 X_3 = 5613691$
$CF = 5711174$	$CF = 6915933$	$CF = 5590640$
$Sx_1 x_2 = 34482$	$Sx_1 x_2 = 21688$	$Sx_2 x_3 = 23051$

$$r_{12} = \frac{34482}{\sqrt{53609.5 \times 26678}} = 34482 / 37818 = 0.9118$$

$$r_{13} = \frac{21688}{\sqrt{53609.5 \times 38624}} = 21688 / 45504 = 0.4766$$

$$r_{23} = \frac{23051}{\sqrt{26678 \times 38624}} = 23051 / 32100 = 0.7181$$

A correlation coefficient table will be helpful in interpretation of the results.

	X1	X2	X3
X1	—	0.9118	0.4766
X2		—	0.7181

The table values of r for 16 degrees of freedom at $P = 0.05$ and $P = 0.01$ are 2.120 and 2.921 respectively. Since the calculated r values are larger than the tabulated values, all three r values are highly significant.

RANK CORRELATION

Sometimes, data consist of two sets of variables measured in the form of ranks. To find the relationship between the ranks, the rank correlation test is used. Spearman (1904) devised this test and it is denoted r_s. This test is used when both X and Y are in an ordinal scale or one is in ordinal and the other is in interval. Similar to correlation coefficient, it is based on the deviations of the rankings on Y from the rankings on X. The following formula is used to find the correlation coefficient:

$$r_s = 1 - \frac{6\Sigma d_i^{\,2}}{n(n^2 - 1)}$$

where r_s = coefficient of rank correlation (subscript s is used to distinguish the rank correlation from the ordinary correlation coefficient).

n = number of pairs of observations

Σ = "sum of"

di = differences between ranks for each pair of observations (Snedecor and Cochran, 1980)

About r_s:

r_s is nonparametric and does not require the assumption that the observations follow normal distribution. Like correlation coefficient (r), the r_s also takes the values from -1 to $+1$. A value of -1 indicates perfect disagreement or nonconcordance, and $+1$ indicates perfect concordance or agreement. The significant levels of r_s worked out by Kendall as quoted by Snedecor and Cochran (1980) can be consulted to learn more about the significance of the r_s. The degree of freedom in this case is $n-2$, where n is the number of pairs. The significance shows the presence of association and nonsignificance implies that the two rankings are independent or not associated.

Example 1: Fourteen applicants appeared for a job interview. Two interviewers interviewed the candidates and ranked them. To find the relationship between the two rankings, the rank correlation test is performed. The data are given as follows:

Interviewer							Applicant							
	1	2	3	4	5	6	7	8	9	10	11	12	13	14
1	1	11	12	2	13	10	3	4	14	5	6	9	7	8
2	4	12	11	2	13	10	1	3	14	8	6	5	9	7
Difference (d)	-3	-1	1	0	0	0	2	1	0	-3	0	4	-2	1
d_i^2	9	1	1	0	0	0	4	1	0	9	0	16	4	1

$d_i^2 = 46$

$$r_s = 1 - \frac{6\Sigma d_i^{\,2}}{n(n^2 - 1)} = 1 - \frac{6 \times 46}{14(14^2 - 1)} = 1 - \frac{276}{2730} = 1 - 0.1011 = 0.8989$$

Test of significance based on *t* statistic:

$$t = \frac{r}{\sqrt{1-r^2}} \times \sqrt{n-2} \text{ with } n-2 \text{ degrees of freedom.}$$

$$r_s = \frac{0.8989}{\sqrt{1-(0.8989)}} \times \sqrt{(14-2)} = \frac{3.2410}{0.4382} = 7.40$$

The *t* value from tables for 12 *df* at $P = 0.05 = 2$ and at $P = 0.01 = 3.055$. The null hypothesis is rejected. Thus there is association between the rankings given by the interviewers.

Example 2: There are five students in math and statistics courses in a class. The rankings are as follows:

Subject	Ranking of students				
	A	B	C	D	E
Math (X)	4	3	1	2	5
Statistics (Y)	4	3	1	2	5
d_i	0	0	0	0	0
Sd_i^2	0	0	0	0	0

$$r^s = 1 - \frac{n\Sigma d_{i2}}{n(n^2-1)} = 1 - \frac{6(0)}{5(5^2-1)} = 1 - 0 = 1$$

The data show a perfect association between the two variables (rankings).

This infers that the rankings of each student are identical.

Procedure:

1. Compute the difference in ranks for each case (rank on X –rank on Y).
2. Square each of the differences.
3. Sum the squared deviations (Σd_i^2).
4. Find 6 times Σd_i^2 (i.e., $6 \Sigma d_i^2$).
5. Square *n* and subtract 1 from this square (i.e, n^2 -1).
6. Calculate $n (n^2-1)$.
7. Find $6\Sigma d_i^2/ n(n^2-1)$.
8. Subtract $6\Sigma d_i^2/ n(n^2-1)$ from 1.
9. The result indicates the rank correlation.

Intraclass Correlation

Sometimes, we wish to study the relationships amongst the individuals (members) within a group. This is referred to as intraclass correlation studies. It is calculated from the data on analysis of variance.

Example: The heights of eight pairs (cm) of plants are recorded to determine the intraclass correlation. The data are as follows.

				Pairs					Total
	1	2	3	4	5	6	7	8	
	71	69	59	65	66	73	68	70	541
	71	72	65	64	60	72	67	68	539
Total	142	141	124	129	126	145	135	138	1080

Calculate: Correction factor (CF) = 1080/16 = 72900
Pairs $SS = (142^2 + \ldots + 138^2)/2 - CF = 73116 - 72900 = 216$
Total $SS = 71^2 + \ldots + 80^2 - CF = 73160 - 72900 = 260$
Error SS = Total SS – pairs SS = 260 – 216 = 44
Form the analysis of variance using ANOVA (Snedecor and Cochran, 1980).

ANOVA

Source of variation	d_f	SS	MS	F	EMS
Between pairs	7	216	30.857	5.61	$\sigma^2 + 2\sigma g^2$
Within pairs	8	44	5.5	σ^2	
Total	15	260			

$\sigma^2 = 5.5$
$\sigma g^2 = (30.857 - 5.5)/2 = 12.678$ where the coefficient of σg^2,
i.e., 2 refers to the number of members in the family.
$r = \sigma g^2/(\sigma g^2 + \sigma^2)$
= 12.678 / 55 + 12.678 = 12.678/ 18.178 = 0.6974.
Since the calculated F value of 5.61 is greater than the table value of $F_{0.05,7,8}$, (3.501) we reject the null hypothesis.

EXERCISES

12.1. Define correlation. What are the methods by which correlation studies are undertaken?

12.2. (a) What do you understand by + or – of a correlation coefficient?

(b) Sketch a graph of the data given as follows. Compute r for the 10 pairs of values.

X	1	2	3	4	5	6	7	8	9	10
Y	2	4	1	5	3	9	10	7	8	6

12.3. A correlation coefficient of 0.45 is based on 45 observations. Can you conclude that this is significant at the 5 percent level?

12.4. Rise in correlation from 0.4 to 0.5 is equivalent to saying an increase in correlation from 0.8 to 0.9. Is this correct?

12.5. What is a scatter diagram for bivariate data? If r is the correlation coefficient between two variables, show the distribution of the pairs of values in appropriate scatter diagrams when $r = 0$, $r > 0$, and $r < 1$.

12.6. Number of advertisements made for the sale of a drink (X) and the number of liters (Y) (in 100 count) sold are given:
a. Find r value.
b. Find the coefficient of determination.

X	5	10	4	0	2	7	3	6
Y	10	12	5	4	1	3	4	8

12.7. Given the following sets of data on X and Y, plot a scatter diagram for each set and calculate the correlation coefficient. Comment on your results.

Set I		Set II		Set III	
X	Y	X	Y	X	Y
1	5	1	1	1	1
3	9	2	9	2	16
		3	2	3	81
		4	7	4	256
		5	6	5	625

12.8. What does r^2 measure? How it is linked with correlation and regression?

12.9. How will you interpret the following correlation coefficients? $+1, -1, 0.9, +0.8$

12.10. How do you define coefficient of determination, r^2? State the possible interpretation of r.

12.11. From the viewpoint of regression analysis, how is correlation coefficient defined?

12.12. Given $n = 57$, $r = 0.412$. Test the hypothesis that rho = greater than or equal to 0.6 at $P = 0.05$ level of significance.

12.13. Does no correlation necessarily imply independence? Explain.

12.14. The following data show the measurements (mm) on length (X) and thickness (Y) of a bark yield. Calculate correlation coefficient.

Specimen	1	2	3	4	5	6	7	8	9	10
X	2.1	2.9	3.2	3.7	4.0	4.4	4.8	5.0	5.4	5.6
Y	0.8	1.2	1.1	1.4	1.9	1.9	1.7	1.9	1.8	2.3

12.15. What information can you draw from the term Sxy in the formula for computation of correlation coefficient?

12.16. Two sets of data are given here. What difference do you observe?

	Set I			Set II	
X		Y		X	Y
1		1		1	2
2		2		2	3
3		2		3	1

12.17. If the calculated correlation coefficient from a sample of 28 observations is equal to 0.8, test the hypothesis that rho = 0.6.

12.18. Why can't correlation be taken as causation between two variables?

12.19. The sign for correlation coefficient and regression coefficient is the same. Why?

12.20. Explain how the scatter diagram will look like (a) when there is positive correlation, (b) when there is negative correlation, and (c) when there is no correlation.

12.21. Correlation between two variables is 0.74. Is it significant at $P = 0.01$, given $n = 8$?

12.22. Define r between two variables. Indicate its interpretations.

12.23. If the correlation coefficient from a sample of 28 observations is equal to 0.8, test the hypothesis that rho = 0.6.

12.24. The correlation coefficient r is determined by using a formula. It is also calculated by taking the square root of the coefficient of determination. Do they give the same results?

12.25. Estimate the intraclass correlation between stem heights (cm) of 10 pairs of plants and test.

1	2	3	4	5	6	7	8	9	10
70	66	65	69	71	72	67	70	69	70
69	67	64	69	73	71	69	70	58	69

REFERENCES

Palaniswamy, K.M. (1990). Design of (field) experiments and crop forecast in agriculture with special reference to rice. Doctoral thesis submitted to the University of Calicut, India.

Snedecor, G.W. and W.G. Cochran (1980). *Statistical methods.* Ames, IA: The Iowa State University Press.

Spearman, C. (1904). General intelligence, objectively determined and measured. *Am. J. Psych.* 15:88.

Usha Rani, P. (1995). Effect of plant attributes on the quality characteristics in chili. *Madras Agric. J.* 82(2):630-634.

Usha Rani, P. (1996). Evaluation of chilli germplasm for capsanthin and capsaicin contents and effect of storage on ground chilli. *Madras Agric. J.* 83(5):288-291.

Usha Rani, P. (2001). Evaluation of germplasm for fruit surface area in chili (*Capsicum annuum* L). *Crop Res.* 21(2):168-173.

Chapter 13

Regression

Scientists in different fields acquire masses of data from their experiments and unless they are properly analyzed and interpreted, the data have little or no value. The science of statistics provides tools for solving problems in fields such as agriculture, medicine, social sciences, engineering, education, bioinformatics, etc. One of the problems they face relates to the relationship between variables and prediction of one variable using another variable or variables.

FUNCTION CONCEPT AND REGRESSION

Consider two variables, X and Y. In mathematics, the relation between X and Y is stated as Y is a function of X. It is written as $Y = f(X)$ where Y is the dependant variable, and X is the independent variable because Y is determined depending upon the X value that we assign. By convention, Y is called the dependent variable and X is the independent variable. If the dependency is established from the given data, we can predict Y for a given X accurately. The study of relationship is called the study of functional relationship between Y and X. The relationship is constant and not subject to error.

Example 1. Study the equation $A = \pi r^2$ where A is a dependent variable and r is an independent variable. A relates to the area of a circle, r = radius of the circle, and π = a constant, whose value is 3.1429. It shows that as radius increases, area also increases proportionately. For a given value of r, we can predict the value of A (area). Here, A is dependent and r is independent. It can be shown in a diagram (see Figure 13.1).

r(cm)	$\pi\ r^2$ (cm²)
0.5	0.7857
1.0	3.14286
1.5	7.07143
2.0	12.57143
2.5	19.64286

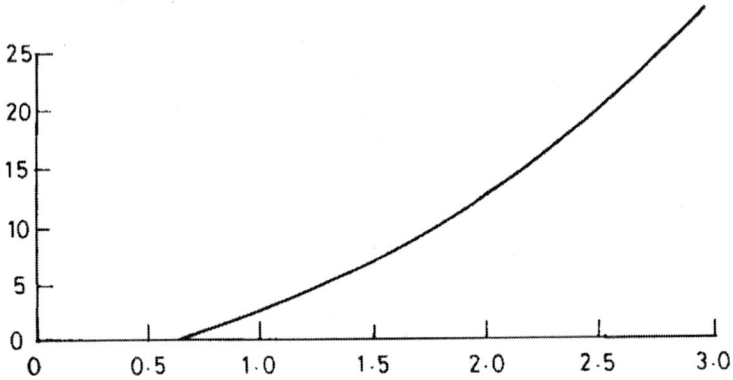

FIGURE 13.1. Perfect relationship between radius and area of a circle.

Example 2. Area of a square = s^2 (s = length of the side).

Functional relationship study in biological sciences is called regression. We will not get exact relationships as the Y variable is subject to influences that are unknown. We study linear relationships between Y and X. Linear means an equation of a straight line of the form $y = a + bx$, where y is the dependent variable and x is the independent variable, and a and b are constants. $Y = a + bx$ is called simple linear regression equation. If the equation contains more than two independent variables, it is called a multiple linear regression equation.

STRAIGHT LINE EQUATION

A linear (straight) relationship between two variables x and y can be expressed by the equation $y = \alpha + \beta x$ for the population data. It is unknown. The main reason for computing the sample regression line is to estimate the population regression line. In $y = \alpha + \beta x$, α and β are constants, and in the sample data the equation is $y = a + bx$ where a and b are the estimates of α and β, respectively. Each sample will give different values of a and b as the sample observations differ, but α and β are fixed ones.

Description of the Straight Line Equation

The straight line or linear equation is $Y = a + bx$ where y = dependent variable and x = independent variable.

a = intercept, i.e., the computed value of y when $x = 0$; or the value where the line intercepts or crosses the y-axis.

b = the regression coefficient. It is the slope of the line as it determines the steepness of the line. It measures the amount of change in y for a unit change in x (see Figure 13.2).

The equation is $y = a + bx$. By increasing one unit in x, we get $y = a + b (x + 1)$. The change in y is therefore: $a + b(x + 1)–(a + bx) = a + bx + b – a – bx = b$. Here, b is represented by a numerical value with a ± (plus or minus) sign depending upon the nature of unit change in x.

Example: (a) A b value of + 0.65 indicates that an increase of one unit of x corresponds to an increase of 0.65 units in y (see Figure 13.3). (b) A b

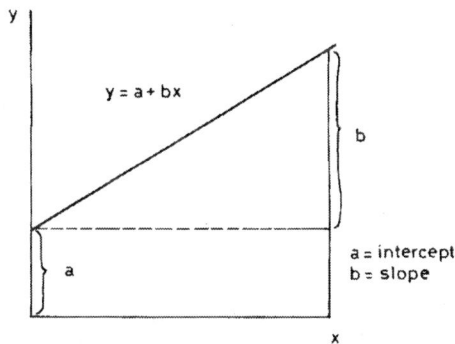

FIGURE 13.2. Linear equation ($h = a + bx$).

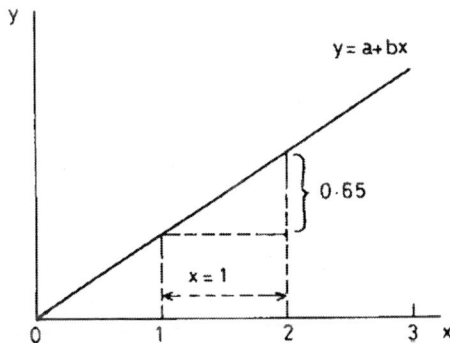

FIGURE 13.3. The meaning of positive slope of 0.65 (unit increase in x results 0.65 units increase in y).

value of -1.5 means that an increase of one unit in x corresponds to a decrease of 1.5 units of y (see Figure 13.4).

Principle in Regression Studies

In regression problems, the values of x are fixed, and they are measured accurately without error. For each fixed x level, the experimenter observes several y values which show variation. For example, if a drug with a certain concentration is administered to several patients and reactions are recorded, we observe variations in reactions. Similarly, in agricultural experiments, for example, when a fixed quantity of fertilizer is applied to a crop in several plots and plot yields are recorded, we observe that plot yields vary from plot to plot. The plot yields show a frequency distribution. This phenomenon is explained with a simple example so that the students understand the regression analysis.

Example. Consider a fertilizer experiment conducted with three doses of N (kg/ha) namely 60, 120, and 180, and the yields (kg/ha) recorded in three plots each. Data are summarized here:

Quantity of N applied (kg/ha)	Yield recorded (t/ha)	Mean yield (kg/ha)
60	2, 3, 4	3
120	3, 4, 5,	4
180	5, 6, 10	7

The data show that when a fixed quantity, e.g., 60 N (kg/ha), is applied to three plots, each plot gives three different yields, 2, 3, and 4 t/ha. Similarly, we get 3, 4, and 5 t/ha for 120 N kg/ha and 5, 6, and 10 t/ha for 180 N kg/ha.

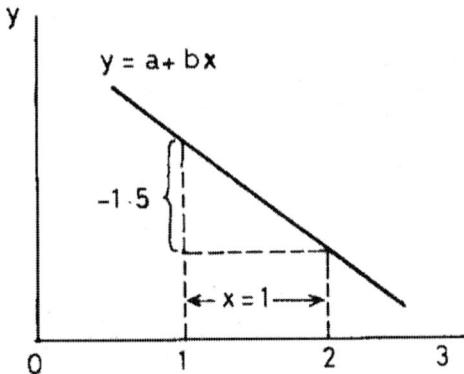

FIGURE 13.4. The slope of -1.5.

For each *x* value, we get a distribution of *y* values, and they are represented in the scatter diagram (see Figure 13.5). Each point in the diagram shows the yield at a fixed level of N. Each distribution will have a mean and variance. Now our objective is to fit a regression line between *y* and *x*. As stated earlier, the *x* values are already fixed as 60, 120, or 180 N (kg/ha). For each *x* value, the mean values of the corresponding *y* values are considered. Thus for *x* = 60, we take the *y* value 3, the average of 2, 3, and 4. Similarly, we take the *y* value 4 (the average of 3, 4, 5) for *x* = 120 and *y* value 7 (average of 5, 6, and 10) for *x* = 180. Thus, when we fit a straight line we consider the *x* and *y* values as shown here.

X	60	120	180
Y	3	4	7

Using these three pairs of values, the linear regression line is fitted. This is the basic principle in regression analysis.

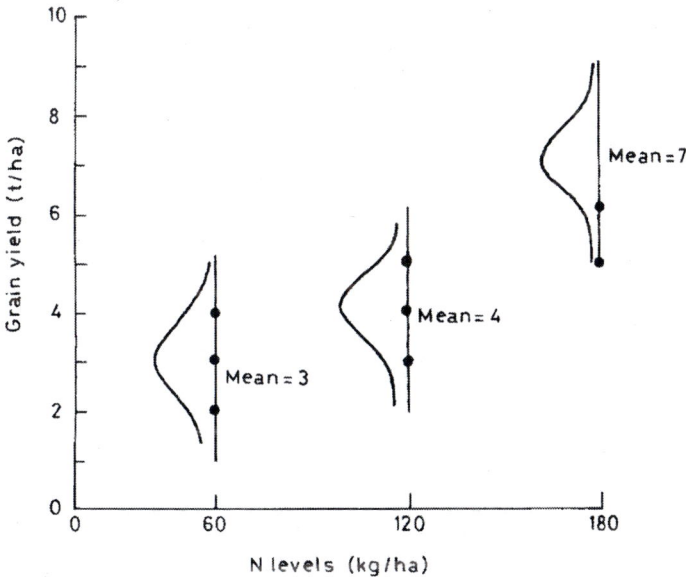

FIGURE 13.5. Scatter diagram showing the distribution of *y* values for each fixed quantity of *x* values.

Principle of Least Squares Employed in Fitting Regression Line

Consider the data from an experiment conducted in an experimental station.

(x) N (kg/ha)	20	30	40	50	60
(y) Yield (t/ha)	3	5	4	6	7

Before we start the analysis, we have to plot the (x, y) points in a scatter diagram. This provides a quick impression of whether there is any relationship between y and x, and whether such a relationship is linear or otherwise (see Figure 13.6).

The scatter diagram shows a linear trend. Now we have to fit a straight line of the type $y = a + bx$ that gives the set of pairs of values very closely. We can obtain several lines. For example, we can get a line by joining the first two points or middle two points or extreme two points or by joining any two points. However, we must set one line that best fits the data points and which gives minimum error in prediction. This line, if obtained following the principle of least squares, will give the best fit.

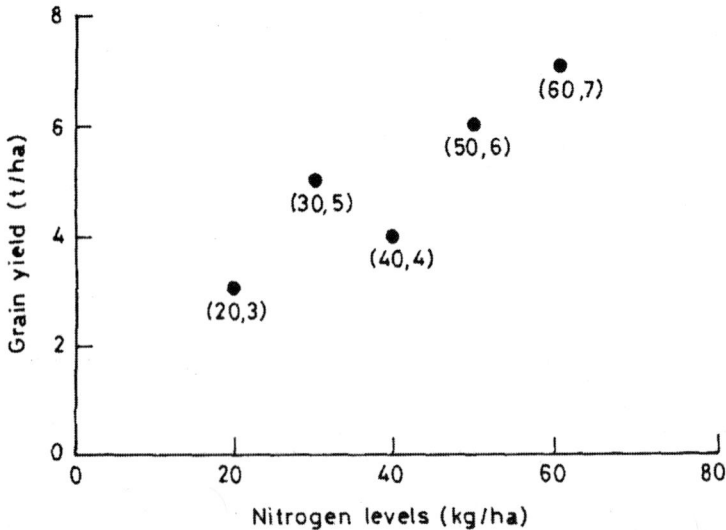

FIGURE 13.6. Scatter diagram showing the relationship between nitrogen and grain production.

Principle of Least Squares

In Figure 13.7, P_1, P_2, P_3, P_4, and P_5 are the observed data points. AB is the fitted line. The vertical distance between P_1 and L_1 (the point on the line) measures the deviation d_1. Similarly, P_2L_2, P_3L_3, P_4L_4, and P_5L_5 measure the deviations at other data points and they are denoted by d_2, d_3, d_4, and d_5, respectively. Some observed points are above the line and some are below the line. Some deviations (d_1, d_2, and d_4) are positive and some (d_3 and d_5), are negative. The sum of these deviations Σd_{is} will be zero. To overcome this positive-negative pattern and to obtain the actual deviations, we first square all these deviations (d_{is}) to make them all positive, and then add them to obtain the least square criterion. This criterion gives the single line of best fit called the least square line. This line, obtained from sample data, $y = a + bx$, is the estimate of the population regression line, $y = \alpha + \beta x$, which is unknown. At X_1 level, the observed point, P_1, is above the line and consequently we observe the error, d_1 positive. Similarly d_2 and d_4 show positive errors and d_3 and d_5 show negative errors. The line obtained following least square criterion gives the minimum value of the sum of squares of the deviations, i.e., Σd_i^2 is minimum. In other words, Σd_i^2 is greater if calculated from any other line. If the fitted line does not relate to the data, the parameters of the line α and β have no significance to the model. If some irrelevant pairs of data are included in the analysis, it will lead to serious errors in the

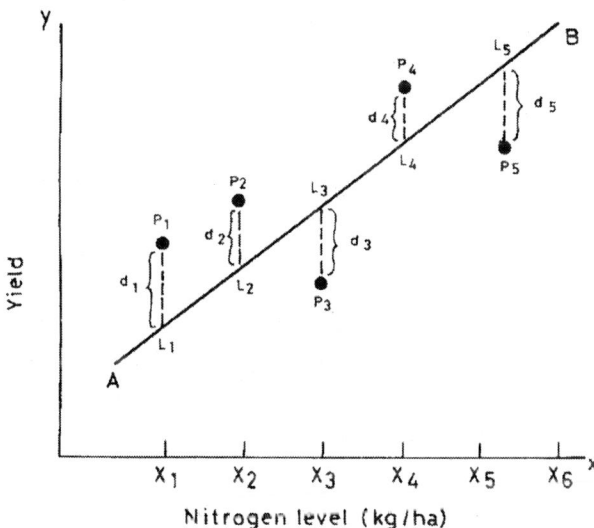

FIGURE 13.7. Figure explaining the principle of least squares.

interpretation of results. Hence, caution is necessary in checking the data before analysis.

Assumptions made in the regression analysis:

1. The distributions of *y* values at each *x* level are normal. The *y* values at each *x* level are independent.
2. The variances of *y* values at each *x* level are equal. This assumption is called the homogeneity of variances.
3. The mean values of *y* for different *x* values lie on a straight line (population regression line).
4. The population parameters α and ß fix the line of regression and they are estimated as a and b in the sample regression line, respectively.
5. Errors have a normal distribution with mean equals zero and common variance.
6. The *X* values are measured without error and *y* values recorded are subject to error.
7. Linearity will hold only for a limited range of *x* values.
8. The *y* value employed in the analysis against each value of *x* is assumed to be the mean values of *y*.

DEFINITION OF ERROR IN REGRESSION ESTIMATES

As explained earlier, for a given *x* value, we get a population of *y* values, and therefore we cannot get exact *y* value for a given *x* though we measure *x* accurately, and thus error occurs. Error is defined as the difference between observed *y* value and the predicted *y* value from the regression line, i.e., $y - \hat{y}$.

The errors are due to two reasons, for example, in the crop yield experiment: (1) inaccurate weighing of the produce, and (2) differences in soil fertility, supply of irrigation water, climatic conditions, pests and diseases, etc., which are often uncontrollable. Also when the experiment is repeated under identical conditions, we fail to get the same results due to unexplainable error. The method of estimation of error is explained in the later part of this chapter.

Method of Fitting the Regression Line from the Sample Data

Example: The data mentioned earlier are considered for fitting the regression line ($n = 5$ pairs).

Y (yield, t/ha)	3	5	4	6	7
X (nitrogen, kg/ha)	20	30	40	50	60

Step 1. Find the totals: $\Sigma x = 200$ $\Sigma y = 25$

Step 2. Find the mean $\bar{x} = \Sigma x/n = 200/5 = 40$ and the mean $\bar{y} = \Sigma y/n = 25/5 = 5$

Step 3. Compute sum of squares of x (Sx^2)

$$Sx^2 \Sigma x^2 - (\Sigma x)^2 / n = [20^2 + 30^2 + 40^2 + 50^2 + 60^2] - (200)^2$$
$$/5 = 9000 - 8000 = 1000$$

Step 4. Compute the sum of squares of y (Sy^2)

$$Sy^2 = \Sigma y^2 - (\Sigma y)^2 / n = [3^2 + 5^2 + 4^2 + 6^2 + 7^2] - (25)^2$$
$$/5 = 135 - 125 = 10$$

Step 5. Compute the sum of production (Sxy)

$$Sxy = \Sigma xy - (\Sigma x)(\Sigma y) / n =$$
$$[(20 \times 3) + (30 \times 5) + (40 \times 4) + (50 \times 6) + (60 \times 7)]$$
$$-[(200 \times 25)]/5 = 1090 - 1000 = 90$$

Step 6. Compute the regression coefficient, b

$$b = Sxy/Sx^2 = 90/1000 = 0.09$$

Step 7. Compute the value of a

$$a = \bar{y} - b(\bar{x}) = 5 - (0.09)(40) = 5 - 3.60 = 1.40$$

Step 8. Substitute the computed values of a and b in the sample regression equation $y = a + bx$, and obtain the required estimated regression equation:

$Y = 1.4 + 0.09X$. Here we are interested in the correlation between y and x.

$r = Sxy / \sqrt{Sx^2 Sy^2} = 90 / \sqrt{(1000 \times 10)} = 0.90$. b can also be calculated using the formula $b = r\sqrt{(Sy^2 / Sx^2)} = 0.9\sqrt{(10/1000)} = 0.09$

The b is estimated as Sxy/Sx^2
Proof: Sxy/Sx^2

$$= [\Sigma(x - \bar{x})(y - \bar{y})] / [(\Sigma(x - \bar{x})^2)]$$
$$= [\Sigma(xy - x\bar{y} - \bar{x}y + \overline{xy})] / [\Sigma x^2 - 2\bar{x}\Sigma x + n\bar{x}^2]$$
$$= [\Sigma xy - \bar{y}\Sigma x - \bar{x}\Sigma y + \Sigma \overline{xy}] / [\Sigma x^2 - 2\bar{x}\Sigma x + n\bar{x}^2]$$
$$= [\Sigma xy - \bar{y}\Sigma x - \bar{x}\Sigma y + n\overline{xy}] / [\Sigma x^2 - 2\bar{x}\Sigma x + n\bar{x}^2]$$
$$= [\Sigma xy - \bar{y}n\bar{x} - \bar{x}n\bar{y} + n\overline{xy}] / [\Sigma x^2 - 2\bar{x}n\bar{x} + n\bar{x}^2]$$
$$= [\Sigma xy - n\overline{xy}] / [\Sigma x^2 - n\bar{x}^2] = [Sxy / Sx^2] = b$$

Use of Estimated Regression Equation

In any regression study, we cannot include many x levels and conduct the experiment. We can predict y value for a particular x value. For example, we want to predict y value for $x = 45$. Substituting $x = 45$ in the equation, we get $y = 1.4 + (0.09)(45) = 5.45$ t/ha.

Relation Between $\bar{x}$ and $\bar{y}$ the Regression Line

The means of x and y are 40 and 5, respectively. Now we substitute the value of $x = 40$, we get $Y = 1.4 + (0.09)(40) = 5$ which is the mean value of y. Thus, the means of x and y lie on the estimated regression line.

Graphic Representation of the Regression Line

In our example, there are five pairs of values namely (20,3), (30,5), (40,4), (50,6), and (60,7), and these points are represented in a graph (see Figure 13.8). The figure thus obtained is called a scatter diagram.

How to Draw a Regression Line

To draw a line we require a minimum of two points. Thus when $x = 30$ the estimated y value is $y = 1.4 + (0.09)(30) = 4.1$ and when $x = 45$, $y = 1.4 + (0.09)(45) = 5.45$. The two points selected are (30,4.1), and (45, 5.45). Using these two points, we draw the line, $y = 1.4 + 0.09\ x$. (see Figure 13.8).

Index of Fit

The fitted line will determine an estimated y ($\hat{Y}$) value on the line. This value is called the estimated or predicted y value for a particular x value. For each value of x, we get a predicted y ($\hat{Y}$). The correlation coefficient between y and $\hat{Y}$ is used to measure the index of fit. It measures the strength of linear relationship between y and x. The r value ranges from -1 to $+1$. If the index

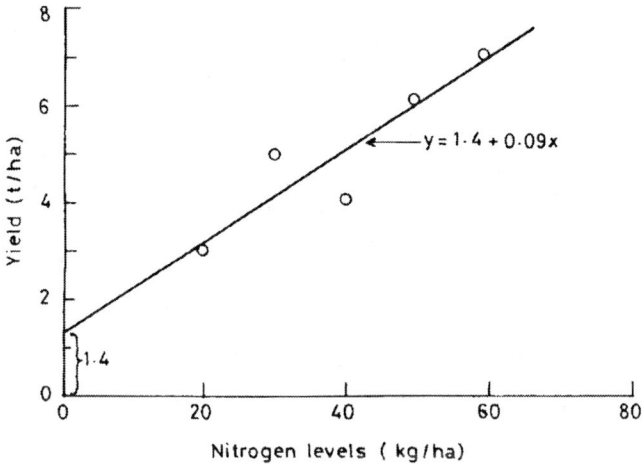

FIGURE 13.8. Scatter diagram showing the fitted simple regression line of *y* on *x*.

of fit is high, the fitted line is also good. The *y* and $\hat{y}$ values in our example are as follows.

X	20	30	40	50	60	Total
Y	3	5	4	6	7	25
$\hat{Y}$	3.2	4.1	5.0	5.9	6.8	25

$Sy^2 = 135 - 125 = 10;\ S\hat{Y}^2 = 133.1 - 125 = 8.1$

$r = Sy\hat{Y} / \sqrt{Sy^2 \times S\hat{Y}^2} = 8.1 / \sqrt{10 \times 8.1} = 8.1 / 9 = 0.9$

Test of significance for *r*:

$t = 0.9 / \sqrt{1 - (0.9)^2} \times \sqrt{(5-2)} = 0.9 / 0.436 \times 1.732 = 3.58$

$t_{3,0.05} = 3.182, t_{3,0.01} = 5.841$ (from *t* tables).

There is a significant correlation between *y* and $\hat{Y}$, and hence there is significant index of fit. The fitted line is good for prediction.

Interpretation of Constants and Variables in the Equation

a = 1.4: It gives an estimated value of grain yield when no nitrogen was applied, i.e., when *x* = 0. It is also called the intercept of the line.

b = 0.09: It measures the rate of increase in y for every addition of one unit of x. For every kg of N application, we can expect an increase of 0.09 kg/ha of grain yield. Increase or decrease is indicted by the algebraic sign of the regression coefficient. In the present example, it is positive (Y = independent variable; x = dependent variable).

PREDICTION AND MEASUREMENT OF ERROR IN PREDICTION

In the equation $y = a + bx$, a and b are only estimates and hence they are subject to error. In order to know the reliability of prediction, we must know the error in prediction. It is measured by the standard error. Standard error of estimate gives a measure of distance between the regression line and the actual data. The error in estimation is the difference between the observed value and the estimated value, i.e, $y-\hat{y}$. To determine the standard error, see Table 13.1.

From Table 13.1, we get the following information.

1. Total variability in $y = Sy^2 = 10$.
2. Variability due to regression or variability explained by regression = $8.10 \Sigma(\hat{y} - \bar{y})^2$.
3. $\hat{y}$ = estimated value of y for the given x values using the regression equation.
4. Out of the total variability of 10, variability explained by regression is 8.10, and the variability not explained by regression is $10 - 8.10 = 1.90$. It is called residual.
5. The error sum of squares may also be computed using the formula, i.e.,

TABLE 13.1. Standard error.

Y	X	$\hat{y}$	$y-\hat{y}$	$(y-\hat{y})^2$	$y-\bar{y}$	$(y-\bar{y})^2$	$\hat{y}-\bar{y}$	$(\hat{y}-\bar{y})^2$
3	20	3.2	−0.2	0.04	−2	4	−1.8	3.24
5	30	4.1	+0.9	0.81	0	0	−0.9	0.81
4	40	5.0	−1.0	1.00	−1	1	0.0	0.00
6	50	5.9	+0.1	0.01	+1	1	+0.9	0.81
7	60	6.8	+0.2	0.04	+2	4	+1.8	3.24
25	200	25.0	0.0	1.90[a]	0	10[b]	0.0	8.10[c]

[a] unexplained variation or error or residual (1.90);
[b] total variability (10);
[c] explained variation regression (8.10)

$$\Sigma d^2 = (1 - r^2) S y^2 = [1 - (0.9)^2] \times 10 = 1.90.$$

Residuals show the amount of inadequacy of the model. Variability explained by the regression in percentage is $(8.1/10) \times 100 = 81$ percent of the total variability in y by x. The error not explained by x is measured as $1 - r^2$ $= 1 - 0.81 = 0.19$, i.e., 19 percent. Thus 19 percent is the proportion of total variability in y not explained by x.

Test of Hypothesis (Testing of Regression Coefficient)

The null hypothesis is H_o: b = 0, i.e., y variable is not dependent on x or y is independent of x and H_1: b $\neq$ 0. The hypothesis is tested by using analysis of variance technique. The total variability is partitioned into variability due to regression and error, i.e.,

$$
\begin{aligned}
\text{Total variability} &= SS \text{ due to regression + deviation} \\
&\quad \text{from regression} \\
&= \text{(explained variation) +} \\
&\quad \text{(unexplained variation)}
\end{aligned}
$$

Total variability in y (total SS) = $Sy^2 = 10$.

Variability explained by regression is computed in several ways:

1. $\dfrac{(Sxy)^2}{Sx^2} = \dfrac{90 \times 90}{1000} = \dfrac{8100}{1000} = 8.1$ or

2. b $Sxy = 0.09 \times 1.90 = 8.10$ or

3. $r^2 Sy^2 = (0.09)^2 \times 10 = 0.81 \times 10 = 8.1.$

Residual SS (unexplained variability) = total SS – regression SS = $10 - 8.10$ = 1.90 (Table 13.2).
The standard error of b is computed as $\sqrt{(\text{Error } SS / df)} = \sqrt{S^2_{yx}} = \sqrt{0.633} = 0.796$. It shows the distance between the regression line and the actual observed point.

Test of significance of b Using t Statistic

The t statistic is t = b/SE_b where SE_b = standard error of b.

TABLE 13.2. ANOVA data.

SV	df	SS	MS	F	F from tables	
					P = 0.05	P = 0.01
Due to regression	1	8.10	8.10	12.9*	10.13	34
Deviation from regression (residuals)	3	1.90	0.633 (S^2_{yx})			
Total	4	10.00				

*As the calculated F (12.9) is greater than the table value of F (10.13), the regression coefficient is significant. The null hypothesis is rejected.

The standard error of b is computed as $SE_b = \sqrt{(s^2 / s_x^2)}$, where s^2 is the error mean square (variance of observations about the linear relationship between y and x). $SE_b = \sqrt{(0.633 / 1000)} = \sqrt{(0.000633)} = 0.025$ $t = (0.09/0.025) = 3.6*$ with n-2. The values of $t_{0.05, 5-2=3}$ and $t_{0.01, 3}$ from t table are 3.182 and 5.841, respectively. Since the calculated t value (3.6) is greater than the table value of t (3.182), b is significant at 5 percent level of probability. If b is not significant, we conclude that y is independent of x, and the computation of regression line is useless.

Verification of the Relationship Between F *and* t *Values*

The relationship is $F = t^2$, i.e., $12.9 = 3.6^2$ (within rounding errors); it may be taken as a check in the calculations.

Suggestions to Improve the Precision of the Estimate

In calculating the t statistic we use the formula $t = b/SE_b$ where $SE_b = \sqrt{(s / sx^2)}$.

It indicates that to obtain the significant t values, we require a smaller standard error. A smaller standard error can be obtained by increasing the spread of X levels. The greater the spread of the X values, the better the line is defined. We can also get lesser error mean square by increasing the error degrees of freedom by involving more levels of X.

Confidence Limits for the Slope of the True Regression Line

We can calculate the confidence limits for the true population regression coefficient ß, which is not known but we can say with confidence that it lies in between certain limits called confidence limits from the sample data.

The following formula is used to compute the confidence limits for ß:

$$b \pm t \times SE_b, \text{ i.e., } b \pm t \times \sqrt{(s^2 / Sx^2)}$$

For example, in grain yield data, the confidence limits for b (0.09) for 95 percent confidence coefficient is determined as follows:

$$0.09 \pm 3.18 \times SE_b$$

The $SE_b = 0.025$, therefore CL $= 0.09 \pm 3.18 \times 0.025 = 0.09 \pm 0.0795 = 0.0105 - 0.1695$.

We are 95 percent confident that ß lies in between these two limits, i.e., $0.0105 < ß < 0.1695$ where the lower limit is 0.0105 and the upper limit is 0.1695 (3.18 is the value of t from tables for 3 degrees of freedom at $P = 0.05$). From the results we can state that on the average, an increase of one unit in fertilizer would be associated with an increase of 0.0105 to 0.1695 t/ha in grain production.

Prediction of Y for a Particular X Value and Standard Error for This Predicted Value

The regression equation obtained in grain yield and fertilizer data is $y = 1.4 + 0.09x$. We can predict Y value for a particular X value and compute the standard error for it. For example, when $x = 50$, the predicted Y value is $1.4 + (0.09)(50) = 5.9$ t/ha. SE for this y value is:

$$\sqrt{\{s^2 [1 / n + (x_0 - \bar{x})^2 / Sx^2]\}}$$
$$= \sqrt{0.633\{1 / 5 + [(50 - 40)^2 / 1000]\}} = 0.4347 \text{ t/ha}.$$

Note: If b is not significant, i.e., Y is not dependent on X, there is no use for prediction.

Confidence Interval for Y *for a Fixed* X *Value*

A confidence interval for *Y* for a particular *X* value, say X_0, is constructed as confidence limits

$$= \hat{y} \pm t_\alpha, (n-2)\sqrt{[1/n + (x_0 - \bar{x})^2 / Sx^2]}$$

For example, x_0 = 45 kg/ha, $\hat{Y}$ = 1.4 + (0.09) (45) = 5.45 t/ha. The confidence limit for this value is 5.45 ± 3.18√0.633/{[(1/5) + (45–40)²]} = 5.45 ± 3.18 (0.796)/(0.474) = 5.45 ± 5.33 = 0.12 – 10.78.

It indicates that the average yield at x_o/45 kg/ha is between 0.12 and 10.78 kg/ha.

Interpretation of r² *(Coefficient of Determination or Correlation Index)*

The coefficient of correlation shows the magnitude of the relationship between *y* and *x*. The square of the correlation coefficient is called coefficient of determination. It is defined as

$$r^2 = 1 - s^2{}_{yx} / Sy^2,$$

where $s^2{}_{yx}$ is the variance around the regression line, *y* = a + b*x*, and Sy^2 is the total variability in *y*. The regression line does not pass through all points in the scatter diagram. The deviation $y - \hat{y}$ gives the variation not explained or accounted for by *x*. Thus,

$$r^2 = 1 - \frac{\text{unexplained variation in } y}{\text{total variation in } y}$$

$$= \frac{\text{total variation} - \text{unexplained variation}}{\text{total variation in } y}$$

$$= \frac{\text{explained variation in } Y}{\text{total variation in } y} = \frac{\text{regression } SS}{\text{total } SS}$$

$$= 8.1/10 = 0.81 = 81 \text{ percent.}$$

However, 0.81 is also the square root of the correlation coefficient ($r = 0.9$). Thus to determine the percentage of total variation in *Y* due to *X*, we must calculate the coefficient of determination, r^2.

When the value of *r* increases, r^2 value also increases, and consequently we are able to account for high percentage of variation in *y* and its dependability on *X*. When $r = 1$, $r^2 = 1$ and we could explain 100 percent variation in *Y*. But it is not possible to get a perfect correlation between two variables in our experiments as the *Y* values are subject to the influence of several factors, which are unknown. When the correlation is low, we get wide scatter of the points around the regression line, and consequently the estimated value will not be accurate. The direction of relationship between *y* and *x* could not be found from r^2 because r^2 will always be positive.

Standard Error and the Data Points

If the observed *y* values are normally independently distributed around the regression line, we can expect that about 68 percent of the points will fall within one SE of estimate above and below the regression line, and about 95 percent above and below 2 SEs of estimate. For example, if we have $Sy^2 = 10$, $s^2_{yx} = 4$, then $r^2 = 1 - 4/10 = 60$ percent. Here the total variability is 10. After fitting the regression line, the residual error or unexplained variation is 4. Thus about 60 percent of the variation in *y* is explained by the relationship between *y* and *x* in the regression line.

Relation Between Correlation and Regression Coefficients

When *y* depends upon *x*, we get $b_{yx} = Sxy/Sx^2$ and $a = \bar{y} - b\bar{x}$ and the fitted regression equation is $y = a + bx$.

When *x* depends upon *y*, we get $b_{xy} = Sxy/Sy^2$ and $a = \bar{x} - b\bar{y}$ and the fitted regression equation is $x = a + by$.

Now,

$$b_{yx} \times b_{xy} = (Sxy/Sx^2) \times (Sxy/Sy^2)$$
$$= [Sxy/\sqrt{(Sx^2 Sy^2)}][Sxy/\sqrt{(Sx^2 Sy^2)}] = r \times r = r^2$$

Thus, this result shows that the correlation coefficient is the geometric mean between the two regression coefficients.

The r^2 value can also be obtained in another method:

$$\begin{aligned}
b_{yx} &= Sxy/Sx^2 = Sxy/(\sqrt{Sx^2}\sqrt{Sx^2})(\sqrt{Sy^2}/\sqrt{Sy^2}) \\
&= Sxy/(\sqrt{Sx^2 Sy^2})\sqrt{Sy^2}/Sx^2 = r(Sy/Sx) \\
b_{xy} &= Sxy/Sy^2 = Sxy/(\sqrt{Sy^2}\sqrt{Sy^2})Sx^2/Sx^2 \\
&= Sxy/\sqrt{Sx^2 Sy^2}(\sqrt{Sx^2 Sy^2}) = r(\sqrt{Sx/Sy}) \\
b_{yx} \times b_{xy} &= r(Sy/Sx) \times (rSx/Sy) = r \times r = r^2
\end{aligned}$$

The two equations $y = a + bx$ (when y = dependent variable and x = independent variable) and $x = a + by$ (when x = dependent variable and y = independent variable) are not the same. The equation $y = a + bx$ is to be used when y is to be predicted from x. The other equation $x = a + by$ is to be used when x is to be predicted from a given value of y.

Computing Regression Coefficient Using the Correlation Coefficient

We know

$$r = \frac{Sxy}{\sqrt{Sx^2\,Sy^2}}$$

$$Sxy = r\sqrt{(Sx^2\,Sy^2)}$$

but

$b = Sxy / Sx^2$ and $b =$
$r(\sqrt{Sx^2\,Sy^2})/\,Sx^2 = r.\,Sx\,Sy\,/\,SxSx = r.\,Sy\,/\,Sx.$

Example: Given $r = 0.456$ $Sy = 38.78$ $Sx = 6.09$ $\bar{y} = 185.821$ $\bar{x} = 37.70$

$b = 0.456 \times (38.78/6.09) = 2.904$
$a = y - bx = 185.8121 - 2.904 \times 37.70 = 76.34$

The required regression equation is $y = 76.34 + 2.904x$.

The formula $b_{yx} = r(Sy/Sx)$ leads to the computation of r from b_{yx}, Sy, and Sx as $rxy = b_{yx}\,Sx/Sy$. This shows that r correlation coefficient is a function of regression coefficient and the ratio of independent and dependent variables. (See Table 13.3.)

Similarities between correlation and regression:

1. The direction of relationship in both the cases is the same, i.e., if the correlation coefficient is negative, the regression coefficient is also negative.
2. The scatter diagram reveals the same trend in both the cases.
3. In regression studies, the total variability in y is accounted for by x. In correlation, the r^2 (coefficient of determination) determines the variation in y. Thus in both the cases, the variability in y is accounted for in terms of x.

4. In correlation, x and y distributions have the same variance. In regression, each distribution of y has the same variance.
5. Only the paired values are used in both the cases; r and r^2 do not have units.

TABLE 13.3. Differences between correlation and regression.

Correlation	Regression
Both x and y are random variables, and they are subject to error.	Only y is a random variable and subject to error. The x values are measured without error, i.e., no sampling variation in x.
x and y are sampled at random after the experiment is laid out.	Experimenter fixes the different x values (levels) before the start of the experiment. Only y values are measured after the experiment is laid.
Joint variation and degree of relationship between x and y are studied. This is done by means of correlation coefficient, which takes the values from −1 to +1.	Dependability of y on x is studied.
With 3 or 4 pairs of values, correlation is not worked out.	Regression analysis can be carried out with 3 or more pairs of values.
In correlation, high values of y are associated with high values of x, low values of y with low values of x, and intermediate values of y with intermediate values of x.	Not so.
Correlation coefficient has no unit of measurement, i.e., it is unit-free.	Unit of measurement is stated.
The range of r values are from −1 to +1.	No range for 'b' values.
Does not assume cause and effect relationship.	Cause and effect relationship between y and x is assumed.
Gives the strength of relationship between y and x.	It does not tell us the strength of relationship between y and x. Points are close to the line when relationship is strong, and points are widely scattered when the correlation is weak.
It refers only to the closeness of the relationship.	It refers the amount of change in y for a unit change in x.
x and y show bivariate normal distribution.	For each value of x, there is a normal distribution.

Extrapolation

Precaution in prediction of *Y* values for extrapolated *X* values: In the present study, the values of *x* ranging from 20 N kg/ha to 60 N kg/ha were included and *y* values estimated. The observed *Y* values were from 3 t/ha to 7 t/ha. The correlation between *Y* and *X* was computed as 0.90. If we predict *Y* values for the *X* values greater than 60 N kg/ha or lesser than 20 N kg/ha (the highest and lowest *X* values included in the experiment), then it is called extrapolation. The fitted linear regression equation, $y = 1.4 + 0.09x$ describes the relation between *Y* and *X* within the range of *X* values. Extrapolation may sometimes give absurd results. For example, observe the regression line $y = 1.4 + 0.09x$ fitted for different *X* levels. The scatter diagram shows the line and the data points. The trend of the line between 0 and 20 N kg/ha and above 60 N kg/ha cannot be predicted exactly as shown in Figure 13.9. Hence extrapolation is to be avoided. If one is interested in the trend in the extrapolated region, conduct the experiment again including the *X* values in the region of interest.

Uses of Regression

1. Regression is used for prediction.
2. One can find the form of relationships between the variables.
3. Regression is used to explain the variation in the dependent variables.
4. Regression is used in trend analysis.
5. The regression technique is used in several situations in the estimation process. A few examples are given as follows for better understanding

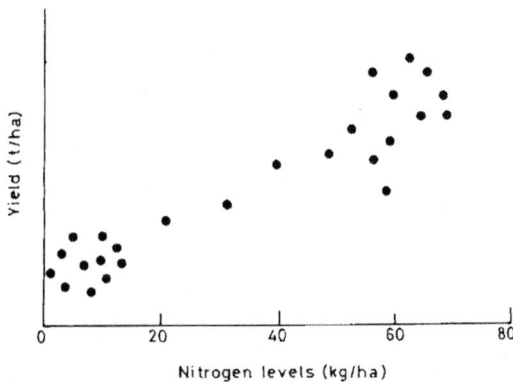

FIGURE 13.9. Diagram explaining the uncertainty in the shape of the line in extrapolation below 20 and above 60 N kg/ha.

of the uses of regression. Researchers can exploit the use of regression in their own areas of research.

a. *Leaf area estimation:* Leaf area (LA) is an important plant trait in plant photosynthesis. However, it is difficult to estimate within the resources available in the experimental stations in developing countries. Furthermore, leaf area must be estimated without destruction of the plant as it will affect the experimental results. For comparative experiments, accurate measurement of leaf area is not essential. Thus, the regression method gives an easy and quick method of estimation. For example, Palaniswamy and Gomez (1974) established the regression equation, leaf area = 0.74 × length of leaf × maximum width of leaf in rice crop. In this case, leaf area is dependent on length and width, and the regression coefficient is 0.74. In addition, they found the correlation coefficient as high as 0.97 between leaf area and length × width of the leaf. Thus, if one wants to estimate leaf area in rice with precision, it can be done with the length of the leaf and its maximum width by using the factor 0.74.

b. Usha Rani (2001) used the regression model $y = k(L \times D)$ where $y =$ fruit surface area in chili (*Capsicum annuum* L.) fruit, $L =$ length of the fruit, $D =$ diameter of the fruit, and k is the regression coefficient. She found the value of $k = 1.6827$. Thus one can estimate the fruit surface area in chili by multiplying fruit length × fruit diameter × 1.6827.

c. Uses of regression are found in studies such as crop yield and fertilizer, age and blood pressure, student scores and amount of time spent studying, nutrition and diseases, etc.

d. Palaniswamy (1990) reported that the linear regression model can be employed in the prediction of grain yield per plant in rice and consequently for crop estimation.

6. Using a regression technique, we can fit a quadratic curve and arrive at an optimum level.

Example 1: The general least square technique can be applied if the levels of X factor are not equally spaced. If X levels are equally spaced, the method of orthogonal polynomials may be used. The method is explained with reference to a problem.

Consider the data where X = nitrogen applied N kg/ha and Y = grain yield (t/ha)

X	0	60	120
Y	3.540	3.600	3.524

Prepare the following table:

x	x'	y	ξ'_1	ξ'_2	$\xi'_1 y$	$\xi'_2 y$
0	0	3.540	−1	+1	−3.540	+3.540
60	1	3.600	0	−2	0	−7.200
120	2	3.524	+1	+1	3.524	+3.524
Total		10.664			−0.016	−0.136

$\lambda = 12_2 = 3\,\bar{x}' = 3/3 = 1$

$\bar{y} = \Sigma y / n = 10.664 / 3 = 3.5547;\ \Sigma \xi_1'^{\,2} = 2;$

$\Sigma \xi_2'^{\,2} = 6;\ \Sigma \xi_1' y = -0.016;\ \Sigma \xi_2' y = -0.136;$

$\xi_1'^{\,2}$, and $\xi_2'^{\,2}$ are given in Fisher and Yates table.

Taking the origin as 0 and unit as 60, the quadratic equation will be of the form

$$y = A + B\xi_1' + C\xi_2'^{\,2}$$
$$A = \bar{y} = 3.5547$$
$$B = C\Sigma \xi_1' y / \Sigma \xi_1'^{\,2} = -0.016 / 2 = -0.008$$
$$C = \Sigma \xi_2' y / \Sigma \xi_2'^{\,2} = -0.136 / 6 = -0.022667$$
$$\xi_1' = \lambda_1 \xi_1 = \lambda_1 (x' - x') = 1(x' - 1)\,\text{(linear equation)}$$
$$\xi_2' = \lambda_2 \xi_2 = \lambda_2 [(x' - \bar{x}')^2 - (n^2 - 1)/12]\,\text{(quadratic equation)}$$
$$\quad = \lambda_2 [(x' - \bar{x}')^2 - (3^2 - 1)/12] = 3(x'^2 - 2x' + 1/3)$$
$$y = a + b'x + cx^2$$
$$\quad = y + B\lambda(x' - \bar{x}') + \lambda_2 C[(x' - \bar{x}')^2 - (n^2 - 1)/12]$$
$$\quad = 3.5547 + (-0.008)(x' - 1) - 0.022667 \times 3[x' + 1 - 2x' - 8/12]$$
$$\quad = 3.5547 - 0.008x' + 0.008 - 0.068001x'^2 - 0.068991$$
$$\qquad + 0.136002x' + 0.045334$$
$$\quad = 3.5400 - 0.068001x'^2 + 0.128002x'$$

Now, putting $x' = x / 60$,

$$\quad = 3.5400 - 0.068001(x^2 / 3600) + 0.128002(x / 60)$$
$$\quad = 3.5400 - 0.000018889x^2 + 0.002133367x$$

This equation gives the optimum: $x_{opt} = 0.002133367/2 \times 0.000018889 = 56.5$.

Example 2: In an experiment conducted with ADT 27 rice variety at 0,40,80,120,160, and 200 N kg/ha, the grain yields recorded were 2526, 3258, 3990, 4107, 4704, and 4535 kg/ha, respectively. Now we will fit a quadratic equation.

x	x= x/40	y	ξ'_1	ξ'_2	$\xi'_1 y$	$\xi'_2 y$
0	0	2526	−5	+5	−12630	+12630
40	1	3238	−3	−1	−9714	−3238
80	2	3990	−1	−4	−3990	−15960
120	3	4107	+1	−4	+4107	−16428
160	4	4704	+3	−1	+14112	−4704
200	5	4535	+5	+5	+22675	+22675
Total	(Σx) 15	(Σy) 23100	$(\Sigma \xi'^2_1)$ 70	$(\Sigma \xi'^2_2)$ 84	$(\Sigma \xi'_1 y)$ 14560	$(\Sigma \xi'_2 y)-$ 5025

$\overline{x}^1/x = 100$

Taking the origin as 0 and unit as 40, the quadratic equation is of the form
$y = A + B\,\xi'_1 + C\,\xi'_2$ where

$$A = \overline{y} = (\Sigma y)/n = 23100/6 = 3850$$
$$B = (\Sigma \xi'_1 y)/(\Sigma \xi'^2_1) = 14560/70 = 208$$
$$C = (\Sigma \xi'_2 y)/(\Sigma \xi'^2_2) = 5025/84 = -59.821$$

Linear function:

$$Y = \overline{y} + B[\lambda_1(x'-\overline{x}')] = \overline{y} + B[2(x-2.5)] = 3850$$
$$+208(2x'-5) = 3850 + 416x' - 1040 = 2810 + 416(x/40)'$$
Or $2810 + 416(x/40) = 2810 + 10.40x$

Quadratic function: It is of the form $Y = A + B\,\xi'_1 + C\xi_2,^2$ where

$$\xi_{1'} = \lambda_1 \xi_1 = \lambda_1(x'-\overline{x}') = 2(x'-2.5) = (2x'-5.0)$$
$$\xi^{2'} = \lambda_2 \xi_2 = \lambda_2[(x'-\overline{x}')^2 - (n^2-1)/12] = 3/2$$
$$[x'^2 + \overline{x}'^2 - .2x'\overline{x}' - (6^2-1)/12]$$
$$= 3/2(x'^2 + 2.5^2 - 2x'2.5 - 35/12] = 3/2[x'^2 + 3.333 - 5.0x']$$

The quadratic equation is

$$A = A + B\xi_{1'} + C\xi_{2'}$$
$$= 3850 + 208(2x' - 5.0) - 59.821 \times 3/2(x'^2 - 5.0x' + 3.3333)$$
$$= 3850 + 416x' - 1040 - 29.9105 \times 3(x'^2 - 5.0x' + 3.3333)$$
$$= 3850 + 416x' - 1040 - 89.7315x'^2 + 448.6575x' - 299.102$$
$$= 2510.8980 + 864.6575(x/40) - 89.7315)(x/40)^2$$
$$= 2510.8980 + 21.6164x - 0.0561x^2 \text{ (the fitted quadratic equation)}$$

The purpose of this experiment is to find the level of X, which maximizes Y, for this particular combination of levels. In order to establish a maximum, at least three levels of X (as in the previous example) must be compared. Four or five levels may be advisable if the range of X is wide and the position of its optimum is not known.

Maximum may be estimated by fitting a parabola to the observed responses. The equation for a fitted parabola is $y = A + Bx + Cx^2$ with $-C$, the optimum is known from

$$dy/dx = a + b + 2cx, \text{ i.e., } b + 2cx = 0, \text{ i.e., } 2cx = -b, \text{ i.e., } x = -b/2c$$

$X_{opt} = (-21.6164)/2(-0.0561) = 192.6595$ kg. The observed and expected Y values for various X levels may be investigated to determine the differences between the observed and expected Y values.

X	Y (observed)	Y (expected)
0	2,526	2,511
40	3,238	3,286
80	3,990	3,881
120	4,107	4,298
160	4,704	4,534
200	4,535	4,590
Total	23,100	23,100

The observed values as well as the predicted values agree to a great extent and hence we can consider the fit as the best fit.

Methods to improve prediction:

1. In regression analysis, we have total variability, variation due to regression and deviation from regression as sources of variation, and degrees of freedom associated with each one of them. The lesser the error means square, the higher the F value leading to significance of dependability of Y on X. Thus the experiment may be conducted in

such a way as to provide more degrees of freedom for error, which can be done by increasing the number of X levels.

2. We cannot completely eliminate error but we can minimize error by modifying the experiment. For instance, in our example, we are able to find the influence of fertilizer on crop yield to an extent of 81 percent. This is because crop yield is not dependent on fertilizer alone. Other factors such as irrigation, temperature, plant protection, etc., may also influence the crop yield. Thus an additional factor or factors may be included, and a new fit attempted, which may perhaps reduce unexplained variation and result in a better fit.

3. Prediction can be improved by studying several variables simultaneously. This technique is called multiple regression.

Regression Through Origin

Using the regression equation, the researchers could predict the unknown value Y for the desired value of X. The simple linear regression model discussed previously is $y = a + bx + e$ where $a = $ intercept, $b = $ regression coefficient, and $e = $ error or the residual about the regression line ($e = y - \hat{y}$) where $\hat{y} = a + bx$. In some situations, it appears appropriate for a regression line to pass through origin (Brownlee, 1984). In such situations, we get the regression model of the type $Y = BX$. This is called nonintercept model. In this model, the value of y is 0 when x variable is 0. If a is found not significant, then $y = bx + e$ model is the appropriate one. The value of a can be tested statistically using t test as $t = (\bar{y} - b\bar{x}) / s_{yx} \sqrt{[(1/n) + (\bar{x}^2 / sx^2)]}$ with n-2 degrees of freedom. If t is not significant, the nonintercept model can be used.

Assumptions in the regression model

1. For each selected x_i, there is a normal distribution of Y.
2. The population values of y corresponding to a selected x_i have a mean ßx, where ß is the parameter.
3. In each population, the standard deviation of y about its mean bx has the same value, σyx.
4. The x_i values are known without error.

Linear Model

The following linear regression model is assumed, $Y = ßx + \varepsilon$. Here, ß is the population regression coefficient to be estimated from the sample, and ε

is a random component distributed normally with mean 0 and variance $\sigma^2 yx$. It is also assumed each pair of observations satisfies the relation

$$Y_i = \beta x_i + \varepsilon_i$$

where Y is NID $(0, \sigma^2)$.

For sample data, the equation is $y_i = bx_i + e_i$. Then an estimate of ith value of the dependent variable is $\hat{y} = bx_i$, where b is estimated using least square by minimizing the residual sum of squares, i.e., $\Sigma(y_i - bx_i)^2$. The b value is estimated by differentiation. Differentiating with respect to b,

$$\Sigma(y_i - bx_i)(0 - x_i) = 0$$
$$\Sigma(y_i - bx_i) - x_i = 0$$
$$\Sigma(-x_i y_i + bx_i x_i) = 0,$$
$$\text{i.e., } \Sigma \xi_i y_i - b\Sigma x_i^2 = 0,$$
$$\text{i.e., } b = \Sigma x_i y_i / \Sigma x_i^2.$$

Thus b is estimated as $b = \Sigma x_i y_i / \Sigma x_i^2$. Then error SS $= s^2 yx = \Sigma y_i^2 - (\Sigma x_i y_i)^2 / \Sigma x_i^2$ where $\Sigma y_i^2 = $ SS of y and $(\Sigma x_i y_i)^2 / \Sigma x_i^2 = $ regression SS with $n - 1$ degrees of freedom. The standard error of the estimate is given by $s = \sqrt{(1/n-1)[\Sigma y_i^2 - (\Sigma x_i y_i)^2] / \Sigma x_i^2}$ (Snedecor and Cochran, 1980). It indicates that if a perfect linear relationship between y and x exists, the standard error becomes 0.

Test for the Significance of the Regression Coefficient

As explained, the total sum of squares and the regression sum of squares may be calculated using ANOVA.

SV	df	SS	MS	F
Regression	1	SSR	MSR=SSR/1	MSR/MSE
Error	$n-1$	SSE	MSE=SSE/$n-1$	
Total	n			

If F in the ANOVA turned out to be significant, the variability in y is explained due to regression. The proportion of variation, i.e., the amount of variation explained by the regression is given be R^2 value. The $R^2 = $ SS due to regression/total SS. If the error in prediction $(y_i - \hat{y}_i)$ is negligible, then $R^2 = 1.00$. When the error is large, R^2 value will be very small. R^2 values are therefore of considerable importance to researchers as they serve as an important measure of adequacy of the predicted equation. The correlation coefficient for a given set of pairs of observations can be calculated and tested by using the t statistic.

Example: Palaniswamy (1990) studied the different methods for estimation of grain yield per plant in IR 22 rice variety in three different fertilizer levels (0, 60, and 120 N kg/ha). The X variable considered was the number of grains in a panicle. Palaniswamy selected the panicle in the tallest tiller of the plant, and the number of grains in the panicle was considered as X variable. The grain yield of the plant was considered as Y variable. He used the regression equation $Y = \beta X$ and estimated the value of the regression coefficient. The regression equations obtained under different N levels were: $Y = 0.1581X$ (0 N kg/ha), $Y = 0.2106X$ (for 60 N kg/ha), and $Y = 0.2714X$ (for 120 N kg/ha). The reported error in the estimation of yield was less than 3 percent, which is an acceptable error in the estimations. The correlation coefficients were 0.9369, 0.9531, and 0.9657 for the three N levels, respectively, and they were highly significant.

EXERCISES

13.1. Explain the terms *correlation* and *regression*.

13.2. Let $y = 2.97 - 0.194x$. Find the estimated y for $x = 4$.

13.3. Given x: 0, 2, 4, and y: 2, 4, 3, (a) calculate the intercept a (b) determine the regression coefficient, and (c) predict the value of y of $x = 4$.

13.4. Given are three scatter diagrams. Find which regression line is most appropriate.

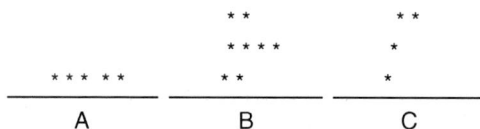

A   B   C

13.5. Find the regression line to the following data. ($Sxy = 16,500$, $Sx^2 = 280,000$, $\bar{x} = 400$, $\bar{y} = 60$).

X	100	200	300	400	500	600	700
Y	40	50	50	70	65	65	80

13.6. Obtain the equation of the line of regression of yield of rice and water supplied from the data given below: (X = amount of water supplied; Y = yield t/ha)

X	12	18	24	30	36	42	48
Y	5.27	5.68	6.25	7.21	8.02	8.71	8.42

Estimate the most probable yield when 40 inches of water is supplied.

13.7. Given: $n = 38$, $\bar{x} = 5$, $\bar{y} = 40$, $Sx^2 = 100$, $Sy^2 = 10,000$, $Sxy = -800$,
   a. Find $y = a + bx$.
   b. Find the r value and coefficient of determination.
   c. Interpret your answers with reference to a and b.
   d. Carry out the test of significance for the regression coefficient (use F and t tests).

13.8. Intensity of pest (X) and production of crop yield in a place are obtained as follows:

Y	23	38	45	36	16	18	39	41
X	7.0	6.5	5.5	6.0	8.0	8.5	6.0	6.5

   a. Draw a scatter diagram.
   b. Find sample regression equation.
   c. Find $\hat{y}$ when $x = 3.5$.
   d. Find standard error of the estimate.

13.9. Why is 100 percent prediction not possible in biological experiments?

13.10. What is the method of least square? Why is it useful?

13.11. What are the two fundamentally different ways of using fitted least squares equations?

13.12. What is the definition of a linear model?

13.13. Can you use the linear model with data that are nonlinear? Explain.

13.14. In regression problems, residuals are always calculated and examined. Why?

13.15. What are some common pitfalls associated with the use of regression analysis?

13.16. In a simple linear regression problem, the following figures were obtained:
   $n = 11$,
   $\bar{x} = 4$, $\bar{y} = 20$, $\Sigma(x - \bar{x})^2 = 64$,
   $\Sigma(x - \bar{x})(y - \bar{y}) = 256$,
   $\Sigma(y_i - \bar{y})^2 = 1600$
   a. Find the regression equation.
   b. Find the estimated variance of y.
   c. Test the hypothesis that $\beta = 0$. Let $\alpha = 0.05$.

    d. Find the coefficient of determination. How do you interpret this value?

13.17. Rainfall and crop yield were recorded in 11 places as follows:

Yield (t/ha)	25	48	123	62	107	119	63	91	152	98	126
Rainfall (in)	37	11	14	44	15	21	33	12	22	29	23

    a. Graph the data.
    b. Assume a linear relationship and using the method of least squares determine the estimated equation.
    c. Show the equation in a scatter diagram.
    d. Calculate the predicted yield when rainfall is 30 inches.

13.18. Why can't correlation be taken as causation between two variables?

13.19. The coefficient of determination explains the proportion of variation explained by regression. Substantiate the statement.

13.20. What are the basic assumptions to be made about the population in regression analysis?

13.21. Test the hypothesis H_o: $\beta = 0$ against $\beta \neq 0$; Use $\alpha = 0.05$.

x	1	1	5	5
y	1	3	2	4

13.22. Given the following data, find the 95 percent confidence interval.

x	10	25	19	42	22	28
y	29	72	40	28	64	87

13.23. The following calculations were made in a regression problem:
$\bar{x} = 20,$
$\bar{y} = 22,$
$\Sigma(x - \bar{x})^2 = 225,$
$\Sigma(y - \bar{y})^2 = 414,$
$\Sigma(x - \bar{x})(y - \bar{y}) = 132,$
$n = 32$
    a. Find the regression line.
    b. Determine its significance.
    c. Draw a scatter diagram.
    d. Verify that the line passes at $(\bar{x}, \bar{y})$ point.

13.24. Given: $n = 38$, $\bar{x} = 5$, $\bar{y} = 40$, $Sx^2 = 100$, $Sy^2 = 10,000$, $Sxy = -800$, compute $y = a + bx$.

13.25. The following data were given in a regression problem,

x	1	2	3	4	5	6
y	6	9	13	12	14	18

verify: (a) $y = 4.601 + 2.114x$; (b) deviation SS = 7.76; (c) regression SS = 78.24; (d) total SS = 86.00; (e) y increases with x; (f) 95 percent confidence interval for $ß = 1.19 - 3.04$; (g) the $\hat{y}$ is 10.943 when $x = 3$; (h) standard error of b = 0.333; (i) standard error for predicted value = 1.513.

13.26. The following data are on age in weeks, x and y mean height in cm of soybean plants,

x	1	2	3	4	5
y	5	17	24	33	41

find the line of regression of y on x. (The line must pass through origin.)

13.27. Given the following data,

x	11	13	15	17	19	21
y	15.2	17.7	19.3	21.5	23.9	25.4

a. Fit the line of the type $y = a + bx$.
b. Test whether the intercept a and slope are significantly different from 0.
c. Find whether the slope differs significantly from a theoretically predicted slope of 1.000.
d. Find 95 percent confidence limits for a and b.
e. Assuming that the regression line can be found for $x = 25$, estimate y value for $x = 25$ and also 90 percent confidence limits for this estimate.
f. Find 95 percent confidence limits for a single estimated value for $x = 25$.

13.28. Given x = nitrogen levels applied and y = yield (t/ha) in a crop,

x	0	40	80	120	160	200	240
y	3	3.2	4.1	4.6	5.0	6.0	7.0

a. Plot the points in a graph.
b. Find the mean and standard deviations for both x and y. Plot the points $\bar{x}, \bar{y}$ on the graph.
c. Find out a and b and mention the regression line $y = a + bx$ and show the line in the graph.

d. Find the deviation of *y* from the corresponding $\hat{y}$ and square and sum these deviations.
e. Find the SS for deviations from regression.
f. Construct a summary table showing the partitioning of the SS and degrees of freedom.

13.29. Given the following data obtained from a regression analysis:

$$\Sigma x = 169, \ \Sigma xy = 3006, \ \Sigma y^2 = 2370, \ \Sigma x^2 = 3856, \ \Sigma y = 130$$

a. Estimate the population regression equation and population correlation coefficient.
b. Does the value of a have any practical significance?
c. What do you understand by the values for b and *r?*
d. Draw the line of regression.
e. Compute the SE for b and test its significance.
f. Can you use the prediction equation to estimate *y* for *x* = 300? Give your reasons.
g. What have you learned from this problem?
h. Do you think there is a strong correlation between these two variables?

13.30. What is the relation between mean of *x,* mean of *y,* and the regression line?

13.31. What do you understand by index of fit in regression studies? How does it relate to prediction value?

13.32. Differentiate between *y* = a + b*x* and *x* = a + b*y.*

REFERENCES

Brownlee, K.A. (1984). Statistical theory and methodology in science and engineering, Second edition. Malabar, FL: Robert E.Krieger Publishing Co. Inc.

Palaniswamy, K.M. (1990). Design of statistical (field) experiments and crop forecast in agriculture with special reference to rice. Doctoral thesis submitted to the University of Calicut, India.

Palaniswamy, K.M. and K.A. Gomez (1974). Length-width method for estimating leaf area in rice. *Agronomy J.* 66(7):430-433.

Snedecor, G.W. and W.G. Cochran (1980). *Statistical methods,* Seventh edition. Ames, IA: The Iowa State University Press.

Usha Rani, P. (2001). Length diameter method for estimating fruit surface area in chili (*Capsicum annuum* L.). *Crop Res.* 21(2):168-173.

Chapter 14

Chi Square Test of Significance

When scientists conduct experiments they collect two types of data: (1) measurement data, e.g., age, income, area, height, weight, rainfall records, barometric measurements, wages, crop yields, burette or pressure gauge readings, or recordings on a potentiometer, etc., where we consider the accuracy of measurements; and (2) enumeration data in which we are concerned with the count.

Examples of enumeration data:
1. Number of plants having green-colored stem
2. Number of blue flowers in an experiment in genetics
3. Number of animals inoculated against anthrax disease
4. Number of animals dead or recovered
5. Number of births in a place
6. Number of deaths in a country
7. Number of telephone poles damaged or undamaged in an area in a given time period
8. Number of patients admitted in hospitals each day in the world
9. Number of people with or without blue eyes
10. Suppose a coin is tossed 100 times. If the coin is fair, we expect 50 heads and 50 tails. We count the number of heads and tails in 100 throws.
11. In genetic problems, the $F2$ plants segregate in a definite ratio, e.g., 3:1, 1:2:1, 9:7, 9:3:3:1, etc. Here we count the plants possessing a particular plant trait.
12. Number of accidents in a day in a particular area
13. Number of students passed or failed in a college in a subject in a particular year
14. Number of children in a family
15. The number of defective items in a consignment

The data obtained by counting are called enumeration data. Enumeration data differ from measurement data in that there is no continuity between counts. For example, the number of plants having pink flowers is noted as 10 or 11 and not 10.2.

In previous chapters, the use of t and F tests of significance were explained. Another important distribution of practical importance in the field of applied statistics is the chi-square statistic. Chi-square is an important discrete distribution (Anderson and Bancroft, 1952).

DEFINITION

χ^2 is a Greek letter pronounced "chi" (kigh) and spelled *chi*. χ^2 is used instead of χ to indicate that χ^2 value is always positive. It is a one-parameter distribution, the parameter being the number of degrees of freedom, which is $n-1$, i.e., the shape of the χ^2 distribution is dependent on the number of degrees of freedom. When the number of degrees of freedom is small, the curve is skewed to the right. As the number of degrees of freedom increases, the curve approaches approximately normal distribution (see Figure 14.1).

The χ^2 test is used to compare two sets of frequency distributions and to verify whether there is significant difference between them. Of the two sets of frequencies, one relates to the observed frequencies calculated from the observed data, and the other relates to frequencies specified by a particular hypothesis or by some assumed model. The chi-square test indicates whether the difference between the two sets of frequencies is real or whether the difference occurs by chance.

The chi-square statistic is calculated by using the formula:

$$\chi^2 = \Sigma(O_i - E_i)^2 / E_i \text{ where}$$

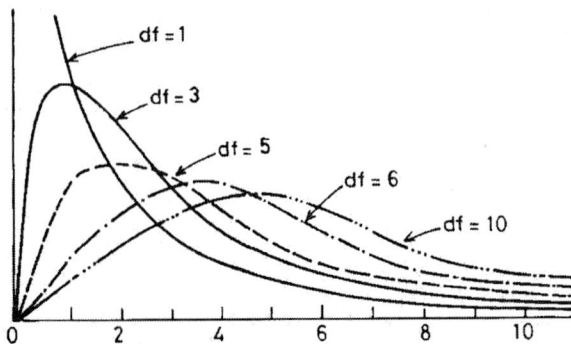

FIGURE 14.1. χ^2 probability curve showing the different shape of the distribution for the different number of degrees of freedom.

O_i = observed frequency in the class i
E_i = expected frequency in the class i
Σ = sum of
Null hypothesis: H_o: $O_i = E_i$ ($i = 1, 2, \ldots k$); H_1: $O_i \neq E_i$ ($i = 1, 2, \ldots k$),

i.e., the observed values conform to some expected values, i.e., the distribution of the observed frequencies does not differ significantly from the distribution of the expected frequencies.

Chi-Square Statistic

In the formula, we consider the difference between a pair of observed and expected frequencies, i.e., $O_i - E_i$. We know $\Sigma O_i = \Sigma E_i = n$. Hence the sum of the deviations is equal to zero. This is avoided by squaring the differences. We get $\Sigma (O_i - E_i)^2$. Then the χ^2 statistic is defined as:

$$\Sigma (O_i - E_i)^2 / E_i, \text{ i.e., (sum of deviation)}^2 / \text{expected frequencies}$$

If $O_i = E_i$, the chi-square value will be zero. It indicates perfect agreement between the frequencies. Rarely $\chi^2 = 0$. The smaller the chi-square value, the smaller the difference. Larger chi-square value indicates disagreement between assumed distribution and the observed values. Thus, the magnitude of the χ^2 value depends upon the amount of differences. A large chi-square value tends to reject the null hypothesis and smaller values of χ^2 tend to confirm the null hypothesis. In chi-square problems, we consider only the number of categories and not the total number of observations. But the total of observed and expected frequencies must be equal.

Chi-Square Value and Test of Significance
(Equal Expected Frequencies)

This is explained with a numerical example: Suppose a coin is tossed 100 times and we observe 65 heads and 35 tails. The numbers 65 and 35 are the observed frequencies. The expected frequencies are computed following a certain theory or hypothesis or assumed model. We assume that the coin is fair. The probability of observing a head is 1/2 and the probability of observing a tail is also 1/2. Thus the expected frequencies in our experiment are 50 heads and 50 tails. The results are tabulated as a 2×2 table.

	Head	Tail	Total
Observed	65	35	100
Expected	50	50	100

Total	115	85	200

$\chi^2 = \Sigma(O_{i-} E_i)^2 / E_i$ with $k - 1$ degrees of freedom $= (65- 50)^2/ 50 + (35 - 50)^2 / 50 = 4.50 + 4 .50 = 9.00$ (without applying the correction for continuity). The degrees of freedom are generally calculated as:

(number of rows $- 1$) (number of columns $- 1$). Therefore in the 2×2 table, we have 2 rows and 2 columns, and the degrees of freedom are given by $(2 - 1) (2 - 1) = 1$. We look for the table value of chi-square for 1 degree of freedom at a percent (5 percent) level of significance $\chi^2_{1, 0.05} = 3.841$. If the calculated χ^2 value is more than the tabulated value at ∞ percent level of significance, we conclude that the fit of the expected frequencies is not equal with that of observed frequencies (see Figure 14.2). The formula can be simplified as $\chi^2 = \Sigma (O_i^2 / E_i) - N$, where $\Sigma O_i = \Sigma E_i = N$.

CORRECTION FOR CONTINUITY

Chi-square is a continuous mathematical function, and when it is calculated from discrete data, some bias is incurred. Due to bias, high χ^2 values result. This is to be corrected by modifying the chi-square values. The modified formula is $\chi^2 = \Sigma (|O_i - E_i| - 0.5)^2 / E_i$.

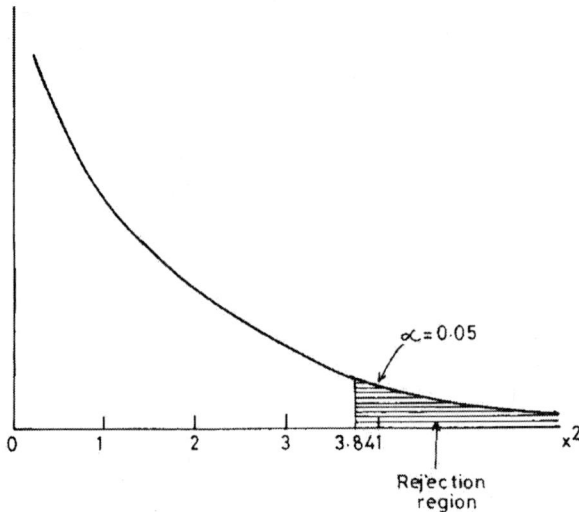

FIGURE 14.2. The χ^2 distribution showing 5 percent critical region.

This correction is called Yates correction. Yates correction is to be used only for 2 × 2 tables (Volk, 1969). The correction relates to subtraction of 0.5 from the absolute difference between the observed and expected frequency.

$$\chi^2 = [\,|\,65\text{–}50|\ - 0.50\ )]^2 /\ 50 + (|\ 35 - 50|\ - 0.5)^2/50 = 14.5^2 /\ 50 + 14.5^2/50 = 8.40.$$

The chi-square values from χ^2 table for 1df at 0.05, 0.01, 0.001 are 3.841, 6.635, and 7.879, respectively (see Appendix, Table A.6). The null hypothesis is rejected. The chi-square value (8.40) is greater than the critical value (3.841) obtained from the table for the accepted level of significance and the appropriate degrees of freedom. Note the computed χ^2 after applying correction for continuity is lesser than the chi-square value computed without applying continuity correction.

TEST OF GOODNESS OF FIT

The test of goodness of fit was developed by Karl Pearson. This test is used to see how well an observed set of data fits an expected set of data, or how well the observed frequency distribution fits with the hypothetical frequency distribution (Denenberg, 1976).

Example 1: In an experiment in the cross 'Red Pigmented Kaliyan' × 'Amarello White' in tobacco *(Nicotiana tobaccum)*, the following frequencies were recorded in F2 for stem color and corolla color.

(CP) stem creamy and corolla pink	69
(Cp) stem creamy and corolla white	25
(cP) stem normal and corolla white	25
(cp) stem normal and corolla pink	5

The theoretical frequencies are expected to be in the ratio 9:3:3:1. The theoretical frequencies are calculated as follows:

Stem creamy and corolla pink	124 × 9/16 = 69.75
Stem creamy and corolla white	124 × 3/16 = 23.25
Stem normal green and corolla white	124 × 1/16 = 7.75
Stem normal green and corolla pink	124 × 3/16 = 23.25

H_o: There is no difference between the set of observed frequencies and the set of expected frequencies. H_1: There is a difference between the observed and expected frequencies.

Form Table 14.1.

The calculated χ^2 of 1.2473 is in the rejection region beyond the critical value of 7.815. The degrees of freedom is 3, i.e., $k - 1$ (4 - 1). The table value of χ^2 for 3 degrees of freedom at 5 percent level of probability is 7.815 (see Appendix Table A.6 and Figure 14.3).

Example 2. A die is cast 60 times and the number of frequencies for each face are as follows:

Face	1	2	3	4	5	6	Total
Frequency	5	10	10	9	9	17	60

The probability of each face appearing at each throw is 1/6. Hence, out of 60 times each face is expected to show ten times. Therefore, the expected frequency in each group is

$$10 \times \chi^2 = \Sigma \, (O_i - E_i)^2 / E_i \text{ with } k - 1 \; df.$$

k (number of groups) $-1 = 6-1 = 5$
O_i = observed frequency in class i ($i = 1 \ldots . 6$)
E_i = expected frequency in class i ($i = 1 \ldots . 6$)

The critical value for five degrees of freedom and for 0.05 level of significance is 11.07. It is found by locating five degrees of freedom in the left margin and moving to the right horizontally, and reading the critical value in the 0.05 column (see Appendix, Table A.6.).

TABLE 14.1. Observed and expected frequencies.

Group	O_i	E_i	$(O_i - E_i)$	$(O_i - E_i)^2$	$(O_i - E_i)^2 / E_i$
1	69	69.75	−0.75	0.5625	0.0081
2	25	23.25	1.75	3.0625	0.1317
3	25	23.25	1.75	3.0625	0.1317
4	5	7.75	−2.75	7.5625	0.9758
Total	124	124	+3.50 −3.50		1.2473

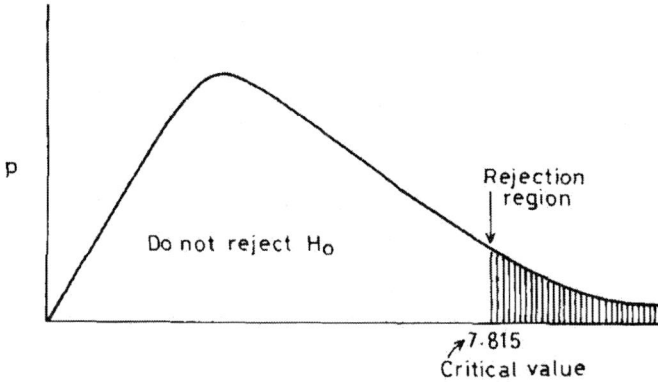

FIGURE 14.3. The χ^2 probability distribution for three degrees of freedom.

Calculation of χ^2:

$\chi^2 = (5-10)^2/10 + (10-10)^2/10 + (10-10)^2/10 + (9-10)^2/10 + (9-10)^2/10 + (17-10)^2/10 = 1/10(25 + 0 + 0 + 1 + 1 + 49) = 76/10 = 7.6$

The result is not significant and hence we conclude that the die is a fair one.

Example 3 (goodness of fit test, unequal expected frequency): In an experiment in a barley cross between two varieties, the following results were reported:

Groups	Obtained No.
Hooded green	885
Hooded cholorina	310
Awned green	292
Awned cholorina	113
Total	1600

Compare the observed data to the expected data in the ratio 9:3:3:1, at 0.05 level of significance. Based on the theoretical model, the expected frequencies would be 900 (1,600 × 9/16), 300 (1,600 × 3/16), 300 (1,600 × 3/16), and 100 (1,600 × 1/16).

Example: The expected frequencies equal the observed frequencies.

$\chi^2 = \Sigma (O_i-E_i)^2 / E_i$
$= (885-900)^2/900 + (310-300)^2/300 + (292-300)^2/300 + (113-100)^2/100$

$= 0.25 + 0.33 + 0.21 + 1.69 = 2.48$. Degrees of freedom is $k-1$, i.e., $4-1 = 3$

The χ^2 value from the table for 3 df at $P = 0.05$ is 7.81. Hence the hypothesis is quite acceptable.

POISSON DISTRIBUTION

Example 4 (Poisson distribution, fitting of Poisson distribution, and test of goodness of fit): Poisson distribution is a good model for the occurrence of some rare events over successive periods of time. It is a discrete probability distribution because it is formed by counting some characters. Some specific examples follow.

1. Number of good seeds and number of bad seeds in a sample
2. Number of accidents occurring per month in an industry
3. Number of typographical errors per page in the typed pages
4. Number of deaths in a place in one year by rare disease
5. Number of defective screws per box of 100 screws
6. Number of cars passing through a street in a particular time period
7. Number of defects (scratches, dents, etc.) observed by a quality inspector in a new piece of cloth produced in a textile mill
8. Number of death claims received per day by an insurance company
9. Number of errors in data entry
10. Number of customers to be served at a restaurant
11. Number of parts produced and number of defective items
12. Number of infected plants in experimental plots containing large numbers of plants
13. Number of men kicked to death by a horse in an army corps each year
14. Number of goals scored by a particular team in a soccer match

The Poisson distribution is described by the following formula:

$$p(x) = \frac{e^{-\gamma}\lambda^2}{x!}$$

$P(x)$ = probability to be computed for a specified value of x
e = base of natural logarithm (it is equal to 2.3182 and $\log_e 10 = 0.4343$)
λ = expected number of occurrences
x = number of occurrences (successes)

The Poisson distribution is a single-parameter distribution; the parameter is mean. Mean is denoted by the Greek letter λ (lambda). λ is estimated by the mean of the data, i.e., $m = \lambda$. The variance and mean are equal in Poisson distribution, i.e., $\sigma^2 = \lambda$. The shape of distribution is skewed on the right. The Poisson distribution occurs when the probability of the events is small and the number of events is large say 100 or more. Thus Poisson distribution describes rare events. A distribution in which frequencies are computed using the previous formula is called a Poisson distribution (Dixon and Massey, 1957).

In the formula, frequencies are computed for various values of x ($x = 0$, 1, 2, ...). The expected frequencies may be considered as the average number of frequencies one would expect for each x value in the long run with repeated sets of, e.g., 200 random observations from the distribution.

Poisson Distribution and Goodness of Fit Test of Significance

Example 1: In an experiment, an entomologist recorded the number of insects on 200 leaves individually and formed the frequency distribution. The data obtained follow:

No. of insects	No. of Leaves
0	110
1	64
2	20
3	6
Total	200

Steps

1. Find the mean.

Mean = $\Sigma f_i x_i / \Sigma f_i$ = $[(110 \times 0) + (64 \times 1) + (20 \times 2) + (6 \times 3)]/200 = 0.61$, i.e., $\lambda = 0.61$

2. Find the probability for each value.

$P(x = 0) = (e^{-\lambda} \lambda^x)/x! = (e^{-0.61} \lambda^0)/0! = 0.5434$
$P(x = 1) = (e^{-\lambda} \lambda^x)/x! = (e^{-0.61} 0.61^1)/1! = 0.3315$
$P(x = 2) = (e^{-\lambda} \lambda^x)/x! = (e^{-0.61} 0.61^2)/2! = 0.1010$
$P(x = 3) = (e^{-\lambda} \lambda^x)/x! = (e^{-0.61} 0.61^3)/3! = 0.0206$
Check: $P(0)+P(1)+P(2)+P(3) = 1$

The cumulative Poisson distribution function is given by $F(x) = \Sigma(e^{-\lambda}\lambda^x)/x!$
   3. Calculate expected frequencies.

$$p_0 \times 200 = 0.5434 \times 200 = 108.7$$
$$p_1 \times 200 = 0.3315 \times 200 = 66.3$$
$$p_2 \times 200 = 0.1010 \times 200 = 20.2$$
$$p_3 \times 200 = 0.0206 \times 200 = 4.8$$
Total 200.0

Remarks: Since the expected total frequency must be equal to 200, the last class frequency was made equal to 4.8. The probabilities can also be calculated using the table of Poisson cumulative distribution.
   4. Form the table for observed and expected frequencies. Before we apply χ^2 test of significance, we must be sure that no class has expected frequencies of less than 5. Accordingly, the groups are formed as:

x	oi	ei	$o_i - e_i$
0	110	108.7	+1.3
1	64	66.3	−2.3
2	26	25.0	+1.0
Total	200	200.0	0

   5. Compute χ^2.

$$\chi^2 = \frac{(110-108.7)^2}{108.7} + \frac{(64-66.3)^2}{66.3} + \frac{(26-25)^2}{25} = 0.15$$

   6. Find the degrees of freedom. For goodness of fit test, the degrees of freedom is $k - r - 1$, where k is the number of rows in the frequency data table and r is the number of parameters estimated in fitting the distribution (Cooper and Weekes, 1983). The degrees of freedom in our problem is $3 - 1 - 1 = 1$ as $k = 3$, $r = 1$.
   7. The χ^2 value from the table for 1 degree of freedom is 3.841. Since the calculated value is less than the table value, the fit is good.
Calculation of λ: Let $\log e^{-0.61} = A$

Taking log, $\log A = -0.61 \times \log_{10} e$
   $= -0.61 \times 0.43429$
$A = $ antilog $(-0.61)(0.43429)$
   $= $ antilog (-0.2649)
   $= -1.7351 = 0.5434$

Inference: When $\lambda = 0.61$, we infer that 54.34 percent of the time x will have the value 0; 87.49 percent of the time x will have the values 0 or 1; 97.60 percent of the time x will have the values 0 or 1 or 2; and 100 percent of the time x will have 0 or 1 or 2 or 3. The table numbers are actually calculated using the formula given (Dixon and Massey, 1957) and this can be used for computation.

Example 6. In an experiment involving spraying treatment for leaf spot in rice, leaves were selected at random and the number of leaf spots counted. The number of leaf spots per leaf was recorded. The distribution of leaf spots was noted, and the data obtained are furnished here:

No. of spots/leaf	0	1	2	3	4	5	6	7	8 or more	Total
Frequency	70	38	17	10	9	3	2	1	0	150

We want to fit the Poisson distribution to the data and apply the test of goodness of fit.

Computation

1. Mean $=[(0 \times 70) + (1 \times 38) + \ldots + (8 \times 0)]/ 150 = 172/150 = 1.147$
2. Probability distribution for the Poisson distribution is given by the formula

$$p(x = x) = (e^{-\lambda} \lambda^x)/x! \ (x = 0,1,2 \ldots 8)$$
$$\text{Compute} = e^{-\lambda} = e^{-1.147} = 0.3177$$

Find the probability for each value	Expected frequencies ($p \times N$)
$P(x=0) = (0.3177)(1.147)^0/0! = 0.3177$	$0.3177 \times 150 = 47.66$
$P(x=1) = (0.3177)(1.147)^1/1! = 0.3614$	$0.3614 \times 150 = 54.66$
$P(x=2) = (0.3177)(1.147)^2/2! = 0.2089$	$0.2089 \times 150 = 31.34$
$P(x=3) = (0.3177)(1.147)^3/3! = 0.0799$	$0.0799 \times 150 = 11.99$
$P(x=4) = (0.3177)(1.147)^4/4! = 0.0229$	$0.0229 \times 150 = 3.42$
$P(x=5) = (0.3177)(1.147)^5/5! = 0.0052$	$0.0052 \times 150 = 0.78$
$P(x=6) = (0.3177)(1.147)^6/6! = 00010$	$0.0010 \times 150 = 0.15$
$P(x=7) = (0.3177)(1.147)^7/7! = 0.0000$	$0.0000 \times 150 = 0$
$P(x=8) = (0.3177)(1.147)^8/8! = 0.0000$	$0.0000 \times 150 = 0$
Total = 1.0000	Total = 150

Check: $P(0)+P(1)+P(2)+P(3) \ldots +P(8) = 1.0000$

Because the expected frequencies in a class should not be less than 5, adjusted frequencies are combined and then χ^2 test applied.

X	0	1	2	3
O_i	70	38	17	25
E_i	47.66	54.66	31.34	16.34

$$\chi^2 = (70 - 47.66)^2 / 47.66 + (38 - 54.66)^2 / 54.66 + (17 - 31.34)^2 / 31.34 + (25 - 16.34)^2 / 16.34 = 10.4716 + 5.0779 + 6.5614 + 4.5897 = 26.7006$$

Degrees of freedom $= k - r - 1 = 4 - 1 - 1 = 2$

The table value of $\chi^2_{0.05,2} = 5.991$ (see Appendix, Table A.6). Since calculated χ^2 value (26.7006) exceeds the table value, we conclude that the observed frequencies do not show a good fit to the Poisson distribution. The null hypothesis is not rejected.

Example 7. Normal distribution, fitting of normal distribution, and testing goodness of fit: Normal distribution is a continuous distribution. The variable x is said to have normal distribution if its density function is of the form

$$f(x) = \frac{N}{\sigma\sqrt{2\pi}} e^{(x-\mu)^2 / 2\sigma^2} \quad (-\infty < x < +\infty)$$

It makes a bell-shaped curve and it is determined by two parameters μ and σ, i.e., mean and standard deviation when $f(x) =$ frequency with which the measurement x repeats.

μ = mean
σ = standard deviation
π = a constant (= 3.1429)
e = base of Naperian logarithm (= 2.7183)
x = any measurement made on a variate

The frequency distribution based on sample data can be fitted to the normal distribution by computing the estimates of the two parameters, μ and σ from the sample. Once the given distribution is known to conform approximately to this pattern, many of the properties of this curve may be exploited to explain the underlying features of the empirical observations. Therefore testing goodness of fit of the data to normal distribution is very important in applied statistics. Furthermore, many of the biological variables such as height of plants, weight of ear heads, crop yields, blood pressure, etc., are known to follow approximately normal distribution. Many events in nature (physical, biological, and psychological measurements) are found to follow normal curve.

The method of fitting data to the normal distribution is illustrated with the numerical example already worked out by Palaniswamy (1990). He studied the nature of distribution of grain yield/plant (g) in IR 22 rice crop by randomly studying a set of 200 plants. The sample data were grouped to form a frequency distribution and tested for normal distribution. Palaniswamy (1990) grouped the 200 observations into 10 mutually exclusive categories and denoted them by O_i. He then computed the expected frequencies for each class ($E_i = 1, 2, \ldots 10$). The procedure for computing the expected frequencies are given as follows.

PROCEDURE FOR FITTING NORMAL DISTRIBUTION

1. Calculate the mean (μ).
2. Calculate the standard deviation (σ).
3. Calculate normal deviate (z) given by $Z = (x-\mu) / \sigma$.
4. Find the proportion of area to the left of z from the normal probability table. (Remarks: when z is negative, the proportion of the area to left of $-z$ is equal to the proportion of the area of the right of $+z$, i.e., 1/2 $(1-\infty)$ is given by $1 - [1/2 (1+\infty)]$. For example when $z = +1.2$, area is 0.88493. The proportion of the area to the left of -1.2 is $1-0.88493 = 0.11507$).
5. Compute the difference of appropriate proportion to get the proportion of area between two limits.
6. Multiply this proportion by total frequency, N (200 in our example) to get the corresponding expected frequencies.

Remarks: The total area in column 5 adds to 1.0000 and the expected frequencies is equal to observed frequencies, i.e., 200.

Form the following table: Mean = 28.02 g., standard deviation = 9.9528 g (see Table 14.2).

Since the excepted frequencies should not be less than 5 in any class we form another table to work out χ^2.

O_i	2	6	12	12	21	27	35	29	26	18	5	7
E_i	2.72	4.30	8.84	15.88	23.12	29.28	31.72	29.28	23.12	15.58	8.98	7.18

We apply the formula $\Sigma (O_i - E_i)^2/E_i$ to work out the chi-square statistic.

TABLE 14.2. Theoretical and observed frequencies.

Class interval	Limits x	$z =$ $(x-\mu)/\sigma$	Area left to z	Area within limits	Theoretical freq. (exp. freq)	Obser. freq.	$\dfrac{(O_i-E_i)^2}{E_i}$
1	2	3	4	5	6	7	
2-6	2	−2.6143	0.0045	0.0136	2.72	2	0.191
6-10	6	−2.2124	0.0136	0.0215	4.30	6	0.672
10-14	10	−1.8105	0.0351	0.0442	8.84	12	1.130
14-18	14	−1.4086	0.0793	0.0794	15.88	12	0.948
18-22	18	−1.0067	0.1587	0.1156	23.12	21	0.194
22-26	22	−0.6048	0.2743	0.1464	29.28	27	0.178
26-30	26	−0.2029	0.4207	0.1586	31.72	35	0.339
30-34	30	+0.1989	0.5793	0.1464	29.28	29	0.008
34-38	34	+0.6008	0.7257	0.1156	23.12	26	0.389
38-42	38	+1.0027	0.8413	0.0779	15.58	18	0.376
42-46	42	+1.4046	0.9192	0.0449	8.98	5	1.764
46-50	46	+1.8065	0.9641	0.0220	4.40	4	0.036
50-54	50	+2.2084	0.9861	0.0094	1.88	0	1.880
54-58	54	+2.6103	0.9955	0.0045	0.90	3	4.990
				1.0000	200.00	200	13.005

$$\chi^2 = (2-2.72)^2/2.72 + (6-4.30)^2/4.30 + (5-8.98)^2/8.98 + \ldots + (7-7.18)^2/7.18 = 0.191 + 0.672 + 1.130 + 0.948 + 0.194 + 0.178 +$$
$$0.339 + 0.008 + 0.359 + 0.376 + 1.764 + 0.005 = 6.164$$

The degrees of freedom will be $k - p - 1$, where p represents number of parameters estimated from the sample, $k =$ number of class intervals used in fitting the distribution. In normal distribution, μ and σ^2 are estimated by $\bar{x}$ and s^2, and the degrees of freedom would be $k - 3$ (Ostle, 1988; Cooper and Weekes, 1983). In our example, $k = 12$, and hence the degrees of freedom is $12 - 3 = 9$. The $\chi^2_{0.05,9} = 16.9$ from the table (see Appendix, Table A.6). Since the calculated value is less than the table value, the χ^2 value is not significant indicating no difference between observed and expected frequencies. Thus the grain yield/plant data follow normal distribution. The fitted equation is reported as:

$$Y = N / 9.9528\sqrt{2\pi e} - (x - 28.02)^2 / 198.1165$$

The observed points and the fitted equation are also shown in Figure 14.4.

Example 8 (binomial distribution—fitting binomial distribution and testing goodness of fit: A trial is conducted giving one or the other of two mutually exclusive categories. Let this trial be repeated n times and now we get n trials. Let us say that the occurrence of an event is called "success" and non-occurrence a "failure." Let p be the probability of success in one trial and q the probability of failure. The trials are independent. In n trials we get 0, 1, 2, ... n successes. We can expect in n trials r success and consequently $n - r$ failures. The r success and $n - r$ failures may occur in any pattern or sequence. The probability of success is the same for each independent trial. The probability of r successes in n trials is given by the expression $p(r) = (^nC_r) p^r q^{n-r} (r = 0,1,....n)$ where, (^nC_r) is the number of combinations of n things taken r at a time. The term (^nC_r) is the binomial coefficient of the rth term; thus the expected frequencies for different values of r can be computed. For example, when $r = 0$, the expected frequency is $(^nC_0) p^0 q^{n-0} \times N$,

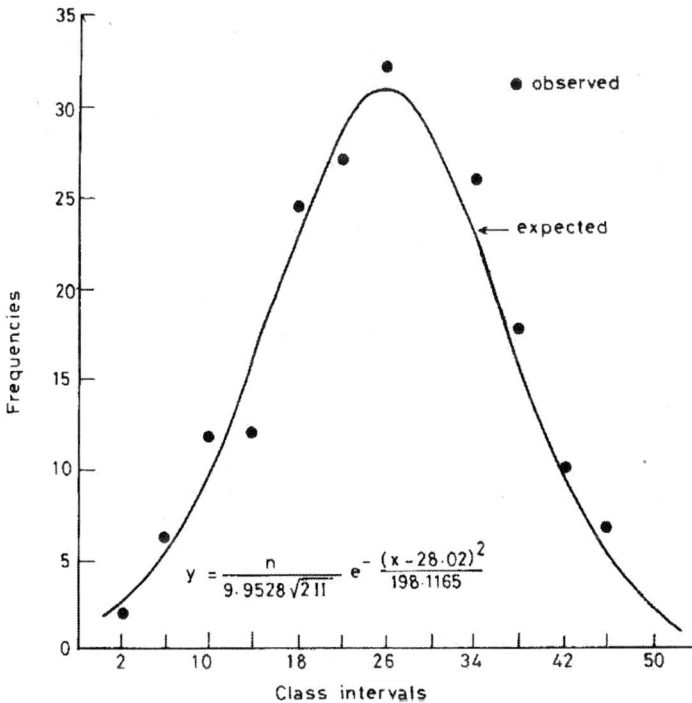

FIGURE 14.4. Observed points and the fitted normal curve for the distribution of grain yield per plant (g) data ($n = 200$).

when $r = 1$, $(^nC_1)\, p^1 q^{n-1} \times N$, etc. In binomial experiments, there are only two possible outcomes such as yes or no, pass or fail, dead or alive, male or female, defective or nondefective. By assigning 1 for success and 0 for failure, the distribution may be made quantitative. The mean (μ) and variance (σ^2) for the binomial distribution are np and npq. When $n = 1$, the binomial distribution is called point binomial. For point binomial, $\mu = p$ and $\sigma^2 = pq$.

Characteristics of the Binomial Variable

1. In n trials, x is number of successes of the random variable.
2. The trials are independent.
3. The trials are identical.
4. There will be only two possible outcomes—one is success and the other is failure.
5. The probability of a success does not vary from trial to trial. The probability of success is denoted by p and the failure by q such that $p + q = 1$.

Suppose we have $p = 0.1$, $q = 0.9$, and $n = 5$. Find the probability for $p(0)$, $p(1)$, $p(2)$, $p(3)$, $p(4)$, and $p(5)$.

Calculation: Formula to be used: $p(r) = (^nC_r)\, p^r q^{n-r}$ ($r = 0, 1, 2, 3, 4, 5$)

$p(0) = (^5C_0)\,(0.1)^0\,(0.9)^{5-0} = 1(0.1)^0(0.9)^5 = 0.59049$
$p(1) = (^5C_1)\,(0.1)^1\,(0.9)^4 = 5(0.1)^1(0.9)^4 = 0.32805$
$p(2) = (^5C_2)\,(0.1)^2\,(0.9)^3 = 10(0.1)^2(0.9)^3 = 0.07290$
$p(3) = (^5C_3)\,(0.1)^3\,(0.9)^2 = 10(0.1)^3(0.9)^2 = 0.00810$
$p(4) = (^5C_4)\,(0.1)^4\,(0.9)^1 = 5(0.1)^4(0.9)^1 = 0.00045$
$p(5) = (^5C_5)\,(0.1)^5\,(0.9)^0 = 1(0.1)^5(0.9)^0 = 0.00001$
Total 1.00000

The data can be shown in a graph (see Figure 14.5).

The probability can be obtained directly from the table of binomial cumulative distribution. From the tables, we read $n = 5$, $r = 0, 1, 2, 3, 4, 5$ as follows.

$p(0) = 0.590$
$p(1) = 0.329\ (0.919\text{--}0.590)$
$p(2) = 0.072\ (0.991\text{--}0.919)$
$p(3) = 0.009\ (1\text{--}0.991)$
$p(4) = 0$
$p(5) = 0$
Total $= 0.991$

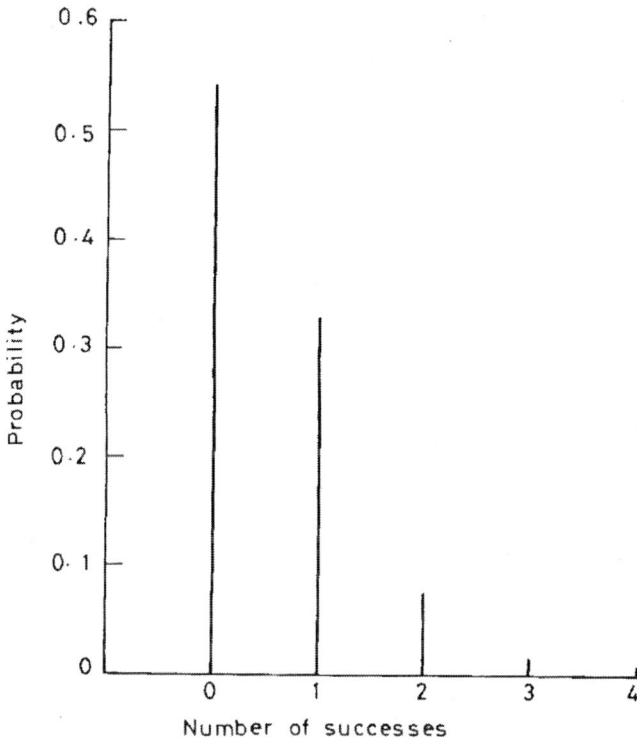

FIGURE 14.5. Distribution of probabilities for different number of successes.

The discrepancy between the computed probability and the probabilities obtained from the table is negligible.

Example 9: Five seeds of corn were sown in 100 petri dishes. The numbers of seeds germinated after a week were as follows:

Number of seeds germinated	0	1	2	3	4	5	Total
Frequency		1	6	14	33	31	15	100

To find whether the data fit binomial distribution, we have to compute the probability and the expected frequencies following binomial distribution.

Solution: Mean = $\Sigma f_i x_i / \Sigma f_i$

$\Sigma f_i x_i / \Sigma f_i = [(0)(1) + (1)(6) + (2)(14) + (3)(33) + (4)(31) + (5)(15)]/100 =$ 332/100 = 3.32,

i.e., $np = 3.32$. Therefore, $p = 3.32/5 = 0.664$ and $q = 1 - p = 1 - 0.664 = 0.336$.

Then we calculate the probabilities and expected frequencies:

Calculation

Formula to be used: $p(r) = (^nC_r)\, p^r q^{n-r}$ $(r = 0, 1, 2, 3, 4, 5)$

$p(0) = (^5C_0)(0.664)^0 \,(0.336)^{5-0} = 1\,(0.664)^0\,(0.336)^5 = 0.0043$
$p(1) = (^5C_1)(0.664)^1 \,(0.336)^4 = 5\,(0.664)^1\,(0.336)^4 = 0.0423$
$p(2) = (^5C_2)(0.664)^2 \,(0.336)^3 = 10\,(0.664)^2\,(0.336)^3 = 0.1673$
$p(3) = (^5C_3)(0.664)^3 \,(0.336)^2 = 10\,(0.664)^3\,(0.336)^2 = 0.3305$
$p(4) = (^5C_4)(0.664)^4 \,(0.336)^1 = 5\,(0.664)^4\,(0.336)^1 = 0.3266$
$p(5) = (^5C_5)(0.664)^5 \,(0.336)^0 = 1\,(0.664)^5\,(0.336)^0 = 0.1290$
Total = 1.0000

Form the following table to work out χ^2 statistic:

Number of germination	0	1	2	3	4	5
Observed frequencies	1	6	14	33	31	15
Expected frequencies	0.43	4.23	16.73	33.05	32.66	12.90

Note that the expected frequencies were obtained by multiplying the probabilities with the total frequencies, i.e., 100. For example, the expected frequency for class 0 is obtained as $0.0043 \times 100 = 0.43$. The chi-square statistic is calculated in the usual manner. The value of chi-square statistic is computed as .043 with degrees of freedom $4 - 1 = 3$. Since the chi-square statistic is not significant, the data follow binomial distribution.

Example 10. In a survey on sex ratio in a village, 1,600 families each with three children were investigated. The following is the frequency distribution of the number of boys.

Number of boys	Number of families observed
0	180
1	600
2	620
3	200
Total	1,600

Assume that the probability of getting a male child is 1/2, i.e., $p = 1/2$. Then the probability of getting male children in the family of three children is ^3C_r $(1/2)^3$. To obtain the expected frequency under the hypothesis for $r = 0, 1, 2,$ and 3, we multiply the corresponding probabilities with the total number of families observed. The expected frequencies formula to be used for finding probabilities:

$p(\mathrm{r}) = (^nC_r)\, p^r q^{n-r}\ (r = 0,1,2,3)$
$p(0) = (^3C_0)\, (1/2)^0\, (1/2)^3 = 1{\times}0.125{\times}1600 = 200$
$p(1) = (^3C_1)\, (1/2)^1\, (1/2)^2 = 3{\times}0.125{\times}1600 = 600$
$p(2) = (^3C_2)\, (1/2)^2\, (1/2)^1 = 3{\times}0.125{\times}1660 = 600$
$p(3) = (^3C_3)\, (1/2)^3\, (1/2)^0 = 1{\times}0.121{\times}1600 = 200$
$\chi 2 = (180-200)2/200 + (600-600)2/200 + (620-600)/600 + (200-200)/2$
$= 0.5000 + 0 + 0.6667 + 0 = 1.6667$

1. The number of classes is 4. The total of the observed and theoretical frequencies agree. The *df* is $4 - 1 = 3$. From χ^2 table, $\chi^2{}_{0.05}$ for 3 *df* = 7.82. The calculated χ^2 value is therefore not significant and hence the observed frequency distribution is consistent with the hypothesis.
2. The probability of a male child is estimated from the given data considering binomial distribution.

Mean $= [0{\times}180) + (1{\times}600) + (2{\times}620) + (3{\times}200)]/1600 = 2440/1600$ $=1.525$, i.e., $np = $ (mean of the binomial distribution) $=1.525$, $p = 1.525/3 = 0.5083$ and $q = 1 - p = 1- 0.5083 = 0.4917$, so that $p + q = 1.0000$.

Fitting binomial distribution when $p = 0.5083$ and $q = 0.4917$ is shown in Table 14.3.

$\chi^2 = (180–190.3)^2/190.1 + (600–590.0)^2/590.0 +$
$(620–609.6)^2/609.6 + (200–210.1)^2/210.1$
$= 106.69/190.3 + 100/590 + 108.6/609.6 + 102.01/210.1$
$= 0.5575 + 01695 + 0.1774 + 0.4855 = 1.3899$

The calculated χ^2 is less than χ^2 value from the table value of

$\chi^2{}_{0.05,\,3} - 7.82$

The observed and expected frequencies agree.

TABLE 14.3. Expected frequencies.

r	^nc_r, i.e., $3C_r$	$p^r q^{n-r}$	$^nc_r, p^r q^{n-r}$	$N \times ^nc_r\, p^r q^{n-r}$
0	1	$(0.5083)^0(0.4917)^3$	1×0.1188	190.3
1	3	$(0.5083)^1(0.4917)^2$	3×0.1229	590.0
2	3	$(0.5083)^2(0.4917)^1$	3×0.1270	609.6
3	1	$(0.5083)^3(0.4917)^0$	1×0.1313	210.1

Contingency Table

In a contingency table, enumeration data are collected simultaneously on two variables. The frequencies of occurrence in the different categories of one variable are shown in columns. Likewise, frequencies of another variable of different categories are shown along the row. The variables are said to be associated when numbers in the cells of the contingency table are not randomly distributed (LeClerg et al., 1962). The number of rows is denoted by r and the columns by c and such a table is called $r \times c$ contingency table (Choi, 1978). By using this table, the compatibility of observed and expected frequencies is tested.

Conditions for Applying χ^2 Test

1. The total of the observed and expected frequencies must be equal.
2. The number of classes must be the same.
3. Chi-square test should not be applied to classes of data in which the expected frequencies are less than 5. In such cases, the expected frequencies must be combined with two or more adjacent groups to bring the expected frequency to above 5 (Chatfield, 1970; Fisher, 1946; Anderson and Bancroft, 1952).
4. Chi-square test should not be used for comparing the data on proportions.

TEST OF INDEPENDENCE (OR TESTING FOR ASSOCIATION)

Association refers to the occurrence of two attributes together. Two events are said to be associated when the presence of one event is dependent upon the presence of the other. Here we deal with the nominal variability, which simply involves the number of occurrences of frequency counts in each category. The degree of relationship is not indicated by the test. In linear correlation, the relationship is indicated by corresponding changes in certain directions of the variables. Here the corresponding changes are not meaningful as the variables occur in frequencies.

Chi-Square Test of Independence in a 2 × 2 Contingency Table

Let there be a population with two variables, *A* and *B*. There are only two categories for each variable. Here the null hypothesis is: No association or relation between *A* and *B*, i.e., *A* and *B* are independent. We study the relationship. No correlation coefficient is used.

First Method

We want to study the relationship between eye color and hair color in a student population. We examine a random sample of 147 students; the results are shown in Table 14.4.

H_0: Row and column variables are independent.
H_1: Row and column variables are associated.

Computation of Expected Cell Frequencies. The probability that one has blue eyes = (71/147). The probability that one has fair hair is 75/147. The probability that an individual has both blue eyes and fair hair (according to multiplication law of probability) (Kohout, 1974) is 71/147 × 75/147. The expected frequency for this group is (71/147) × (75/147) × 147 = 0.1483 × 0.510 × 147 = 36.224. The expected frequency = (column total/grand total) × (row total/grand total) × grand total. The other expected frequencies in the cells are computed by subtraction. The concerned expected frequencies are given in brackets.

$$\chi^2 = (40 - 36.224)^2/36.224 + (35 - 38.776)^2/38.776 + (31 - 34.776)^2/34.776 + (41 - 37.224)^2 37.224 = 0.3936 + 0.3677 + 0.4100 + 0.3830 = 1.5543$$

TABLE 14.4. 2 × 2 contingency table.

Haircolor	Blue eyes	Not blue eyes	Total
Fair hair	40 (36.224)	35 (38.776)	75
Not fair hair	31 (34.776)	41 (37.224)	72
Total	71	76	147

The degrees of freedom is $(r-1)(c-1) = (2-1)(2-1) = 1$. The χ^2 value for 1 *df* at 5 percent level of significance from χ^2 table is 3.84. Since the calculated χ^2 is lesser than the tabulated value, H_o is not rejected, i.e., there is no association between the two attributes studied. In other words, two factors are independent.

Second Method (Shortcut Method)

In a fourfold problem, the difference between $(o_i - e_i)$ observed and expected frequencies is the sum for all cells, the formula for chi-square is

$$\chi^2 = (o_i - e_i)^2 \, \Sigma(1/e_i) = (3.776)^2 \, (1/36.224 + 1/38.776 + 1/34.776 + 1/37.224) = 14.258 \times 0.1090 = 1.5543 \text{ (as before)}$$

In other words, the chi-square value is equal to the common difference squared times the sum of the reciprocals of the four expected frequencies.

Third Method (Without Correction Factor) (Box et al., 1978)

$$d^2 = \frac{N(ad - bc)^2}{(a+b)(a+c)(b+a)(c+d)} \text{ where}$$

$N = 147$; $ad = 1640$; $bc = 1085$; $a+b = 75$; $a+c = 71$; $b+d = 76$; $c+d = 72$
$\chi^2 = 147(1640 - 1085)^2/75 \times 71 \times 76 \times 72 = 45279675/29138400 = 1.554$
    (as before within rounding errors)

Fourth Method with Correction Factor

We can also compute χ^2 without finding expected frequencies using the formula,

$$\chi^2 = N \, \{/ad - bc/ - N/2\}^2/\{(a+b)(a+c)(b+d)(c+d)\}$$
$$= 147(1640 - 1085 - 147/2)^2/75 \times 71 \times 76 \times 72$$
$$= 34088810/29138400 = 1.169$$

χ^2 *Computation for any Contingency Table (for 2 × 4 Table)*

Example 1: Table 14.5 gives the distribution of 884 fields (selected in a random sample of paddy crop) classified according to (1) type of manure, and (2) type of irrigation. Test whether the type of manure is independent of supply of irrigation.

The column variable is method of manuring. The column totals show the distribution of this variable, i.e., there are 534 fields with no manure, 304 fields with farm yard manure (FYM), 14 with oil cakes, and 32 with other types of manures. The row variable consists of two categories of irrigation and the row totals show the number of fields irrigated (224) and unirrigated (660). Using the cell frequencies, we can study their association by χ^2 test. The expected frequencies for each cell are calculated as before and they are shown in brackets.

$$\chi^2 = (121 - 135.31)^2/135.31 + \ldots + (18 - 23.89)^2/23.89 =$$
$$204.7761/135.31 + 15.761/77.03 + 19.803/3.55 + 34.692/8.11 +$$
$$204.778/398.69 + 15.761/226.97 + 19.803/10.45 + 34.692/23.89$$
$$=1.513 + 0.205 + 5.578 + 4.277 + 0.5124 + 0.069 + 1.895 + 1.452 =$$
$$15.503$$

The degrees of freedom for a $r \times c$ table is $(r–1)(c–1) = (2–1)(4–1) = 3$. χ^2 value for $\chi^2_{3,0.05} = 7.815$. Since the computed χ^2 value exceeds the table value at 5 percent level, we reject the H_o and conclude that irrigation and manure are associated.

Example 2: In a study, performance in an examination in a statistics course (pass or fail) and sex (male or female) are entered in a $r \times c$ table. Our objective is to find whether a relationship exists between the two criteria of classification. In the study, the following data are shown in Table 14.6. Figures in parentheses indicate the expected frequencies for each cell.

$$\text{Chi-square} = (24 - 19.8)^2/19.8 + (31 - 35.2)^2/35.2 + (8 - 12.2)^2/12.2$$
$$+ (26 - 21.8)^2/21.8 = 0.88 + 0.50 + 1.45 + 0.81 = 3.64$$

TABLE 14.5. 2 × 4 contingency table.

	No manure	FYM	Oil cakes	Others	Total
Irrigation	121 (135.31)	81 (77.03)	8 (3.55)	14 (8.11)	224
No irrigation	413 (398.69)	223 (226.97)	6 (10.45)	18 (23.89)	660
Total	534	304	14	32	884

Note: Numbers in parentheses indicate expected frequencies for each cell.

FYM = farm yard manure

TABLE 14.6. 2 × 2 contingency table.

	Pass	Fail	Total
Male	24 (19.8)	31 (35.2)	55
Female	8 (12.2)	26 (21.8)	34
Total	32	57	89

The degrees of freedom $(r-1)(c-1) = 1$. The $\chi^2_{3,0.05} = 3.84$ (from table). The result is not significant. It indicates that sex and performance in the studies are independent of each other, i.e., they are not related. Therefore, it gives sufficient grounds to accept the null hypothesis.

Testing Two Population Proportions

By Z Test

Example 1. For the study, we have 32 students who passed and 57 students who failed. There are two populations, one of passed students and the other failed. H_o: proportion of pass = proportion of fail, i.e., difference between the two proportions is zero.

Given $p_1 = 24/32 = 0.75$ $p_2 = 31/57 = 0.544$ $p = 55/89 = 0.6180$ $q = 0.3820 = (1-p)$; $N_1 = 32$ $N_2 = 57$ $N = N_1 + N_2 = 89$,

$$= \frac{p_1 - p_2}{\sqrt{\{pq(N_1 + N_2)\}/(N_1 N_2)}} = \frac{p_1 - p_2}{\sqrt{(pqN)/(N_1 N_2)}}$$

$$= \frac{0.75 - 0.544}{\sqrt{[(0.6180) \times (0.3820) \times (89)]/(32 \times 57)}}$$

$z^2 = 3.686$, which agrees with the value given previously for χ^2. The z value can be obtained by taking the square root of chi-square value.

Limitations on the Use of Chi-Square

1. The χ^2 test is to be used for the data expressed in terms of frequencies. Chi-square test should not be used for data expressed in proportions or other derived measures (Dayton and Stunkard, 1971).
2. For data collected as repeated measurements, χ^2 test should not be applied. For analysis for such data, one may consult Siegel (1956).

3. For 2 × 2 tables, χ^2 is to be applied for samples larger than 40, otherwise a serious misinterpretation of the data may arise (Denenberg, 1976; Dayton and Stunkard, 1971).
4. The group to which the observations are classified is mutually exclusive.
5. All observations are independent of all other observations.
6. It is necessary to use Yates correction when χ^2 is computed with 1 *df.*
7. The minimum expected frequency in a class should not be less than 5. We can combine adjacent classes with a frequency below 5 in order to reach or exceed this value.

PEARSON'S COEFFICIENT OF CONTINGENCY

Pearson developed a coefficient to show the degree of association. It is denoted by the letter C.

$$C = \frac{\sqrt{\chi^2}}{\sqrt{(N + \chi^2)}}$$

$C = 0$ if there is no association. It takes maximum value of 1 and it is used to compare association in two tables of the same group.

Example: In a sample study about gender and educational performance in two schools, A and B, the following data were observed (see Table 14.7). Is there any association between sex and success? If so, in which school is it greater?

Procedure: The expected frequencies were computed as before and they are shown in parentheses.

1. Calculate χ^2 values: χ^2 value for school A = 1.714
   χ^2 value for school B = 4.862. χ^2 value from tables is 3.84.
2. Inference: In school A, there is no association and in school B there is evidence of association.
3. Compute coefficient of contingency

$$C: (C = \sqrt{\chi^2 / (N + \chi^2)})$$

4. For school A:

$$C = \sqrt{1.714 / (100 + 1.714)} = 0.13$$

For school B:

$$C = \sqrt{4.862 / (100 + 4.862)} = 0.22$$

TABLE 14.7. Gender and educational performance.

	School A			School B		
	Male	Female	Total	Male	Female	Total
Pass	18 (15)	12 (15)	30	16 (11.25)	9 (13.75)	25
Fail	32 (35)	38(35)	70	29 (33.75)	46 (41.25)	75
Total	50	50	100	45	55	100

As C value in school B is greater, there is greater association between sex and educational performance.

Phi Coefficient

The association between two variables studied in a 2×2 contingency table is measured by phi coefficient (ϕ).

Computation: Consider the following data on sex and education.

Gender		Education	
Sex	Pass	Fail	Total
Male	24	31	55
Female	8	26	34
Total	32	57	89

Phi coefficient is computed as

$$= (ad - bc) / \sqrt{(a + b)(c + d)(a + c) + (b + d)}$$
$$\phi = (24 - 26 - 31 \times 8) / \sqrt{55 \times 34 \times 32 \times 57} = (624 - 248) / \sqrt{3410880}.$$
$$= 376 / 1846.9 = 0.2034$$

The phi coefficient takes values from 0 to 1. Zero indicates no correlation. It can be used for comparisons between two variables in different populations. The phi coefficient measures the correlation between two traits. The χ^2 statistic from the data is calculated:

Gender		Education	Total
	Pass	Fail	
Male	24(19.78)	31 (35.22)	55
Female	8(12.22)	26 (21.78)	34
	32	57	89

Figures in parentheses are expected frequencies.

$$\chi^2 = (24.0-19.78)^2/19.78 + (31-35.22)^2/35.22 + (8-12.22)^2/12.22 + (26-21.78)^2/21.78$$
$$= 0.9003 + 0.5056 + 1.4573 + 0.8176 = 3.6808$$

From χ^2 value, the phi coefficient can also be calculated using the formula

$$\phi = \sqrt{\chi^2 / n} = \sqrt{3.6808 / 89} = 0.2034$$

The association between two variables is also given by V. It is called Cramer's V.

$$V = \sqrt{\chi^2 / n \times df} = \sqrt{3.6808 / 89 \times 1} = 0.2034$$

The contingency coefficient, C, for the data, is calculated as

$$C = \sqrt{(\chi^2 / n + \chi^2)} = \sqrt{3.6801 / (89 + 3.6808)} = \sqrt{3.6808 / 92.6808}$$
$$= 0.1994$$

Summary for the data: Phi coefficient = 0.2034
Coefficient of contingency = 0.1993
Cramer's V coefficient = 0.2034
$\chi^2 = 3.6808$

The results show that the traits are not associated. The relationship is also low as indicated by the values of different coefficients. We may interpret these coefficients as we interpret the χ^2 value. For more details on phi coefficient, see Tai (1978).

HETEROGENEITY CHI-SQUARE ANALYSIS

In ten F3 families of a cotton cross 'Burma Ghost' × 'Kanpur White', segregating for the character leaf shape, the following numbers of plants in the narrow and broad classes were recorded.

					Families					
Number	114	96	88	68	54	83	113	95	100	117
Narrow	12	28	24	19	34	17	17	32	34	18
Broad	4	12	7	10	16	7	5	8	7	5

Now we have to test for agreement with 3:1 ratio for each family and hetero-geneity between families. Also, we have to test whether all the families are in agreement regarding the ratio of narrow to broad.

H_o: The ten families are homogeneous
H_1: The ten families are heterogeneous

To test for heterogeneity, first calculate the expected frequencies for each family for the ratio 3:1 and compute chi-square statistics for each family. The results are:

Family number	Narrow	Broad	Total	χ^2 value	df
114	12	4	16	0.0000	1
96	28	12	40	0.5333	1
68	24	7	31	0.0967	1
58	19	10	29	1.3908	1
54	34	16	50	1.3067	1
83	17	7	24	0.2222	1
113	17	5	22	0.0606	1
95	32	8	40	0.5334	1
100	34	7	41	1.3734	1
117	18	5	22	0.1304	1
Total	235	81	316	5.6480	10

Chi-square analysis for the pooled data for all families for the totals of 235 (narrow) and 81 (broad) is calculated from the expected frequencies of 237 and 79 (based on the 3:1 ratio) as $0.0169 + 0.0506 = 0.0675$.
Now the total chi-square value is 5.6480 with 9 *df:*

chi-square for totals $= 0.0675$ (for pooled data) with 1 *df*
heterogeneity chi-square $= 5.6480 - 0.0675 = 5.5805$ with 8 *df.*

Table value of chi-square for 8 *df* at $P = 0.05$ is 15.51. Since the calculated chi-square value is less than the tabulated value, heterogeneity is not signif-icant, we decide in favor of homogeneity of families, i.e., the families segre-gate in the ratio 3:1. Therefore, do not reject the null hypothesis (Zar, 1974).

Test for Equality of Two Variances

Testing of Sample Variance with That of a Known Variance

Palaniswamy (1990) studied the distribution of grain weight per panicle in the tallest two tillers in IR 22 rice variety grown without nitrogen application and established the variance as $0.1477g^2$, which can be considered as population variance σ^2_0. A sample of size 20 is taken and the observations recorded in the sample are as follows:

2.4323	2.7662	2.6487	2.3612	3.1165
2.6854	2.4317	2.4474	2.4491	2.4430
2.4855	2.9894	3.1812	3.2927	2.4320
2.5329	3.1625	2.3934	2.9920	2.6664

The test statistic χ^2 is distributed as

$$\chi^2 = \frac{(n-1)s^2}{\sigma^2_0} \text{ with } n-1 \text{ } df, \text{ if } \sigma^2 = \sigma^2_0$$

The variance of the sample is estimated as 0.2221 g^2. Now we have to find out whether the population variance is maintained. The calculated value of the test statistic is

$$\chi^2 = \frac{(20-1)(0.0968)}{0.1477} = 12.45.$$

The critical value for a two-sided test is 30.14 (see Appendix, Table A.6) at $P = 0.05$. Because the calculated value is less than the tabular value, we do not reject the null hypothesis.

Another sample taken randomly from each plant gave the following measurements (g):

2.2006	2.1265	1.9455	3.1875	2.7151
2.5853	2.8345	2.6691	2.9971	2.4750
1.7887	2.4943	2.2631	2.0543	2.5351
2.2734	2.6673	1.9981	2.2326	1.7081

This sample gave a variance of 0.2221. The chi-square statistic is $[(20-1) \times (0.2221)]/0.1477 = 28.58$, which indicates that the random sample shows more variability. In comparative experiments, the random selection method for estimating the plant traits in different treatments is not advisable.

Sometimes we have to compare the variabilities of two populations. Common practice for plant breeders is to raise crops in fields for seed purposes. Let there be two fields cultivated for seed purposes in a crop. The breeder would prefer the field showing smaller variance in the traits, for example, plant height. We test the hypothesis that $\sigma_1^2 = \sigma_2^{\,2}$ (H_o). We take samples of n_1 and n_2 from the fields and find variances s_1^2 and s_2^2. Then the ratio s_1^2 / s_2^2 follows F distribution with (n_1-1) and (n_2-1) degrees of freedom if $\sigma_1^2 = \sigma_2^{\,2}$. If s_1^2/s_2^2 is greater than f_{n1-1}, n_{2-1} at $P = 0.05$, we reject H_o of equal variances. Here we take larger sample variance in the numerator and smaller mean square in the denominator.

Testing of Homogeneity of More Than Two Variances Using χ^2 Statistic

Testing homogeneity of variances is important in applied statistics. Scientists, particularly agricultural researchers, conduct their experiments across several years or seasons, and several locations and analyze the data using the ANOVA technique. Experiments are conducted in different crops, soils, weather conditions, etc. The experimenters are interested in studying the interaction of their treatments with the latter factors to arrive at a common recommendation. The average response depends upon these factors. If the experimental errors computed in the ANOVA are not homogenous, the average response cannot be determined. The assumption is that the error variances in different experiments are the estimates of the variance of the same population. Hence, testing of homogeneity of variances is imperative in the analysis of the data. Several methods are available to test the homogeneity of variances (Bartlett, 1937; Bishop and Nair, 1939; Stevens, 1936). The test recommended by Bartlett (1937) is illustrated with a numerical example.

Example: An experiment was conducted for four years in an agricultural research station to evaluate four different varieties of sorghum *(Sorghum vulgare)*. The design adopted was a randomized complete block design with five replications. The experiments were analyzed individually. It is assumed that the error variances are the estimates of the variance of the same population. The error sum of squares, error mean square, and the degrees of freedom are used in the chi-square test of significance. See Table 14.8.

Check

1. $27900.96 \times 12 = 334811.60$ (degrees of freedom are equal in each case)
2. $15.19111 \times 12 \times 2.30259 = 419.748$

Find pooled mean square:

$(s^{-2})\, s^{-2} = (1/\Sigma k_i)\, (\Sigma k_i s_i^2)$
$= [(12 \times 7491.71) + (12 \times 2830.93) + (12 \times 10793.80) +$
$(12 \times 6784.52)]/(12 + 12 + 12 + 12)$
$= 12 \times 27900.96/48 = 6975.24$

Check: $27900.96\,/4 = 6975.24$ (as in Table 14.8). Then, $\Sigma k_i \times \log_e s^2 = 48 \times 3.8435 \times 2.30259 = 424.80022$
The χ^2 statistic (Bartlett, 1937) is given by

$$\chi^2 = \frac{1}{C}\Sigma k_i \times \log_e s^2 - \Sigma(k_i \log_e s_i^2) \text{ with } k - 1 \text{ degrees of freedom}$$

where
k = number of mean squares being compared
C = Correction factor = $1 + 1/[3(k-1)]\} [\Sigma (1/k_i) - 1/ (1/\Sigma k_i) - 1/ \Sigma k_i]$
k_i = degrees of freedom associated with each mean square
Σk_i = total degrees of freedom
s_i^2 = individual mean square
s^{-2} = pooled mean square computed as $[1/\Sigma k_i] [(\Sigma k_i s_i^2)]$

TABLE 14.8. Obtaining pooled error.

Year df (k_i)	Error *SS*	Error *MS*	$\log_e s_i^2$	$f_i \log_e s_i 2$
1 12	89900.50	7491.71	3.87460×2.30259	12×8.92165
2 12	33971.20	2830.93	3.45194×2.30259	12×7.94840
3 12	129525.70	10793.80	4.03302×2.30259	12×9.28639
4 12	81414.20	6784.52	3.83155×2.30259	12×8.82249
Total 48	334811.60	27900.96	15.19111×2.30259	$12 \times 34.97893 =$ 419.74716

Calculation of C

$$C = 1 + 1/3(n-1) \, (\Sigma 1/k_i - 1/\Sigma k_i = 1 + 1/9 \, (4/12 - 1/48)$$
$$= 1 + 1/9(15/18) = 149/144$$

Calculation of χ^2

$$\chi'^2 = [(\Sigma k_i \log_e s^2 - \Sigma k_i \log s_i^2\} = 424.80022 - 419.74716 = 5.05306$$
$$\chi^2 = \chi'^2/C = 5.05306/(149/144) = 4.883$$

Now referring to the chi-square table for 3 degrees of freedom (four individual mean squares are involved), we get the critical value as 7.82 at $P = 0.05$. Hence the calculated χ^2 value is not significant at $P = 0.05$. Therefore we can conclude that the four experimental error mean squares estimate the common population variance σ^2. The error mean squares are therefore homogeneous, and hence we can consider the pooled variance of 6975.24 in the comparison of treatments.

EXERCISES

14.1. Define the terms *observed frequency, expected frequency,* and *test of independence.*

14.2. Expand the factor $(a + b)^5$. Write down the binomial coefficients.

14.3. What is the term containing b10 in the expansion $(a + b)^{15}$?

14.4. What are the two general requirements for use of chi-square goodness of fit test?

14.5. In an orchard, there are 1,040 female trees out of the total of 2,000 trees. Do you think the sex ratio is 1:1?

14.6. In a random sample of ten babies, there are nine boys and one girl, Our assumption is boys and girls are equally likely. Apply χ^2 goodness of fit test using correction factor and without using correction for continuity.

14.7. How does the term association in chi-square problems differ from that of measurement data?

14.8. What conditions should the expected frequencies satisfy for the application of χ^2 test of significance?

14.9. In an experiment on the cross 'Red Pigmented Kaliyan' × 'White Pigmented Amaralo 15' in *Nicotiana tobaccum,* the following frequencies were obtained in F2 in respect to segregation for stem color and corolla color:

Stem creasy and corolla pink CP = 69
Stem creasy and corolla white Cp = 25
Stem normal and corolla pink cP = 25
Stem normal and corolla white cp = 5

Test whether the segregation ratio 9:3:3:1 for CP, Cp, cP, and cp fit the data.

14.10. In an experiment on genetics, Mendel crossed two types of peas and counted the seeds as shown here. Data include Yellow round 315; Yellow wrinkled 101; Green round 108; Green wrinkled 32. Mendal's law of segregation states that the frequencies should be in the ration 9:3:3:1. Does the data support the theory?

14.11. In an experiment in rice crop, four varieties were kept in a green-house. Insects were kept in a box and on release they go to any variety. The number of insects choosing the varieties A, B, C, and D are given here: A = 100, B = 200, C = 300, and D = 400. Test whether the probability of selection of the varieties by the insects is a 0.25. What additional information will you get from the experiment?

14.12. In a breeding experiment, one observes 1,500 seedlings. According to theory, the seedlings should segregate in the ratio 1:2:1. The observed frequencies in the three types were 401, 780, and 389. Do you think that the observed frequencies deviate from those of expected as per law of segregation?

14.13. In a university, three professors taught a statistics course and awarded the following grades to the students. Do the professors differ in grading their students?

Professor	Grade				
	A	B	C	D	E
I	25	44	132	23	27
II	39	56	216	28	33
III	19	46	182	24	21

14.14. The department of plant breeding in a university released three varieties of corn (C1, C2, and C3) and wanted to know the performance from the farmers. A survey was conducted on 150 randomly selected farms. The data obtained are given here:

	Variety C1	Variety C2	Variety C3
No. of farmers preferred	68	43	39

Find out whether the varieties are equally preferred by the farmers. (Ans. χ^2 = 9.01).

14.15. Given here is a 2 × 2 contingency table showing sex and the results of an examination (pass or fail).

Sex	Result		Total
	Pass	Fail	
Male	891	569	1460
Female	666	290	956
Total	1557	859	2416

What is the percentage of pass for male? female? Also find the expected frequency in each cell.

14.16. Sex and blood group data for 400 persons are given.

Sex	Blood Group			
	O	A	B	AB
Male	100	40	45	10
Female	10	35	55	5

Analyze the data and state your inference.

14.17. In a survey on smoking in a village, the interviewed persons gave the following data:

Habit	Sex		Total
	Male	Female	
Smoking	136	72	208
Nonsmoking	88	104	192
Total	224	176	400

Compute χ^2, phi coefficient, and Yule's Q Coefficient and interpret your results.

14.18. In a cattle farm, twenty-two animals were identified with the same disease and with same degree. An inoculation was administered to ten and the remaining 12 were kept as control. The results are shown here:

Treatment	Results		Total
	Recovered	Dead	
Inoculated	7	3	10

Not inoculated (control)	3	9	12
Total	10	12	22

Hint: $\chi^2 = 2.82$.

14.19. In an experiment, the following contingency table was obtained. We want to test the association between row and column classification:

	Column 1	Column 2	Column 3
Row 1	9	34	53
Row 2	16	30	25

    a. State the null hypothesis and also the alternate hypothesis.
    b. Show the rejection region in a graph.
    c. Find the expected counts for each cell assuming association.
    d. Interpret your results.
    e. Convert all the observations to percentage and comment.

14.20. A problem with four groups gave a χ^2 value of 1.04. Is it significant? Give your reasons (the df is $4 - 1 = 3$). (Hint: A value with 3 $df = 1.04$ is not significant. A value would be significant at $P = 0.05$ and a value of 12.5 would be significant at both 5 percent and 1 percent, respectively.)

14.21. An anthrax immunization project on goats yielded the following data:

Treatment	Died	Survived	Total
Inoculated	2	10	12
Not inoculated	6	6	12

Is vaccination effective? ($\chi^2 = 1.69$)

14.22. From a table, 200 digits were taken randomly and their frequencies are shown in the table here:

Digit	0	1	2	3	4	5	6	7	8	9
Frequency	22	21	16	20	23	15	18	21	19	25

Do you think the digits are distributed equally? (Ans: $\chi^2 = 4.2$)

14.23. The count data of the incidence of an event are 0, 0, 2, 2, 1, 1, 2, 0, 1, 0, 1, 2. Given $np = 17$. Do these data follow Poisson distribution?

14.24. In an experiment, 671 plants with green foliage and 569 with yellow foliage were observed. This is a backcross showing 1:1 ratio. Do you

think that the observed frequencies are compatible with that of theory? ($\chi^2 = 8.39$)

14.25. The lengths (in cm) of 50 radishes from a uniformity trial are given here:

31	30	23	20	23	20	28	28	24.5	26.5
23	18	20	24	17	29	37	31	27	22.5
28	22	23	22	29	20	30	25	22.5	19.5
22	25	19	18	21	23	23	27.5	26.5	23.5
23.5	25	35	23	29	17	23	32	23.5	34

Do you think that the lengths of the radishes are normally distributed?

14.26. In an extensive orchard there are 1,000 trees grown under two different conditions, shaded and unshaded. The trees are low yielding and
high yielding; categories are as follows:

	Shaded	Unshaded	Total
Low yielding	250	195	445
High yielding	350	205	555
Total	600	400	1000

Do you think shading had any effect on the yields of trees?

14.27. Length of ear and grain number in 3,000 wheat ear heads are given here:

	Number of grains/ear			Total
Length(cm)	20-29	30-39	40-49	
7-8.9	40	30	10	80
9-10.9	50	40	20	110
11-12.9	30	50	30	110
Total	120	120	60	300

Are the ear head lengths and grain numbers independent?

14.28. In a village, 2,000 families from two groups, A and B, were randomly selected and the numbers of families taking tea and not taking
tea were recorded.

Habit	A	B	Total
Drinking tea	1240	160	1400
Not drinking tea	560	40	600
Total	1800	200	2000

Find whether drinking tea is associated with groups.

14.29. Some entomologists investigated yellow, short-leaved, and spruce pines in a certain forest to see how many were being seriously attacked by insects. Investigation of 250 trees of each species gave the following results:

Species	Seriously damaged	Not damaged	Total
Yellow	58	192	250
Short leaved	80	170	250
Spruce	78	172	250

Are insects attacking one of the species more than the others?

14.30. Fit a normal distribution for the data of daily milk yield for a herd of cows:

Milk yield (kg)	1-3	3-5	5-7	7-9	9-11	11-13	13-15	15-17
No. of cows	10	95	150	375	260	110	85	15

14.31. A rice breeder states that out of 100 seeds, 90 will germinate, 10 will fail to germinate. State your hypothesis about the breeder's claim and perform a test of hypothesis.

14.32. Given the following probability distribution:

$$p(x) = {}^5C_{x(0.7)} {}^x (0.3)^{5-x} \; (x = 0, \, 1 \ldots 5)$$

   a. State whether x is a discrete or continuous random variable.
   b. State the name of the probability distribution.
   c. Find the mean and standard deviation of x.

14.33. Define the term *point binomial.*

14.34. Discuss the important characteristics of binomial distribution.

REFERENCES

Anderson, A.L. and T.A. Bancroft (1952). *Statistical theory in research.* New York: McGraw Hill Book Company Inc.

Bartlett, M.S. (1937). Properties of sufficiency and statistical tests. *Proc. Roy. Soc., London, Series A.* 160:268-282.

Bishop, D.J and U.S. Nair (1939). A note on certain methods of testing for the homogeneity of a set of estimated variances. *Suppl. J. Roy, Statis. Soc.* 6:89-99,

Box, G.E.P, W.G. Hunter, and J.S. Hunter (1978). *Statistics for experimenters—An introduction to design, data analysis and model building.* New York: John Wiley and Sons.

Chatfield, C. (1970). *Statistics for technology.* London: Chapman and Hall Ltd.

Choi, S.C. (1978). *Introductory applied statistics in science.* Englewood Cliffs, NJ: Prentice Hall Inc.

Cooper, R.A. and A.J. Weekes (1983). *Data, models and statistical analysis.* New Delhi, India: Heritage Publishers.

Dayton, C.M. and C.L. Stunkard (1971). *Statistics in problem solving.* New York: McGraw Hill Book Company.

Denenberg, V.H. (1976). *Statistics and experimental design for behavioral and biological researchers—An introduction.* Washington, DC: Hemisphere Publishing Corporation.

Dixon, W.J. and F.J. Massey (1957). *Introduction to statistical analysis.* New York: McGraw-Hill Book Co., Inc.

Fisher, R.A. (1946). *Statistical methods for research workers,* Tenth edition. Edinburgh and London: Oliver and Boyd Ltd.

Kohout, F.J. (1974). *Statistics for social scientists.* New York: John Wiley and Sons, Inc.

LeClerg, E.L., W.H. Leonard, and A.G. Clark. (1962). *Field plot technique.* Minneapolis, MN: Burgess Publishing Company.

Ostle, B. (1988). *Statistics for research—Basic concepts and techniques for research workers.* Ames, IA: The Iowa State University Press.

Palaniswamy, K.M. (1990). Design of statistical (field) experiments and crop forecast in agriculture with special reference to rice. Doctoral thesis submitted to the University of Calicut, India.

Siegel, S. (1956). *Non-parametric statistics for the behavioral sciences.* New York: McGraw-Hill.

Stevens, W.I. (1936). Heterogeneity of a set of variances. *J. Gen.* 33:398-399.

Tai, S.W. (1978). *Social science statistics: Its elements and applications.* Santa Monica, CA: Goodyear Publishing Company, Inc.

Volk, W. (1969). *Applied statistics for engineers.* New York: McGraw-Hill Book Co.

Zar, J.H. (1974). *Biostatistical analysis.* Englewood, NJ: Prentice Hall, Inc.

PART II:
EXPERIMENTAL DESIGN

Chapter 15

Experimental Design

Intensive research occurs in almost all areas around the world, and researchers spend enormous amounts of money, time, labor, and expertise under different experimental situations. This shows the importance of experimental design. Researchers also run their experiments to find out something that is unknown (Mclean and Anderson, 1974). For example, in agriculture, scientists deal with several factors that contribute to higher production in crop plants. Crop production is influenced by factors such as seed, cultivation practices, irrigation methods and levels of irrigation, rainfall, temperature, humidity, pests and diseases, soil types and soil fertility levels, sunlight, solar radiation, farm management practices, different types of fertilizers, different doses of fertilizers, and also different combinations of two or more of these factors. Thus selection of a proper experimental design to meet the different objectives of an experiment is very complicated. Furthermore, scientists conduct their experiments in different types of media such as fields, laboratories, and greenhouses. For example, in agriculture itself, there are many disciplines, e.g., plant breeding, agronomy, soil science, agricultural chemistry, plant pathology, entomology, seed technology, plant physiology, economics, agricultural engineering, horticulture, vegetable crops, plantation crops, spices, fisheries, forestry, animal science, feeds and fodders, food science, agricultural extension, agricultural statistics, plant biochemistry, agricultural biotechnology, genetics, bioinformatics, etc. At present, interdisciplinary research is being done to tackle several problems at once to save time and expenditure. For example, in the plant breeding and genetics area, breeders conduct research on the evaluation of varieties for high yield, wide adoptability, resistance to specific factors such as pests and diseases, cold, temperature, soil salinity, water-logging conditions, drought conditions, cropping systems, grain quality, nutritional values, etc., in collaboration with other specialist scientists. Agronomists concentrate on time of sowing/planting, seed rate, spacing, doses of fertilizers and methods of application to different crops, water requirements, irrigation methods, weed control, land utilization (farm management), crop rotation, multiple cropping, mixed cropping, cropping systems, harvesting techniques, etc.

Experiments in other sciences are done in a similar manner. In some types of experiments, several treatments are compared simultaneously. Experiments may be repeated for confirmation of the results either in the same place or in different locations. Thus, design of an experiment for a specific situation is of central importance. Ill-designed investigations lead to false conclusions. Researchers realize the significance and importance of design because of the complexity of the experiments. Researchers should consult a statistician prior to the commencement of an experiment regarding several aspects of the design to obtain the maximum amount of information at minimum cost. The principles of experimental design apply to all areas of research equally.

EXPERIMENTAL DESIGN

Experimental design involves applied mathematics and applied statistics, and hence readers are expected to have basic knowledge in mathematics and statistics. The design of an experiment includes formulation of specific objectives of the experiment, the specific variables to be measured during the course of the experiment and method or methods of collection of various types of data to achieve the specific objectives, description of the treatments to be tested, the design selected and full details of experimental layout and experimental units, outline of the analysis of the data relevant to the design model selected, tabulation of data, statistical analysis, and drawing of inferences and conclusions. Design also deals with the different methods by which the treatments are placed in the experimental units. Some of the literature, i.e., Cox (1958), Finney (1955), Cochran and Cox (1957), Federer (1967), and Hicks (1973) deal with experimental design without involving much mathematics. We will define some of the terms used in the experimental design.

EXPERIMENTAL UNIT

The unit of material to which a treatment is applied is called an experimental unit or plot. A plot may be an area of land on which a crop is grown. It may be a patient, hospital, a piece of animal tissue, etc. Any quantitative measurement obtained from a plot is called yield or an observation. In an experimental unit, imposed treatment effects, and controls are recorded and compared. The unit differs from one science to another. For example, in agriculture, the units are plots, pots, petri dishes, animals, etc. If the researcher is interested in the effect of a fertilizer in a crop of corn grown in pots, then the pots would be the experimental unit. The important characteristic of the unit is that it exhibits variation from one unit to another, and this variation masks the real effect of the treatments. If the experimenter thinks that the treatment effects are slight or low, then he or she should take larger numbers of units to each treatment group to obtain a significant treatment mean square in the analysis of variance. Hence, the experimenters must understand the causes for the existence of such variability. This is explained with reference to agriculture later in this chapter. In the next section, we define the terms *accuracy* and *precision,* which are commonly used in statistics.

ACCURACY

Suppose we have a population consisting of four observations ($N = 4$) in a yield of a crop (kg/plot) in four different plots $X_1 = 15$, $X_2 = 17$, $X_3 = 18$, and $X_4 = 22$. Then we know the population mean μ (mu) is

$$\Sigma X_i \,/\, N = (15 + 17 + 18 + 22) \,/\, 4 = 72 \,/\, 4 = 18 \text{ kg/plot.}$$

If we take a sample of size $n = 2$, e.g., X_1 and X_3, then the sample mean is

$$(X_1 + X_3) \,/\, 2, \text{ i.e., } (15 + 18)/2 = 33/2 = 16.5 \text{ kg /plot.}$$

Then we say that the sample mean is inaccurate. If the sample consists of X_3 and X_4, then the sample mean is $(18 + 22)/2 = 40/2 = 20$, which is less accurate. Suppose our sample observations are X_2 and X_3, then the sample mean is $(17 + 18) = 35/2 = 17.5$, which is very close to the population mean μ. Thus accuracy refers to the nearness or closeness of the measurements obtained to their true value, i.e., the population mean. As μ is unknown, the actual accuracy is not assessable. Accuracy also refers to the absence of the bias.

PRECISION

Suppose we have a population of four values, and from it we take samples of size 2. We get $4C2 = [(4 \times 3)/2 \times 1] = 6$ samples and consequently 6 estimates of the same population. The extent to which these derived estimates differ from one another is measured by the standard error. This is called precision. The precision of any one estimate is given by $s \,/\, \sqrt{n}$, where

s = standard deviation and n = size of the sample. $s / \sqrt{n}$ is called standard error of the estimate. Lack of precision may also happen due to faulty technique in taking samples, etc.

BIAS

In our first example, when the sample elements are X_3 and X_4, the estimated sample mean is $(18 + 22)/2 = 20$. The difference between the estimated sample mean and the population mean is called bias. Thus in this sample, the bias is $20 - 18 = 2$. If there is no bias, we get 100 percent accurate results.

STUDY OF VARIABILITY AMONG PLOTS

Example 1

Consider ABCD is a field with a uniform corn crop partitioned into four equally sized plots (see Figure 15.1). The observed plot yields are 10, 12, 8, and 14 kg/plot. Under the given soil and fertility conditions of the experiment, each plot has a certain true yield while the yield recorded is only the estimate of the true yield being subject to error. The observed observations, namely 10, 12, 8, and 14 kg/plot, are not the same though we expect the same yield in each plot. The cause or causes for such variation are not easily explainable. The unknown causes may be due to inherent variability in the soil fertility, variability among the several meteorological parameters that

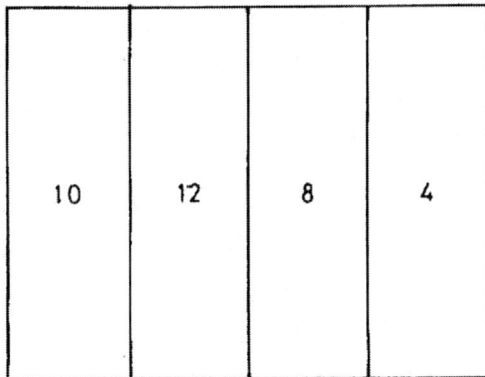

FIGURE 15.1. Figure showing the variation in yield (kg/plot) recorded in equal size plots and in the same crop.

exist in the experimental units, and other unknown causes. These variations are not only uncontrollable by the experimenter but also could not be completely eliminated. This variation is expressed in terms of variance (s^2), which is calculated as $\Sigma(x_i - \bar{x})^2 / (n-1) = \{(10-11)^2 + (12-11)^2 + (8-11)^2 + (14-11)^2 = 20/3 = 6\,2/3$ kg^2. The larger the deviations the more the variance.

Example 2

Consider ABCD is a field with two varieties of rice V1 and V2 in equally sized plots (see Figure 15.2). Two varieties are randomly allotted to the plots. The V1 gives yields 5, 6, 4, and 5 kg/ plot with a mean of 5 kg/plot and V2 gives 8, 10, 7 and 7 kg/plot with a mean of 8 kg/plot. The figures are shown in the dot diagram in Figure 15.3. The figures indicate (1) variation between varieties, and also (2) variation within varieties. This diagram therefore helps to show the central tendency and spread of the observations. It shows that V1 and V2 differ in their mean values. When we get more numbers of observations, we can study the distribution of V1 and V2 by forming a histogram, frequency polygon, and frequency curve, etc., and also the spread. The variation due to (1) differences between the varieties is called "known cause" as the experimenter used two varieties to study the comparison of these two treatments, and (2) the differences *within* a variety are of "unknown causes." The known causes are due to deliberate inclusion of two different varieties (also called treatments) and the unknown causes are unexplainable. The unknown causes may be minimized by proper selection of an appropriate design and following the principles of design. In animal science experiments, the known causes due to breed, diet, housing, location, etc., can be controlled but we cannot control the differences due to genetic constitution of the animals. In agriculture, by selecting appropriate design, the errors due to unknown causes may be minimized.

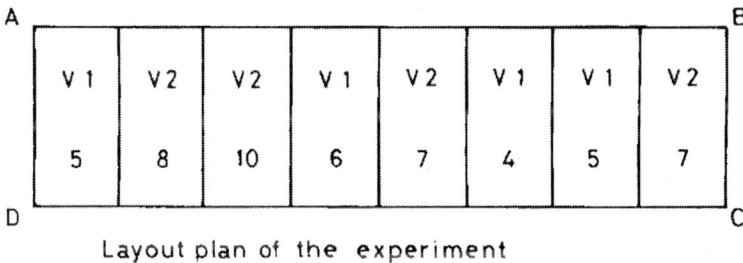

Layout plan of the experiment

FIGURE 15.2. The variation in plot yields recorded by different varieties. (Figures indicate grain yield in kg/plot.)

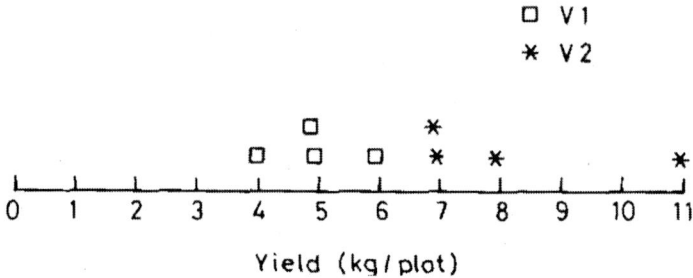

FIGURE 15.3. Dot diagram showing the yield figures of the two varieties, V1 and V2.

TREATMENT

Suppose an experimenter is interested in the evaluation of two drugs, i.e., in finding out of the efficiency of one drug over the other, then these two drugs in the experiment are called treatments.

Example 1

Suppose, we want to study the gain in weight in rats by administering three materials, namely no sucrose, liquid sucrose, and solid sucrose. Then the treatments in the experiment are (1) no sucrose, (2) liquid sucrose, and (3) solid sucrose. The rats are experimental units. The treatment with no sucrose can be termed as the control treatment.

Example 2

If a physician studies the effect of four treatment combinations of a drug on the cure of a disease in human beings, then the four combinations are called treatments.

Treatment Effect

How do we measure the treatment effect? In agriculture, scientists conduct experiments to determine the effect of fertilizer on crop production in the experimental units called plots. Researchers select sufficient numbers of replications and conduct the experiments. In one set, the fertilizer is applied and in another set, no fertilizer is applied. The difference in yields gives the treatment effect.

When assessing treatment effects, one has to consider several aspects, which are important and interesting. For example, in fruit crops, one has to consider not only the weight of the crop but also the number of fruits. In experiments involving application of fertilizers, cost as well as optimum dose must be taken into consideration. A variety of plant attributes, which are responsible for manifestation of treatment effect, must be considered. For example, in chilies (*Capsicum annuum* L.), while evaluating fruit yields of 73 genotypes, other characters such as fruit diameter, pericarp weight, seed weight, pedicel weight, number of fruits/plant, number of seeds/fruit, number of primary and secondary branches, etc., should also be considered as they show positive correlation with fruit yield (Usha Rani, 1996).

Fixation of Treatments

The treatments included in the experiment play an important role in the success of the experiment. Investigators should not forget to include a control treatment in the experiment. The control treatment does not receive any treatment. In medicine, it is called a placebo. The main purpose of a control is to provide a baseline for comparison with the experimental figures. The treatment effect is to be measured in the entire area of the experimental unit or a random sample taken from the experimental unit.

EXPERIMENTAL ERROR

When the same treatment is applied in two or more experimental units and treated alike in all respects, the yields as well as all other characters are expected to give identical values. But under actual conditions, we do not get the same or identical values. This is mainly due to variability that exists in the experimental units namely plots, human beings, rats, etc. Thus, failure of the same treatment to give identical results in different experimental units is called experimental error or simply error. If the experimental units exhibit greater variability, there will be greater error and consequently the treatment effects are completely or partially obscured. As a result, the error will lead to results that are not true. On the contrary, if the experimental error is small or negligible, we can detect even small differences among the treatments. The importance of the error should be understood by researchers who conduct experiments in controlled conditions also. Those who conduct their experiments in controlled greenhouses may feel that there is no experimental error in their studies. This is not correct. They should know it is unlikely that two plants in the same treatment grown side by side in pots will give exactly the same growth under identical conditions. Results also do not give the same measurements on other plant attributes. These differences are the result of unknown causes. Such a variation is called residual variation. This variation follows normal distribution.

Examples

1. If a corn crop is harvested from two plots of equal size or equal numbers of plants of a treatment in an experiment, the yields will seldom be equal.
2. The weights of two fruits of the same age harvested from a tree always differ.

Importance of Study of Experimental Error

In any ideal experiment, the observed yield or other variables of interest should reflect the effect of the treatment alone. However, certain unknown and uncontrollable factors contribute to error. In crop experiments, the main source of error is the inherent differences in soil fertility among the units (Goulden, 1959). The error can be minimized by adopting optimum plot size and shape. If the size is too small, even small errors will get magnified when the measurements are expressed in standard units. If the size is too large, one cannot manage the experiments as it involves more work and expenditure. Hence, the necessity for information on optimum size of the plot for the different types of experiments and crops.

Reasons for the Study of Plot Size and Shape

An experimental unit is measured by its size or area and hence, study of plot size is important for the following reasons:

1. It will help the experimenter to accommodate large number of varieties (treatments) in one experiment.
2. It will help the experimenter to make use of the experimental materials in an economic way particularly when they are in short supply.
3. Very small plots may not give reliable estimates of the treatments.
4. Large plots will involve increased cost.
5. When the plot size is optimum, supervision of the experiment can be done efficiently and it will lead to reliable estimates of the characters.
6. Optimum plot size and shape vary from crop to crop.

Method of Studying Optimum Plot Size and Shape
(Uniformity Trial for Grain Yield in Rice Crop)

Because the size and shape of a plot play a significant role in the control of experimental error, consequently increasing precision and minimizing cost, a uniformity trial is conducted. This is explained with rice crop. A large field representing the experimental station is selected and one crop (of all the same variety) is grown uniformly till harvest. A suitable small plot, called a basic unit, is fixed. The whole area is divided into as many basic units as possible. Finally, the yield is recorded separately for each basic unit, and analyses are done to determine the optimum size. Palaniswamy (1990) conducted studies in rice crop. In two selected fields, rice crop (*Oryza sativa* L.) was grown uniformly under two fertilizer levels, one with 0 fertilizer and another with 120 N kg/ha. The area of the experiment was 48.00 m by 9.60 m at the center of the field. The crop was planted at a spacing of 0.40 m by 0.20 m (Figure 15.4).

An area of 0.40 m by 0.20 m was taken as the basic unit (x) at harvest, and the whole experimental area was divided into 5,760 (120 by 48) basic units, and the grain yield recorded separately for each basic unit. The adjacent basic units were then combined to form plots of various sizes and shapes. The grain yields of the unit plots were then combined to form different plot sizes and shapes of plots consisting of varying numbers of basic units in order to study the amount, and pattern of variability. There were 120 basic units along the west-east direction and 48 across the north-south direction (see Figure 15.4). The coefficient of variation for each size and shape was calculated (see Table 15.1). The coefficients of variation were then plotted against the respective plot sizes. The goodness of fit between CV and plot size were determined by computing R^2 values. The optimum plot size was found at maximum curvature.

Fairfield Smith's (1938) equation $Y = ax^b$, where y is the coefficient of variation corresponding to plot size x and a and b were unknown parameters that were fitted to the data relating to the unfertilized and fertilized trials (applied in Table 15.1). The observed and predicted CVs for different plot sizes and the fitted equations are given in Table 15.1. Theoretically, the optimum plot size arrived at was about 3.20 m^2. The R^2 values, 0.9769 in unfertilized trial and 0.9740 in fertilized trial, were highly significant. They showed that about 97 percent of the soil variability is influenced by plot size. Thus plot size is an important character to be considered in laying out field experiments. The coefficients of variation did not reduce after a certain plot size (see Figures 15.5, 15.6, 15.7) and hence, there is no use in increasing the plot size after a certain plot size. Federer (1967) suggested that while

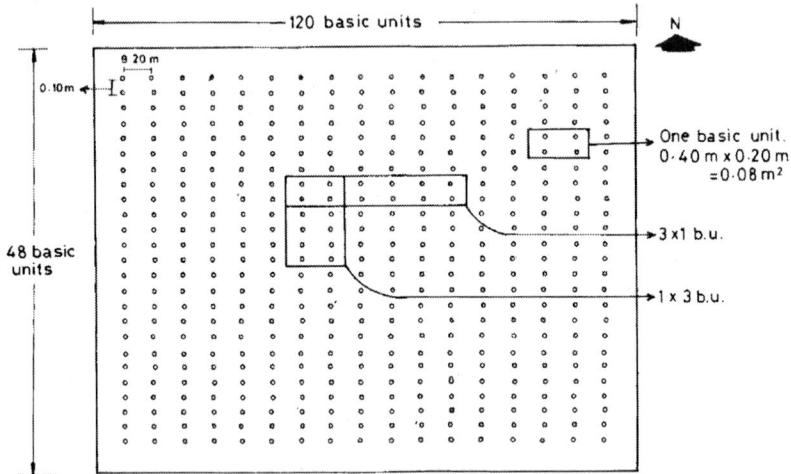

FIGURE 15.4. Field plan of uniformity trial with rice variety IR 20 grown with 20 cm spacing in between lines and 10 cm within lines.

fixing optimum plot size, practical considerations should also be taken into consideration. Very small plots do not give reliable estimates of the treatment responses (Sukhatme, 1947). Considering all practical aspects, an optimum plot size of 6.40 m^2 to 7.20 m^2 was recommended.

The coefficients of variation for different plot sizes and shapes are shown in Table 15.2.

Plot shapes did not exhibit much difference. For example, plot shapes of the same size, 1 by 12, 12 by 1, 2 by 6, 6 by 2, 3 by 4, and 4 by 3 recorded CVs of 9.1 percent, 9.7 percent, 9.3 percent, 9.4 percent, 9.4 percent, and 9.3 percent, respectively. They did not show much difference (see Table 15.2). However, slightly lesser variability was observed in rectangular plots. Rectangular shapes have been reported to be efficient (Murray, 1950; Brim and Mason, 1959; Torrie et al., 1958; Gopani et al., 1970). In tomato, Currence (1947) and Palaniswamy et al. (1975) observed that rectangular plots showed lesser variability and recommended rectangular-shaped plots for experiments in tomato. Fisher (1935) observed that each plot must represent the whole block and the plots must lie side by side as narrow strips. The differences between long narrow plots when they are placed side by side are usually less than those between square plots (Goulden, 1959).

TABLE 15.1. Number of basic units, size and shape of the plots, variance, and CV for the uniformity trial data. (Character: grain yield) Nitrogen levels: 0 and 120 N kg/ha.

No. of basic units (b.u)	Size and shape of the plot	0 N kg/ha			120 N kg/ha		
		Var/b.u	CV(%)	Mean CV(%)	Var/b.u	CV(%)	Mean CV(%)
1	1×1 (0.08)	3110.29	22.67	22.67	5339.31	23.86	23.86
2	1×2	1762.32	17.07		2995.55	17.65	
	2×1 (0.16)	1718.10	16.85	16.96	2807.41	17.21	17.43
3	1×3	1237.16	14.30		2111.10	14.92	
	3×1 (0.24)	1309.47	14.71	14.51	1988.86	14.48	14.70
4	1×4	1020.96	12.99		1448.75	13.58	
	4×1	1095.44	13.46		1664.23	13.25	
	2×2 (0.32)	1059.89	13.18	13.21	1638.23	13.14	13.32
5	5×1 (0.40)	993.51	12.81	12.81	1460.77	12.41	12.41
6	1×6	695.38	10.72		1260.25	11.53	
	6×1	889.83	12.13		1286.78	11.65	
	2×3	763.69	11.23		1223.13	11.36	
	3×2 (0.48)	848.56	11.84	11.48	1206.29	11.28	11.46
8	1×8	607.44	10.08		1022.64	10.39	
	8×1	776.18	11.33		1111.97	10.83	
	2×4	646.37	10.34		1070.93	10.63	
	4×2 (0.64)	734.34	11.02	10.68	1034.99	10.45	10.58
9	3×3 (0.72)	620.78	10.13	10.13	932.42	9.92	9.92
10	5×2	674.03	10.55		908.66	9.79	
	10×1 (0.80)	718.72	10.50	10.53	1011.88	10.33	10.06
12	1×12	438.59	8.51		790.31	9.13	
	12×1	612.52	10.06		897.15	9.73	
	2×6	466.06	8.78		822.40	9.31	
	6×2	608.65	10.03		830.16	9.36	
	3×4	536.19	9.41		830.16	9.36	
	4×3 (0.96)	545.96	9.50	9.38	814.01	9.27	9.36
15	5×3	503.73	9.12		717.67	8.70	
	3×5 (1.20)	584.25	9.83	9.48	807.01	9.23	9.07
16	1×16	306.18	7.11		636.58	8.19	
	2×8	407.79	8.21		690.21	8.53	
	8×2	547.30	9.51		749.19	8.89	
18	4×4 (1.28)	473.20	8.84	8.42	731.97	8.79	8.60
	3×6 (1.44)	387.48	8.80		661.55	8.35	
	6×3	449.18	8.62	8.31	670.38	8.41	8.38

TABLE 15.1 *(continued)*

No. of basic units (b.u)	Size and shape of the plot	0 N kg/ha			120 N kg/ha		
		Var/b.u	CV(%)	Mean CV(%)	Var/b.u	CV(%)	Mean CV(%)
20	5×4	443.21	8.56		656.74	8.32	
	10×2	506.03	9.15		650.48	8.28	
	20×1 (1.60)	541.73	9.46	9.06	718.94	8.71	8.44
24	1×24	252.80	6.49		534.65	7.37	
	24×1	444.35	8.57		607.38	8.00	
	2×12	310.10	7.16		567.00	7.73	
	12×2	449.85	8.62		598.84	7.95	
	3×8	349.05	7.60		581.94	7.83	
	8×3	409.59	8.23		608.57	8.01	
	4×6	353.28	7.64		596.32	7.93	
	6×4 (1.920)	391.41	8.04	7.79	609.06	8.01	7.85
30	5×6	324.68	7.33		543.46	7.57	
	10×3	378.64	7.91		542.69	7.57	
	15×2	435.78	8.49		534.81	7.51	
	30×1 (2.40)	447.04	8.60	8.08	560.84	7.69	7.59
32	2×16	221.41	6.05		455.90	7.60	
	4×8	314.57	7.21		528.47	7.47	
	8×4 (2.56)	355.42	7.66	6.97	564.95	7.72	7.60
36	3×12	266.55	6.64		490.98	7.20	
	12×3	328.89	7.37		511.78	7.35	
	6×6 (2.88)	289.23	6.91	6.97	509.37	7.33	7.29
40	5×8	298.59	7.02		477.35	7.10	
	10×4	331.90	7.41		505.34	7.30	
	20×2	415.57	8.29		479.74	7.11	
	40×1 (3.20)	380.40	7.93	7.66	544.93	7.58	7.27
45	15×3 (3.60)	316.74	7.24	7.24	453.80	6.92	6.92
48	1×48	133.35	4.59		388.02	4.40	
	2×24	184.76	5.53		382.26	6.35	
	24×2	340.12	7.50		405.67	6.54	
	3×16	185.05	5.53		403.09	6.52	
	4×12	244.94	6.36		439.23	6.81	
	12×4	291.82	6.94		477.08	7.28	
	6×8	256.29	6.51		468.08	7.05	
	8×6 (3.84)	269.21	6.67	6.20	471.95	7.06	6.50
60	5×12	234.54	6.23		396.08	6.46	
	10×6	246.73	6.39		422.91	6.68	

No. of basic units (b.u)	Size and shape of the plot	0 N kg/ha Var/b.u	CV(%)	Mean CV(%)	120 N kg/ha Var/b.u	CV(%)	Mean CV(%)
	15×4	282.23	6.83		425.98	6.70	
	20×3	307.43	7.15		417.60	6.64	
	30×2	356.99	7.68		368.58	6.23	
	60×1 (4.80)	313.76	7.20	6.91	442.22	6.83	6.59
64	4×16	173.84	5.36		367.17	6.22	
	8×8 (5.12)	243.67	6.35	5.86	442.94	6.83	6.53
72	3×24	164.05	5.21		351.62	6.09	
	24×3	241.04	6.31		346..27	6.04	
	6×12	208.44	5.87		390.74	6.42	
	12×6 (5.76)	216.51	5.96	5.84	400.84	6.50	6.26
80	5×16	161.87	5.17		328.84	5.89	
	10×8	231.87	6.19		394.10	6.45	
	20×4	275.99	6.75		394.66	6.45	
	40×2 (6.40)	300.76	7.05	6.29	381.89	6.35	6.29
90	15×6	211.76	5.92		359.12	6.15	
	30×3 (7.20)	248.56	6.41	6.17	312.55	5.74	5.95
96	2×48	109.01	4.24		305.61	5.60	
	4×24	151.94	5.01		330.49	5.90	
	24×4	209.27	5.88		334.10	5.94	
	8×12	197.24	5.71		363.30	6.19	
	12×8 (7.68)	203.52	5.80	5.33	375.97	6.30	6.00
120	5×24	143.61	4.87		300.96	5.63	
	10×12	193.74	5.66		329.23	5.89	
	15×8	199.52	5.74		334.90	5.94	
	20×6	212.59	5.93		329.25	5.89	
	30×4	226.40	6.12		299.99	5.62	
	40×3	206.29	5.84		331.58	5.91	
	60×2	254.22	6.48		316.54	5.78	
	120×1 (9.60)	234.52	6.23	5.86	295.55	5.58	5.78
128	8×16 (10.24)	129.99	4.64	4.64	315.97	5.77	5.77
144	3×48	105.02	4.17		294.40	5.57	
	6×24	124.18	4.53		302.09	5.64	
	12×12	161.57	5.17		310.48	5.72	
	24×6 (11.52)	145.29	4.90	4.69	271.56	5.35	5.57
160	10×16	134.94	4.72		277.60	5.41	
	20×8	201.13	5.77		313.27	5.75	

TABLE 15.1 *(continued)*

No. of basic units (b.u)	Size and shape of the plot	0 N kg/ha			120 N kg/ha		
		Var/b.u	CV(%)	Mean CV(%)	Var/b.u	CV(%)	Mean CV(%)
	40×4 (12.80)	181.22	5.47	5.32	326.46	5.87	5.68
180	15×12	164.16	5.21		270.31	5.34	
	30×6	161.45	5.17		237.20	5.00	
	6×30 (14.40)	168.14	5.27	5.21	269.98	5.34	5.23
192	4×48	94.66	3.96		278.07	5.42	
	8×24	119.77	4.45		295.85	5.59	
	24×8	142.91	4.86		267.84	5.32	
	12×16 (15.36)	106.01	4.19	4.37	266.27	5.30	5.41
240	5×48	91.81	3.90		255.20	5.19	
	10×24	119.82	4.45		256.21	5.20	
	15×16	113.69	4.33		224.35	4.86	
	20×12	176.62	5.40		251.92	5.15	
	30×8	154.54	5.05		236.15	4.99	
	40×6	127.00	4.58		267.55	5.31	
	60×4	145.52	4.90		264.81	5.28	
	120×2 (19.20)	201.16	5.77	4.80	189.37	4.47	5.06
288	6×48	78.33	3.60		256.26	5.20	
	12×24	91.42	3.89		248.42	5.12	
	24×12 (23.04)	112.88	4.32	3.94	203.87	4.64	4.99
320	20×16	114.02	4.34		212.32	4.73	
	40×8 (25.60)	119.47	4.44	4.38	269.28	5.33	5.03
360	15×24	98.75	4.04		212.37	4.73	
	30×12	138.63	4.79		178.07	4.33	
	60×6	101.26	4.09		209.33	4.70	
	120×3 (28.80)	121.13	4.47	4.35	146.38	3.93	4.42
384	8×48	80.08	3.64		266.26	5.30	
	24×16 (30.72)	73.71	3.49	3.57	162.00	4.13	4.72
480	10×48	85.91	3.77		229.78	4.92	
	20×24	115.10	4.36		200.11	4.59	
	30×16	91.40	3.89		142.75	3.88	
	40×12	114.45	4.35		209.18	4.70	
	60×8	91.45	3.89		220.94	4.83	
	120×4 (38.40)	104.63	4.16	4.07	153.77	4.03	4.49
576	12×48	68.88	3.37		228.52	4.91	
	24×24 (46.08)	67.67	3.34	3.36	153.71	4.03	4.46
640	40×16 (51.20)	69.01	3.38	3.38	176.36	4.31	4.31

No. of basic units (b.u)	Size and shape of the plot	0 N kg/ha			120 N kg/ha		
		Var/b.u	CV(%)	Mean CV(%)	Var/b.u	CV(%)	Mean CV(%)
720	15×48	75.56	3.53		197.02	4.56	
	30×24	89.19	3.84		132.96	3.74	
	60×12	93.87	3.94		170.29	4.24	
	120×6 (57.60)	60.90	3.17	3.62	102.49	3.29	3.96
960	24×40	91.67	3.89		189.05	4.47	
	40×24	64.55	3.27		172.79	4.27	
	60×16	40.01	2.57		133.63	3.75	
	120×8 (76.80)	57.64	3.09	3.21	112.95	3.45	3.99
1152	24×48 (92.16)	54.86	3.01	3.01	138.15	3.82	3.82
	30×48	74.28	3.50		122.53	3.60	
1440	60×24	50.19	2.88		140.40	3.85	
	120×12 (115.20)	62.90	3.22	3.20	66.43	2.65	
	30×48	62.06	3.20		181.95	4.38	3.37
1920	120×16 (153.60)	10.65	1.33	2.27	16.25	2.85	2.85
2880	60×48	40.38	2.58		165.14	4.17	
	120×24 (230.40)	16.56	1.65	2.12	22.61	1.54	2.86
5760	120×48 (460.80)	0	0	0	0	0	0

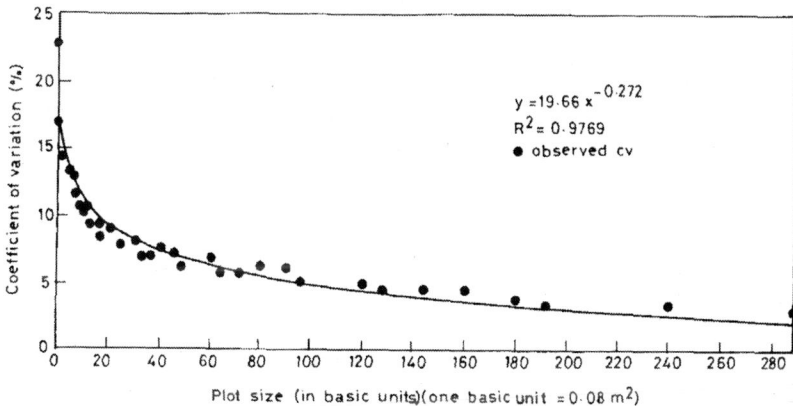

FIGURE 15.5. Coefficient of variation in relation to plot size for the uniformity data on the grain yield for IR 20 rice grown with N at 0 kg/ha.

Optimum Plot Sizes for Different Crops

The optimum plot size varies from crop to crop. For example, in soybean, Brim and Mason (1959) estimated the plot size as 86 sq. ft (3.6 times the basic units of 3 by 8 feet); in Egyptian cotton, Galal and Fittough (1973) estimated optimum plot size as 18 to 29 m^2; in tomato, Palaniswamy et al. (1975) fixed the optimum plot size as 9.00 m^2 (20 times the basic unit); and in groundnut, Gopani et al. (1970) arrived the optimum plot size as 20 m^2 to 30 m^2. Kittock et al. (1986) conducted uniformity trials to fix the optimum number of plants to estimate the mean plant height in cotton in the United States. Considering each plant as a basic plot and using the CV method, they concluded four to eight plants per plot as optimum size to estimate mean plant height in a plot. These data confirm that optimum plot sizes are not the same for different crops.

Uniformity Trial for Estimation of Number of Ear-Bearing Tillers/Plant in Rice Crop

The uniformity trial conducted by Palaniswamy (1990) in rice crop without fertilization was used to study the optimum number of plants required to estimate the number of tillers/plant. The number of tillers/plant is an important biometric and yield component trait in rice contributing to crop yield and hence, this character is invariably recorded in rice experiments. Each plant was considered as basic plot. Palaniswamy obtained $180 \times 86 =$ 17,280 basic plots in the experiment. The CV for each size and shape was computed (see Table 15.2) and studied graphically (Figure 15.7). Palaniswamy concluded that five plants per plot were to be considered for estimation of this character. However, considering the highly variable nature of the character under different soil and fertility conditions, plant spacing, varieties, seasons, cultural practices, etc., he recommended selecting ten plants per plot for its estimation.

EXERCISES

15.1. Define *experimental error.* State the main sources for error. Suggest methods to increase the accuracy of an experiment.

15.2. You are required to fix a sample plot for estimation of weed counts in an experiment. State the procedure for location of the plot.

15.3. Explain the terms *accuracy* and *precision* with suitable examples.

15.4. What is an experimental unit? Give some examples in different areas of research.

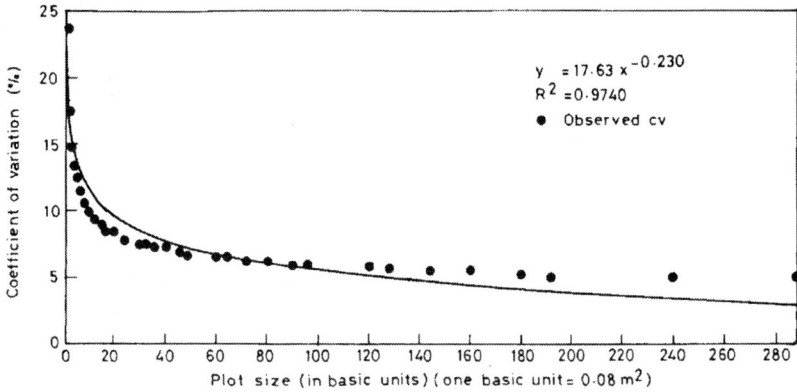

FIGURE 15.6. Coefficient of variation in relation to plot size for the uniformity data on the grain yield for IR 20 rice grown with N at 120 kg/ha.

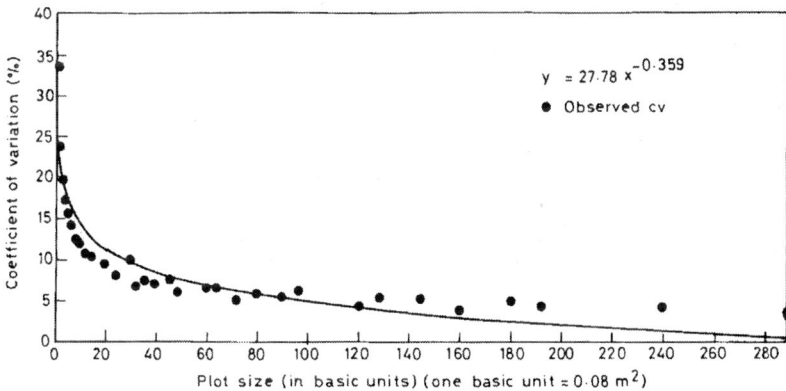

FIGURE 15.7. Relationship of coefficient of variation on plot size based on the uniformity data on the number of tillers per plant in IR 20 rice grown with N at 0 kg/ha.

15.5. Explain what is meant by experimental design. Why is it important in research?
15.6. What is uniformity trial? How is it useful in agricultural research?
15.7. What is treatment? How will you measure the treatment effect?
15.8. What measure of dispersion is used in uniformity trials?

TABLE 15.2. Coefficient of variation for the different sizes and shapes of the plots in respect of the character, number of ear-bearing tillers in rice crop in unfertilized uniformity trial.

No. of plants	Shape/Size of the plot	CV(%)	Mean CV(%)
1	1×1	33.62	33.62
2	1×2	23.98	
	2×1	23.44	23.71
3	1×3	19.79	
	3×1	19.40	19.60
4	1×4	17.25	
	4×1	17.17	
	2×2	16.92	17.11
5	5×1	15.67	15.67
6	1×6	14.35	
	6×1	14.66	
	2×3	14.13	
	3×2	13.95	14.27
8	1×8	12.53	
	2×4	12.43	
	4×2	12.59	12.52
9	3×3	11.80	
	9×1	12.74	12.27
10	5×2	11.45	
	10×1	12.07	11.76
12	1×12	10.58	
	12×1	11.57	
	2×6	10.40	
	6×2	10.65	
	3×4	10.27	
	4×3	10.61	10.68
15	5×3	9.69	
	15×1	10.82	10.26
16	1×16	9.23	
	2×8	9.10	
	4×4	9.42	
	16×1	10.43	9.55
18	3×6	8.80	
	6×3	9.11	
	9×2	9.36	9.09
20	5×4	8.57	

No. of plants	Shape/Size of the plot	CV(%)	Mean CV(%)
	10×2	8.88	
	20×1	9.90	9.12
24	1×24	7.77	
	2×12	7.86	
	12×2	8.54	
	3×8	7.62	
	4×6	8.15	
	6×4	8.04	8.00
30	5×6	7.51	
	10×3	7.65	
	5×2	7.95	
	10×1	8.57	7.92
32	1×32	6.30	
	2×16	6.94	
	4×8	7.08	6.77
36	3×12	6.75	
	6×6	7.07	
	9×4	7.12	
	12×3	7.41	
	18×2	7.73	
	36×1	8.49	7.43
40	5×8	6.59	
	10×4	6.88	
	20×2	7.43	6.97
45	15×3	6.92	
	45×1	8.46	7.69
48	1×48	4.92	
	2×24	5.84	
	3×16	5.89	
	4×12	6.30	
	6×8	6.07	
	12×4	6.67	5.95
54	9×6	6.37	
	18×3	6.76	6.57
60	5×12	5.89	
	10×6	6.10	
	15×4	6.14	
	20×3	6.40	
	30×2	6.64	

TABLE 15.2 *(continued)*

No. of plants	Shape/Size of the plot	CV(%)	Mean CV(%)
	60×1	7.93	6.52
64	2×32	4.78	
	4×16	5.59	5.19
72	3×24	5.13	
	6×12	5.61	
	12×6	5.97	
	9×8	5.56	
	18×4	6.10	
	36×2	6.38	5.79
80	5×16	5.25	
	10×8	5.34	
	20×4	5.87	5.49
90	15×6	5.61	
	30×3	5.87	
	45×2	6.37	
	90×1	6.97	6.21
96	1×96	3.65	
	2×48	3.75	
	3×32	4.13	
	4×24	4.97	
	6×16	4.91	
	12×8	5.19	4.43
108	9×12	5.14	
	18×6	5.55	
	36×3	5.73	5.47
120	5×24	4.64	
	10×12	5.02	
	15×8	4.89	
	20×6	5.31	
	30×4	5.30	
	60×2	5.90	5.18
128	4×32	3.87	3.87
135	45×3	5.69	5.69
144	3×48	2.79	
	6×24	4.38	
	9×16	4.63	
	12×12	4.93	
	18×8	4.73	
	36×4	5.24	4.45

No. of plants	Shape/Size of the plot	CV (%)	Mean CV(%)
160	5×32	3.60	
	10×16	4.53	
	20×8	4.65	4.26
180	15×12	4.70	
	30×6	4.92	
	45×4	5.22	
	60×3	5.22	
	90×2	5.26	
	80×1	5.75	5.18
192	2×96	2.71	
	4×48	2.99	
	6×32	3.33	
	12×16	4.40	3.36
216	9×24	4.12	
	18×12	4.54	
	36×6	4.84	4.50
240	5×48	2.63	
	10×24	4.09	
	15×16	4.24	
	20×12	4.50	
	30×8	4.31	
	60×4	4.80	4.10
270	45×6	4.82	
	90×3	4.92	4.87
288	3×96	2.29	
	6×48	2.34	
	9×32	3.15	
	12×24	3.98	
	18×16	4.10	
	36×8	3.34	3.20
320	10×32	3.11	
	20×16	4.10	3.61
360	15×24	3.91	
	30×12	4.28	
	45×8	4.26	
	60×6	4.45	
	90×4	4.65	
	180×2	4.45	4.33
384	4×96	2.15	

TABLE 15.2 *(continued)*

No. of plants	Shape/Size of the plot	CV(%)	Mean CV(%)
	12×32	2.94	2.55
432	9×48	2.07	
	18×24	3.71	
	36×12	4.23	3.34
480	5×96	2.00	
	10×48	2.11	
	15×32	2.87	
	20×24	3.74	
	30×16	3.84	
	60×8	3.90	3.08
540	45×12	4.28	
	90×6	4.33	
	180×3	4.31	4.31
576	6×96	1.92	
	12×48	1.90	
	18×32	2.73	
	36×16	3.83	2.60
640	20×32	2.74	2.74
720	15×48	1.74	
	30×24	3.52	
	45×16	3.90	
	60×12	3.96.	
	90×8	3.88	
	180×4	4.12	3.52
864	9×96	1.55	
	18×48	1.49	
	36×24	3.53	2.19
960	10×96	1.64	
	20×48	1.64	
	30×32	2.55	
	60×16	3.63	2.37
1080	45×24	3.60	
	90×12	3.98	
	180×6	3.85	3.81
1152	12×96	1.57	
	36×32	2.52	2.05
1440	15×96	1.48	
	30×48	1.31	
	45×32	2.65	

No. of plants	Shape/Size of the plot	CV(%)	Mean CV(%)
	60×24	3.35	
	90×16	3.60	
	180×8	3.52	2.65
1728	18×96	1.35	
	36×48	1.25	1.30
1920	20×96	1.33	
	60×32	2.41	1.87
2160	45×48	1.44	
	90×24	3.36	
	180×12	3.75	2.85
2880	30×96	1.08	
	60×48	0.91	
	90×32	2.35	
	180×16	3.51	1.96
3456	36×96	1.11	1.11
4320	45×96	1.33	
	90×48	0.56	
	180×24	3.38	1.76
5760	60×96	0.80	
	180×32	2.52	1.66
8640	90×96	0.04	
	80×48	0.62	0.33
17280	180×96	0	0

15.9. Given two treatments how will you measure (a) the variability in experimental units, and (b) the treatment effect?

15.10. Why is optimum plot size in an experiment recommended? Give some examples of different plot shapes having the same plot size.

15.11. Different plot sizes are necessary for different crops. Substantiate this statement.

REFERENCES

Brim, C.A. and D.D. Mason (1959). Estimation of optimum plot size for soybean yield trials. *Agron. J.* 51(6):330-334.

Cochran, W.G. and G.M. Cox. (1957). *Experimental design,* Second edition. New York: Wiley; London: Chapman and Hall.

Cox, D.R. (1958). *Planning of experiments.* New York: Wiley; London: Chapman and Hall.

Currence, T. M. (1967). Studies related to field plot technique with tomato. *Proc. Amer. Soc. Hort. Sci.* 290-296.

Federer, W.T. (1967). *Experimental design.* New York: Macmillan.

Finney, D.J. (1955). *Experimental design and its statistical basis.* Chicago, IL: University of Chicago Press.

Fisher, R.A. (1935). *The design of experiments.* Edinburgh and London: Oliver and Boyd.

Galal, U.E and H.A. Abou-el-Fittough (1973). Estimation of optimum plot size and shape for Egyptian cotton yield trials. *Plant Breed. Abstr.* 43(6):363.

Gopani, D.D., M.M. Kabaria, and Vaishnani (1970). Size and shape plots in field experiment on groundnut. *Indian J. Agric. Sci.* 40:1004-1010.

Goulden, C.H. (1959). *Method of statistical analysis.* New York: John Wiley and Sons, Inc.

Hicks, C.R. (1973). *Fundamental concepts in the design of experiments.* New York: Holt, Rinehart and Winston.

Kittock, D.L., C.J. Gain, R.A. Selly, and B.B. Taylor (1986). Samples needed for estimation of plant height of pima cotton. *Agron. J.* 78(3):546-547.

Mclean, R.A. and V.L. Anderson (1974). *Design of experiments—A realistic approach.* New York: Marcel Dekker, Inc.

Murray, D.B. (1950). A uniformity trial with swamp rice. *Trop. Agric.* 27(4-6):105-107.

Palaniswamy, K.M. (1990). Design of statistical (field) experiments and crop forecast in agriculture with special reference to rice. Doctoral thesis submitted to the University of Calicut, India.

Palaniswamy, K.M., S. Thamburaj, P. Kamalanathan, P. Gnanamurty, and A. Shanmughasundaram (1975). Estimation of optimum plot size and shape for field experiments in tomato (*Lycopersicon esculantum* Mill). *Madras Agric J.* 62(3):110-113.

Smith, M.F. (1938). An empirical law describing heterogeneity in the yields of agricultural crops. *J. Agric. Sci.* 28:1-29.

Sukhatme, P.V. (1947). The problem of plot size in large-scale yield surveys. *J. Amer. Stat. Assoc.* 42:291-310.

Torrie, J.H., D.R. Sahmidt, and G.M. Tempo. (1958). Estimates of optimum plot size and shape and replicate number for forage yield of Alfalfa-Bromegrass mixtures. *Agron. J.* 50:258-260.

Usha Rani, P. (1996). Evaluation of chili *(Capsicum annuum L.)* germplasm for capsanthin and capsaicin contents and effect of storage on ground chili. *Madras Agric. J.* 83(5): 288-291.

Chapter 16

Analysis of Variance

FUNDAMENTAL CONCEPTS

The analysis of variance (ANOVA) is a statistical tool with many uses. This technique is used in many areas of research. The term *analysis of variance* was first introduced by R. A. Fisher in the 1920s to deal with the problems in the analyses of agronomic data in agriculture. The basic concepts involved in ANOVA should be understood by researchers, and those who intend to undertake research. The basic component in ANOVA is the sum of squares, usually written as *SS*. Its uses are manifold in the analysis of data; it is used in the computation of variance, standard deviation, standard error of the mean, and the standard error of difference between two means. In particular, it is used in the estimation of precision of the experiments. The SS is computed using the formula $\Sigma(x - \bar{x})^2$ (definition formula) or $\Sigma x_i^2 - (\Sigma x)^2 / n$ (computational formula). In the working formula, Σx_i^2 is the sum of squares of the original observations, and $(\Sigma x)^2 / n$ is the correction term to obtain the sum of the deviation squares of the observations from their mean.

In analysis of variance, the total SS is analyzed and partitioned into different components according to the experimental model. The models are of linear type. There is a corresponding breakdown of the total degrees of freedom into degrees of freedom for each component sum of squares. The total degrees of freedom in a linear model are always $(n-1)$ where n is the total number of observations.

Example: The mathematical linear model for completely randomized design is

$$Y_j = \mu + T_i + \varepsilon_{ij}$$

where $i = 1, 2, \ldots t; j = 1, 2 \ldots r$.
Y_{ij} = observed value of the *jth* replication of the *ith* treatment
μ = grand mean (general mean)
T_i = treatment effect
ε_{ij} = error effect of the *jth* replication of the *ith* treatment

Since we are dealing with samples, the model is written as

$$y_{ij} = m + t_i + e_{ij} \, (i = 1,2\ldots t) \text{ and } (j = 1,2\ldots r), \text{ where}$$

y_{ij} = observed value of the *jth* replication of the *ith* treatment in the experiment

m = general mean

t_i = *ith* treatment effect

e_{ij} = error effect of the *jth* replication of the *ith* treatment. The total variation among the y_{ijs} is partitioned into components of (a) between groups SS, and (b) within groups SS.

If we repeat the experiment under similar conditions, the population parameters μ and T_i, remain unchanged but the sample values of m and t_i vary from one experiment to another. Scheffe (1959) gives the following definition for the analysis of variance. The analysis of variance is a statistical technique for analyzing measurements depending upon several kinds of effects operating simultaneously to decide which kinds of effects are important to estimate the effects.

Example: Two treatments (X1 and X2) (two different varieties of wheat) are evaluated in a completely randomized block design with four replications.

The yield data (kg/plot) are tabulated here:

Rep/treatments	X1	X2	Total
1	5	8	13
2	6	10	16
3	4	7	11
4	5	7	12
Total	20	32	52 (grand total)

Null hypothesis is H_o: $\overline{X}1 = \overline{X}2$, i.e., $\overline{X}1 - \overline{X}2 = 0$
Alternate hypothesis is H_1: $\overline{X}1 \neq \overline{X}2$
Computations:

1. Compute totals for X1, X2, replication 1, replication 2, replication 3, replication 4, and grand total. The sample mean, under the H_o, is an unbiased estimate of the population mean μ, if null hypothesis is true the sample mean is equal to grand mean.

2. Total SS: $\Sigma x_i^2 - (\Sigma x)^2 / n = 5^2 + 6^2 + \ldots + 7^2 - 52^2 / 8$

$\qquad = 364 - 338 - 26$

3. Between treatment SS: $\Sigma T_i^2 - (\Sigma T_i)^2 / r = (20^2 + 32^2) / 4 - \text{CF} =$

$\qquad 356 - 338 = 18$

4. Within treatment SS:

SS within treatments = SS within X1 + SS within X2

SS within X1 $= \{\Sigma X1^2 - (\Sigma X1)^2 / r\} = (5^2 + 6^2 + 4^2 + 5^2)$

$\qquad -20^2 / 4 = 102 - 100 - 2$

SS within X2 $= \{\Sigma X2^2 - (\Sigma X2)^2 / r\} = (8^2 + 10^2 + 7^2 + 7^2)$

$\qquad -32^2 / 4 = 262 - 256 - 6$

Hence within SS = 2 + 6 = 8

or within SS = total SS – SS between treatment SS = 26 –18 = 8,

i.e., total SS = between treatment SS + within treatment SS

26 = 18 + 8

The total SS is equal to the sum of between treatment SS and within treatment SS.

COMPUTATION OF VARIANCE

Variance is defined as $[SS/(n - 1)]$, i.e., sum of squares/degrees of freedom. Therefore we find the degrees of freedom for each, i.e., for total SS, between treatments SS, and within treatment SS. Total degrees of freedom is the total number of observations minus one, i.e., in our example it is $8 - 1$ = 7. Between treatment degrees of freedom is the number of treatments in the experiment minus one, i.e., $2 - 1 = 1$. The within treatment degrees of freedom is the total number of observations minus the number of treatments, i.e., $8 - 2 = 6$ or it is the total of number of degrees of freedom within treatment 1 $(4 - 1 = 3)$ and the number of degrees of freedom within treatment 2 $(4 - 1 = 3)$, i.e., $3 + 3 = 6$. Then calculate variance estimates for the two sources of variation namely between groups (treatments) and within groups (treatments), by dividing the corresponding SS by the concerned degrees of freedom.

Variance between groups is 18/1 = 18
Variance within groups: 8/6 = 1.33

TABLE 16.1. ANOVA.

Source of variation (SV)	Degrees of freedom (*df*)	Sum of squares (*SS*)	Mean square (*MS*)	*F*	*F* from tables at	
					P = 0.05	*P* = 0.01
Between treatments	1	18	18 (B)	13.55*	5.99	13.74
Within treatments (Error)	6	8	1.33 (C)			
Total	7	26				

*indicates significant at *P* = 0.05

FIND THE F RATIO VALUE

The *F* ratio is obtained by dividing the treatment mean square by the within treatment mean square, i.e., $18/1.33 = 13.5$. The fact that the between treatment mean square is 13.5 times the within treatment mean square seems to indicate that the variability *amongst* the treatment is much greater than that *within* groups. The *F* ratio with 1 and 6 degrees of freedom from *F* table are 6.99 and 13. 74 for *P* = 0.05 and *P* = 0.01, respectively. These are the values given in the *F* table for significant levels of 5 percent and 1 percent, respectively (see Appendix Table A.4). As the calculated *F* value is greater than the table value of *F*, we conclude that there is significant difference between the two treatment means at *P* = 0.05. The results obtained are shown in Table 16.1.

In Table 16.1, the expected value of error mean square (EMS) is σ^2, which can also be estimated. The expected mean square for treatments [E(B)] is

$$\sigma^2 + r\sigma^2 = E(C) + r\sigma^2.$$

As per H_o: $r\sigma^2 = 0$, E(B) = E(C), if B/C is greater, it indicates that the treatment effect is present. Thus the analysis of variance serves two purposes (1) to test the hypothesis that the treatments are equal, and (2) to estimate variance components.

DEGREES OF FREEDOM

We can also verify the degrees of freedom corresponding to the various SS as follows:

Total SS = Treatment SS + Error SS (in terms of SS)

In terms of degrees of freedom, we have, i.e.,

$$n - 1 = (t - 1) + t(r - 1), \text{ i.e., } n = t - 1 + rt - t + 1 = rt,$$

i.e., the SS between treatments and within treatments give the total SS, and the same is true in degrees of freedom also. In the ANOVA, we calculate F ratio

F = (between treatment mean square)/within treatment mean square

= (treatment mean square + error mean square)/error mean square

i.e., if the null hypothesis is true, i.e., the treatment effect is zero, we get

$$= \frac{0 + \text{error mean square}}{\text{error mean square}} = 1$$

i.e., the null hypothesis is expected to be 1. If the null hypothesis is not true, then

$$F = \frac{\text{treatment effect } + \text{ experimental error}}{\text{experimental error}} > 1$$

When the calculated F ratio is greater than the F ratio value from F table, it indicates that the differences between treatments are greater than would be expected by chance. The computed F value lies in the rejection region and hence the null hypothesis is rejected.

An F value of unity or less is always nonsignificant and in such a situation, it would not be necessary to consult an F table. According to Fisher (1935), if the treatment differences in the experiment fail to satisfy the F test of significance, then there is no need to proceed in further analyses as all differences observed in the experiment are due to chance.

F DISTRIBUTION

F distribution is explained with the F curve. The acceptance and rejection regions for the example discussed previously are shown in Figure 16.1. The shape of the F curve is determined by the degrees of freedom associated with the F ratio. The F ratio depends upon the numerator degrees of

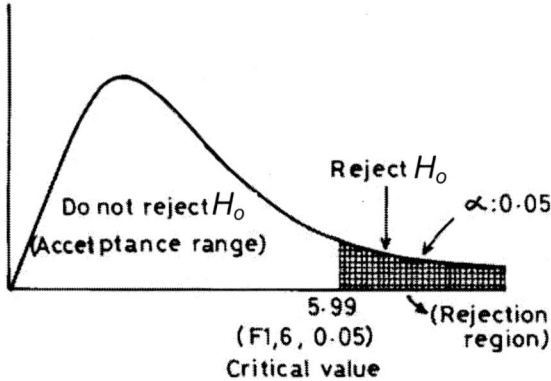

FIGURE 16.1. Distribution of F for $df_1 = 1$, $df_2 = 6$, and $\alpha = 0.05$.

freedom (n_1) and denominator degrees of freedom (n_2). For a fixed probability level say $\alpha = 0.05$, we compare the F value calculated in the ANOVA with the table value of F. In our example, the F ratio for 1 and 6 degrees of freedom at $\alpha = 0.05$ is 5.59 and it means that an ordinate drawn at 1 and 6 = 5.59 will divide the sampling distribution of F at a point where proportion of the area under the curve to the right is 0.05, i.e., 5 percent (see Figure 16.1).

For a fixed a level, the calculated F value is compared with the table value of F. If the calculated F falls in the rejection region, the null hypothesis is rejected. It shows that there is evidence that the difference in the treatment means is significant at 5 percent level. If the calculated F value is less than the table value of F, i.e., if the calculated F value falls in the acceptance region, there is no significant evidence to say that the treatment means differ.

However, sometimes we are wrong in our statement. We expect to be wrong 5 percent of the time when the experiment is repeated several times. The F test shows only that there is some difference between the means. If the stipulated assumptions are satisfied, the ANOVA technique may be applied in several areas of research.

SUMMARY OF THE ANOVA TECHNIQUE

1. State the null hypothesis and alternate hypothesis.
2. The null hypothesis states that there is no difference in yielding capacity of the two varieties (treatments). The alternate hypothesis is that at least one treatment differs significantly from the other.

3. Calculate the sample means ($\overline{V}1 = 5$ and $\overline{V}2 = 8$).
4. If the null hypothesis is true, $\overline{V}1$ and $\overline{V}2$ means are close to the grand mean $(\overline{V}1 + \overline{V}2)/2 = (5 + 8)/2 = 13/2 = 6.5$.
5. Find SS.
6. Find between treatment variance and error variance.
7. Determine level of significance (say $\alpha = 0.05$).
8. Find F: F = (between treatment variance)/pooled variance = $18/1.33 = 13.5$.
9. If calculated F is greater than the table F ratio value, conclude the existence of significant difference.

EXERCISES

16.1. When there is no treatment effect, we expect the F ratio value is 1. Why?
16.2. Describe the similarities between an F ratio and t statistic.
16.3. Analysis of variance technique is to be used when the treatments are more than two instead of several t tests. Why?
16.4. Look at the data obtained from an experiment conducted in completely randomized design, and tell what value should be obtained for the variance between treatments. Calculate SS between treatments and mean square between treatments:

Replication	Treatment I	Treatment II
1	1	2
2	4	5
3	0	0
4	3	1
Total	8	8

16.5. What is the F value?
16.6. What two purposes does the analysis of variance serve?
16.7. Draw F curve, and show acceptance and rejection regions when the number of treatments are 4, and error degrees of freedom is 8 for $\alpha = 0.05$.
16.8. Given the following ANOVA ($r = 10$). Fill in the blanks.

SV	df	SS	M
Between treatments	3	_____	_____
Within treatments	_____	_____	_____
Total	39	153	_____

16.9. A researcher reports an *F* ratio with 3, 30 for ANOVA. How many treatment conditions were compared in the experiment? How many subjects participated in the experiment?

16.10. Given the following:

Treatments	T1 ($n_1 = 5$)	T2 ($n_2 = 5$)	T3 ($n_3 = 5$)
SS within each tr.	45	25	50

Total SS = 325

Do these data provide evidence of any significant mean difference among the treatments?

16.11. What are the basic ideas of the analysis of variance? You are given three treatments and you have to conduct the experiments in four replications each. How you will decompose the data? What are the parts? Give your ANOVA.

16.12. What do you understand by residuals? How you will calculate them?

16.13. What is the concept of ANOVA?

16.14. What is the procedure to set up ANOVA table when two treatments are studied with six replications in completely randomized design?

REFERENCES

Fisher, R.A. (1935). *The design of experiments*. London: Oliver Boyd; Edinburgh: Tweeddale Court.

Scheffe, H. (1959). *The analysis of variance*. New York: John Wiley and Sons, Inc.

Chapter 17

Principles of Experimental Design

Experiments are conducted in many areas of research involving high costs, time, materials, and labor, and hence the results obtained from these experiments must be as accurate as possible. The aim of the researcher is to detect even small differences that exist among treatments, and also to measure the precision of the estimates for the purpose of reliability. Reliable results save expenditures. Hence, it is imperative to know the basic principles of experimental design, which are common to all areas of research. Following are three basic principles of experimental design:

1. Replication
2. Randomization
3. Local control

They are discussed in detail in this chapter.

REPLICATION

The first basic principle is replication. Replication of an experiment is defined as the repetition of the experiment. If a treatment is allotted to r experimental units in a design, it is replicated r times. The necessity of replication is that the results of a single replication (trial) cannot be trusted because of, e.g., soil variability, drainage, insect pests, and other known and unknown factors (Cox, 1968). Even with the best of experimental control, results vary from trial to trial (Lorenzen and Anderson, 1993). The relationship between standard error of a treatment mean (SEm) and replication is given by

$$SEm = s / \sqrt{r},$$

where s = standard deviation and r = number of replications. The standard error of difference between two treatment means is given by

$$\sqrt{2s^2 r}.$$

These two formulae indicate that to increase the precision of the experiment, the number of replications should be increased (Armitage and Remington, 1967). If there is no replication, experimental error cannot be estimated in any other way (Goulden, 1959), and consequently the precision of the estimates cannot be determined. Suppose there are two treatments in an experiment without replication, the difference between the two treatments may either be real or by chance. Replication helps to detect the real difference.

Replication and Standard Error

One can reduce the standard error of a treatment mean by increasing the number of replications. However, by increasing the number of replications beyond a certain number, we cannot get further reduction in the standard error. Thus, there is no benefit in increasing the number of replications beyond a certain number. Figure 17.1 shows that there is practically no reduction in standard error beyond four replications.

Furthermore, too many replications increase the costs of experiments considerably. The number of replications to be fixed in an experiment depends upon several factors such as number of treatments, nature of experimental units, the degree of precision required, etc. Generally, the number of replications should be fixed in a design in such a way that we get 10 to 12 error degrees of freedom in the analysis of variance. Experience shows that it is wise to have four to five replications in field experiments. In medical science, Armitage and Remington (1967) suggested the number of hospitals

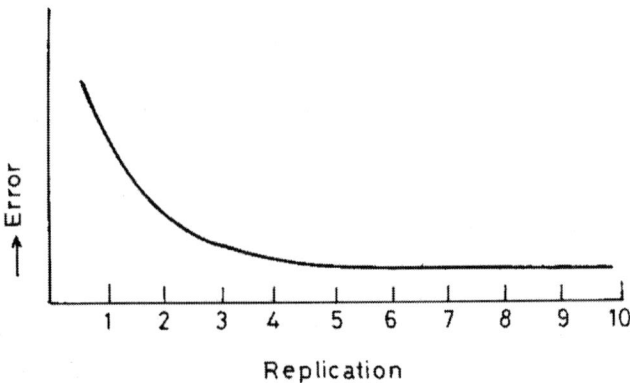

FIGURE 17.1. Relationship between number of replications and standard error.

can be considered as replications in clinical trials to compare certain medical treatments. Fisher's diagram may help determine the significance of replication in the experiment (see Figure 17.2).

RANDOMIZATION

Randomization is an important feature of modern experimentation. In any experiment, the treatments are placed in experimental units in each replication. The problem faced by the experimenter is how to place the treatments in the given experimental units. Fisher devised a method called randomization for the allotment of treatments to the experimental units. In the randomization procedure, the treatments are allotted at random to the units. In animal science experiments, Emmens (1948) points out that haphazard selection of experimental units may lead to systemic differences in weights between the animals chosen. Randomization guarantees a valid estimate of error irrespective of the nature of the experimental error in the individual units (Cochran, 1976). The importance of randomization can be explained with certain examples. Suppose in a varietal trial, several varieties of a crop are compared in a design with replications. Placement of a low-yielding variety in a highly fertile unit will certainly overestimate its yielding capacity and give unreliable results. Similarly, capacity of a high-yielding variety if placed in a low fertile plot will be underestimated. However honest the experimenter is, he or she may likely favor one treatment over another in the allotment of treatments in the experimental units. Ogawa (1974) stressed that randomization is the only way of ensuring that comparisons between treatments are not biased. Furthermore, a number of variables cannot be

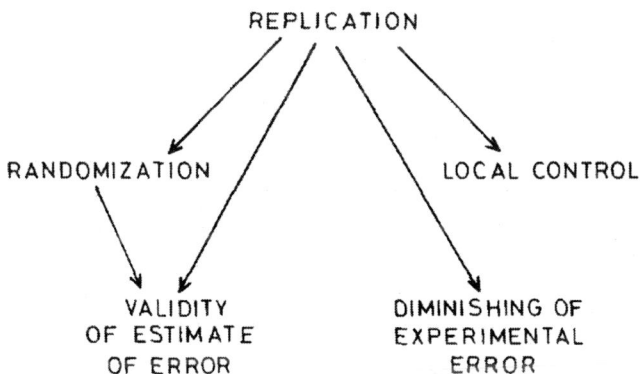

FIGURE 17.2. Diagram showing the relationship between replication, randomization, and local control. (*Source:* LeClerg et al., 1962.)

controlled. Randomization will tend to average out the effect of these uncontrolled variables. By randomization, one can infer that errors of measurement are independent, a common assumption in most statistical analyses (Hicks, 1973).

Reasons for Randomization

Suppose we want to compare five varieties (treatments) of a crop in a randomized complete block design with four replications. In one replication, the five treatments can be allotted in 5!, i.e., we can expect 120 possible results. Hence, for four replications in the experiment, we can arrange treatments in many possible ways. Yet the results we obtain from our experiment are one among them and hence the random allotment of treatments is the only choice. By randomization, we are avoiding the personal or subjective bias, which may be conscious or unconscious. Randomization procedures differ from one design to another and there are several methods of randomization. We can use random tables (see Appendix, Table A.1), coins, dice, etc., depending upon the number of treatments. Randomization is the basis for the design of experiments but it does not eliminate variation caused by extraneous factors (Yates, 1964; Finney, 1960).

LOCAL CONTROL

Local control is a device recommended to reduce the experimental error. Generally, the plots (experimental units) adjacent to each other tend to be similar in fertility levels. Based on this principle, the adjacent plots are grouped into a replication and the treatments within a replication are compared under uniform fertility conditions. Control of local variation due to soil, moisture, shelter, or anything else can be achieved by devices such as blocks or perhaps rows and columns as in Latin square design (Pearce, 1985).

Recommendations to improve precision of the experiment:

1. The precision of the experiment can be increased by increasing the number of replications.
2. Precision can be improved by careful selection of treatments and experimental units, and also by taking additional measurements, and using covariance techniques, and by selecting appropriate experimental designs.
3. In any field, the experimental area should be located in the center of the field and the area around the experiment should be covered by

plantings of the same crop. Plants adjacent to blank spaces tend to benefit from the additional habitat factors such as water, light, etc.

4. In row-planted crops, the border rows around the experimental units should be discarded during harvest to avoid border effect. If the experimenter suspects more border effect, it is advisable to discard two border rows on all sides of the experiment. By discarding the border rows, the treatment effects are estimated more precisely.

5. If the experiment involves removal of plants for estimation of plant attributes during the course of experiment, it is advisable to provide larger plot sizes and also separate the experimental area for sampling purposes.

6. If harvesting takes place, it is advisable to carry out the harvest of the experimental units replicationwise, i.e., the experimental units in a replication should not be harvested on two different days.

7. In comparative trials, the plant population should be equally maintained in all the experimental units. Often, we come across missing hills due to the death of plants, nongermination, nonestablishment of the plants, attack of pests and diseases, and other reasons. Under such situations, the missing hills should be replanted as soon as possible.

8. In some types of experiments such as fertilizer trials, the experimental units within a replication have to be separated by bunds. Under such situations, the width of the bunds, which demarcates one plot from the other, should be decided depending upon the crop, soil conditions, etc., prevailing at each experimental station.

9. When laying out experiments, avoid any experimental unit near the shade to avoid shade effect.

10. During the course of the experiment, any cultural practices such as weeding, irrigation, manuring, etc., should be applied uniformly to all the plots.

11. The bunds that separate the experimental units in the replication should always be kept intact so that flow of water from one plot to another can be prevented. This is important especially when the experiment involves fertilizer applications or fertilizer treatments.

12. Care should be taken when deciding the width of the bunds depending upon the nature of the experiment. Probably larger (wider) bunds must be maintained in fertilizer trials.

13. Many of these suggestions are applicable also for trails conducted in greenhouses.

Estimation of Plant Attributes in Experiments

In comparative experiments, the main objective is to detect significant differences among yields of different treatments. This also estimates the plant attributes or yield components that contribute to crop yields. Plant attributes must therefore be estimated as accurately as possible and with precision. Improper and unrepresented selection of samples from different treatments lead to biased estimates. The sample size and the sampling procedure for selection of plants vary from crop to crop. This is explained with reference to rice crop.

Example: In rice crop experiments, besides recording grain yields in different treatments, experimenters invariably record yield components and other plant characters, e.g., plant height, leaf area, length of the panicle, number of grains per panicle, number of ear-bearing tillers per plant or number of panicles per plant, and grain weight per panicle. These components contribute to the yield and possess high positive correlation with grain yield. Palaniswamy (1990) studied variability among the different tillers in rice in two varieties, and three nitrogen (N) levels (0, 60, and 120 kg/ha). A large amount of variation was observed in these characters (see Table 17.1).

Palaniswamy (1990) identified the tillers in a plant based on their heights (see Figure 17.3), and observed variation in the same plant with respect to the four characters studied. The coefficient of variation, i.e., the variability among the tallest tillers is lesser than the less tall (shorter) ones. These studies show that when we select tillers for biometric observations from different treatments, we should select the tallest tiller uniformly in the selected plants (hills) to get better estimates for comparative purposes. The distribution of the characters in different tillers is tabulated in Table 17.1.

ANALYSIS OF DATA

Analysis of the data depends upon the design model employed. Computer programs are recommended for accurate computations. If computer facilities are not available, one can refer to the method of calculations in Hicks (1973), Ostle (1963), Steel and Torrie (1960), and also those discussed in this book.

Method of Selection of a Sample Plot for Sampling Purpose in an Experimental Plot

Sometimes, experimenters wish to fix a small area or a portion of the experimental plot for sampling purposes. Such situations arise in herbicide

TABLE 17.1. Distribution of the different attributes on the different tillers in the same plant in two rice varieties grown under 0, 60, and 120 N kg/ha level.

N	Tiller	IR22				IR662			
		X1	X2	X3	X4	X1	X2	X3	X4
0	1	85.9	21.6	109	2.59	86.9	25.3	101	2.65
	2	83.9	21.2	100	2.38	85.0	24.9	97	2.52
	3	83.8	21.2	98	2.30	83.8	24.8	97	2.50
	4	80.1	21.2	98	2.27	82.3	24.5	90	2.34
	5	79.4	21.0	91	2.14	81.6	24.1	87	2.27
	6	78.5	20.7	88	2.00	80.0	23.8	79	2.06
	7	77.6	20.3	80	1.88	78.4	23.0	74	1.90
	8	76.1	20.0	77	1.80	77.8	23.4	75	1.93
	9	75.5	20.0	76	1.77	76.7	23.0	70	1.80
	10	74.0	19.2	71	1.66	75.4	22.3	64	1.73
	11	74.0	19.8	69	1.68				
60	1	96.4	21.5	107	2.56	90.1	24.8	98	2.62
	2	93.4	21.2	104	2.47	88.2	24.6	97	2.86
	3	92.3	21.1	102	2.45	86.9	24.2	92	2.44
	4	91.1	20.8	98	2.22	86.3	24.2	91	2.39
	5	90.0	20.8	95	2.75	85.4	24.1	90	2.36
	6	88.6	20.7	94	2.74	84.5	23.9	87	2.30
	7	89.9	20.7	94	2.24	83.3	23.3	81	2.30
	8	88.0	20.5	89	2.11	82.1	23.4	80	2.09
	9	86.9	20.2	84	2.00	80.9	23.3	79	2.08
	10	86.1	19.9	84	2.00	80.6	23.2	79	2.04
	11	84.2	19.8	80	1.88	79.8	22.5	73	1.93
	12	83.9	19.5	70	1.64	78.5	22.4	67	1.73
	13	80.6	19.0	66	1.55				
120	1	98.6	21.5	108	2.58	96.6	25.3	100	2.72
	2	97.1	21.5	107	2.56	94.8	24.8	95	2.57
	3	95.9	21.6	101	2.41	93.6	24.6	95	2.58
	4	94.8	21.3	101	2.38	92.3	24.7	92	2.47
	5	94.1	21.4	101	2.36	91.0	24.2	88	2.36
	6	93.2	21.2	98	2.30	90.1	24.0	86	2.28
	7	93.0	21.0	93	2.22	88.9	23.6	83	2.21
	8	92.4	21.0	94	2.20	87.5	23.5	79	2.12
	9	91.7	21.2	94	2.20	86.8	23.2	79	2.08

TABLE 17.1 *(continued)*

N	Tiller	IR22				IR662			
		X1	X2	X3	X4	X1	X2	X3	X4
	10	90.7	20.9	89	2.11	85.6	23.0	77	2.03
	11	88.9	20.6	81	1.93	84.7	23.3	75	1.99
	12	87.8	20.0	77	1.79	84.3	22.4	69	1.80
	13	85.6	20.2	75	1.80	83.3	22.1	68	1.71

X1 = Mean height of the tillers (cm)
X2 = Mean length of the panicle (ear head)
X3 = Mean number of grains/panicle
X4 = Mean weight of grains in the panicle

Note: Tillers panicles are described according to their positions (based on their heights in descending order).

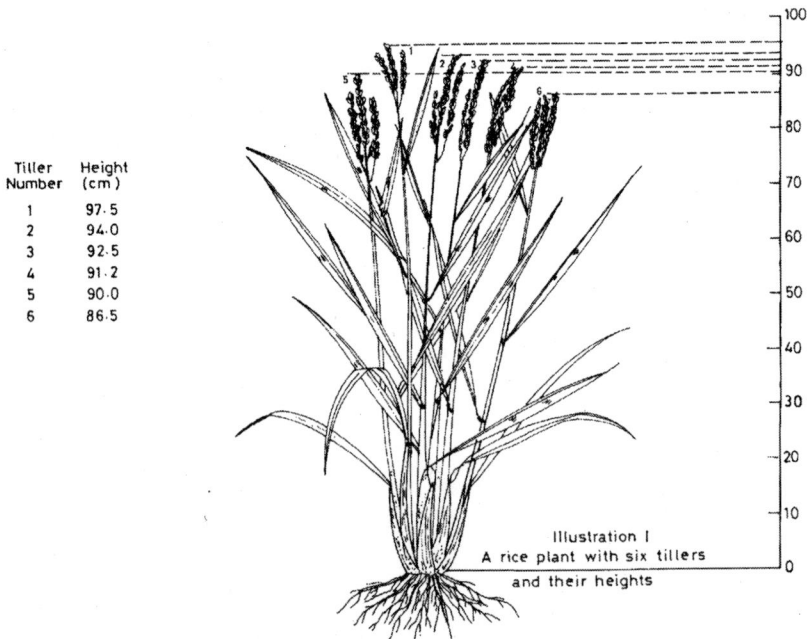

Tiller Number	Height (cm)
1	97.5
2	94.0
3	92.5
4	91.2
5	90.0
6	86.5

Illustration I
A rice plant with six tillers
and their heights

FIGURE 17.3. Differences in height of the tillers in the same plant.

experiments in which weed weights have to be recorded, crop estimation experiments, entomological experiments involving counts of infested plants, etc. A random selection of a small area in the experimental plot should be fixed in order to get an unbiased estimate of the character under study. The procedure is explained with an example. Let ABCD be the rectangular experimental unit (see Figure 17.4). We have to select a plot of size 4 feet by 2 feet for recording observations or for sampling purposes. D is the southwest corner of the plot. Let the length of the plot $DC = x$ feet and width $AD = y$ feet. Select two random numbers, one between 0 and $(x - 4$ feet), and another between 0 and $(y - 2$ feet). Let the selected random numbers be M and N. Here we are considering $(x - 4$ feet) and $(y - 2$ feet) so that the plot should not fall outside the plot. From D, move M feet along DC, and then N feet toward north parallel to DA. Let this point be P (Figure 17.4). Then from P, move 4 feet toward east parallel to DC and then again come to P. Then move 2 feet toward north parallel to DA to a point marked S. Then form the rectangle $PQRS$. $PQRS$ is the required sample plot to be used for recording observations or for sampling purposes to obtain unbiased estimate(s) of the character or characters of interest.

EXERCISES

17.1. State fundamental principles of experimental design. Briefly explain each one of them with suitable examples.
17.2. Describe the role of replication.
17.3. How will the principle of local control minimize the error?

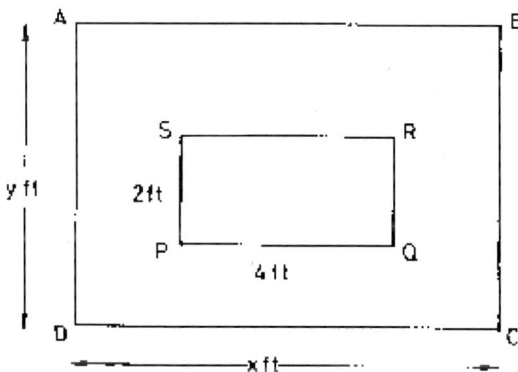

FIGURE 17.4. Procedure for fixing a plot size of 4 feet x 2 feet in an experimental plot for sampling purposes.

17.4. You are required to fix a sample plot for estimation of weed counts in an experiment. State the procedure for location of the plot.
17.5. What is experimental unit? Give some examples in different areas of research.
17.6. Give some recommendations for increasing the precision of the experiment in your field of research.
17.7. How do you fix a plot of 3 feet by 2 feet in a field measuring 10 feet by 30 feet for estimation of plant attributes?
17.8. Explain briefly Fisher's diagram in relation to experimental design.
17.9. What is the reason for recommending randomization of treatments?
17.10. State some methods to increase the precision in crop research.

REFERENCES

Armitage, P. and R.D. Remington (1967). *Experimental design—Statistics in endocrinology.* London: The MIT Press.
Cochran, W.G. (1976). *Early development of techniques in comparative experiments on the history of statistics and probability.* New York: Marcel Dekker, Inc.
Cox, D.R. (1968). *Planning of experiments.* New York: John Wiley and Sons.
Emmens, C. W. (1948). *Principles of biological assays.* London: Chapman and Hall.
Finney, D. J. (1960). *An introduction to the theory of experimental design.* Chicago: University of Chicago Press.
Goulden, C.H. (1959). *Method of statistical analysis.* New York: John Wiley and Sons, Inc.
Hicks, C.R. (1973). *Fundamental concepts in the design of experiments.* New York: Holt, Rinehart and Winston.
LeClerg, E.L., W.H. Leonard, and A.G.Clark (1962). *Field plot technique.* Minneapolis, MN: Burgess Publishing Company.
Lorenzen, T.J. and V.L. Anderson (1993). *Design of experiments.* New York: Marcel Dekker, Inc.
Ogawa, J. (1974). *Statistical theory of the analysis of experimental designs.* New York: Marcel Dekker, Inc.
Ostle, B. (1963). *Statistics in research,* Second edition. Ames, IA: Iowa State University Press.
Palaniswamy, K.M. (1990). Design of statistical (field) experiments and crop forecast in agriculture with special reference to rice. Doctoral thesis submitted to the University of Calicut, India.
Pearce, S.C. (1985). *Agricultural experimentation in a developing country—A celebration of statistics.* The ISI-Centenary Volume. New York: Springer-Verlag.
Steel R.G.D. and J.H. Torrie (1960). *Principles and procedures of statistics.* New York: McGraw-Hill.
Yates, F. (1964). Sir Ronald Fisher and the design of experiments. *Biometrics,* 20:307-321.

Chapter 18

Completely Randomized Design

Researchers require appropriate experimental design to test their hypotheses. Investigations should be conducted as efficiently as possible to detect even small differences among the treatments. Several designs are available. Thus researchers should consult a statistician before formulating the design and starting the experiment. Completely randomized design (CRD) is a basic design. It is also called a one-way analysis of variance model. Sometimes, it is also referred to as a one-way classification or one-way layout model. In this design, the treatments are completely assigned at random to the experimental units without blocking. This is useful when the experimental units are uniform or homogeneous. When the experimental units are not uniform, one should use an alternate design.

DESCRIPTION OF THE DESIGN

Fixed Effects Model: When the Number of Replications for the Treatments Are Equal

Example 1: An agronomist wishes to test the yielding capacity of three wheat varieties (treatments), V1, V2, and V3, in a field trial. The agronomist also wishes to have four replications for each treatment. Thus 3 by 4 = 12 experimental units. The variable under comparison is grain yield.

Procedure for the Layout of the Experiment

Let ABCD be the field available for the trial (see Figure 18.1). Since we require 12 plots (experimental units), the field ABCD is divided into 12 plots of equal size. Each plot is numbered serially from 1 through 12 as shown in Figure 18.1.

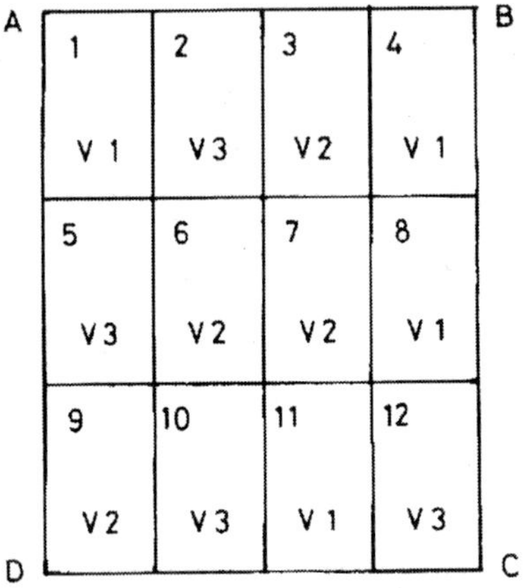

FIGURE 18.1. Layout plan of completely randomized design for testing three treatments each with four replications.

Randomization Procedure

The treatments are allotted to the plots in a random manner. This randomization procedure ensures that each and every plot has an equal chance of getting a treatment. Since there are 12 plots (two-digit number), we consult the two-digit random number table (see Appendix, Table A.1), and select 12 random numbers (Hader, 1973). Let the random numbers selected be 4, 8, 11, 1, 6, 3, 7, 9, 12, 2, 5, and 10. The V1 treatment is allotted to the plots having numbers 4, 8, 11, and 1; V2 to the 6, 3, 7, and 9; and finally V3 to 12, 2, 5, and 10. When selecting random numbers, if a random number, say 4, was drawn a second time, it would be disregarded. Finally, the field plan will look like the plan shown (see Figure 18.1). In this design, the treatments are allotted to the plots completely at random and hence the term *completely randomized design.* The plot yields (kg/plot) recorded are shown in Table 18.1 under each treatment for analysis purpose.

The distribution of plot yields in each treatment can be studied by plotting the points in a graph. This graph is called a dot diagram (see Figure 18.2).

The dot diagram gives a rough idea about the distribution of observed values in each treatment. V1 records yields 3, 4, 3, and 2 in replications I, II, III, and IV, respectively, with a total of 12 and mean of 3. If all four plots are uniform and all other extraneous factors do not affect the plots, each plot should have recorded a value of 3. The soil fertility variation and other external factors unknown to the experimenter cause the differences in yields

TABLE 18.1. Grain yield recorded (kg/plot) by the treatments.

Replication	Variety		
	V1	V2	V3
I	3	6	15
II	4	8	15
III	3	7	14
IV	2	7	12
Total	12	28	56
Mean	3	7	14
Treatment Effect	−5 (3-8)	−1 (7-8)	+6 (14-8)

Note: Grand total = 96; grand mean = 8.

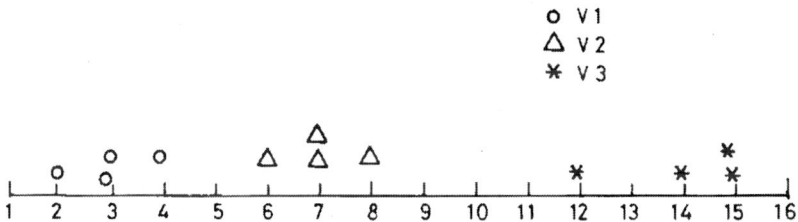

FIGURE 18.2. Dot diagram for the distribution of plot yields in each treatment.

in the different plots. These differences contribute to error and are measured by taking deviations of the observed values from their means. Thus the distributions of errors in the four plots in V1 are 0 (3–3), +1 (4–3), 0 (3–3), and −1 (2–3) (note the sum of errors is zero). Variability is therefore expressed as sum of squares (SS) as mentioned in Chapter 15. Thus SS for V1 is 0 + 1 + 0 + 1 = 2. Similarly, we can find the errors and error SS for V2 and V3 as shown in Table 18.2.

Suppose we remove the experimental errors. Then we can expect the same plot yields under each variety in all the replications as shown in Table 18.3.

Now we find differences among the varieties and this variation is the combined effect of varieties and random effects. If there is no variation among varieties, we should then get the value 8, the grand mean in each

TABLE 18.2. Experimental errors and error sum of squares.

	Errors for				Error sum of squares for		
Replication	V1	V2	V3	Replication	V1	V2	V3
I	0	−1	+1	I	0	1	1
II	+1	+1	+1	II	1	1	1
III	0	0	0	III	0	0	0
IV	−1	0	−2	IV	1	0	4
Total	0	0	0		2	2	6
							Total: 10

TABLE 18.3. Yield figures after removing errors.

	Variety		
Replication	V1	V2	V3
I	3	7	14
II	3	7	14
III	3	7	14
IV	3	7	14
Total	12	28	56
Mean	3	7	14

Note: Grand total = 96; grand mean = 8.

plot. The deviation between variety means and grand mean is due to the treatment and random effects. This variation is calculated by SS as:

$$[(3-8)^2 + (3-8)^2 + ... + (14-8)^2] = 248,$$
$$\text{i.e., } [4(3-8)^2 + 4(7-8)^2 + 4(14-8)^2] = 248.$$

The total variability among all 12 observations is the sum of squared deviations of each sample observation (x_{ij}) and the grand mean $(\bar{x})$. This variability is given by the SS as

$$(\Sigma x_i - \bar{x}), \text{ i.e.,}$$
$$(3-8)^2 + (4-8)^2 ... + (4-8)^2 ... + (14-8)^2 + (12-8)^2 = 258.$$

The analysis is carried out according to the design. In completely randomized design, the total variation in the sample observations is expressed in terms of total SS, which is divided into its components: between treatment SS and within treatment SS, i.e.,

Total SS = between treatment SS + within treatment SS
Thus: 258 = 248 + 10 (Montgomery, 1991)
Total df (11) = 2 [(among treatments) $(t-1)$] + 9
[within treatments t $(r-1)$]

Assumptions: In a completely randomized design, the following assumptions are made:

1. Each observation is assumed to follow the linear additive model namely

$Y_{ij} = \mu + T_i + \varepsilon_{ij}$ where

μ = population mean
T_i = treatment effect
ε_{ij} = random error. For the sample data, we have the model
$y_{ij} = m + t_i + e_{ij,}$ where
y_{ij} = observed value of *ith* variety grown in *jth* replication;
m = general mean, which is constant in each observation;
t_i = *ith* treatment effect;
e_{ij} = experimental error effect contributed by *jth* replication receiving *ith* treatment. Experimental error is a random variable whose mean is zero and a constant variance, σ^2. The experimental errors are normally independently distributed with mean zero and common variance, e.g., σ^2 for every treatment. Thus we obtain the pooled variance following the assumption of homogeneity of variances. Distribution of errors independently means the value of experimental error in one observation does not affect the value of another error term. Homogeneity of variances can be tested using Bartlett's test. The within error sum of squares of the three varieties (V1 = 2, V2 = 2, and V3 = 6) adds to the error sum of squares of 10 as shown in the ANOVA. It satisfies the additivity assumption.

2. Observations in each treatment follow normal distribution, and because of this assumption the test of significance for the differences among the treatment means is carried out. According to the model,

$y_{ij} = m + t_i + e_{ij}.$

Each observation and its components in the sample are given in Table 18.4.

TABLE 18.4. The additive model on which each observation is made.

Variety	Replication	Observed value	General mean *(m)*	Treatment effect	Error effect $(y_{ij}-m-t_i)$
V1	I	3	8	−5	0
	II	4	8	−5	+1
	III	3	8	−5	0
	IV	2	8	−5	−1
V2	I	6	8	−1	−1
	II	8	8	−1	+1
	III	7	8	−1	0
	IV	7	8	−1	0
V3	I	15	8	+6	+1
	II	15	8	+6	+1
	III	14	8	+6	0
	IV	12	8	+6	−2

Note that $y_{ij}-m$ measures the total deviation, which is equivalent to treatment deviation and error.

In completely randomized design, the variability is caused by only one source, namely the treatments and hence this design is also called one-way classification or single factor or single classification design.

Null hypothesis (H_o): The null hypothesis before the start of the experiment is that there is no difference among the treatment means, i.e.,

$$\mu_1 = \mu_2 = \mu_3 = 0, \text{ and } \Sigma T_i = 0, \Sigma \varepsilon_{ij}^{2}$$

is normally independently distributed with mean 0 and variance σ^2. If the null hypothesis is true, the treatment mean square and within treatment mean square are both estimates of common variance σ^2. This is true as we are sampling from the same population. Figure 18.3 explains this principle:

Since the treatment SS, within treatment SS, and the total SS are based on different degrees of freedom, they are not easily explainable and comparable, and therefore they are expressed as variances (mean squares) by dividing by respective degrees of freedom. This technique is called analysis of variance (ANOVA) as the total variability is partitioned into two components, namely between treatment variability and within treatment according to the linear model of the design. Finally, the analysis of variance table is formed as shown in Table 18.5.

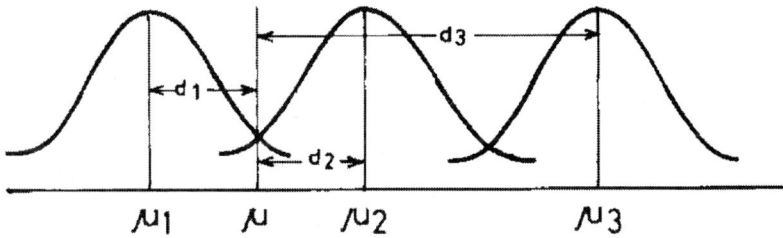

FIGURE 18.3. Explains the null hypothesis.

TABLE 18.5. ANOVA example.

Source of variation (SV)	Degrees of freedom (df)	Sum of squares (SS)	Variance or mean square (MS)	F	F from tables 5%	F from tables 1%
Between varieties	2	248	124	111.7**	4.26	8.02
Experimental error	9	10	1.11			
Total	11	258				

**Significant at P = 0.01

Between treatment variance or mean square is given by: between treatment SS / $t-1$, where t is the number of treatments. It provides an unbiased estimate of variance σ^2 of the population of observations. The within treatment mean square or simply experimental error or error mean square is obtained by dividing the error SS by its degrees of freedom $n-t$ (in this case $12-3 = 9$). It is another unbiased estimate of the quantity σ^2. These two variances are independent. The ratio s^2 (between)/ s^2 (within), i.e., $124/1.11 = 111.7$ follows F distribution with n_1 and n_2 degrees of freedom when H_o is true. Here $n_1 = 2$ and $n_2 = 9$. The F values from F table for n_1 and n_2 degrees of freedom at $P = 0.05$ and $P = 0.01$ levels of probability are 4.26 and 8.02, respectively. (See Appendix, Table A.4.)

F RATIO AND ITS SIGNIFICANCE

The reader must be familiar with the use of the F table as it is frequently used in experimental design. The F value is a statistic. The F values are given in the table called the F table for different probability levels and degrees of freedom (see Appendix, Table A.4). We are most concerned with two levels 0.05 and 0.01, and n_1 and n_2 degrees of freedom; n_1 represents degrees of freedom associated with the larger mean square, and n_2 the smaller mean square. The values given in the F table across the top row refer to the degrees of freedom associated with the larger mean square (n_1), and the column shows degrees of freedom (n_2) of the smaller mean square.

USING THE F TABLE

First locate the column marked n_1 degrees of freedom for greater mean square (numerator), and then go down under this column until you reach the degrees of freedom associated with error variance, i.e., n_2. In our example it is 9. Then go across this row until you reach the degrees of freedom associated with treatment mean square n_1. In our example n_1 is 2. At this junction, find the critical F value for n_1 and n_2 (2, 9) degrees of freedom. It is 4.26 at 0.05 probability level, and 8.02 at 0.01 level of probability. If the null hypothesis is true, i.e., $\mu_1 = \mu_2 = \mu_3 = 0$, the treatment mean square and the error mean square should be the same as both estimates σ^2 (this is because we are sampling from the same population), and consequently the F ratio value is equal to 1. If the computed F value is higher than the table value of F, it indicates that among treatment mean square is significantly greater than the within treatment mean square. We infer that the treatment means differ significantly and consequently our hypothesis of equality of variety means is rejected, i.e., there is evidence that the treatment means differ significantly.

F *Distribution*

The shape of the F curve is determined by the degrees of freedom associated with F ratio. As stated earlier, the F ratio depends upon the numerator degrees of freedom and denominator degrees of freedom. We want the F value to the right of which a certain proportion of the area under the curve falls. The $F_{2, 9, 0.05} = 4.26$ means that an ordinate drawn at $F_{2, 9} = 4.26$ will divide the sampling distribution of $F_{2, 9}$ at a point, when the proportion of the area under the curve to the right is 0.05, i.e., 5 percent ($\alpha \times 100 = 0.05 \times 100$) of the area under the curve falls at $F_{2, 9} = 4.26$. For $\alpha = 0.01$, $F_{2, 9} = 8.02$, i.e., 1 percent of the sampling distribution of $F_{2, 9}$ falls to the right of

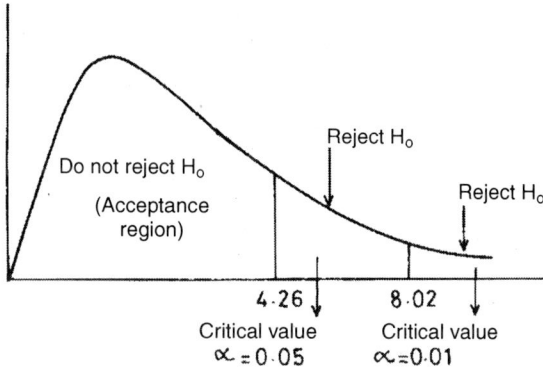

FIGURE 18.4. Sampling distribution of F for $df_1 = 2$, $df_2 = 9$, at 0.05 and 0.01 levels of significance. Areas of rejection and nonrejection are shown.

this point. These F values are considered when the null hypothesis is true, i.e., when the population means are equal. The sampling distribution of F is shown in Figure 18.4. Since the observed F is greater than table F at 5 percent and 1 percent, the data discarded the null hypothesis and we believe that real differences exist among the treatment means. For a fixed α, calculated F value is compared with the table value of F.

ASSUMPTIONS

For the validity of the F test in ANOVA, the following assumptions are made.
1. The observations are independent.
2. The parent population from which observations are taken is normal.
3. The treatment and error effects are additive.
4. Errors are homogeneous.

Distribution of Error or Residual Variation

One of the assumptions is that errors are distributed with mean zero and with a common variance. In our example, the residuals are distributed as follows:

Residual variation of	Replication I	Replication II	Replication III	Replication IV	Total
V1	0	+1	0	−1	0
V2	−1	+1	0	0	0
V3	+1	+1	0	−2	0

For better understanding, the residuals are exhibited in the dot diagrams in Figures 18.5, 18.6, 18.7, and 18.8. If the null hypothesis is true, the observations follow normal distribution with mean zero and variance σ^2.

Comparison of Treatment Means

When the null hypothesis is rejected, we infer that differences between treatments exist, but it does not specify where the differences are. Differences among the means are tested by the least significant difference (LSD) test.

LEAST SIGNIFICANT DIFFERENCE

The statistic t is defined by
$$t = d\,/\,SE_d = (\bar{x}_1 - \bar{x}_2)\,/\,[\sqrt{s^2\,(1/n_1 + 1/n_2)}], \text{ where}$$
$\bar{x}_1$ and $\bar{x}_2$ are the two means whose difference is tested.
s^2 = error variance and n_1 and n_2 are the sample sizes of the treatments X1 and X2.

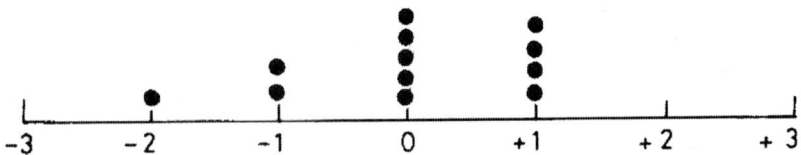

FIGURE 18.5. Overall residuals in the experiment are shown in the dot diagram.

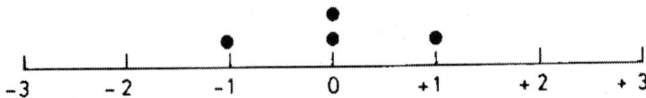

FIGURE 18.6. Dot diagram showing the distribution of residuals of V1 treatment.

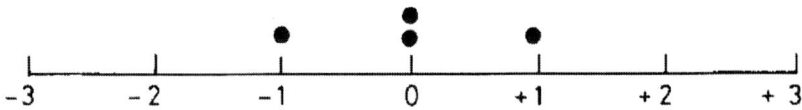

FIGURE 18.7. Dot diagram showing the distribution of residuals of V2 treatment.

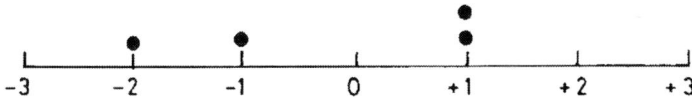

FIGURE 18.8. Dot diagram showing the distribution of residuals of V3 treatment.

When $n_1 = n_2 = r$, then $t = (\bar{x}_1 - \bar{x}_2) / \sqrt{2s^2 / r}$, i.e., $(\bar{x}_1 - \bar{x}_2) / \sqrt{(2s / r)} \geq t_\alpha$

We call $[t_\alpha \sqrt{(2s^2 / r)}]$ least significant difference (LSD). If we want to declare the difference between any pair of means significant, then the difference between the means should be equal to or greater than LSD. In our example,

$$\text{LSD} = t_\alpha \sqrt{2s^2 / r} = 2.262 \sqrt{(2 \times 1.11) / 4} = 1.68$$

where 2.262 is the table value of $t_{0.05,9} = 2.262$) (see Appendix, Table A.5).

Testing Treatment Differences (Conclusion)

After computing LSD, arrange the treatment means in decreasing order

Treatment means	$\bar{V3}$	$\bar{V2}$	$\bar{V1}$
Mean values	14	7	3

Since the difference between any two means is greater than LSD, the three varieties differ significantly from one another. V3 records the highest yield and an increase of 100 percent and 366.7 percent over V2 and V1, respectively.

Standard error of the mean $= s / \sqrt{r} = [\sqrt{(1.11)} / \sqrt{4}] = 0.53$ kg/plot.

Coefficient of variation: $CV = (s / \bar{x}) \times 100 = [(\sqrt{1.11}) / 8] \times 100$
$$= 13.1 \text{ percent}$$

Effect of Variability within Treatments and **F** Ratio

Within treatment variability in an experiment is very important as it can obscure the variance between treatments and, consequently, the experiment will not reveal the real (significant) difference. This is explained with a suitable example to understand the importance of variability within the treatment in different replications in an experiment. It further shows the importance of experimental units and their uniformity in all respects (except treatments), which are the main causes for unknown variability.

Example: Consider the same experiment with the results shown in Table 18.6.

Suppose experiment 1 was conducted by X and experiment 2 by Y. The ANOVA for X has already been given in our example. The ANOVA of experiment 2 is given in Table 18.7.

TABLE 18.6. Results of two experiments.

	Experiment 1				Experiment 2		
	Varieties				Varieties		
Rep.	V1	V2	V3	Rep.	V1	V2	V3
I	3	6	15	I	7	3	19
II	4	8	15	II	2	15	8
III	3	7	14	III	2	7	24
IV	2	7	12	IV	1	3	5
Total	12	28	56	Total	12	28	56
Mean	3	7	14	Mean	3	7	14

TABLE 18.7. Resulting ANOVA for experiment 2.

Source of variation	df	SS	MS	F	Table F 5 percent	1 percent
Between varieties	2	248	124	3.1ns	4.26	8.02
Within varieties	9	360	40			
Total	11	608				

In these two experiments, X and Y reported the same mean values for V1, V2, and V3. In the experiment conducted by X, the variability within treatments (error mean square = 1.11) is lesser than the error mean square of 40 reported by Y. X's data showed significant results and those of Y did not. The F ratio in X's experiment was high (111.7) as within the treatment mean square, i.e., the experimental error is less. The standard error of the mean of the experiment conducted by Y was 3.16 and the CV 79.1 percent, which were higher when compared with those of X's experiment.

Precision and Statistical Significance

Significant effect depends not only on the estimated treatment effect but also on precision of the experiment. The treatment effect is based on normal distribution. The LSD is at least twice the SE and three times the SE at 5 percent and 1 percent level of probability to declare significant results

COMPUTATION PROCEDURE FOR COMPLETELY RANDOMIZED DESIGN

This procedure is explained with reference to the example mentioned earlier.

Step 1. Compute the totals and means of the treatments (varieties). If V1, V2, and V3 indicate variety total and means, then V1 = 12, V2 = 28, V3 = 56 and V1 = 3, V2 = 7, and V3 = 14; grand total (GT): V1 + V2 + V3, i.e., 12 + 28 + 56 = 96 and grand mean (GM) = 96 = 12 = 8.
Step 2. Compute correction factor (*CF*): $CF = GT^2/n = 96^2/12 = 768$. The correction factor corrects simple sum of squared quantities to the sum of squared deviations, i.e., SS and n = total number of observations.
Step 3. Compute variety sum of squares (VSS):

$$VSS = (V1^2 + V2^2 + V3^2)/r - CF = (12^2 + 28^2 + 56^2)/4 - CF$$
$$= 1016 - 768 = 248$$

r = number of replications; the divisor 4 relates that each of the values (12, 28, and 56) were obtained by the total of four observations.

Step 4. Find the total SS (TSS): TSS = $3^2 + \ldots + 12^2 - CF = 1026 - 768$ = 258

Step 5. Find the error sum of squares (ESS). Error SS = $TSS - VSS$ = 258 − 248 = 10
Step 6. Form the ANOVA. See Table 18.8.
Step 7. Find mean squares for treatments and error by dividing with the degrees of freedom associated with them.
Step 8. Find the F value.

$$F = \frac{\text{Mean square for treatments}}{\text{Mean square for error}}$$ with n_1 and n_2 df. Here it is 2, 9.

Step 9. Extract the F ratio value from F table. They are 4.26 and 8.02 for 0.05 and 0.01 levels of probability, respectively.
Step 10. If the computed F is equal to or greater than the table value of F, the difference in the means of the treatments is statistically significant. If F is significant at $P = 0.05$, it is significant, and if F is significant at $P = 0.01$, the differences are highly significant. A single * indicates significance at $P = 0.05$; double ** is significant at $P = 0.01$ probability level; *** indicates significance at $P = 0.001$.

How do we find the divisor for calculating the treatment sum of squares?
1. When calculating treatment sum of squares, each treatment total is squared and divided by the number of values involved in the sum of the treatment total.
2. The sum of squares will not be negative.
3. The standard deviation calculated from the error mean square (square root of error SS) should not be higher than the range of the observations.

TABLE 18.8. Computed ANOVA.

Source of variation	df	SS	MS	F	Table F 5 percent	Table F 1 percent
Treatments	2	248	124	111.7**	4.26	8.02
Error	9	10	1.11			
Total	11	258				

** = significant at P = 0.01

CRD with Unequal Numbers of Replications for Treatments

Sometimes, the experimenter may not be in a position to accommodate an equal number of replications for each treatment due to shortage of experimental materials, want of experimental units, etc. Under such circumstances, conduct the experiment with an unequal number of replications. An example follows.

Example 1: Let there be four treatments T1, T2, T3, and T4, with 6, 5, 8, and 7 replications, respectively, for a wheat crop varietal trial. The plot yields (kg/plot) are recorded as follows:

Replication	Treatments			
	T1	T2	T3	T4
I	2	7	3	10
II	3	8	1	12
III	4	4	2	8
IV	3	10	1	5
V	1	6	2	12
VI	5		4	10
VII			2	9
VIII			3	
Total	18	35	18	66
Mean	3.00	7.00	2.25	9.43

Grand total: 137.

Number of replications = $n_1 = 6$, $n_2 = 5$, $n_3 = 8$, $n_4 = 7$, $n = \Sigma n_i = 26$

Procedure for Computation

Compute $CF = GT^2/n = 137^2/26 = 721.88$

$$\text{Treatment SS} = T1^2/n_1 + T2^2/n_2 + T3^2/n_3 + T4^2/n_v - CF$$
$$= 18^2/6 + 35^2/5 + 18^2/8 + 66^2/7 - CF = 54.00$$
$$+245.00 + 40.50 + 622.29 - 721.88$$
$$= 961.79 - 721.88 = 239.91$$
$$\text{Total SS} = 2^2 + ... + 9^2 - CF = 1035.00 - 721.88 = 313.12$$

The ANOVA results are as follows.

Source of variation	df	SS	MS	F
Treatments	3	239.91	79.97	24.02**
Error	22	73.21	3.33	
Total	25	313.12		

** significant at $P = 0.01$

As F is significant, the treatment means can be compared using the least significant difference (LSD) test. The LSD at the 5 percent level is given by:

For Comparing T1 and T2

$$\text{LSD} = t \times \text{SEd} = 2.074 \times \sqrt{3.33(1/6+1/5)} = 2.29$$

For comparing T1 and T3: $\text{LSD} = 2.2074 \times \sqrt{3.33(1/6+1/8)} = 2.04$

For comparing T1 and T4: $\text{LSD} = 2.2074 \times \sqrt{3.33(1/6+1/7)} = 2.11$

For comparing T2 and T3: $\text{LSD} = 2.074 \times \sqrt{3.33(1/5+1/8)} = 2.16$

For comparing T2 and T4: $\text{LSD} = 2.074 \times \sqrt{3.33(1/5+1/8)} = 2.22$

For comparing T3 and T4: $\text{LSD} = 2.074 \times \sqrt{3.33(1/8+1/7)} = 1.96$

Conclusion

Treatments:	T4	T2	T1	T3
Mean values:	9.43	7.00	3.00	2.25

Comparing Means at P = 0.01

LSD for comparing two means:

t- value for comparing two means at $P = 0.01$ for 22 *df* is 2.819 (see Appendix, Table A.4).

For comparing T1 and T2: $\text{LSD} = 2.819 \times \sqrt{3.33(1/6+1/5)} = 3.11$

For comparing T1 and T3: $\text{LSD} = 2.819 \times \sqrt{3.33(1/6+1/8)} = 2.78$

For comparing T1 and T4: $\text{LSD} = 2.819 \times \sqrt{3.33(1/6+1/7)} = 2.86$

For comparing T2 and T3: $\text{LSD} = 2.819 \times \sqrt{3.33(1/5+1/8)} = 2.93$

For comparing T2 and T4: $\text{LSD} = 2.819 \times \sqrt{3.33(1/5+1/7)} = 3.01$

For comparing T3 and T4: $\text{LSD} = 2.819 \times \sqrt{3.33(1/8+1/7)} = 2.66$

Conclusion

Treatments:	T4	T2	T1	T3
Mean values:	9.43	7.00	3.00	2.25

Differences between pairs of means may also be shown in a table format:

Treatment	Means	T3 2.25	T1 3.00	T2 7.00	T4 9.43
T3	2.25	—	0.75	4.75**	7.18*
T1	3.00	—	—	4.00*	6.43*
T2	7.00	—	—	—	2.43
T4	9.43	—	—	—	—

*significant at $P = 0.05$; **significant at $P = 0.01$

The cells with asterisks indicate that the corresponding differences are statistically significant. Treatments indicated by a line do not differ. Treatments with no line do differ. Thus, treatments 2 and 4 differ from treatments 3 and 1, but there is no difference between treatments 2 and 4 and also between 3 and 1.

The standard errors of the means are calculated as follows:

Treatment	n	Treatment mean	SEm (kg/plot)
1	6	3.00	$\sqrt{3.33/6} = 0.65$
2	5	7.00	$\sqrt{3.33/5} = 0.36$
3	8	2.25	$\sqrt{3.33/8} = 0.82$
4	7	9.43	$\sqrt{3.33/7} = 0.69$

Example 2: Because of the importance of the design with unequal replications, a second example with five treatments is explained.

A simple analysis of variance is shown with five treatments M1, M2, M3, M4, and M5 (different methods of cultivation in corn crop with unequal number of replications). The final yield data are shown here.

Replication	M1	M2	M3	M4	M5	Total
I	13	23	24	18	21	
II	15	19	22	17	22	
III	11	19	28	21	17	
IV	15	23	26	15	17	
V	16	—	24	—	20	
VI	—	—	27	—	21	
Number of replications:	5	4	6	4	6	25
Total	70	84	151	71	118	494
Mean	14.0	21.0	25.2	17.8	19.7	

GM = 19.8

Sufficient space was not available for M1, M2, and M4, and so we have fewer replications. The average yield varies from 14.0 to 25.2 kg/plot. The experimenter wants to test whether the methods of cultivation are equal. This is stated in the form of null hypothesis as $\mu_1 = \mu_2 = \mu_3 = \mu_4 = \mu_5$. The variance among the means of methods of cultivation is compared with the variance within the methods. If the variation among the methods is significantly greater than the variance within the methods, the added variance must be due to real difference among the methods rather than the chance factor.

Computation: $CF = 494^2/25 = 9761.44$

> *Step 1.* Total SS $= 13^2 + 15^2 + \ldots + 21^2 - 494^2/25 = 10224 - 9761.44$
> $= 462.56$
> *Step 2.* SS among the treatments: $70^2/5 + 84^2/4 + 151^2/6 + 71^2/4 + 118^2/6$
> $- CF = 10,125.08 - 9761.44 = 363.64$
> (see Table 18.9)
> *Step 3.* SS within the treatments $= 462.56 - 363.64 = 98.92$

We know that the experimental error is an unbiased estimate of the quantity, σ^2. The F value for 0.05 and 0.01 levels of probability are 4 and 20 *df* are 2.87 and 4.43, respectively. Since the computed F exceeds both these levels, the means are said to differ significantly at $P = 0.01$. The treatment MS is greater than the within mean square and hence our null hypothesis is rejected. Significant differences exist among the means. To determine as to where they are, an LSD test is applied.

LSD is calculated as:

$$LSD = t_{20df,0.05} \times SEd = 2.086\sqrt{s^2/n_1 + s^2/n_2}$$

LSD for testing M1 and M2 $= 2.086 \times \sqrt{(4.95/5 + 4.95/4)} = 3.11$
Mean of M1 $= 14.00$ and mean of M2 $= 21.00$. Since the difference is greater than LSD, M2 records significantly higher yield than M1.

$$\text{M1 versus M3: LSD} = 2.086 \times \sqrt{(4.95/5 + 4.95/6)} = 2.81$$

M1 mean $= 14.00$ and M3 mean $= 25.2$: M3 is significantly higher than M1.

$$\text{M1 versus M4: LSD} = 2.086 \times \sqrt{4.95/5 + 4.95/4} = 3.11$$

TABLE 18.9. ANOVA results.

Source of variation	df	SS	MS	F	Table F 5 percent	Table F 1 percent
Among treatments	4	363.64	90.91	18.37**	2.87	4.43
Within treatments	20	98.92	4.95			
Total	24	462.56				

M1 mean = 14.00 and M4 mean = 17.8. M4 is significantly greater than M1.

M2 versus M3: LSD = $2.086 \times \sqrt{4.95/4 + 4.95/6} = 2.22$

M2 mean = 21.00. M3 mean = 25.2. M3 is significantly higher than M2.

M2 versus M4: LSD = $2.086 \times (2 \times 4.95)/4 = 4.89$

(Since M2 and M4 had the same number of replications i.e., 4.) M2 mean is 21.00 M4 mean is 7.8. The difference is less than LSD and hence M2 and M4 are equal: M2 = M4. M2 versus M5: LSD = 2.22; therefore M2 = M5.
  M3 versus M4 LSD = 2.22. Therefore, the difference 7.4 is significant.
  M3 versus M5: LSD = 2.086 . M3 is significantly higher than M5.
M4 and M5: LSD = 2.22; therefore M 4 = M5.

Advantages and Disadvantages of Completely Randomized Design

Advantages

1. We can test a number of treatments.
2. Replications for different treatments may be kept unequal for practical considerations. For example, in agricultural experiments, the supply of seed for certain varieties may be sufficient for only lesser numbers of plots and in such situations, fewer replications may be used for such varieties. Sometimes, the control treatment needs to be more accurately assessed than the other treatments to compare the control with new treatments. Under such situations, the control treatment may have more replications.
3. Layout of the design is easy.
4. Statistical analysis is easy. Even if a treatment value is lost (damage caused by pests, diseases, birds, etc.), the analysis can be carried out by discarding the value or values that are lost.
5. We get more error degrees of freedom and thereby we get lesser error mean square which will increase the F value.

Disadvantages

1. Since randomization is unrestricted, systematic error, if any present, will render our estimate of variance (σ^2) an invalid estimate of random variation.
2. CRD is suitable only if the experimental units are uniform or homogeneous.

Relation Between t and F Values When the Number of Treatments are Two

Sometimes the researcher conducts the experiment with only two treatments. The researcher can test the difference between the treatment means either by t or F test as there is a relation between t and F statistics when only two treatments are compared. The relation is $t = \sqrt{F}$ or $t^2 = F$. It is explained with an example.

Example: Consider an experiment with two treatments and four replications conducted in completely randomized design give the following results $(n_1 = 4, n_2 = 5)$

Replication	Treatment A	Treatment B
I	4	4
II	2	6
III	4	3
IV	5	5
V	—	7
Total	15	25
Mean	3.75	5.00

Computations

Grand total (GT) = 40
Treatment totals: A = 15, B = 25
Correction factor (CF): $40^2/9 = 177.78$
Treatment SS: $15^2/4 + 25^2/5 - CF = 161.25 - 177.78 = 3.47$
Total SS: $4^2 + 2^2 + \ldots + 7^2 - CF = 196 - 177.78 = 18.22$
(see Table 18.10)

TABLE 18.10. ANOVA.

Source of variation	df	SS	MS	F	Table F 5 percent	1 percent
Treatments	1	3.47	3.47	1.647ns	5.59	12.25
Error	7	14.75	2.107			
Total	8	18.22				

t test $t = (\bar{A} - \bar{B}) / \text{SEd} = (5.00 - 3.75) \sqrt{s^2(1/n_1 + 1/n_2)} = 1.25$

$\sqrt{2.107}(1/4 + 1/5) = 1.25/0.974 = 1.28336$

$t = 1.28336$, $t^2 = (1.28336)^2 = 1.647$, thus $t^2 = F$.

Since the calculated F (1.647) fails to exceed the table F ($F_{1,7,0.05}$), 5.59, we do not reject the null hypothesis and conclude that no evidence was found to suggest that the treatments made a difference in the population.

EXERCISES

18.1. Briefly describe the completely randomized design.

18.2. Write the mathematical model for CRD.

18.3. Mention the advantages and disadvantages of CRD.

18.4. An experimenter is interested in the testing of two drugs on guinea pigs. He selects 18 animals and injects drug A. Drug B is very costly and he could select only 9 animals. After injection, he gave a pain-sensitivity test and obtained the following results. Should he reject the null hypothesis at $P = 0.05$ level?

Drug A: 69 68 70 73 69 68 67 69 73 69 70 68 73 68 71 72 71 70
Drug B: 88 88 89 82 86 84 75 82 81

18.5. Ten varieties of rice were tested for their quality. 1,000 grain weights were recorded in 10 samples and the data were analyzed in a CRD model. Results are given in the analysis of variance table:

Source of variation	df	SS	MS	F	Table F	
					5 percent	1 percent
Between varieties	9	352.36	39.15	195.75	1.93	2.61
Within varieties	90	17.72	0.20			
Total	99	370.08				

Mean values of the varieties are as follows:

$V1 = 16.22$, $V2 = 19.31$, $V3 = 13.30$, $V4 = 14.82$, $V5 = 18.35$
$V6 = 16.90$, $V7 = 16.45$, $V8 = 16.91$, $V9 = 13.56$, $V10 = 17.88$

Find standard error, standard error of difference, and group the varieties based on LSD.

18.6. In a completely randomized design, five wheat varieties were tested with ten replications for each. The data (yield in kg/plot) are given:

Var/Rep.	I	II	III	IV	V	VI	VII	VIII	IX	X
V1	22	20	25	23	29	31	26	28	24	24
V2	27	29	21	35	31	31	33	28	30	25
V3	25	30	32	26	23	27	25	30	33	28
V4	37	33	36	40	34	32	29	34	33	38
V5	42	38	41	45	43	48	44	45	39	42

1. Test the homogeneity of variances.
2. Write the ANOVA table and test for differences in the mean values of varieties.
3. Compare the average mean values in V4 and V5 with the three remaining varieties using Scheff's multiple comparison procedure. Use 95 percent confidence coefficient [consider the contrast: $1/2 (\mu_4 + \mu_5) - 1/2 (\mu_1 + \mu_2 + \mu_3)$].

18.7. Four kinds of feed (F) were each fed to a lot of 5 baby chicks. The gains in weight (g) were noted as follows.

Feed/chick	Gain (g)				
	1	2	3	4	5
F1	55	49	42	21	52
F2	61	80	30	89	63
F3	42	60	70	65	90
F4	150	140	170	120	150

Analyze the data and test the equality of the means of the feed.

18.8. In a germplasm bank in an experimental station, four new types of rice varieties were selected and compared. There were only 23 plots available and the yields were recorded as given:

Variety	Plot yield (kg)						
V1	30	74	46	58	62	38	
V2	50	38	66	62	44	58	80
V3	18	56	34	24	66	52	
V4	88	78	60	76			

1. It is suggested that the yielding capacity of the varieties are the same. Examine this suggestion and report your conclusion. State the model and your assumptions.

2. Test whether the yields of V1 and V2 are the same.
3. In this experiment, V1 is a standard variety and V2, V3, and V4 are new varieties. V3 gives the worst yield and is discarded. Another experiment is planned with the plots available to be allotted randomly between the varieties V1, V2, and V4 with V2 and V4 having the same number of replications. Comparison between V1 and V2 and between V1 and V4 are of particular interest. How would you plan the experiment?

18.9. A factory employs a number of workers. The manager wants to test whether the workers take equal time for their lunch breaks. He recorded the time taken (minutes) by four workers on different days. The observations are given as follows:

Worker	Time (min)											
W1	31	19	39	22	28	32	36	40	20	22	29	33
W2	33	32	29	29	32	25	28	22	34	36		
W3	28	26	29	29	24	25	36	32	36	27	26	26
W4	24	22	30	22	30	30	28	30				

Do you think that the time taken by the workers is chance variation?

18.10. Twenty-four rats were selected to study the effect of four different types of diets D1, D2, D3, and D4. The diets were given for a two-week period and after the experimental period, the gains in weights were recorded. Using $P = 0.01$ as the level of significance, test the null hypothesis that the diets have no different effects on weights.

Diets	Weights (g)					
D1	20	18	18	18	16	15
D2	14	20	18	18	16	16
D3	30	31	31	28	28	24
D4	21	19	18	17	16	15

18.11. The amounts of food (kg) consumed by a pet during different months of the year are given. Test whether there is any difference in the amounts consumed in the different months.

18.12. A chemist collected soil samples from four different soil types and estimated the moisture content. Ten determinations were made in

March	June	September	December
6.9	6.8	7.0	7.1
7.1	6.6	6.8	7.6
7.0	6.6	6.0	7.3
6.9	6.4	7.0	7.8

each soil type. Calculate the ANOVA to obtain an estimate of sampling variation within a soil, calculate the SEm of moisture content of a soil, and the SE of difference between two soil means. Test the hypothesis that there is no variation of soil moisture content between different soils

Soil Type	i	II	III	IV	V	VI	VII	VIII	IX	X
					Replication					
A	11.7	12.3	15.1	10.5	8.3	9.2	13.0	10.8	9.4	15.3
B	7.0	9.2	3.1	6.7	4.5	7.0	11.6	5.7	5.8	5.3
C	8.7	9.2	8.0	3.2	10.1	10.5	7.3	7.7	8.1	5.3
D	15.3	7.2	14.0	9.3	6.7	8.1	11.5	10.0	6.9	3.7

18.13. Blood sugar was estimated from ten people and two determinations were made. The results are given here.

Source of variation	df	SS	MS	F	Table F 5 percent 1 percent
Among people	9	1710	190		
Within	10	100	10		

a. What model has been used?
b. Should the null hypothesis be rejected?
c. Compute the intraclass correlation.

18.14. A, B, and C are the new fertilizers and D is the standard one. An experiment was laid out to find the efficacy of the new fertilizers over the standard in equally sized plots in wheat crop. The yields were recorded as follows:

A: 29.8, 32.2, 31.4, 33.2, 32.5, 32.4   B: 29.1, 31.0, 28.0, 26.4, 27.4, 30.0
C: 26.8, 31.3, 31.5, 28.1, 27.2, 32.5   D: 30.8, 24.9, 30.9, 30.6, 31.6, 26.0

Is there any difference in the yield produced by these fertilizers? Use $P = 0.05$.

18.15. In a varietal trial in rice, the experimenter reported the following results:

Source of variation	df	SS	MS	F	Table F 5 percent 1 percent
Among varieties	—	2392	—		
Within varieties	20	3600	—		
Total					

Complete the analysis of variance table and show that the SEm is equal to 6. The means of the varieties are: V1 = 80, V2 = 85, V3 = 60, V4 = 82, and V5 = 93. Carry out the LSD test and write your conclusions.

18.16. Three diets, D1, D2, and D3, were formulated, and their effect on weight loss was examined on people of the same age, sex, and other activities. After a certain period, the weight loss (kg) was recorded. Test the differences in weight loss due to diets.

D1: 2, 4, 3, 3, 3, 4
D2: 4, 5, 6, 4, 5, 6, 4, 2
D3: 7, 6, 9, 9, 8, 7, 6, 10, 12

18.17. In an experimental station, three fertilizers (F1, F2, and F3) were in common use. A new fertilizer (F4) was tested with the old ones to compare their efficacy. Fertilizers were allotted randomly to the plots. The plot yields (kg) are as given:

F1: 18, 22, 19, 21, 23, 20, 19   F2: 19, 26, 21, 17, 20, 22, 21
F3: 24, 21, 23, 22, 19, 20, 20   F4: 18, 19, 20, 17, 19, 20, 18

   a. What is the model used in the experiment?
   b. Test for differences in the mean yields.
   c. What are the point estimates for the different fertilizers?
   d. Compare F1, F2, F3 versus F4.

18.18. The plot yields (kg) of three varieties of oats in an experiment are:
   V1: 22, 19, 22, 24, 17, 21, 19, 25, 15, 22, 19
   V2: 24, 23, 25, 24, 23, 30, 21, 22, 34, 22, 23
   V3: 26, 22, 24, 20, 22, 17, 22, 16, 9, 18, 20
   a. Plot the data.
   b. Are the data symmetrical about their mean values? (If not symmetrical, the assumption of normality will not hold.)
   c. Is there any value that does not represent the same population? (If you find any outlier you must find out the reason.)
   d. Do you find that the variability in each set of observations is the same? (If not, assumption of the same variance is not correct.)

18.19. What is the usual model for a one-way classification analysis? What are its possible shortcomings?

18.20. Five treatments were each assigned to six experimental units. The total SS was 3240. The variance of each treatment mean was equal to 12. Reproduce the ANOVA table and test the hypothesis of equal treatment means. Let $P = 0.05$.

18.21. Six rice plants of a certain variety were grown with fertilizer and six plants of the same variety were grown without fertilizer. The heights (cm) were recorded at harvest stage as:

| Without fertilizer | 90 | 100 | 99 | 98 | 99 | 97 |
| With fertilizer | 100 | 105 | 110 | 99 | 102 | 120 |

a. Test the hypothesis: H_o: $\mu_1 = \mu_2$ by ANOVA method
b. Test H_o: $\mu_1 = \mu_2$ by a t test (paired)
c. What is the relationship between F and t statistics?

Test the null hypothesis using $P = 0.01$ that the fertilizer has no effect on the heights.

18.22. An agricultural experimental station wishes to find out the yields of five varieties of soybean crop and conduct an experiment in completely randomized design with four plots each. The yields recorded are given in kg/plot.

A	B	C	D	E
4	7	5	11	4
6	13	4	10	8
4	10	4	9	6
10	12	9	14	9

Can we conclude that the mean yields of the soybean varieties are the same $(P = 0.05)$?

18.23. Six varieties of rice were tested. The following data are the yields/plot (kg). Test for homogeneity of means.

Variety					Yields/plot (kg)				
	I	II	III	IV	V	VI	VII	VIII	IX
1	7	4.5	7.4	7.4	5.0	5.9	6.4	6.3	5.0
2	8	5.3	5.4	4.0					
3	6.9	6.8	7.6	8.1	9.0	9.0	11.2	7.4	5.7
4	9.0	8.2	7.7	8.5	7.2	9.0	9.5	8.5	
5	5.9	9.2	8.7	8.8	7.9	6.8			
6	4.9	8.2	7.9	6.9	9.0	7.9			

18.24. Why is the assumption of normality to be made in ANOVA? If the experiment is properly randomized, is this assumption necessary?

18.25. Given four treatments A, B, C, and D, in CRD design with five replications, show the point estimates of the treatments in a figure. The data relate to the plot yields of wheat varieties.

Treatments		Yield (t/ha) Replications			
	I	II	III	IV	V
A	62	37	44	58	63
B	48	70	64	52	81
C	77	65	68	61	74
D	49	64	42	54	59

18.26. Draw a layout plan for testing 12 soybean varieties with three replications. Assume CRD. Show how you will randomize the treatments.

18.27. When will you resort to unequal number of replications in CRD?

REFERENCES

Hader, R .J. (1973). Random number table. *The Am. Statistician,* 27:82.

Montgomery, D.C. (1991). *Design and analysis of experiments,* Third edition. New York: John Wiley and Sons.

Chapter 19

Randomized Complete Block Design

TWO-WAY CLASSIFICATION

The purpose of any experiment is to detect even small differences that exist among the treatments under comparison. This is possible only when the experimental units used in the experiment are homogeneous. Experimental units such as plots (used in agricultural field experiments), animals, and human beings (used in medical research), etc., show variation among themselves. The degree and pattern of variation depend upon several factors such as soil fertility, stand of the crop, moisture content in the soil, temperature, sunshine, microorganisms, alkalinity, salinity, etc., over which experimenters have little or no control. Generally, plots adjacent to each other are more alike than plots that are farther away (Fisher, 1966). Detailed information about the nature and variation of soil can be obtained from uniformity trials for different characters. When such plots are used in experiments, the observed treatment differences are not only attributable to real treatment effects but also differences in the experimental units (plots). Similar situations occur in animal experiments. Animals exhibit variation in age, sex, initial weight, height, health conditions, breed, genetic constitution, and these factors mask the real effect of the treatment.

Example: Suppose one is interested in studying the effect of a drug in animals. Because animals differ in age, breed, constitution, etc., they may respond differently to the same drug and same dose. Thus, when animals are selected at random, the variability in the animals causes experimental error and the differences in the treatments are overlooked. To increase sensitivity, the sample animals must be divided into homogeneous groups (or strata or blocks). This principle leads to the formation of randomized complete block design (RCBD). The heterogeneity in the experimental units, although not completely eliminated, can be reduced or minimized by following the method of local control or blocking.

Randomized complete block design is commonly used in agricultural experiments. In this design, the field is divided into homogeneous blocks. Each block is again subdivided into plots. The number of plots is equal to the number of treatments, i.e., all treatments are included in each block. Thus each block is considered as an experiment or replication. Each block is compact, and the plots within a block, as they are adjacent to one another, are more alike than the plots farther away. Variations between the units within blocks contribute to the experimental error. Block to block variation is eliminated. The number of replications (blocks) depends upon the number of treatments and the error degrees of freedom. The number of replications is fixed in such a way that the design provides a minimum of 12 degrees of freedom for error.

THE SPECIFICS OF RANDOMIZED COMPLETE BLOCK DESIGN

Layout Procedure

Formation of Blocks

Block formation depends upon the fertility gradient in the field. If the fertility gradient is from north to south, then the blocks are formed across the gradient as shown in the plan (see Figure 19.1). The size and shape of the plot depends upon several factors such as crop, nature of treatments, availability of experimental area, cost of the experiment, availability of skilled personnel, etc. Suppose an experimenter wishes to compare yield potential of six varieties V1, V2, V3, V4, V5, and V6 of rice crop in four replications, a total of $6 \times 4 = 24$ plots are required. As there are six treatments, there will be six plots in each block. In this design, variation due to blocks is removed from error variation (Hicks, 1993). In animal or medical experiments, one may wish to study the effect of diet, ration, drug, breed, etc. In such situations, one should consider initial weight as the criterion for formation of blocks. Sometimes, age or breed may be the criterion for formation of blocks. Littermates may also be used as blocks, in which case each mate receives a treatment. When an experiment is conducted in several locations or laboratories, the locations or laboratories may be used as blocks.

Mead and Curnow (1983) reported that in experiments with plants in pots in greenhouses or laboratories where different treatments are applied to different pots, geographical blocks of similarly situated plots will usually prove useful. In medical trials in which the experimental unit is usually an individual person, factors such as age of the person, sex, height or weight, social class, medical history, and racial characteristics can be used as a basis for blocking. Hicks (1993) observed that blocking in an experiment is a

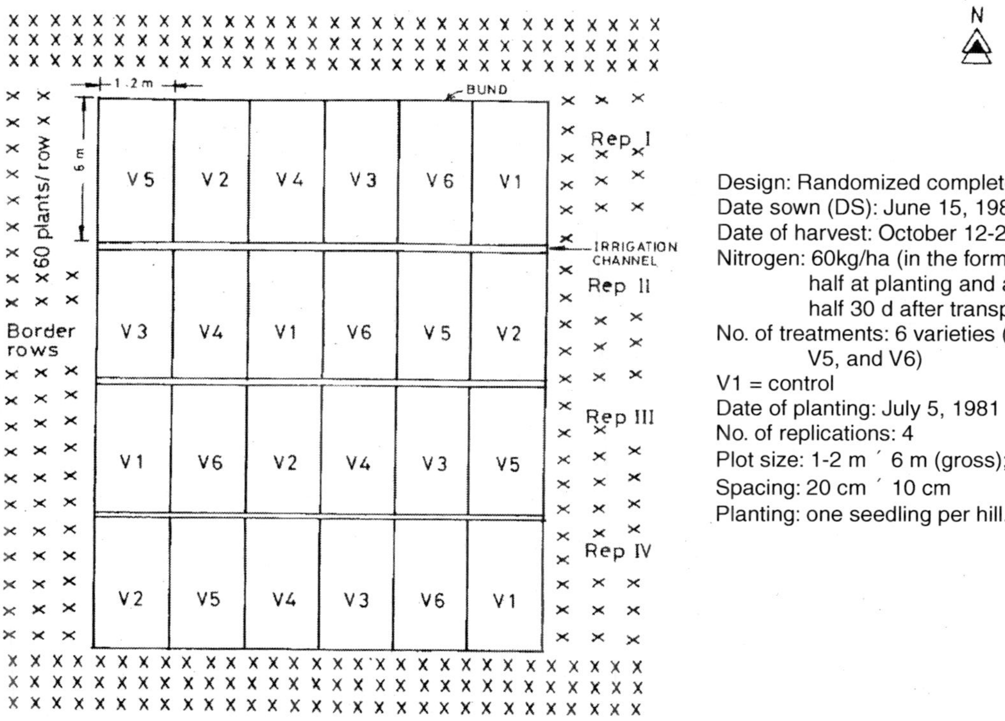

FIGURE 19.1. Model field layout plan for the study potential of different rice varieties in an agricultural research station.

Design: Randomized complete block
Date sown (DS): June 15, 1981
Date of harvest: October 12-20
Nitrogen: 60kg/ha (in the form of urea)
    half at planting and another
    half 30 d after transplanting
No. of treatments: 6 varieties (V1, V2, V3, V4,
    V5, and V6)
V1 = control
Date of planting: July 5, 1981
No. of replications: 4
Plot size: 1-2 m ´ 6 m (gross); 0-8 m ´ 5-8 m (net)
Spacing: 20 cm ´ 10 cm
Planting: one seedling per hill.

very useful procedure to reduce the experimental error. This design can be used in a situation in which the same experimental units (animals, people, parts, etc.) are measured before the experiment and then again at the conclusion of the experiment. The experimental units are the blocks and the block effects can be removed to reduce the experimental error. Anderson and McLean (1974) stated that the main reason for blocking is that the experimenter does not have a sufficient number of homogeneous experimental units available to run a completely randomized design with several observations per treatment combination. The data obtained from RCBD were analyzed as a completely randomized design. Any effects that should have been attributed to blocks would end up in the error term of the model. Thus another reason for blocking is to remove a source of variation from the error.

Randomization Procedure

There are six treatments and four replications in our example. First, randomly allocate the six treatments in the six plots in Block 1 (replication 1). Then allocate randomly six treatments in Block II. Similarly, randomize the treatments in Blocks III and IV separately and independently. In this way, better comparisons can be made between treatments as a more homogeneous environment is made to test the treatments in each block. Randomization is done either by drawing lots or using random number tables.

Example 1: The plan for three replications and four treatments (A, B, C, and D) is given:

Block 1	D	B	A	C
Block 2	A	B	C	D
Block 3	C	A	B	D

Example 2: For a field layout plan for evaluating six varieties of rice in four replications see Figure 19.1.

Suggestions to Improve Accuracy

When forming blocks, it is better to use rectangular shaped plots; the length is formed along the direction of the gradient.

1. In the middle of the experiment, one may intend to study some technique. It can be done by imposing the technique in all the plots within a block uniformly.

2. All agricultural operations should be performed uniformly in all plots within a block. Some agricultural operations are: irrigation, fertilizer application, weeding, intercultivation, harvesting, plant protection, etc.
3. Operations such as harvesting of the crop, recording of biometric observations, etc., should be done in all plots in a replication on the same day.
4. Discard one border row on all sides of each experimental plot to avoid border effect.
5. Use net plot size for calculating the plot yield.
6. Convert the plot yield into kg/ha and then analyze the data.
7. Plant the seedlings in the outskirts of the experimental field, i.e., the space outside the experimental plots should not be kept unplanted (vacant space).
8. A small bund in between the experimental plots to separate the treatments is necessary.
9. Form the irrigation channel in between the blocks.
10. After harvest, dry the produce and use dry weight for statistical analysis.
11. If different varieties are under comparison, the produce must be adjusted to the same moisture level.

Assumptions of the Model

1. Each observation (X_{ij}) is considered as a random independent sample of size one from the total observations.
2. Errors are normally independently distributed with mean zero and common variance, σ^2.
3. Block and treatment effects are additive, i.e., there is no interaction between blocks and treatments.

Example 1. Numerical example: The plot yields (kg/plot) recorded in the six treatments (V1, V2, V3, V4, V5, and V6) and in four replications are tabulated before the start of the statistical analysis.

Step 1. Form the treatment × replication table (see Table 19.1).

Step 2. Compute the following:

1. Correction factor (CF) $= (GT)^2 \, / \, rt = 230^2 \, / \, 4 \times 6 = 2204.2$
2. Replication SS: $[(R1^2 + R2^2 + R3^2 + R4^2)] / t - CF = (57^2 + 58^2 + 61^2 + 54^2)/6 - CF = 2208.3 - 2204.2 = 4.1$

3. Variety SS = $(V1^2 + V2^2 + V3^2 + V4^2 + V5^2 + V6^2)/r$–CF = $(50^2 + 44^2 + 28^2 + 36^2 + 44^2 + 28^2)/4$–CF = 2309.0–2204.2 = 104.8
4. Total SS = $\Sigma X^2 = -$ CF = $10^2 + 15^2 + \ldots 6^2 + 9^2 -$ CF = 2358 – 2204.2 = 153.8
5. Error SS: Total SS – Rep. SS – Treatment SS = 153.8 – 4.1 – 104.8 = 44.9.

Step 3: Form the ANOVA table and compute mean squares and F value (see Table 19.2).

The F value for treatments is significant at the 1 percent level. It is customary to denote values that are significant at 5 percent by one asterisk (*), those at 1 percent by two asterisks (**), those at 0.1 percent by three asterisks (***), and those not significant as "ns". The actual probability (*P* values) are printed in computer programs. By using blocking, there is no gain as blocks do not vary. The ANOVA shows that differences between treatments do exist. However, it does not tell which variety or treatment is best and by how much. To perform tests of significance, we have to compute LSD.

TABLE 19.1. Treatments × replications.

| Replications | Treatments | | | | | | Total |
	V1	V2	V3	V4	V5	V6	
I	10	12	7	8	12	8	57 (RI)
II	15	11	6	10	11	5	58 (RII)
III	15	11	7	10	12	6	61 (RIII)
IV	10	10	8	8	9	9	54 (RIV)
Total	50	44	28	36	44	28	230

TABLE 19.2. ANOVA results.

| SV | df | SS | MS | F | Table value of F | |
					0.05	0.01
Replication	3	4.1	1.37	0.46^{ns}	3.29	5.42
Treatments	5	104.8	20.96	7.01^{**}	2.90	4.56
Error	15	44.9	2.99			
Total	23	153.8				

^{ns} = not significant
** = significant at $P = 0.01$

Test of Significance and Comparisons of Treatment Means

To test the null hypothesis, i.e., H_o = no difference among the variety means, i.e., $\overline{V1} = \overline{V2} = \overline{V3} = \overline{V4} = \overline{V5} = \overline{V6}$, the F test is used.

$$F = \frac{\text{mean square for treatments}}{\text{mean square for error}} = \frac{20.96}{2.99} = 7.01$$

Since the observed F exceeds 4.56 ($F_{5, 15, 0.01}$ at $P = 0.01$; see Appendix, Table A.4), the test is significant, which indicates that treatment means differ significantly at $P = 0.01$. We shall compare the treatment means at $P = 0.05$ level of significance.

Comparison of Means

We use the least significant difference (LSD) test:
1. Compute LSD at $P = 0.05$:

$$\text{LSD}_{0.05} = [(t_{\text{error df}, 0.05}) \times \sqrt{(2s^2 / r)}] = 2.131 \times \sqrt{(2 \times 2.99)}/ 4 = 2.61$$
kg/plot

where 2.131 is the value of t for $P = 0.05$ probability level and error degrees of freedom obtained from table of t (see Appendix, Table A.5).
2. Arrange the mean values of the treatments in descending order.

Treatments	V1	V5	V2	V4	V3	V6
Mean values	12.5	11.0	11.0	9.0	7.0	7.0

3. Any two means connected by the same line or followed by a common letter are not significantly different at 5 percent level of probability.
4. Determine SEm $= \sqrt{s^2 / r} = \sqrt{2.99 / 4} = 0.86$.
5. Determine the coefficient of variation:

$$\text{CV (percent)} = (s / \bar{x}) \times 100 = [\sqrt{2.99}) / 9.58] \times 100 = 18 \text{ percent.}$$

Conclusion: There is no difference in the yield of the varieties V1, V5, and V2. Likewise, there are no differences among V4, V3, and V6. The V6 (standard or control) gave the lowest yield of 7.00 kg/plot. V1 gave about 78 percent higher yield than the standard. The F test for blocks is not significant and hence formation of blocks has not increased the efficiency of the design.

Efficiency of RCBD over CRD

The precision of two designs is judged based on error variance. Fisher (1966) and Cochran and Cox (1957) have developed a formula to find out the relative efficiency of RCBD over CRD. The formula is as follows:

$$\text{Relative efficiency (RE)} = \frac{\text{Error MS (CRD)} (n_1 + 1)(n_2 + 3)}{\text{Error MS (RCBD)} (n_2 + 1)(n_1 + 3)}$$

where error MS (CRD) = error mean square for CRD
error MS (RCBD) = error mean square for RCBD
n_1 = df associated with error MS for RCBD
n_2 = df associated with error MS for CRD.

We know EMS for RCBD from the experiment and EMS for CRD is unknown. However, it can be estimated using the following formula:

$$\text{EMS (CRD)} = \frac{n_b E_b + (n_1 + n_e)E_e}{n_b + n_1 + n_e}$$

where E_b = block SS, E_e = error SS, $n_{b, nt,}$ and n_e are the degrees of freedom associated with blocks, treatments, and error MS in RCBD.

$n_b = 3$ $E_b = 1.37$
$n_t = 5$ $E_e = 2.99$
$n_e = 15$

$$\text{estimated error MS for CRD} = \frac{[(3)(1.37) + (5 + 15)2.99]}{3 + 5 + 15} = \frac{63.91}{23}$$
$$= 2.78$$

$$\text{RE} = \frac{[2.78(15 + 1)(18 + 3)]}{[2.99(18 + 1)(15 + 3)]} \times 100 = \frac{934.08}{1022.58} \times 100 = 91.3$$

RCBD is not efficient compared with CRD as there is no substantial reduction in the soil heterogeneity by formation of blocks.

Advantages

In RCBD, smaller error mean square is expected due to formation of blocks and treatments effects are estimated more precisely. The design will be efficient in agricultural experiments as the fields usually possess gradient and fertility differences. Since all the treatments are included in each block, the treatments are compared equally in each block. Each block is also considered as a separate experiment.

1. Even if the whole block is affected, analysis can be done using the data from the remaining blocks.
2. Since each block forms an experiment, blocks may be formed non-contiguously also, depending upon the availability of experimental area.
3. If any plot in a block is affected adversely by pests, birds, rat damage, etc., the value of that plot may be considered as a missing value. The missing value can be estimated using the missing plot technique, and using the estimated value, analysis can be done and conclusions made.
4. The experimenter can have reasonably more treatments and replications.
5. If any extra replication is required for a particular treatment, we can provide the extra replication for that treatment in each block.
6. Statistical analysis is simple and the design is easy to understand.

Disadvantages

1. If more treatments are included, the block size becomes large and consequently the treatments in the blocks are not compared efficiently. As the block size increases, the within block variability tends to increase.
2. If any missing value occurs, it must be estimated and hence the analysis becomes complicated.
3. It will be efficient only if the variability in experimental units exists in one direction. If fertility variation exists in two directions, Latin square design is to be used.

LINEAR MODEL FOR RANDOMIZED
COMPLETE BLOCK DESIGN

The basic assumption in RCBD is that any observation in the experimental unit is represented by a linear statistical model, i.e.,

Any observation = general mean + treatment effect + block effect + error effect

In symbols, $Y_{ij} = \mu + Ti + \beta j + E_{ij}$, $i = 1, 2 \ldots t$, and $r = 1, 2 \ldots r$
where

$\quad Y_{ij} =$ value of the observation made in jth block of the ith treatment
$\quad \mu =$ general mean
$\quad Ti =$ true effect of ith treatment
$\quad \beta j =$ true effect of jth block
$\quad E_{ij} =$ error effect of the observation made in the jth block of the ith treatment.

Population values are unknown and are estimated from a sample experiment and then the model is written as

$\quad y_{ij} = m + t_i + b_j + e_{ij}$ are the estimates of μ, Ti, βj, and E_{ij}, respectively.

The model is explained with reference to a numerical example. Consider the data obtained from RCBD with four treatments and three replications as shown in Table 19.3.

TABLE 19.3. Treatment and two-way replications.

Treatment	Blocks 1	Blocks 2	Blocks 3	Total	Mean	Treatment effect
V1	6 (y_{11})	4 (y_{12})	8 (y_{13})	18	6	0
V2	2 (y_{21})	4 (y_{22})	3 (y_{23})	9	3	−3
V3	10 (y_{31})	12 (y_{32})	14(y_{33})	36	12	6
V4	2 (y_{41})	4 (y_{42})	3 (y_{43})	9	3	−3
Total	20 By1$_j$	24By2$_j$	28By3$_j$	72Σy_{ij}	6	0
Mean	5	6	7	72	6	
Block Effect	−1	0	+1			

y_{ij} indicates ith treatment in jth block; for example $y_{11} =$ indicates first treatment in first block = six

The distribution of e_{ijs} is shown as follows,

Treatment	Block		
	1	2	3
V1	+1	−2	+1
V2	0	+1	−1
V3	−1	0	+1
V4	0	+1	−1

From these data, each observed value is the total of general mean, block effect, treatment effect, and error effect. For example:

$$Y_{ij} = Y_{11} = 6 = m + t_i + b_j + e_{ij} = 6 + (0) + 1 + (-1)$$

The sum of block effects add to zero. Also, the sum of treatment effects add to zero.

Assumptions

1. Treatment and replication effects are additive.
2. The errors are assumed normally and independently distributed with mean zero and variance σ^2, i.e., NID $(0, \sigma^2)$.

When Blocking will be Effective in RCBD

1. Within block heterogeneity should be as low as possible.
2. Between block sum of squares should exceed the treatment sum of squares.

MISSING VALUES

In agricultural experiments, some of the yields may be lost due to birds, rats, pests, diseases, improper germination, etc., which affect the expected yield as in other experimental units. In such cases, these values can be estimated and analysis carried out after substituting the estimated value. The formula for estimating one missing value is given in the following.

$$\hat{Y} = (Tt + Rr - G) / t - 1)(r - 1), \text{ where}$$

$\hat{Y}$ = estimated value for the missing plot

t = number of treatments
r = number of replications
T = total of the treatment whose missing value is to be estimated from other blocks
R = total of the block in which missing value occurs
G = total of all observations excluding the missing value.

The estimated value is substituted and then the analysis is done as usual. Since one observation is lost, the error degrees of freedom and the total degrees of freedom should be reduced by one. The SE of difference between the mean of the treatment with a missing value and the mean of any other treatment is given by $\sqrt{\{s^2[(2/r)+t/(r-1)(r-1)]\}}$ where s^2 is error variance in the analysis of variance.

Numerical example: An experiment was conducted to determine the efficacy of four treatments (four wheat varieties) with four replications in RCBD. The plot yields (kg/plot) are given here:

Replications	Treatments				Total
	A	B	C	D	
I	8	6	6	4	24
II	7	7	5	4	23
III	9	9	8	3	29
IV	12	9	8	8	37
Total	36	31	27	19	113

Analysis: Correction factor (CF) = $113^2/16 = 798.06$

Rep. SS = $(24^2 + 23^2 + 29^2 + 37^2)/4 - $ CF = $3315/4 - 798.06 =$
$828.75 - 798.06 = 30.69$
Treatment SS: $(36^2 + 31^2 + 27^2 + 19^2)/4 - $ CF = $3347/4 - $ CF =
$836.75 - 798.06 = 38.69$
Total SS: $(8^2 + \ldots + 8^2) - $ CF = $879.00 = 80.94$

See Table 19.4.

TABLE 19.4. ANOVA results.

Source of variation	df	SS	MS	F	Table F 5 percent	Table F 1 percent
Replication	3	30.69	10.23	8.00	3.86	6.99
Treatment	3	38.69	12.90	10.08**	3.86	6.99
Error	9	11.56	1.28			
Total	15	80.94				

Now assume the value in treatment B in replication III (value 9) is missing and we have to estimate it. The data will be as follows. (The symbol X indicates the missing observations.)

Replication	Treatments A	B	C	D	Total
I	8	6	6	4	24
II	7	7	5	4	23
III	9	X	8	3	20
IV	12	9	8	8	37
Total	36	22	27	19	104

Now $G = 104$, $t = 4$, $T = 22$, $r = 4$, $R = 20$, and $G = 104$. By substituting these values in the equation,

$$\hat{y} = (tT + rR - g) / (T - 1)(R - 1),$$

we get

$$[4(22) + 4(20) - 104] / [(4 - 1)(4 - 1)] = 7.11.$$

After inserting the estimated missing value, the analysis can be carried out. The data after substituting the estimated value are as follows:

Replication	Treatments A	B	C	D	Total
I	8	6	6	4	24
II	7	7	5	4	23
III	9	**7.11**	8	3	**29.11**
IV	12	9	8	8	37
Total	36	**29.11**	27	19	**111.11**

Analysis: CF $111.11^2/16 = 771.59$; Rep. SS $= 802.24 - 771.59 = 30.65$

Tr. SS $= 808.34 - 771.59 = 36.75$.

Total SS $= 76.95$. See Table 19.5.

Both replications and treatments are significant at $P = 0.01$. Note that the total *df* is 14 and error *df* is only 8.

For a mean of a treatment with a missing unit and that of any other treatments =

$$\sqrt{s^2\{2/r + t/r(r-1)(t-1)\}} = \sqrt{\{1.19(2/4 + 4/(4\times3\times3))\}} = 0.853$$

Different Number of Replications Within Blocks

Sometimes, we wish to compare several new treatments with a standard or control or check. In such cases, we can use more numbers of plots for the control treatment within each block than the new treatments. By doing so, we reduce the standard error for comparing the new treatment with the control treatment. For example: we have six new treatments and one control treatment and four replications, we require a total of $7 \times 4 = 28$ plots. The SEd for comparing any two treatment means is $\sqrt{2s^2/4} = s\sqrt{0.5}$. On the other hand, if we have six new treatments and three plots for the control treatment in each replication, we require $9 \times 3 = 27$ plots. Therefore, SEd for comparing the new treatment with the control is $\sqrt{s^2/3 + s^2/0}$ $= s\sqrt{(3+1)/9} = s\sqrt{(4/9)} = s\sqrt{0.44}$. Thus if our object is to compare the new treatment with the control, the second design is better than the first.

MODIFICATIONS IN THE ANALYSIS

Treatment SS is calculated as: $T1^2/r_1 + T2^2/r_2 + .. Tt^2/r_t - (\Sigma x)^2/rt$, where $r_1, r_2 \ldots$ are the replications for treatments $T1, T2 \ldots$, respectively. When

TABLE 19.5. ANOVA results.

Source of variation	df	SS	MS	F	Table *F*	
					5 percent	1 percent
Replication	3	30.65	10.22	8.59	3.86	6.99
Treatment	3	36.75	12.25	10.29**	3.86	6.99
Error	9	9.55	1.19			
Total	15	76.95				

randomizing the treatments in a block, we consider the total of new treatments and control plots together.

Example: Suppose the new treatments are designated as 1, 2, 3, 4, 5, 6, and the control treatment is designated C. We plan to provide four plots for the control in a block. Thus, we now have 10 treatments. Take 10 random numbers. Let them be 1, 2, 3, 4, 5, 6, 7, 8, 9, 0. Now allot 1 to 6 to new treatments and 7, 8, 9, and 0 to control. The layout plan will be as follows.

Layout plan:

Block 1	C	C	2	C	4	C	6	3	5	1
Block 2	C	3	C	5	6	4	C	2	1	C
Block 3	3	C	6	C	5	1	C	4	C	2
Block 4	1	2	C	C	5	C	6	4	C	3
Block 5	C	3	C	2	C	5	6	C	4	1

$$\text{Block SS} = (B1^2 + B2^2 + B3^2 + B4^2 + B5^2)/5 - (\Sigma x)^2/50$$
$$\text{TR SS} = (T1^2 + T2^2 + T3^2 + T4^2 + T5^2 + T6^2)/5 + Tc^2/20 - (\Sigma x)^2/50$$
$$\text{Total SS} = \Sigma x^2 - (\Sigma x)^2/n$$

Where T1 to T6 are the totals for the six new treatments, and TC is the total for the control.

Some Hints for Calculations for the ANOVA

1. To find the divisor for calculating treatment SS, one must be familiar with the divisor to be used. We must divide the squared value of a treatment by the number of units included in the total of the treatment.
2. Since the treatment mean square estimates the same quantity as the error mean square, this enables us to use F test for overall treatment difference.
3. Check: The SS will not be negative.
4. Total SS = treatment SS + block SS + error SS.
5. The square root of error mean square in the ANOVA should not be higher than the range of observations.

Comparison Between RCBD and CRD

Numerical example: It is important to know the difference between RCBD and CRD as these two are often used. This is explained with a numerical example for better understanding. There are four treatments, A, B, C, and D in RCBD and CRD with four replications each. Data are shown here and in Table 19.6.

TABLE 19.6. ANOVA results for comparison of CRD and RCBD.

CRD

Source of variation	df	SS	MS	F	Table F 5 percent
Trt	3	27.7	9.2	2.10ns	3.49
Error	12	53.2	4.4		
Total	15	80.9			

RCBD

Source of variation	df	SS	MS	F	Table F 5 percent
Rep	3	38.6	12.9	8.1	3.86
Trt	3	27.7	9.2	5.8*	3.86
Error	9	14.6	1.6		
Total	15	80.9			

CRD				RCBD			
C	A	C	A	B	C	A	D
9	14	12	17	14	12	17	13
A	A	D	D	D	C	B	A
13	13	11	13	11	12	14	14
D	B	B	B	A	D	B	C
9	14	14	8	13	13	11	10
B	C	D	C	C	D	B	A
11	10	11	12	9	9	8	13

Model: $y_{ij} = \mu + T_i + \varepsilon_{ij}$ Model: $y_{ij} = \mu + \alpha_i + \beta_j + \varepsilon_{ij}$
$(i = 1, 2, 3, 4)\,(j = 1, 2, 3, 4)\,(i = 1, 2, 3, 4)\,(j = 1, 2, 3, 4)$
$H_o: \mu_1 = \mu_2 = \mu_3 = \mu_4$ $H_o: \mu_1 = \mu_2 = \mu_3 = \mu_4$
H_o: not rejected H_o: rejected

In CRD, the null hypothesis could not be rejected. But in RCBD, the null hypothesis is rejected, as this design allows removal of block-to-block differences and reduces error, from 4.4 to 1.6, to test the hypothesis that the treatments are all the same. In RCBD, the block-to-block differences are significant. Complete randomization is not done in RCBD. Similar results would be obtained in other areas of research.

Regression in RCBD

A research trial with 0, 60, and 120 N kg/ha conducted in RCBD in rice crop with five replications gave the following plot yields in an experimental station. Analyze the data and partition the treatment sum of squares into linear and quadratic sum of squares.

Plot Yields in Different Replications

Replications	Treatment			Total
	0	60	120	
I	3	7	10	20
II	7	9	11	27
III	2	4	7	13
IV	2	6	10	18
V	6	9	12	27
Total	20	35	50	105

Total SS = 144.00; replication SS = 48.67; treatment SS = 90.00 (see Table 19.7)

TABLE 19.7. ANOVA results.

SV	df	SS	MS	F	Table value of F 0.05	0.01
Rep	4	48.67	12.17		3.84	7.01
Treatments	2	90.00	45.00	67.2**	4.46	8.65
Error	8	5.33	0.67			
Total	14	144.00				

A significant difference exists among the treatment means. We can apply the LSD test and compare the means. Suppose we hypothesize that yield is linearly dependent upon fertilizer. We can carry out the trend analysis. In this case, the treatment sum of squares can further be partitioned into sum of squares for linear regression with one degree of freedom and sum of squares for curvature, i.e., the quadratic component with one degree of freedom. The table of coefficients (c) for obtaining the linear and quadratic components of the treatment sum of squares, when treatments are equally spaced are given in the following table for four treatments:

Comparison	Number of treatments 1	2	3	c_i^2
Linear	−1	0	1	2
Quadratic	+1	−2	1	6

Form the following table:

Treatments	0	60	120
Tr. Total	20	35	50
Linear	−1	0	+1
Quadratic	+1	−2	+1

Using the orthogonal coefficients from the table, we calculate the sum of squares due to linear and quadratic components.

$$\text{SS linear component} = [(-1)(20) + (0)(35) + (+1)(50)]^2 /$$
$$[n\Sigma c_1^2 = 5 \times 2 = 10] = (-20 + 50)^2 / 10 = 900 / 10 = 90$$

The SS is equal to SS for treatments. This indicates that the SS for the quadratic component must be equal to zero. This further shows that the

TABLE 19.8. ANOVA results.

Source of variation	df	SS	MS	F	Table F 5 percent	Table F 1 percent
Replication	4	48.67	12.17		3.84	7.01
Treatments	2	90.00	45.00	67.2 **	4.46	8.65
Linear	(1)	90.00	90.00	134.30**	5.32	11.26
Quadratic	(1)	0.00	0.00			
Error	8	5.33	0.67			

trend of the treatment means can be exactly represented by a straight line within the nitrogen levels included. Here our interest is to find out whether the trend is significant or not. Linear and quadratic trends have to be tested. We are not interested in finding an equation between yield and fertilizer levels. Table 19.8 shows that the linear trend is significant.

EXERCISES

19.1. Briefly explain the method of analysis of the two-way classification model. Assume equal number of observations per cell and both factors are quantitative. Give the dummy analysis of variance table.

19.2. What is randomized complete block design? When do you think that it is necessary to use the randomized complete block design? What is the usual model for a two-way ANOVA? What are its shortcomings? How can you check if the model is not adequate?

19.3. Two treatments (treated and control) were compared in RCBD in soybean variety and the heights (cm) were measured in 12 replications. The data are given here.

	Replications											
	1	2	3	4	5	6	7	8	9	10	11	12
Treated plants	27	30	19	18	17	16	18	19	22	23	22	25
Control	14	15	20	18	15	13	20	18	18	20	12	11

a. Compute the ANOVA.
b. Find the relation between the values of F and t.

19.4. Four varieties of introduced corn were grown in equally sized plots in five locations and the yields (kg/plot) recorded. They are given here:

			Location		
Variety	1	2	3	4	5
V1	60	64	64	65	66
V2	63	63	66	67	65
V3	61	64	65	63	65
V4	55	56	60	58	59

a. Give the linear model and the assumptions.
b. Conduct the analysis of variance.
c. Are there differences in yield among the means of the varieties?

19.5. In a RCBD with five treatments and four blocks, the ANOVA results follow.

SV	MS
Blocks	24.06
Treatments	43.81
Error	12.36

Determine whether the treatment effects are different at 5 percent or 1 percent level of significance.

19.6. In a fodder crop experiment, the plot yields of a crop harvested on different days are as follows:

Days	Replications			
	I	II	III	IV
4	21.2	21.4	12.0	17.2
10	19.3	17.4	24.5	30.3
16	22.8	29.0	18.5	24.5
22	26.0	34.0	33.0	30.2
34	43.5	37.0	25.1	23.4
40	32.1	30.5	35.7	32.3

a. Compute the ANOVA.
b. Fit a linear regression of yield on harvesting time and partition the treatment sum of squares into linear regression and residuals.
c. Interpret the results.

19.7. The following is a RCBD. Five varieties of wheat in five replications were tried. The plot yield data are given here:

Variety	Replication				
	I	II	III	IV	V
V1	8.2	6.2	4.8	4.7	6.2
V2	3.3	3.1	3.6	2.6	2.3
V3	2.0	2.9	1.7	2.5	3.8
V4	3.3	3.2	2.3	2.9	2.9
V5	1.0	1.6	1.2	1.3	1.5

a. Run the ANOVA for RCBD for the data.
b. Compute the residuals and correlation coefficient between observed yields and the corresponding residuals.
c. Analyze the experiment without using the classification of blocks and comment.

19.8. An investigation was conducted to compare three methods for producing a certain commodity. The output on eight successive days from the methods is given here:

Methods	Days							
	1	2	3	4	5	6	7	8
M1	9	11	15	8	12	10	10	12
M2	8	16	14	14	12	13	11	16
M3	16	19	21	19	22	22	18	23

a. State any assumption you make to test the differences on the mean outputs of the three methods (assume $P = 0.05$).
b. Compare the means of methods M1 and M2 and the mean of M3.
c. Use Scheffe's multiple comparison method. Also use the LSD method. Compare these two methods. Plot the residuals of the three methods. Are the assumptions satisfied?

19.9. Moisture contents (percent) of four soil types were determined. Five samples were taken in each soil. Calculate the analysis of variance to obtain an estimate of the sampling variation within a soil and hence calculate the standard error of the mean moisture content of a soil and the standard error of difference between two soil means. Test the hypothesis that there is no variation of soil moisture content among the soils. Write down your conclusions. Soil moisture content in the different soils types (A, B, C and D) are as follows:

A	B	C	D
13.5	8.2	16.8	8.0
10.6	8.4	10.5	4.2
15.0	9.1	11.5	8.2
15.2	9.3	12.7	9.8
16.5	6.4	9.8	7.8

19.10. An experiment was carried out to determine whether there was any significant difference between the strengths of four materials. The data collected are shown here (measured in standard units). The design was RCB with days as blocks.

Blocks	Materials			
	1	*2*	*3*	*4*
1	20.7	23.8	21.5	25.2
2	19.1	21.5	22.9	25.5
3	21.4	23.2	22.4	24.8
4	22.5	22.5	22.2	24.6
5	19.2	23.1	23.2	23.1

Analyze the data and determine whether the strengths of the materials are different from one another. Find 95 percent confidence limits on the average strength of the materials.

19.11. In an agricultural research station, three laborers were asked to perform a certain task. The time (minutes) was recorded for the completion of the task.

	Time taken (min)		
Time of work	*1*	*2*	*3*
8-9 a.m.	25	26	28
12-1 p.m.	35	38	40
4-5 p.m.	42	53	49

a. What is the model for this situation?
b. Form the ANOVA table and test for the time.
c. Provide point estimates for the parameters.
d. Compare the time effects pairwise.

19.12. Four different types of rations were tested in an experiment. For each ration four pigs were used. The gain in weight (lb) was recorded after some time.

Ration	Gain in weight (lb) in replications			
	1	2	3	4
1	8.5	17.8	11.6	14.4
2	15.6	16.0	16.0	19.0
3	6.1	7.1	9.3	7.5
4	13.7	12.0	8.9	10.9

State the null hypothesis. Compare the gain in weights due to different rations. What is the SE of a mean? Compute the CV in the experiment.

19.13. A violent movie was watched by eight people. Their pulse rates were recorded before and after the movie.

Before	97	100	90	89	90	86	92	89
After	99	112	99	103	102	90	105	100

Use ANOVA and test the pulse rate means before and after the movie. Carry out the analysis using paired t test and find the relationship between the t statistic and the F statistic. Calculate 95 percent confidence limits for the difference in mean pulse rates.

19.14. State the similarities and differences between one-way and two-way classification models.

19.15. Why is randomized complete block design called a two-way classification model?

19.16. The effect of temperature on the preservation of apples was tested. The number of rotten apples was noted for apples in five batches and reported as follows. Find the differences among the temperatures on the quality of apples:

Batch	Temperature (°F)			
	50	55	60	65
1	8	5	7	10
2	14	10	3	5
3	12	8	6	5
4	9	9	5	9

REFERENCES

Anderson, V.L. and R.A. McLean (1974). *Design of experiments—A realistic approach.* New York: Marcel Dekker Inc.

Cochran, W.G. and G.M. Cox (1957). *Experimental design,* Second edition. New York: Wiley.

Fisher, R.A.(1966). *The design of experiments,* Eighth edition. London: Tweeddale Court.

Hicks, C.R. (1993). *Fundamental concepts in the design of experiments.* New York: Saunders College Publishing

Mead, R. and R.N. Curnow (1983). *Statistical methods in agricultural and experimental biology.* New York: Chapman and Hall.

Chapter 20

Group Comparisons

The analysis of variance technique may be applied to a great variety of complex experiments. In many experiments, we study the effects of two or more factors simultaneously. In such experiments, we can get some additional information if we use a group analysis technique. Every degree of freedom in a statistical problem can be used to provide information of some statistical interest. A procedure for subdivision of the sum of squares of treatments, each of which has one degree of freedom, is explained in this chapter. We compare one treatment with the other. If there are two treatment means, $\overline{T}1$ and $\overline{T}2$, then $\overline{T}1 - \overline{T}2$ is a comparison or a contrast. If there are three treatments, a pairwise comparison of two treatments can be made in three different ways, e.g., $\overline{T}1 - \overline{T}2$, $\overline{T}1 - \overline{T}3$, and $\overline{T}2 - \overline{T}3$. Also, we can get a comparison average of the two means with the third one $[(\overline{T}1 + \overline{T}2)]/2 - \overline{T}3$ (where $\overline{T}1$, and $\overline{T}2$ are similar but differ considerably from $\overline{T}3$), $[(\overline{T}1 + \overline{T}3)]/2 - \overline{T}2$ and $[(\overline{T}2 + \overline{T}3)]/2 - \overline{T}1$. In these situations, we can make use of group comparison analysis; it is convenient to work with the totals of treatments when the numbers of replications are equal in each treatment. We will define some of the terms that are commonly used.

COMPARISON

Suppose there are k treatments in an experiment, each based on an equal number of replications. Consequently, we get k totals T1, T2, T3 ... Tk, respectively. Therefore, a combination $C = l_1T_1 + l_2T_2 + l_3T_3 + \ldots + l_kT_k$ (Σl_iT_i) is called a *comparison* or contrast among the treatments, if $\Sigma l_i = 0$. In other words, a contrast is defined by $C = l_1T_1 + l_2T_2 + l_3T_3 + \ldots + l_kT_k$ where $\Sigma l_iT_i = 0$.

If C is a contrast, then $C^2/r \Sigma l_i^2$ is a part of the sum of squares of the treatments associated with one degrees of freedom, i.e., $[(l_1T_1 + l_2T_2 + l_3T_3 + \ldots + l_kT_k)^2]/\{r [l_1^2 + l_2^2 + l_3^2 + \ldots + l_k^2]\}$, where $\Sigma l_i = 0$ associated with 1 *df*.

ORTHOGONAL AND NONORTHOGONAL CONTRASTS

Orthogonal comparisons are explained with a numerical example. Let $C1$, $C2$, and $C3$, be three comparisons and $T1$, $T2$, $T3$, and $T4$ be treatment totals.

$$C1 = (-3)T1 + (-1)T2 + (+1)T3 + (+3)T4$$
$$C2 = (+1)T1 + (-1)T2 + (-1)T3 + (+1)T4$$
$$C3 = (+1)T1 + (+1)T3 + (-1)T3 + (-1)T4$$

Calculate the sum of the products of the corresponding coefficients of the pairs of contrasts namely, $C1$ and $C2$, $C1$ and $C3$, and $C2$ and $C3$. In $C1$ and $C2$, the sum of the products of the corresponding coefficients is

$$(-3)(+1) + (-1)(-1) + (+1)(-1) + (+3)(+1) = 0.$$

In $C1$ and $C3$, the sum of the products of the corresponding coefficients is

$$(-3)(+1) + (-1)(+1) + (+1)(-1) + (+3)(-1) = -8.$$

In $C2$ and $C3$, the sum of the products of the corresponding coefficients is

$$(+1)(+1) + (-1)(+1) + (-1)(-1) + (+1)(-1) = 0.$$

Orthogonality is defined for any two comparisons in terms of the coefficients l_is. In any two comparisons, if the sum of the products of the corresponding coefficients is equal to zero, then these two comparisons are said to be orthogonal. Thus, the comparisons $C1$ and $C2$ as well as $C2$ and $C3$ are orthogonal and $C1$ and $C3$ are not orthogonal. In addition to comparing simple differences between pairs, it is often necessary to examine linear combinations of means. For example, let there be three treatments namely A (0 N kg/ha), B (60 kg N/ha), and C (120 N kg/ha). Here we can compare the average of the two N treatment means with the control. For this, a comparison or contrast is used (Cox, 1987; Ostle, 1963).

PARTITIONING OF TREATMENT SUM OF SQUARES

Let there be t treatments in an experiment. The experiment yields $t-1$ degrees of freedom. The treatments' sum of squares may be partitioned to give $t-1$ independent comparisons each based on one degrees of freedom. Each of these separate comparisons can be tested for significance individually by experimental error. For this test, the F test need not be significant. Orthogonal components are additive, i.e., the sum of parts is equal to the total. Each component in an orthogonal set covers a different portion of the total variation; nonorthogonal components are not additive.

Numerical example: A study of grain yield in rice crop in a fertilizer trial with different manures was conducted in a randomized complete block design with four replications at a rice research station. The results relating to plot yields (lb/plot) are furnished here:

	Replication				
Treatments N/ac	1	2	3	4	Total
Control (T_1)	20.1	19.1	19.0	19.5	77.7
Ammonium sulphate (T_2)	22.5	23.0	23.6	25.2	94.3
Green leaf (T_3)	24.7	25.5	23.9	23.6	97.7
Ammonium sulphate + green leaf (T_4)	24.4	26.8	25.9	25.9	103.0
Total	91.7	94.4	92.4	94.2	372.7

Analysis: Correction factor = $372.7^2/16 = 8681.581$
Replication SS = $91.7^2 + \ldots + 94.2^2 /4 - CF = 1.33$
Treatment SS: $77.8^2 + \ldots + 103^2/4 - CF = 89.43$
Total SS = 99.47
(see Table 20.1)
From the data, we have $r = 4$, $t = 4$, control (T1) = 77.7

$(T_2) = 94.3$, $T_3 = 97.7$, $T_4 = 103.0$
SS treatments = 89.43, MS $_{error} = 0.96$.

TABLE 20.1. ANOVA results.

Source of variation	df	SS	MS	F	Table F 5 percent	Table F 1 percent
Replication	3	1.33	0.44		3.86	6.99
Treatments	3	89.43	29.81	31.05**	3.86	6.99
Error	9	8.71	0.96			
Total	15	99.47				

Note: Table value of F for 3,9 and at $P = 0.05$ and $P = 0.01$, see Appendix, Table A.6.

The effectiveness of the three sources of fertilizer application is evaluated. This experiment is considered as a random sample from the population of interest. Consider the following comparisons:

Contrast	Control (T_1)	Ammonium Sulfate (T_2)	Green leaf (T_3)	Ammonium sulfate + green leaf (T_4)
C1	−3	+1	+1	+1
C2	0	−1	+1	0
C3	0	−1	−1	+2
Treatment totals	77.7	94.3	97.7	103.0

C1 = control versus fertilizers SS: $[(-3)(77.7) + (+1)(94.3) + (+1)(97.7) + (+1)(103.0)]^2/[r(-3)^2 + (1)^2 + (1)^2 + (1)^2] = (-233.1 + 94.3 + 97.7 + 103.0)^2/48 = (295-233.1)^2/48 = 61.9^2/48 = 79.82$.
C2 = Amm. sul. versus green leaf SS $= [(-1)(94.3) + (+1)(97.7)]^2/[r(1^2+1^2)] = 3.4^2/8 = 1.44$.
C3 = Amm. sul. and green leaf versus Amm. sul. + green leaf $= [(-1)(94.3) + (-1)(97.7) + (+2)(103)]^2/[r(-1)^2 + (-1)^2 + (+2)^2] = 14^2/24 = 8.17$.

To test these comparisons within ANOVA procedure, the treatment SS as per comparisons are tested as follows (see Table 20.2):

In this test we are able to make particular comparisons instead of the simple comparisons of all possible pairs of treatments. The mean of control treatment is 77.7 kg/plot, whereas the mean of the fertilized plots is 98.3. Thus the difference is significant. There is no difference between ammonium sulfate and green leaf plots. Ammonium sulfate combined with green

TABLE 20.2. ANOVA results.

Source of variation	df	SS	MS	F	Table F 5 percent	1 percent
Replication	3	1.33	0.44			
Treatments	3	89.43	29.81	31.05**	3.86	6.99
C1 (control vs. others)	(1)	79.82	79.82	83.14**	5.12	10.56
C2 (Amm. sul vs. Gr. leaf)	(1)	1.44	1.44	1.50ns	5.12	10.56
C3 (Amm.sul. and Gr. leaf vs. Amm. sul + green leaf)	(1)	8.17	8.17	8.51*	5.12	10.56
Error	9	8.71	0.96			
Total	15	99.47				

leaf application is better than separate applications. We can also form orthogonal comparisons in another way.

SECOND SET OF ORTHOGONAL COMPARISONS

	Control	Amm. sul.	Green leaf	Amm. sul.+ green leaf
C1	+1	−1	+1	−1
C2	+1	0	−1	0
C3	0	+1	0	−1
Treatment totals	77.7	94.3	97.7	103.0

C1 = control + green leaf versus amm. sul + green leaf:

$$[(+1)(77.7) + (-1(94.3) + (+1)(97.7) + (-1)(103.0)]^2/[r\{(+1)^2 + (-1)^2 + (1)^2 + (-1)^2\}] = [77.7{-}94.3 + 97.7{-}103.0]^2/16 =$$
$$(175.4{-}197.3)^2/16 = 21.9^2/16 = 29.97$$
C2 = control versus green leaf SS = $[(+1)(77.7) + (-1)(97.7)]^2/[r(1^2 + 1^2)] = -20.0^2/8 = 50.00$
C3 = amm. sul. versus green leaf = $[(+1)(94.3) + (-1)(103.0)^2] = -8.7^2/8 = 9.46$

To test these comparisons within the ANOVA procedure, the treatment SS as per comparisons are tested in Table 20.3.
All comparisons are significant.

TABLE 20.3. ANOVA results.

Source of variation	df	SS	MS	F	Table F 5 percent	Table F 1 percent
Replication	3	1.33	0.44			
Treatments	3	89.43	29.81	31.05**	3.86	6.99
C1	(1)	29.97	29.97	31.21**	5.12	10.56
C2	(1)	50.00	50.00	52.08**	5.12	10.56
C3	(1)	9.46	9.46	9.8*	5.12	10.56
Error	9	8.71	0.96			
Total	15	99.47				

*significant at P = 0.05
**significant at P = 0.01

PRACTICAL UTILITY OF CONTRASTS

In a greenhouse under controlled conditions Palaniswamy (1998) studied purslane (*Portulaca oleracea* L.). The essential fatty acids, alpha-linolenic acid (ALNA) and linoleic acid (LA), which are very essential for good health, were evaluated. Palaniswamy (1998) studied the effect of nitrogen management for enhancement of these fatty acids and included the treatments nitrate nitrogen and ammonium nitrogen in different concentrations. Specific comparisons among the different nitrate nitrogen to ammonium nitrogen ratios (A = 1:0, B = 0.75:0.25, C = 0.5:0.5, and D = 0.25: 0.75) on essential fatty acids are as follows:

Character	A versus (B+C+D)/3		B versus D		C versus (B+D)/2	
Alpha-linolenic acid	8.075	19.440**	12.82	17.99*	27.51	15.41**
Linoleic acid	1.505	3.220**	2.19	3.02*	4.56	2.603**

From the analysis of such specific comparisons, researchers can draw very useful inferences:

1. When all contrasts are mutually orthogonal, the addition of the sum of squares of the different contrasts is equal to the treatment sum of squares. This is similar to the case for the degrees of freedom.
2. The number of replications for each treatment is the same.
3. The contrasts are determined using treatment totals rather than the treatment means for the sake of easy computations. We will get the same results when we compute the sum of squares using the treatment totals or treatments means.
4. It is assumed that the contrasts are formulated before the start of the experiment.
5. All assumptions underlying the analysis of variance are satisfied.
6. Since the sum of squares of the contrasts add up to the total of the treatment sum of squares, all the variability in the treatment sum of squares is explained by the contrasts.
7. When the number of comparisons is large, the number of decisions that go wrong is also relatively large.
8. The total variation of the treatments may be subdivided in many different ways and contrasts should be constructed depending upon the nature of the experimental variables and the purpose of the experiment. For example, in the previous experiment two different sets of comparisons are shown.

EXERCISES

20.1. Given $T1 = 1267.3$, $T2 = 1341.2$, $T3 = 1247.2$, $T4 = 1368.2$, and $r = 12$, find SS due to C1 $= T1 + T3$ versus $T2 + T4$ C2 $= T1 - T3$ and C3 $= T2 - T4$.

20.2. A study was conducted to test two drugs. The treatments are (1) placebo, (2) drug 1 alone, (3) drug 1 + diet I, (4) drug 2 alone, (5) drug 2 + diet II, and (6) drug 1 + drug 2 + diet I + diet II. To study the effectiveness on patients, state how you will perform the experiment. State the orthogonal contrasts that you can make such as placebo versus drugs + diets, drug 1 versus drug 2, etc.

20.3. Given $t = 5$ and $r = 4$, design a CRD. The treatment totals are $T1 = 197.4$, $T2 = 199.0$, $T3 = 211.3$, $T4 = 215.8$, and $T5 = 186.5$. The sum of squares for treatments, error, and total are 136.8, 13.0, and 149.8, respectively. Establish four different contrasts and test them by using the analysis of variance technique.

20.4. In an experiment, one standard variety (A) and three new strains (B, C, D) of rice were evaluated in a CRD with five replications in equally sized plots. The plot yields are given in Table 20.4. Analyze the data for the contrast (a) A versus B, C and D, (b) B and C versus D and (c) B versus C.

20.5. In an entomological experiment, five methods of pest control treatment were tried along with a standard one without any application. The treatments are C = control, standard method of treatment, A1= chemical A with low concentration, A2 = chemical A with higher concentration, B1 = chemical B with low concentration, B2 = chemical B higher concentration. Calculate the following contrasts.
   a. Does control differ from others?
   b. Does the standard method differ from others?
   c. Are A1 and A2 different?
   d. Are B1 and B2 different?
   e. Are A1 and A2 different from B1 and B2?
   f. Find the contrasts along with the divisors.

20.6. What is a contrast? How is it useful in the analysis of variance?

20.7. What are the points to be remembered in the analysis using contrasts?

20.8. What do you understand by orthogonal contrasts? Give two examples.

20.9. There are two factors, A and B. In A, there are three levels, a_1, a_2, and a_3 and in B there are three levels b_1, b_2, and b_3. An experiment was conducted with six combinations of the treatments in a randomized

TABLE 20.4. ANOVA results for completely randomized design in rice crop varieties.

| Variety | Replications | | | | |
	I	II	III	IV	V
A	32	25	28	17	17
B	26	22	24	28	35
C	19	23	24	24	20
D	36	39	36	44	40

complete block design. Show how you will partition the treatment sum of squares. The experimenter is interested in (a) A versus B, (b) a_1 versus a_3, (c) b_1 versus b_3, (d) a_1 versus $a_2 + a_3$, and (e) b_1 versus $b_2 + b_3$. Can you find another set of contrasts?

20.10. The researcher wants to test nitrogen in three different sources, each with two levels. The researcher wants to test two light treatments for all three nitrogen sources. Set up seven different contrasts and show how you will partition the treatment sum of squares having seven degrees of freedom.

20.11. Set up a problem from your own area of specialization involving comparisons.

20.12. Define group comparison with suitable examples.

REFERENCES

Cox, C.P. (1987). *A handbook of introductory statistical method.* New York: John Wiley and Sons.

Ostle, B.I. (1963). *Statistics in research.* Ames, IA: The Iowa University Press.

Palaniswamy, U. (1998). Nitrogen management studies to enhance the essential fatty acids in purslane (*Portulaca oleracea* L.). Part of doctoral thesis submitted to the University of Connecticut.

Chapter 21

Multiple Comparison Procedures

Because of its simplicity the analysis of variance technique is used to test the differences among the treatment means when the number of treatments is more than two. Experimenters set up a null hypothesis H_o. For example, $H_o: \mu_1 = \mu_2 = \mu_3 = \ldots = \mu_t$, i.e., there are no differences among the population means of the t treatments. Multiple range tests are used for testing the differences between two means. In the analysis of variance, the test of hypothesis is based on the F ratio statistic. If F test is significant, the null hypothesis is rejected, and we conclude that the means differ significantly. At least two means differ significantly. The experimenter may want to know which pairs of treatment means differ significantly or to contrast one treatment effect with the average of some other treatment effects or to estimate a treatment effect. The different tests, e.g., LSD, Duncan's New Multiple Range test, Dunnett's test, Tukey's w-procedure test, Newman-Keuls test, and Scheff's test are explained with suitable illustrations.

LEAST SIGNIFICANT DIFFERENCE TEST

This test was designed by Fisher. The treatment means have to be compared using LSD when the F test in the analysis of variance table (ANOVA table) is significant. The LSD for a given probability level can be calculated and the means compared (Bancroft, 1968; Carmer and Swanson, 1973). The procedure for the least significant difference test has been described by Goulden (1939), Davies (1949), and others. LSD at 5 percent level of probability can be determined as $LSD_{0.05} = t_{0.05} \times SE_d$ (Steel and Torrie, 1960), where $t_{0.05}$ is the value from t table for the error degrees of freedom (see Appendix, Table A.5.), and the given level of significance, SE_d, is the standard error of difference between two means, which equals , where s^2 is the error mean square (EMS) in the ANOVA and r is the number of replications.

Relation Between t and LSD

We have already defined the statistic t as

$$t = (\bar{x}_1 - \bar{x}_2) / SE_d,$$

i.e., $\bar{x}_1 - \bar{x}_2 = t \times SE_d$, i.e., if the observed difference between two means exceeds LSD, the difference is declared as significant.

Advantages of LSD Test

LSD is used in detecting true difference among the means and applied only after F test in the ANOVA is significant. Only one value is involved in the comparison. It is easy to compute.

Disadvantages of LSD Test

The disadvantage of LSD is that if the overall F test is not significant, none of the individual pairs of means can be compared. When it is used indiscriminately to test all possible differences among several means, certain differences will be significant but not at the level of significance chosen, i.e., instead of making comparisons at the 5 percent level, some treatments are compared at lower levels of significance.

Example 1: In an experiment with six treatments (six different methods of irrigation) in sugarcane crop conducted in Latin square design, Table 21.1 was obtained. The F ratio value from F table for 5 and 20 degrees of freedom at $P = 0.05$ is 2.71 (see Appendix, Table A.6). As the F calculated (7.62) is greater than F table value (2.71) and also F at $P = 0.01$ (4.10), the treatment differences are highly significant. H_o is rejected.

Calculation of LSD and Comparison of Means Using LSD

Standard error of the mean $SE_m = \sqrt{(2.423 / 6)} = \sqrt{(2.423)}/2.45 = 1.557/2.45 = 0.63$

$LSD = \sqrt{(2s^2 / r)} = 2.086 \times \sqrt{[(2 \times 2.423) / 6]} = 2.086 \times 0.898 = 1.87$

TABLE 21.1. ANOVA results for six irrigation methods in sugarcane.

Sources of variation	df	SS	MS	F	F from tables (5 percent)
Rows	5	10.774	2.155	0.89	2.71
Columns	5	11.524	2.305	0.95	2.71
Treatments	5	92.339	18.468	7.62**	2.71
Error	20	48.467	2.423		
Total	35	163.104			

($t = 2.086$ is the value of t at $P = 0.05$ and 20 degrees of freedom from the table of t (see Appendix, Table A.5).

Arrange the means in descending order:

Treatments:	M5	M1	M4	M3	M0	M2
Mean values:	19.48	19.29	19.19	18.79	15.87	15.73

Note: Any pair of treatment means that differ in absolute value by more than 1.87 would imply that the corresponding pairs of population means are significantly different (Montgomery, 1991).

Conclusion: The treatments M5, M1, M4, and M3 can be grouped together, i.e., any pair in this group does not differ significantly. However, they gave significantly higher production than M0 and M2. No difference between M0 and M2 exists.

We can also set up confidence limits for the means as follows:

Confidence limits: $\overline{X} \pm t \times SE$ (at $P = 0.05$)
$\overline{X} \pm 2.086 \times 0.63 = \overline{X} \pm 1.31$

		Confidence limits	
Treatment	Mean	Upper limit	Lower limit
M5	19.48	20.79	18.17
M1	19.29	20.60	17.98
M4	19.19	20.50	17.88
M3	18.79	20.10	17.48
M0	15.87	17.18	14.56
M2	15.73	17.04	14.42

Example 2: Procedure for comparing means using t test and F test. Data collected from an experiment in RCBD with four treatments and four replications follow.

			Treatment			
Replication	I	II	III	IV	Total	Mean
A	47	52	62	51	212	53
B	50	54	67	57	228	57
C	57	53	69	57	236	59
D	54	65	74	59	252	63
Total	208	224	272	224	928	

$LSD = t \times SE_d = 2.262 \times \sqrt{(2 \times 7.78)/4} = 4.46$ (t at $P = 0.05, 9 = 2.262$, Standard of error of difference between two means $= 1.972$; see Table 21.2).

TABLE 21.2. ANOVA results.

SV	df	SS	MS	F	F from tables P = 0.05	P = 0.01
Rep	3	574	191	24.55**	3.86	6.99
Tr	3	208	69.3	8.91**	3.86	6.99
Error	9	70	7.78			
Total	15	852				

Treatment means:

63	59	57	53
D	C	B	A

t test: Comparing pairs of means at a time:
1. D versus B $t = (\underline{D} - \underline{B})/1.972 = (63 - 57)/1.972 = 6/1.972 = 3.04*$
2. D versus A $t = (\underline{D} - \underline{A})/1.972 = (63 - 53)/1.972 = 10/1.972 = 5.07*$
3. C versus B $t = (\underline{C} - \underline{B})/1.972 = (59 - 57)/1.972 = 2/1.972 = 1.014^{ns}$
4. C versus A $t = (\underline{C} - \underline{B})/1.972 = (59 - 53)/1.972 = 6/1.972 = 3.04*$
5. B versus A $t = (\underline{B} - \underline{A})/1.972 = (57 - 53)/1.972 = 4/1.972 = 2.028^{ns}$
6. D versus C $t = (\underline{D} - \underline{C})/1.972 = (63 - 59)/1.972 = 4/1.972 = 2.028^{ns}$

Note that t test as well as F test give the same results.

$F = t^2$ verification:

1. B versus A: SS $= (228^2 + 212^2)/4 - 440^2/8 = 24232 - 24200 = 32$

    $F = 32/7.78 = 4.11^{ns}$ as $F_{1, 9, 0.05} = 5.12$
    (where 7.78 is the EMS from ANOVA)
    $(F = 4.11, t = 2.028,$ i.e., $F = t^2)$

2. C versus A: SS $= (236^2 + 212^2)/4 - 448^2/8 = 25160 - 25088 = 72$

    $F = 72/7.78 = 9.25$ $(F = 9.25, t = 3.04,$ i.e., $F = t^2)$

3. C versus B: SS $= (236^2 + 228^2)/4 - 464^2/8 = 26920 - 26912 = 8$

    $F = 8/7.78 = 1.03^{ns}$ $(F = 1.028, t = 1.014,$ i.e., $F = t^2)$

4. D versus A: SS $= (252^2 + 212^2)/4 - 464^2/8 = 27112 - 26912 = 200$

    $F = 200/7.78 = 25.7$ $(F = 25.7, t = 5.07, F = t^2)$

5. D versus B: SS $= (252^2 + 228^2)/4 - 480^2/8 = 28872 - 28800 = 72$

    $F = 72/7.78 = 9.25$ $(F = 9.25, t = 3.04,$ i.e., $F = t^2)$

6. D versus C: SS = $(252^2 + 236^2)/4 - 488^2/8 = 29800 - 29768 = 32$

$F = 32/7.78 = 4.11^{ns}$. ($F = 4.11$, $t = 2.028$, i.e., $F = t^2$)

The relationship $F = t^2$ has been verified in all comparisons between any two treatments. The t test for each difference leads to the same inference as obtained in LSD test. Thus the researchers can employ ANOVA for comparison of treatments when the number of treatments is more than two.

DUNCAN'S NEW MULTIPLE RANGE TEST

Duncan (1955) developed this test and, because of its simplicity, it is commonly used by researchers. This test can be used whether or not the F test is significant. Duncan's New Multiple Range test is best for testing differences in all pairs of means (Wine, 1964). In this test, each treatment mean is compared with every other treatment mean. It is similar to Newman-Keuls test but the least significant ranges R_p are different. This test is explained using the data given in Example 2.

Procedure

Step 1: Find the SE of the mean as $= \sqrt{(s_e^2 / r)} = \sqrt{(\text{error mean square} / r)}$.

Step 2: Extract the significant studentized ranges (SSR) for 5 percent and 1 percent levels from the table taking into consideration the number of means involved for the comparison and the error degrees of freedom on which the standard error is based (see Appendix, Table A.10).

Step 3: The SSRs are then multiplied by the standard error of the mean to get the least significant ranges (LSR).

Step 4: Rank the means.

Step 5: Test the differences. Each difference is declared significant if it exceeds the corresponding LSR. (For unequal replications: in the equation SE of the mean $= \sqrt{s_e^2 / r}$, replace r by harmonic mean r_h of (r_i) where, $r_h = t/\Sigma 1/r_i$ for unequal replications (Montgomery, 1991). The test criterion for Duncan's test is to compare each difference with R_p.

Procedure

1. Arrange all the means in the increasing order of magnitude.
2. Compute R_p for each size of subset of the means.

3. Compute d_{ij} for all the $k\,(k-1)/2$ pairs of the means.
4. Declare d_{ij} significant when $d_{ij} > R_p$.
5. Any two or more means not significantly different from each other are underlined.

A numerical example is shown in Table 21.3.
Standard error of the mean $\sqrt{(7.78/4)}/1.395$:

P	2	3	4
SSR 5 percent	3.20	3.34	3.41
SSR 1 percent	4.60	4.86	4.99
LSR 5 percent R_p	4.464	4.659	4.757
LSR 1 percent R_p	6.417	6.779	6.961

The critical difference R_p of Duncan's test for p means known as shortest significant difference is computed as $R_p = (r\,\alpha, p, v)\sqrt{EMS/r}$ where, r, p, v are from the tables for $\alpha = 0.05$ and 0.01 for various values of p and error degrees of freedom.
    Step 1: Compute $SEm = \sqrt{EMS/r} = \sqrt{7.78/4} = 1.395$.
    Step 2: Compute $R_p = (r\,\alpha, p, v) \times SE$.

Mean differences	Rp	p
$d_{ij} = (\bar{D}-\bar{A}) = 63-53 = 10^*$	4.757	4
Since $(\bar{D}-\bar{A}) > R_p$, the difference d_{ij} is significant.		
$d_{ij} = (\bar{D}-\bar{B}) = 63-57 = 6^*$	4659	3
Since $(\bar{D}-\bar{B}) > R_p$, the difference is significant.		
$d_{ij} = (\bar{D}-\bar{C}) = 63-59 = 4^{ns}$	4.464	2
Since $(\bar{D}-\bar{C}) < R_p$, the difference (4) < 4.464 is not significant.		
$d_{ij} = (\bar{C}-\bar{A}) = 59-53 = 6^*$	4.659	3
Since $(\bar{C}-\bar{A}) = 6 > 4.659$, the difference is significant.		

TABLE 21.3. ANOVA results.

SV	Df	SS	MS	F	F from tables 5 percent	1 percent
Replications	3	574	191.33	24.59	3.86	6.99
Treatments	3	208	69.3	8.91**	3.86	6.99
Error	9	70	7.78 (s^2)			
Total	15	852				

$d_{ij} = (\overline{C} - \overline{B}) = 59 - 57 = 2^{ns}$ 4.464   2

Since $(\overline{C} - \overline{B}) = 2 < 4.464$, the difference is not significant.

$d_{ij} = (\overline{B} - \overline{A}) = 57 - 53 = 4^{ns}$ 4.464   2

Since $(\overline{B} - \overline{A}) = 4 < 4.464$, the difference is not significant.

Note that the LSD test and Duncan's New Multiple Range test give the same conclusions. These results are shown as follows. The means are arranged in descending order.

D	C	B	A
63	59	57	53

DUNNETT'S TEST

This test was designed by Dunnett (1955). In experiments, researchers invariably include a control as one of the treatments in order to evaluate the new treatments in comparison with the control (Dunnett, 1964). For example, in an experiment involving fertilizers, one treatment may be without any fertilizer application, which is considered as the control. Under such situations Dunnett's Test is the best choice. Dunnett's least significant difference is used for testing the difference. The test procedure is as follows: This test is explained using the data given in Example 1.

Step 1. Calculate Dunnett's least significant difference: Dunnett's least significant difference is = Dunnett's *t* value from the table of Dunnett's *t* for the number of treatments, error degrees of freedom, and given level of significance × SE_d.

Step 2: Compute

$$SE_d = \sqrt{S^2(1/r_1 + 1/r_2)}$$

where r_1 and r_2 are the number of replications for control and other treatments, respectively. If $r_1 = r_2$, then the

$$SE_d = SE_d = \sqrt{2s^2/r}.$$

In the given example under LSD test, r (replications) for control and r (replications) for other treatments are the same, i.e., 6, $s^2 = 2.423$ with 20 degrees of freedom (see Table 21.1).

$$SE_d = \sqrt{(2 \times 2.423)} / 6 = 0.89 \text{ (as before)}$$

LSD (Dunnett) = $D_t \times SE_d = 2.81 \times 0.89 = 2.50$ (where D_t = Dunnett's t value, i.e., 2.81 for $P = 0.05$ and error degrees of freedom for two sided— see Appendix, Table A.8).

Step 3: Treatment means are (highest to lowest):

M5	M1	M4	M3	M0	M2
19.48	19.20	19.19	18.79	15.87	15.73

The mean of the control treatment M0 = 15.87. Thus every treatment which is greater than 15.87 + 2.50 = 18.37 is significantly greater than the control. Thus, the treatments M5, M1, M4, and M3 are significantly different from control treatment. It is assumed that all populations have the same variance, σ^2 and are normally distributed. The values of D_α, k, and v are given in Table A.8 (see Appendix).

Comparison of differences can also be made using LSD (Dunnett) values of 2.50:

Treatment	M5	M1	M4	M3	M0
Mean	19.48	19.20	19.19	18.79	15.87

M5 – M0 = 19.48 – 15.87 = 3.61* (as 3.61 > 2.50)
M1 – M0 = 19.20 – 15.87 = 3.33* (as 3.33 > 2.50)
M4 – M0 = 19.19 – 15.87 = 3.32* (as 3.32 > 2.50)
M3 – M0 = 18.79 – 15.87 = 2.92* (as 2.92 > 2.50)

TUKEY'S w – PROCEDURE TEST

Tukey (1949) proposed a multiple comparison procedure in which all pairs of treatment means in the experiment can be compared. Like the LSD test, a single value is used for judging the significance of all the differences (Steel and Torrie, 1960). The least significant difference in Tukey's test is given by honestly significant difference (HSD).

HSD $= q_\alpha (p_1 n_2) \times SEm$, where $SEm = s\sqrt{r}$ (q_α to be obtained from Appendix, Table A.9) where

s^2 = error mean square in the ANOVA with n_2 degrees of freedom;
r = number of replications for each treatment;
$q_{0.95,p1,n2}$ = the tabulated value for studentized range at $P = 0.05$;
p = number of treatments; and
n_2 = degrees of freedom associated with error mean square.

Example: The LSD test in Example 1 is used to explain Tukey's w-procedure test. We have $t = 6$, $MSE = 2.423$, $\alpha = 0.05$, degrees of freedom = 20, $q_{0.95,6,20} = 4.45$, which is the value from (see Appendix, Table A.9). HSD = $4.45 \times = 2.83$ (4.45 is the value from Appendix, Table A.9). To declare the difference between any two means as significant, the difference should exceed the HSD of 2.83.

Comparison of means using Tukey's test:

M5	M1	M4	M3	M0	M2
19.48	19.20	19.19	18.79	15.87	15.73

The results are similar to those reported in LSD test based on t value. However, 2.83 is greater than the least significant difference based on t (1.87). Hence, one can expect smaller numbers of pairs to be found as significant. In Tukey's test, all treatments must have equal numbers of replications whereas in LSD test, the numbers of replications may differ. It is also assumed that all populations have the same variance and are normally distributed.

Comparison of Pairs of Means

We have six treatments and we have to perform 15 pairwise comparisons as shown:

M5 versus M1: M5 − M1 = 19.48 − 19.20 = 0.28[ns]
M5 versus M4: M5 − M4 = 19.48 − 19.19 = 0.29[ns]
M5 versus M3: M5 − M3 = 19.48 − 18.79 = 0.69[ns]
M5 versus M0: M5 − M0 = 19.48 − 15.87 = 3.61*
M5 versus M2: M5 − M2 = 19.48 − 15.73 = 3.75*
M1 versus M4: M1 − M4 = 19.20 − 19.19 = 0.01[ns]
M1 versus M3: M1 − M3 = 19.20 − 18.79 = 0.41[ns]
M1 versus M0: M1 − M0 = 19.20 − 15.87 = 3.33*

M1 versus M2: M1 − M2 = 19.20 − 15.73 = 3.47*
M4 versus M3: M4 − M3 = 19.19 − 18.79 = 0.40[ns]
M4 versus M0: M4 − M0 = 19.19 − 15.87 = 3.32*
M4 versus M2: M4 − M2 = 19.19 − 15.73 = 3.46*
M3 versus M0: M4 − M0 = 18.79 − 15.87 = 2.92*
M3 versus M2: M3 − M2 = 18.79 − 15.73 = 3.06*
M0 versus M2: M0 − M0 = 15.87 − 15.73 = 0.14[ns]

Drawing lines under the means that are not significantly different appears to be a good presentation of results.

NEWMAN-KEUHLS TEST OR STUDENT-NEWMAN-KEUHLS TEST

This test was devised by Newman (1939). It allows comparison of all possible pairs of means in a sequential manner and keeps a constant level for comparison of all pairs of means (Newman, 1939; Keuhls, 1952). This is a multiple range test.

Example 1: In an experiment conducted in CRD design with five treatments and six replications each, the following ANOVA was obtained. The data relate to plot yields in kg of different varieties of rice crop. The ANOVA is shown in Table 21.4.

H_o: $\mu1 = \mu2 = \mu3 = \mu4 = \mu5$
H_1: $\mu1 \neq \mu2 \neq \mu3 \neq \mu4 \neq \mu5$
$F_{4,25,0.05} = 2.76$ H_o: Rejected.
$SE_m = \sqrt{(9.77 / 6)} = \sqrt{1.63} = 1.28$

Procedure for comparisons of means:

1. Arrange the means in increasing order of magnitude.
2. Tabulate pairwise differences.
3. Find q = standardized ranges based on error degrees of freedom (25) and number of treatments (5) and at 5 percent significant ranges (see Appendix, Table A.9).

TABLE 21.4. ANOVA results for plot yields (Kg) of varieties of rice crop.

SV	df	SS	MS	F
Treatments	4	2193.44	548.4**	56.13**
Experimental error	25	244.13	9.77	
Total	29	2437.57		

P	2	3	4	5
$q [0.05, _{(p,n^2)}]$	2.919	3.532	3.901	4.166
	$(q_{2,25,0.05})$	$(q_{3,25,0.05})$	$(q_{4,25,0.05})$	$(q_{5,25,0.05})$

q = difference between two means / SEm (values of q are to be obtained from Appendix, Table A.9). Upper percentage points of studentized range, $q_\alpha = (\bar{x}_{max} - \bar{x}_{min})/s\bar{x}$.

Tabulate the results as shown here for comparison (see Table 21.5).

Comparison Between Treatment 5 and Treatment 1

$q = (\text{Tr } 5 - \text{Tr1})/\text{SE} = (58.3 - 32.1)/1.28 = 20.47 > 4.166$ (tabular q value).

Hence we reject the null hypothesis H_o: $\mu 5 = \mu 1$. To compare $\mu 5$ and $\mu 1$, consider a range of five means, i.e., $p = 5$. Since $20.47 > 4.1666$ (q 0.05, 24) reject H_o.

To compare Tr 5 and Tr 2,

$q = (58.3 - 40.2) / 1.28 = 14.06 > 3.901$ (q 0.05, 24, $p = 4$).

Since $14.06 > 3.901$, reject H_o. Just like $t = \bar{x}_1 - \bar{x}_2)/\text{SE}$, q value is calculated by dividing the difference between two means by standard error. If q is $\geq$ the critical q from tables, we reject the null hypothesis.

Procedure: First compare the largest treatment (Tr 5) against the smallest (Tr 1), then the largest (Tr 5) against the next smallest (Tr 2) and so on. Continue like this until the largest (Tr 5) is compared against the next largest (Tr

TABLE 21.5. Comparison of treatment means using Newman-Keuhls test.

Comparison	Difference	SE	q	q from Table	H_o	Decision
5 versus 1	58.3 − 32.1 = 26.2	1.28	20.47	4.166	5 = 1	Reject
5 versus 2	58.2 − 40.2 = 18.1	1.28	14.14	3.901	5 = 2	Reject
5 versus 3	58.3 − 41.1 = 17.2	1.28	13.44	3.532	5 = 3	Reject
5 versus 4	58.3 − 44.1 = 14.2	1.28	11.09	2.919	5 = 4	Reject
4 versus 1	44.1 − 32.1 = 12.0	1.28	9.38	3.901	4 = 1	Reject
4 versus 2	44.1 − 40.2 = 3.9	1.28	3.05	3.532	4 = 2	Accept
4 versus 3	44.1 − 41.1 = 3.0	1.28	2.34	2.919	4 = 3	Accept
3 versus 1	41.1 − 32.1 = 9.0	1.28	7.03	3.532	3 = 1	Reject
3 versus 2	41.4 − 40.2 = 1.2	1.28	0.94	2.919	3 = 2	Accept
2 versus 1	40.2 − 32.1 = 8.1	1.28	6.33	2.919	2 = 1	Reject

4). Then compare the second largest (Tr 4) with the smallest (Tr 1), the second largest (Tr 4) with the next smallest (Tr 2) and so on.

A line is drawn under the appropriate sample means to show the similarity between the population means as shown here:

Conclusion:

Treatment	1	2	3	4	5
Mean values	32.1	40.2	41.1	44.1	58.3

Example 2: Newman-Keuhls Test: The details of an experiment conducted in CRD design with four treatments (A, B, C, D) with six replications and the yield data of five types of varieties of corn yield/plot (kg) are shown here:

		Treatments		
Replication	A	B	C	D
I	6	10	3	10
II	2	8	7	4
III	5	11	6	6
IV	4	7	4	6
V	6	7	8	7
VI	7	9	6	8
Total	30	52	34	41
Mean	5.0	8.7	5.7	6.8

Grand total: $\Sigma y_{ij} = 157$. Uncorrected SS $\Sigma y_{ij}{}^2 = 1141$; CF $= 157^2/24 = 1027.0$

Total SS $= 1141 - 1027 = 114.0$

Tr SS $= 1073.5 - 1027.0 = 46.5$

Error SS $= 114.0 - 46.5 = 67.5$; F from table for 3,20 df and for $P = 0.05 = 3.1$.

See Table 21.6.

Arrange the means in rank order:

B	D	C	A
8.7	6.8	5.7	5.0

$SE = \sqrt{EMS / r} = \sqrt{3.4 / 6} = 0.75$

$R_{k-1} = (q_{\alpha(k-1),df})$ SE $\alpha = 0.05$

$R_k = R4 = (q_{0.05,4,20})$ SE $= 3.96 \times 0.75 = 2.97$

TABLE 21.6. ANOVA results.

SV	df	SS	MS	F
Treatments	3	46.5	15.5	4.6
Error	20	67.5	3.4	
Total	23	114.0		

$R_{k-1} = R3 = (q_{0.05,3,20})SE = 3.58 \times 0.75 = 2.68$
$R_{k-2} = R2 = (q_{0.05,2,20})SE = 2.95 \times 0.75 = 2.21$
$B - A = 8.7 - 5.0 = 3.7$ versus $2.97*$
$B - C = 8.7 - 5.7 = 3.0$ versus $2.68*$
$B - D = 8.7 - 6.8 = 1.9$ versus 2.21^{ns}
$D - A = 6.8 - 5.0 = 1.8$ versus 2.68^{ns}
$D - C = 6.8 - 5.7 = 1.1$ versus 2.21^{ns}
$C - A = 5.7 - 5.0 = 0.7$ versus 2.21^{ns}

Conclusion: (1) B is significantly greater than A and C; (2) All others are not significantly different.

SCHEFFE'S OR S TEST

Scheffe's (1953) suggested this method for making any and all comparisons on a set of k means. We do not need to plan the comparisons in advance (Edwards, 1971) and it can be applied after the results have been obtained and studied (Federer, 1967). In this test, the multiplier is obtained from the ordinary F table, which is equal to

$$S = [\sqrt{(t-1)} \times F_{\alpha,(t-1),\text{error } df}],$$

where t = number of treatments in the experiment, α = level of significance, and error df = the number of degrees of freedom associated with the error variance. For comparison of two means, the standard error of the comparison is equal to $\sqrt{2_s{}^2 / r}$; where, $s/\sqrt{r}$ is the standard error of the mean. This is described in an example.

Example 1. There are three treatments with ten replications in a CRD design. The data follow:

					Replications						
Tr	I	II	III	IV	V	VI	VII	VIII	IX	X	Total	Mean
1	10	9	14	15	8	12	14	17	20	13	132	13.2
2	13	9	17	21	17	15	9	16	19	23	159	15.9
3	21	17	15	14	25	25	27	18	21	22	205	20.5

Grand total = GT = 496, CF = 496/30 = 8200.53
Tr. SS = $(132^2 + 159^2 + 205^2)/10 - CF = 272.5$
Total SS: $10^2 + 9^2 + \ldots + 22^2 - CF = 8964 - 8200.5 = 763.5$.
See Table 21.7.
Calculations: Standard error of difference between two means

$$= \sqrt{2s^2 / r}$$
$$= \sqrt{(2 \times 18.18)/10} = \sqrt{3.836} = 1.91$$

Value of F from the table of F: $F_{\alpha, (t-1), \text{error } df} = F_{0.05, 2, 27} = 3.35$ (see Appendix, Table A.6).
   Calculate the multiplier, $S =$
$[\sqrt{(t-1) \times F_{\alpha, (t-1), \text{error } df}} = \sqrt{(3-1)(3.35)} = 2.588]$.
   Standard error of difference for comparing two means = $S \times 1.91 = 2.588 \times 1.91 = 4.94$.
   If difference between any two means exceeds 4.94, the two means are said to differ significantly.
   Conclusion: Arrange the means as follows:

Tr 3	Tr 2	Tr 1
20.5	15.9	13.2

Example 2: There are five treatments whose mean values are given:

Treatments:	1	2	3	4	5
Means:	32.1	40.2	41.1	44.1	58.3

These means were recorded in CRD with five treatments and each with six replications. The ANOVA results are given in Table 21.8.

TABLE 21.7. ANOVA results.

SV	df	SS	MS	F	F from tables 5 percent	F from tables 1 percent
Treatment	2	272.5	136.25	7.49**	3.35	5.49
Error	27	491.0	18.18 (s^2)			
Total	29	763.5				

TABLE 21.8. ANOVA results in CRD.

SV	df	SS	MS	F
Treatments	4	2193.44	548.4**	56.13**
Error	25	244.13	9.77 (s^2)	
Total	29	2437.57		

Calculations:

SEd for comparing two means = $\sqrt{(2 \times 9.77 / 6)} = 1.805$

Value of F from F table for $F_{0.05,\,4,\,25\,=\,2.76} = 2.76$

Multiplier $S = \sqrt{(t-1) \times F}_{\,0.05,4,25=2.76} = \sqrt{(4 \times 2.76)} = 3.32$

$S\text{–}SE_d$ for comparing two means = $3.32 \times 1.805 = 5.99$

If the difference between any two means exceeds 5.99, the two means are said to differ significantly.

Conclusion:

Tr	5	4	3	2	1
Means	58.3	44.1	41.1	40.2	32.1

EXERCISES

21.1. Explain the LSD test.

21.2. The means of six treatments are given:
T1 = 13.4; T2 = 24.3; T3 = 44.4; T4 = 54.8; T5 = 69.6; T6 = 86.8; LSD = 25.99. Compare the means and give your conclusions.

21.3. What is Duncan's New Multiple Range test?

21.4. What are the important methods available for testing the treatment means?

21.5. There are four treatments in an experiment. The results are given here ($r = 4$, design CRD). Apply Dunnett's Test using Treatment 1 as control.

Treatment 1	38	38	38	39
Treatment 2	40	41	42	39
Treatment 3	43	44	45	46
Treatment 4	43	47	48	45

21.6. In an analysis of an experiment, the following results were reported.

Source of variation	df	MS
Treatment	7	503.9
Error	40	141.6

The treatment means are: T1 = 172; T2 = 178; T3 = 182; T4 = 185; T5 = 165; T6 = 176; T7 = 168; T8 = 162. Apply Duncan's New Multiple Range test and draw conclusions.
Use the following information: SEm = 4.86.

Value of p	2	3	4	5	6	7	8
SSR (P = 0.05)	2.86	3.01	3.10	3.17	3.22	3.27	3.38
LSR	13	15	15	15	16	16	16

21.7. Apply Tukey's w-procedure test to the treatment means in problem 21.6.
21.8. What is the difference between Tukey's test and LSD test regarding the number of replications?
21.9. Can Duncan's test be applied in all circumstances?
21.10. Can LSD test be applied when F is not significant?
21.11. The values of six treatments recorded in a completely randomized design are as follows:

Treatment	1	2	3	4	5	6
Mean	14.6	16.7	19.6	15.7	24.5	22.4

Treatment 1 is control. Use Dunnett's Test to find whether the control treatment differs significantly from the others. Given error mean square = 30.
21.12. Write a short description on the meaning of confidence limits for a population mean.
21.13. In a CRD, we have eight treatments with six replications each and the error mean square is 60.5 with 40 *df*. Use Duncan's New Multiple Range test with $\alpha = 0.01$ to investigate the difference between the means. The mean values follow:

1	2	3	4	5	6	7	8
17.7	33.7	48.6	49.4	53.1	59.3	63.3	70.0

21.14. Describe briefly each of the terms: *least significant difference, multiple comparison, Dunnett's Test.*

21.15. What is the main use of the various tables containing the critical values? Explain with suitable examples.

21.16. An ANOVA table obtained in an analyis of an experiment is given here:

SV	df	SS	MS	F
Replication	3	0.54	0.18	
Treatments	9	17.22	1.91	9.27
Error	27	5.57	0.21	
Total	39	23.33		

The mean values of the ten treatments are as follows:

Tr	1	2	3	4	5	6	7	8	9	10
Mean	2.25	2.50	2.75	3.00	3.25	3.00	4.00	3.75	4.25	4.00

a. Apply LSD test and compare the means.
b. Consider Treatment 1 as the control. Apply Dunnett's Test and state your results.

21.17. In a fertilizer trial, the following five treatments were evaluated: A = control; B = ammonium sulfate = 15 kg/ac; C = ammonium phosphate = 30 kg/ac; D = super phosphate = 15 kg/ac; E = super phosphate = 30 kg/ac. Grain yield per plot is given here.

	A	B	C	D	E
	2.02	2.62	2.30	1.75	1.55
	1.65	2.32	1.62	1.90	3.00
	2.17	2.75	3.22	2.20	3.95
	3.60	3.10	3.12	3.55	3.32
	3.00	3.70	3.32	2.70	3.43
Total	12.44	14.49	12.58	12.10	15.25

Analyze the data after identifying the design. Apply different kinds of tests of significance for the differences amongst the treatment means and state your inferences.

REFERENCES

Bancroft, T.A. (1968). *Topics in intermediate statistical methods.* Ames, IA: Iowa State University Press.

Carmer, S.G. and M.R. Swanson (1973). Evaluation of ten pair-wise multiple comparison procedures by Monte-Carlo methods. *J. Amer. Stat. Assoc.* 68:66-74.

Davies, O. L. (1949). *Statistical methods in research and production,* Second edition. London: Olover and Boyd.

Duncan, D.B. (1955). Multiple range and multiple *F* tests. *Biometrics,* 11:1-42.

Dunnett, C.W.A. (1955). A multiple comparison procedure for comparing several treatments with a control. *J. Amer. Stat. Assoc.* 50:1096-1121.

Dunnett, C.W. (1964). New tables for multiple comparisons with a control. *Biometrics,* 20:482-491.

Edwards, A.L. (1971). *Experimental design in psychological research.* New Delhi: AmerInd Publishing Co. PVT.

Federer, W.T. (1967). *Experimental design, theory and application.* New Delhi: Oxford and IBH Publishing Co.

Goulden, C.H. (1939). *Method of statistical analysis,* First edition. New York: John Wiley and Sons. Inc.

Keuhls, M. (1952). The use of the "studentized range" in common with an analysis of variance. *Euphytica,* 1:112-122.

Montgomery, D.C. (1991). *Design and analysis of experiments,* Third edition. New York: John Wiley and Sons.

Newman, D. (1939). The distribution of range in samples from a normal population, expected in terms of an independent estimate of standard deviation. *Biometrika,* 31:20-30.

Scheffe, H. (1953). A method for judging all contrasts in the analysis of variance. *Biometrika,* 40:87-104.

Steel, R.G.D. and J.H. Torrie (1960). *Principles and procedures of statistics with special reference to biological sciences.* New York: McGraw-Hill.

Tukey, J.W. (1949). Comparing of individual means in the analysis of variance. *Biometrics,* 5:99-114.

Wine, R.W. (1964). *Statistics for scientists and engineers.* New Delhi: Prentice-Hall of India, (Private) Ltd.

Chapter 22

Latin Square Design

In randomized complete block design, the blocks (replications) are formed across a fertility gradient to minimize the experimental error in one direction and to compare the treatments with increased precision. Sometimes the soil fertility and other variations occur in both directions. In such situations, the blocks (replications) may be formed in both directions. Such a design is called Latin square (LS) design. This design is expected to further reduce the experimental error.

The important features of Latin square design:

1. The number of treatments, rows, and columns are equal.
2. Each row and column contains one treatment only once, thus each row as well as each column is a replication containing all the treatments in the experiment.

Example: A plan for comparing five treatments in a 5 × 5 Latin square design is shown in Figure 22.1. C1, C2, C3, C4, and C5 are columns; R1, R2, R3, R4, and R5 are rows. A, B, C, D, and E are the treatments.

Example 1. We have four job applicants (A, B, C, D) and we want to test their efficiency. We select the Latin square design. We select four problems and four machines. If an applicant works the same problem in different machines, then there would be problem practice. Similarly, if an applicant works on the same machine, there would be machine effect. We select equal numbers of candidates, machines, and problems. Each candidate uses each machine once and each problem once. In Latin square design, each row represents a machine, and each column a problem, and the letters represent the candidates. It is shown in Figure 22.2.

Example 2. Sometimes, the field gradient is in the same direction. Blocks and order in blocks may be used to remove the effects of the fertility gradient. The plan would be like the one that follows:

A	B	C	D

Replication 1

B	C	D	A

Replication 2

C	D	A	B

Replication 3

D	A	B	C

Replication 4

Columns

	C1	C2	C3	C4	C5
R1	A	B	C	D	E
R2	B	C	D	E	A
R3	C	D	E	A	B
R4	D	E	A	B	C
R5	E	A	B	C	D

Rows

FIGURE 22.1. 5 × 5 Latin square.

Problems

	1	2	3	4
1	A	B	C	D
2	B	C	D	A
3	C	D	A	B
4	D	A	B	C

Machines

FIGURE 22.2. 4 × 4 Latin square design.

Here the rows serve as blocks and positions in the rows serve as columns (Cochran and Cox, 1957).

Example 3. Consider five treatments, A, B, C, D, and E. They have to be tested in a laboratory that can accommodate only five technicians. Suppose we expect variation from one day to another. A group of five technicians randomly assigned to each day's testing could be considered as forming a block. If any day-to-day variation is of some importance, then the design will remove this source of variation from the estimate of experimental error. Suppose we anticipate that the hour of testing is also a source of error, then the five candidates are tested each day at five different times. Variation between times could also be considered by assigning the candidates once in each hour. The plan will be as shown:

Day   *Hours*
                  7   8   9   10   11
Mon.   B   E   D   A   C

Tues.	C	A	B	D	E
Wed.	D	B	C	E	A
Thurs.	E	C	A	B	D
Fri.	A	D	E	C	B

Each treatment occurs once and only once in each row and in each column.

Example 4: Variation in milk yield in the four different quarters of a cow's udder can be studied in a 4 × 4 Latin square design. Order of milking can be used as rows and time of milking as columns and the letters A, B, C, and D (as treatments), which are the four quarters of the udder, respectively. The design will be as follows (see Figure 22.3):

In a biological assay of vitamins, the 4 × 4, 5 × 5, and 6 × 6 Latin squares will be most useful. The typical Latin square designs before randomization are as follows:

4 × 4				5 ×5					6 × 6					
D	A	B	C	D	E	A	B	C	F	A	B	C	D	E
C	D	A	B	C	D	E	A	B	E	F	A	B	C	D
B	C	D	A	B	C	D	E	A	D	E	F	A	B	C
A	B	C	D	E	A	B	C	D	C	D	E	F	A	B
				A	B	C	D	E	B	C	D	E	F	A
									A	B	C	D	E	F

In a Latin square design, the experimenter is not interested in the effects of rows and columns. As shown in the analysis table, the variation remaining after removal of rows, columns, and treatments was used as the experimental error. When no interaction exists among rows, columns, and treatments, the test is quite valid. The randomization procedure for different Latin squares has been given by Fisher and Yates (1963), and Cochran and Cox (1957).

The ANOVA table for a Latin square is in Table 22.1.

Order of milk

		1	2	3	4
	4 a.m.	D	A	B	C
Time	10 a.m.	C	D	A	B
	4 p.m.	B	C	D	A
	10 p.m.	A	B	C	D

FIGURE 22.3. 4 × 4 Latin square design for milk yield.

TABLE 22.1. ANOVA results for a Latin square.

Source of variation	df	SS	MS	F	Table F 5 percent	1 percent
Rows	$t-1$	$\Sigma i^2/t - CF$				
Columns	$t-1$	$\Sigma C_j^2/t - CF$				
Treatments	$t-1$	$\Sigma T_k^2/t - CF$				
Error	$(t-1)(t-2)$	By subtraction				
Total	t^2-1	$\Sigma x_{ij}^2 - CF$				

Anderson and McLean (1974) emphasized that Latin square design was satisfactory for agricultural experiments but not for engineering, social sciences, and other areas. Pearce (1985) remarked that rows and columns were not necessarily better than RCBD, and if Latin square design is laid out on land of unknown fertility, there would be a serious failure of the model which assumes row and column effects to be additive.

An experiment was conducted with four different rice varieties (A, B, C, D) to study their performance in a Latin square design. The layout plan and the yields recorded are given here:

Rows		1	2	3	4	Total
		Columns				
1		A	B	C	D	
		4	8	3	8	23(R1)
2		B	C	D	A	
		10	3	14	5	32(R2)
3		C	D	A	B	
		4	10	4	9	27(R3)
4		D	A	B	C	
		16	3	9	2	30(R4)
Total		34	24	30	24	
		C1	C2	C3	C4	112 (GT)

Calculations

1. Find the variety (treatment) totals by forming the table as follows

			Treatments	
	A	*B*	*C*	*D*
	4	8	3	8
	5	10	3	14
	4	9	4	10
	3	9	2	16
Total	16	36	12	48
Mean	4	9	3	12

GT = 112

2. Calculate column totals, row totals, treatment totals, and grand total (GT).
3. Compute correction factor

   (CF): $CF = (GT)^2 / t^2 = 112^2/16 = 784$

4. Total $SS = 4^2 + 5^2 + \ldots + 10^2 + 16^2 - CF = 1046 - 784 = 262$
5. Column $SS = (C1^2 + C2^2 + C3^2 + C4^2)/t - CF = (34^2 + 24^2 + 30^2 + 24^2)/4 - CF = 18.0$
6. Row $SS = (R1^2 + R2^2 + R3^2 + R4^2)/t - CF = (23^2 + 32^2 + 27^2 + 30^2)/4 - CF = 11.57$
7. Treatment $SS = (A^2 + B^2 + C^2 + D^2)/t - CF = (16^2 + 36^2 + 12^2 + 48^2)/4 - CF = 1000 - 784 = 216.0$
8. Form the ANOVA table (see Table 22.2)

Test of Significance

F = (treatment mean square/error mean square) = $72.00/2.75 = 26.18$. Since the computed F (26.18) exceeds 9.78 (table value of F at $P = 0.01$: See Appendix, Table A.6), the treatment means differ significantly at 1 per-

TABLE 22.2. ANOVA results.

Source of variation	*df*	*SS*	*MS*	*F*	Table *F* 5 percent	Table *F* 1 percent
Rows	3	11.5	3.83	1.39[ns]	4.76	9.78
Columns	3	18.0	6.00	2.18[ns]	4.76	9.78
Treatments	3	216.0	72.00	26.18**	4.76	9.78
Error	6	16.5	2.75			
Total	15	262.0				

[ns] = not significant; ** significant at $P = 0.01$

cent level of significance. Since the F values for rows and columns are not significant, formation of blocks along rows or columns did not give any appreciable reduction in error variability.

Calculation of least significant difference: (LSD) at $P = 0.05$ for comparison of treatment means

$$LSD = t_{0.05,6} \times \sqrt{2s^2 / t} = 2.447\sqrt{(2 \times 2.75) / 4} = 2.87.$$

Comparison of treatment means:

Treatments: T4   T2   T1   T3
Mean values: 12.0   9.0   4.0   3.0

Inference: The variety 4 recorded the highest yield, which is significantly different than the other varieties. The difference between T1 and T3 is not significant, and therefore it is denoted by a line under the two means.

LINEAR MODEL

The assumed linear model for a Latin square design is as follows:

$$Y_{ijk} = m + r_i + c_j + t_k + e_{ijk}$$

where, Y_{ijk} = any observation = general mean + row effect + column effect + treatment effect + error effect, where

Y_{ijk} = value of the observation in the ith row, jth column, and kth treatment

m = general mean
r_i = effect of ith row
c_j = effect of jth column
t_k = effect of the kth treatment
e_{ijk} = experimental error effect from the ith row, jth column of the kth treatment estimated by the residual.

Example: The model is explained with a numerical example for better understanding of the various effects. The data obtained from an experiment are shown in Table 22.3:

The row effect and column effect are calculated as in the previous case. For example, the column effect in Column 1 is column mean minus grand mean $(8.50 - 9.00 = -0.5)$, and the row effect in Row 1 is mean of row effect in Row 1 minus the grand mean $(9.25 - 9.00 = 0.25)$.

TABLE 22.3. Experimental data.

Row	C1	C2	C3	C4	Total	Mean	Row Effect
1	D 12	A 4	B 10	C 11	37	9.25	+0.25
2	C 9	D 10	A 6	B 7	32	8.00	−1.00
3	A 6	B 4	C 12	D 14	36	9.00	0.00
4	B 7	C 12	D 12	A 8	39	9.75	0.75
Total	34	30	40	40	144		
Mean	8.50	7.50	10.0	10.0	9.00		
Column effect	−0.5	−1.5	+1.0	+1.0			

The treatment effects are calculated as follows:

Rows	Treatment			
	A	B	C	D
1	4	10	11	12
2	6	7	9	10
3	6	4	12	14
4	8	7	12	12
Total	24	28	44	48
Mean	6	7	11	12

Grand total = 144 Grand mean = 9.00

Tr. effect: −3.00 − 2.00 + 2.00 + 3.0

Error effect is estimated after deducting the row, column, and treatment effects from the total effect for each observation according to the model.

Error effect $(e_{ijk}) = y_{ijk} - $ GM − row effect − column effect − treatment effect.

For example, $e_{113} = +0.25 = 12 - 9 - 0.25 - (-0.5) - 3.0$.

Distribution of errors and error sum of squares in each observation (see Table 22.4).

Now we will perform statistical analysis from the original data as follows.

Calculations

1. Row totals: R1 = 37 ; R2 = 32; R3 = 36 ; R4 = 39; total: 144
2. Column totals: C1 = 34; C2 = 30; C3 = 40; C4 = 40; total : 144
3. Treatment totals: A = 24; B = 28; C = 44; D = 48; total: 144
4. $\Sigma R_i = \Sigma C_i = \Sigma T_i = \Sigma y_{ijk} = 144$ (check)

5. General mean: $GT/t^2 = 144/16 = 9$
6. Correction factor $(CF) = GT^2/t^2 = 144^2/16 = 20736/16 = 1296$
7. Total $SS\ (TSS) = \Sigma y_{ijk}^2 - CF = 1440 - 1296 = 144$
8. Row $SS = (R1^2 + R2^2 + R3^2 + R4^2)/6 - CF = 1302.5 - 1296.0 = 6.5$
9. Col. $SS = (C1^2 + C2^2 + C3^2 + C4^2)/6 - CF = 1314 - 1296 = 18.0$
10. Tr. $SS = (T1^2 + T2^2 + T3^2 + T4^2)/6 - CF = 1400 - 1296 = 104$
11. Error $SS = TSS - $ Row $SS - $ Col. $SS - $ Tr. $SS = 144 - 6.5 - 18.0 - 104 = 15.5$
12. Now form the analysis of variance table as shown in Table 22.5.

Additivity assumption: Note that the error sum of squares within each treatment (0.6250 in treatment A, 6.1250 in treatment B, 6.1250 in treatment C, and 2.6250 in treatment D) add to the total error sum of squares as obtained in the ANOVA.

Test of Significance

The analysis indicated a significant difference among treatments but not among rows and columns. The nonsignificant rows or columns indicates that formation of rows or columns failed to improve the precision of the experiment. To test the null hypothesis that there is no difference between treatment means (variety means), $F = 13.44$, is compared with $F_{0.05} = 4.76$ and $F_{0.01} = 9.78$. At 0.01 level of significance, the null hypothesis is rejected.

TABLE 22.4. Error and error sum of squares for treatments.

Errors				Error sum of squares			
Treatments				Treatments			
A	B	C	D	A	B	C	D
+0.25	−0.75	+1.75	−1.25	0.0625	0.5625	3.0625	1.5625
−0.50	+0.50	0	0	0.2500	0.2500	0	0
+0.50	−1.50	0	+1.00	0.2500	2.2500	0	1.0000
−0.25	+1.75	−1.75	+0.25	0.0625	3.0625	3.0625	0.0625
0	0	0	0	0.625	6.125	6.125	2.625
	Total: 0				Total error $SS = 15.5000$ ei_{jk}^2		

TABLE 22.5. ANOVA results.

Source of variation	df	SS	MS	F	Table *F* 5 percent	Table *F* 1 percent
Rows	$(t-1)=3$	6.5	2.167	0.84^{ns}	4.76	9.7
Columns	$(t-1)=3$	18.0	6.00	2.33^{ns}	4.76	9.7
Treatments	$(t-1)=3$	104.0	34.676	13.44^{**}	4.76	9.7
Error	$(t-1)(t-2)=6$	15.5	2.58			
Total	15	144.0				

^{ns} = not significant; ** significant at $P=0.01$

Comparison of Means

Calculate $SEm = \sqrt{s^2} / \sqrt{n} = \sqrt{2.58} / \sqrt{4} = 0.803$. LSD to test the difference between two means is

$$t \times SEd = 2.447 \times \sqrt{(2 \times 2.58 / 4)} = 2.447 \times 1.135 = 2.78.$$

(The value of t at $P_{0.05,6}$ is 2.447, from the t tables.)
Arrange the means in descending order and find the differences between the means:

Treatments:	D	C	B	A
Means:	12.0	11.0	7.0	6.0

Inference: There is no difference between D and C, or between B and A.

Efficiency of Latin Square over CRD

Relative efficiency $(RE) = [R + C(t-1)E] / (t+1)E$

where R = row mean square = 2.167; C = column mean square = 6.00; and E = error means square = 2.58.

$$RE = [2.167 + 6.000 + (3 \times 2.58)] / (5 \times 2.58) = [15.907 / 12.90] \times 100$$
$$= 123 \text{ percent.}$$

Efficiency of Latin Square over RCBD Using Rows As Block

$$RE = \left[C + (t-1)E\right] / (t \times E) = \left[6.00 + (3 \times 2.58)\right] / (4 \times 2.58) = \left[13.74 / 10.32\right] \times 100 = 133 \text{ percent.}$$

Efficiency of Latin Square over RCBD Using Columns As Block

$$RE = \left[R + (t-1)E\right] / (t \times E) = \left[2.167 + (3 \times 2.58)\right] / (4 \times 3.58) = (9.907 / 10.32) \times 100 = 96 \text{ percent.}$$

ADVANTAGES AND DISADVANTAGES OF LATIN SQUARE DESIGN

Advantages

1. Columns as well as rows serve as blocks, and hence the design minimizes the soil heterogeneity in both directions.
2. Experiments may be laid out in compact square blocks.
3. It is used in many fields such as agriculture, industry, medicine, etc.

Disadvantages

1. Number of degrees of freedom: For any efficient design we require a minimum of 12 degrees of freedom for error in the ANOVA. The error degrees of freedom for different sizes of Latin square designs are as follows.

SV			Sizes		
	2 x 2	3 x 3	4 x 4	5 x 5	6 x 6
Rows	1	2	3	4	5
Columns	1	2	3	4	5
Treatments	1	2	3	4	5
Error	0	2	6	12	20
Total	3	8	15	24	35

The error degrees of freedom for 2 × 2, 3 × 3, 4 × 4, 5 × 5, and 6 × 6 are 0, 2, 6, 12, and 20, respectively. Thus, Latin square will not be efficient when the treatments are four or less than four. In our example, we used 4 × 4 Latin square which had six degrees of freedom for error. The test is insensitive, and hence Latin square design for smaller than 5 × 5 square should not be used. The number of degrees of freedom may be increased when the number of treatments is less. For example, when $t = 4$, two 4×4 Latin squares may be used. Then the ANOVA results will be as follows:

Source of variation	df
Between squares	1
Rows within squares	6
Columns within squares	6
Treatments	3
Error	15
Total	31

2. For larger numbers of treatments, the row size increases, which causes heterogeneity among the plots within the blocks, and hence, for larger numbers of treatments, this design is not suitable.
3. It can be used only when the number of rows, number of columns, and number of treatments are equal.
4. Latin square is not useful for fewer than four treatments nor more than eight (Kempthorne, 1952).

How do we compute different SS when the number of squares is 2 and $t = 3$?

Between squares: $(G1^2 + G2^2)/9 - (\Sigma x)^2/18$ (with 1 degree of freedom)
Row SS = (row SS for square 1 + row SS for square 2) with 4 degrees of freedom
Col. SS = (column SS for square 1 + column SS for square 2) with 4 degrees of freedom
Tr. SS = $[T1^2 + T2^2 + T3^2]/6 - [(\Sigma x)^2/18]$ with 2 degrees of freedom
Error df = total df – squares df – row df – col. df – treatment df., i.e., $(17 - 1 - 4 - 4 - 2 / 6)$

G1 and G2 are the totals of square 1 and 2, respectively. T1, T2, and T3 are the totals of six units, three in each square.

Partition of Degrees of Freedom When We Use Several Squares

ANOVA showing the degrees of freedom for various sources of variation when more squares are used in an experiment. This information will be useful for statistical analyses when researchers use several squares in their experiments.

Sources of variation	df	MS
Between squares	$S-1$	S
Rows within squares	$S(k-1)$	R
Columns within squares	$S(k-1)$	C
Treatments	$k-1$	T
Treatments × squares	$(S-1)(k-1)$	TS
Error	$S(k-1)(k-2)$	E
Total	Sk^2-1	

The treatment × squares sum of squares may be pooled with error sum of squares, if there is no treatment × square interaction.

Example: An experiment was conducted to evaluate four different treatments in a rice crop trial. For want of contiguous area, two 4 × 4 Latin squares were used. The yield data are shown in Table 22.6 and Table 22.7: Carry out the analysis and draw conclusions.

Form the square × treatments table as shown in Table 22.8.

Treatment totals: A = 42.0; B = 44.0; C = 50.0; D = 54; GT = 190
Treatment means: $\bar{A}=5.25$; $\bar{B}=5.50$; $\bar{C}=6.25$; $\bar{D}=6.75$
CF = $190^2/32 = 1128.13$
Total $SS = (7^2 + 7^2 + \ldots + 6^2 + 6^2) - CF = 53.87$
SS for squares = $(104^2 + 86^2)/16 - CF = 10.12$
SS for treatments: $(42^2 + 44^2 + 50^2 + 54^2)/8 - CF = 11.37$

TABLE 22.6. Square 1 data.

Row/col.	C1	C2	C3	C4	Total
R1	D 7	A 7	C 7	B 7	28
R2	B 5	C 6	A 5	D 7	23
R3	A 6	B 7	D 7	C 6	26
R4	C 8	D 8	B 6	A 6	27
Total	26	28	25	25	104

TABLE 22.7. Square 2 data.

Row/col.	C1	C2	C3	C4	Total
R1	A 6	D 7	C 8	B 5	26
R2	B 2	A 4	D 5	C 4	16
R3	D 7	C 6	B 5	A 6	24
R4	C 5	B 6	A 3	D 6	20
Total	21	23	21	21	86

TABLE 22.8. Square x Treatment table.

Tr	Sq 1	Sq 2	Total
A	23	19	42
B	25	19	44
C	27	23	50
D	29	25	54
Total	104	86	190

SS sq × treatments SS: $(23^2 + 25^2 + \ldots + 23^2 + 25^2)/24 - CF = SS \text{ sq.} - SS$ Tr. = 0.38

SS columns w/*n* squares: $(26^2 + 28^2 + 25^2 + 25^2 + 21^2 + 23^2 + 21^2 + 21^2)/4 - (104^2 + 86^2)/16 = 1.25$

SS rows within squares: $(28^2 + 23^2 + 26^2 + 27^2 + 26^2 + 16^2 + 24^2 + 20^2)/4 - (104^2 + 86^2)16 = 18.25$

Form the analysis of variance table as in Table 22.9.

LATIN SQUARE AND TREND ANALYSIS

Example: The effect of super phosphate on sugarcane crop was studied in an experimental farm using Latin square design. The treatments were 0, 40, 60, 120, and 160 lb P_2O_5/acre as super phosphate. The variety used was CO 42. Plot size (net) 33 ft × 33 ft. The layout plan and the yield in kg/plot is as follows:

B 269	D 262	E 322	C 320	A 202
D 164	B 248	C 200	A 258	E 229
E 210	A 159	D 185	B 224	C 225
A 217	C 257	B 210	E 158	D 151
C 189	E 230	A 220	D 218	B 205

The following analyses are to be carried out: (1) comparison of means; (2) partition the treatment SS with linear, quadratic, cubic, and quartic analyses; (3) one missing value occurs in Tr. A in the third row. Find the missing value.

Analysis 1

1. Find treatment totals.

TABLE 22.9. ANOVA results.

Source of variation	df	SS	MS	F	Table F (5 percent)
Squares	1	10.12	10.12	10.25	
Treatments	3	11.37	3.79	3.95	3.49
Sq × Tr	3	0.38	0.13	<1	
Col within sq	6	2.25	0.38	<1	
Rows within sq	6	18.25	3.09	3.17	
Error	12	11.50	0.96		
Total	31	53.87			

2. Form the following table.

	A	B	C	D	E	
	202	269	320	262	322	
	258	248	200	164	229	
	159	224	225	185	210	
	217	210	257	151	158	
	220	205	189	218	230	
Total	1056	1156	1191	980	1149	Grand total = 5532

Find row totals and column totals.

B 269	D 262	E 322	C 320	A 202	1375
D 164	B 248	C 200	A 258	E 229	1099
E 210	A 159	D 185	B 224	C 225	1003
A 217	C 257	B 210	E 158	D 151	993
C 189	E 230	A 220	D 218	B 205	1062
Total 1049	1156	1137	1178	1012	5532

Find row totals and column totals.

B 269	D 262	E 322	C 320	A 202	1375
D 164	B 248	C 200	A 258	E 229	1099
E 210	A 159	D 185	B 224	C 225	1003
A 217	C 257	B 210	E 158	D 151	993
C 189	E 230	A 220	D 218	B 205	1062
Total 1049	1156	1137	1178	1012	5532

Analysis 2

1. Find correction factor (CF) = $5532^2/25$ = 1224120.9
2. Row SS = $(R1^2 + R2^2 + R3^2 + R4^2 + R5^2)/5 - CF$ = $(6218325)/5 - CF$ = $1243665.6 - CF$ = 19544.7
3. Column SS = $(C1^2 + C2^2 + C3^2 + C4^2 + C5^2)/5 - CF$ = $1228266.8 - CF$ = 4145.9
4. Treatment SS = $(A^2 + B^2 + C^2 + D^2 + E^2)/5 - CF$ = $(6150554/5) - CF$ = $1230110.8 - CF$ = 5989.9
5. Total SS = $1271278 - CF$ = 47157.9

Partition of treatment SS using orthogonal polynomials.
Form the following table:

SS	A	B	Treatment totals C	D	E	Divisor
	1056	1156	1191	980	1149	
Linear	-2	-1	0	1	2	10
Quadratic	2	-1	-2	-1	2	14
Cubic	-1	2	0	-2	1	10
Quartic	1	-4	6	-4	1	70

SS linear $= [(-2)(1056) + (-1)(1156) + 0 + (1)(980)$
$+ (2)(1149)]^2/r\Sigma c_i^2$
$= (10.00)^2/ \ 5 \times 10 = 2.00$

SS quadratic $= [(2)(1056) + (-1)(1156) + (-2)(1191) + (-1)(980)$
$+ (2)(1149)]^2/5 \times 14$
$= (-108)^2/70 = 11664/70 = 166.6$

SS cubic $= [(-1)(1056) + (2)(1156) + (0)(1191) + (-2)(980)$
$+ (1)(1149)]^2/5 \times 10$
$= 445^2/50 = 198025/ \ 50 = 3960.52$

SS quartic $= [(1)(1056) + (-4)(1156) + (6)(1191) + (-4)(980)$
$+ (1)(1149)]^2/5 \times 70$
$= (1056 - 4624 + 7146 - 3920 + 1149)/5 \times 70$
$= (9351 - 8544)^2 = 807^2/350 = 1860.7$

Since F is not significant, the treatment differences are not significant (see Table 22.10).

TABLE 22.10. ANOVA results.

Source of variation	df	SS	MS	F	Table F (5 percent)
Rows	4	19544.7	4886.1 (s^2r)	3.35	
Columns	4	4145.9	1036.5 (s^2c)	< 1	
Treatments	4	5989.9	1497.4 (s^2t)	1.27ns	
Linear	(1)	2.0	2.00	<1	
Quadratic	(1)	166.6	66.60	<1	
Cubic	(1)	3960.5	3960.50	2.72ns	
Quartic	(1)	1860.7	1860.70	1.27ns	
Error	12	17477.4	1456.40 (s^2e)		
Total	24	47157.9			

Compare Latin square design with randomized complete block design by taking rows as blocks: The efficiency of Latin square design as related to corresponding randomized complete block design is given by the formula

$$\left[\frac{(k-1)(k-2)+1}{(k-1)(k-2)+3}\right]\left[\frac{(k-1)^2+3}{(k-1)^2+1}\right]\left[\frac{s^2c+(k-1)s^2e}{ks^2c}\right] \text{ where,}$$

k = number of treatments
s^2c = column mean square
s^2e = error mean square

$$\left[\frac{(5-1)(5-2)+1}{(5-1)(5-2)+3}\right]\left[\frac{(5-1)^2+3}{(5-1)^2+1}\right]\left[\frac{1036+(5-1)1456.4}{5\times1456.4}\right]$$

$(13/15)(19/17)(0.9433) = 0.9128 = 91.3$ percent

If columns are replicates, the efficiency of Latin square relative to randomized complete block design is obtained using the following formula:

$$\left[\frac{(k-1)(k-2)+1}{(k-1)(k-2)+3}\right]\left[\frac{(k-1)^2+3}{(k-1)^2+1}\right]\left[\frac{s^2r+(k-1)s^2e}{ks^2e}\right]$$

$$\left[\frac{(5-1)(5-2)+1}{(5-1)(5-2)+3}\right]\left[\frac{(5-1)^2+3}{(5-1)^2+1}\right]\left[\frac{4886.1+(5-1)1456.4}{5\times1456.4}\right]$$

$$\left[\frac{(4\times3)+1}{(4\times3)+3}\right]\left[\frac{(16+3)}{(16+1)}\right][1.4710]=(13/15)(19/17)(1.4710)=1.4248=$$

142.5 percent

Columns as replicates, the Latin square is efficient by 142.5 percent.

Estimation of missing value: Assume that treatment A in Row 3 and Column 2 is missing. The missing value is 159. Assume this value as X.

If the missing value is X, then, $X = \dfrac{r(R+C+T)-2G}{(t-1)(t-2)}$

General formula for estimation of missing value in Latin square design: If any observation is lost, it may be estimated using the formula,

$$y = \frac{r(R+C+T)-2G}{(r-1)(r-2)} \text{ where}$$

y = estimated value for the missing observation,
r = number of replications,
R = total of all observations in the row where missing value occurs,
C = total of all observations in the column where missing value occurs,
T = total of the missing treatment occurring in other rows and columns, and
G = grand total of all the known values.

$r = 4$; $R = 844$; $C = 997$, $T = 897$; $t = 897$; and $G = 5373$.

$$X = \left[\frac{5(844+997+897)-2\times5373}{4\times3=12}\right] = (13690-10746)/12 =$$

$2944/12 = 245.3$

Grand total = 5373 + 245 = 5618.

Statistical Analysis

$CF = GT^2/n = GT^2/25 = 5618^2/25 = 1262476.9$

Row *SS:* 1279648.0 − *CF* = 17171.1
Column *SS:* 1269512.4 − *CF* = 7035.5

Treatment *SS:* 1267916.4 − *CF* = 5439.5
Total *SS:* 1306022 − *CF* = 43545.1
(See Table 22.11)

Substitute the estimated value and then carry out the analysis of variance as in randomized complete block design. The standard error of difference for comparing two means in one of which a missing value occurs is

$$\sqrt{[V_E\{(2/r) + [1/(r-1)(r-2)]\}]}.$$

Latin Square and Study of Contrasts Among the Treatments

Example: There are five treatments, A, B, C, D, E studied in Latin square design. The yield data are given. Study the contrasts of interest. See Table 22.12.
Row totals: R1 = 142.80; R2 = 139.34; R3 = 152.72; R4 = 157.83; R5 = 145.43

Treatments	A	B	C	D	E	
Total	107.02	181.10	105.53	151.80	192.67	738.12 (GT)
Mean	21.40	36.22	24.12	30.36	38.53	

Compute the following contrasts:
1. A, B versus C, D, E
2. A versus B
3. C versus D, E
4. D versus E

TABLE 22.11 ANOVA results.

Source of variation	df	SS	MS	F	Table F 5 percent	Table F 1 percent
Rows	4	17171.1	4292.7	3.4	3.36	5.67
Columns	4	7035.5	1758.8	1.4	3.36	5.67
Treatments	4	5439.5	1359.8	1.1	3.36	5.67
Error	11	13899.0	1263.5			
Total	23	43545.1				

TABLE 22.12. 5 × 5 Latin square design.

Row	Column				
	1	2	3	4	5
1	C 18.30	B 35.25	D 30.32	A 16.08	E 42.85
2	D 28.05	E 36.16	A 17.25	C 25.90	B 31.98
3	A 25.12	D 28.55	B 37.10	E 38.27	C 23.68
4	B 40.25	C 22.60	E 41.15	D 31.68	A 22.15
5	E 4.24	A 26.42	C 15.05	B 36.52	D 33.20
Total	145.96	148.98	140.87	148.45	153.86

Contrast treatments

	A	B	C	D	E	Value	Divisor
Tr. totals	107.02	181.10	105.53	151.80	192.67		
C1	−3	−3	2	2	2	35.64	30x5
C2	−1	+1	0	0	0	74.08	2x5
C3	0	0	−2	1	1	133.41	6x5
C4	0	0	0	−1	+1	40.87	2x5

Analysis

$CF = 738.12^2/25 = 21792.845$

Row SS: $(142.8^2 + \ldots + 145.43^2)/5 - CF = 21838.212 - CF = 45.367$

Column SS: $(145.96^2 + \ldots + 153.86^2)/5 - CF = 21810.802 - CF = 17.957$

Tr. SS: $(107.2^2 + \ldots + 192.67^2)/5 - CF = 23110.406 - CF = 1317.561$

SS for contrasts:

C1 = $35.64^2/150 = 8.468$
C2 = $74.08^2/10 = 548.784$
C3 = $133.41^2/30 = 593.274$
C4 = $40.87^2/10 = 167.035$
Total = 1317.561
See Table 22.13.

TABLE 22.13. ANOVA results.

Source of variation	df	SS	MS	F	Table F 5 percent	Table F 1 percent
Rows	4	45.367	11.341	< 1	3.26	5.41
Columns	4	17.957	4.489	< 1	3.26	5.41
Treatments	4	1317.561	329.390	19.54	3.26	5.41
C1 (AB) vs (C+D+E)	(1)	8.468	8.468	< 1	4.75	9.33
C2 A vs B	(1)	548.784	548.785	32.56**	4.75	9.33
C3 C vs (D,E)	(1)	593.274	593.274	35.20**	4.75	9.33
C4 D vs E	(1)	167.035	167.036	9.91**	4.75	9.33
Error	12	202.249	16.854			
Total	24	1583.134				

Conclusion: Contrasts 2 and 3 show large differences. Also C4 shows significant difference.

EXERCISES

22.1. How do the principles of experimentation satisfy a Latin square design?

22.2. Fill in the blanks in the following ANOVA table for a Latin square design.

Source of variation	df	SS	MS	F	Table F 5 percent	Table F 1 percent
Rows	_____	72	_____	2		
Columns	_____	_____	36			
Treatments		180	_____			
Error	6	_____	12			
Total						

22.3. What is Latin square design? How does it differ from randomized complete block design?

22.4. What is the underlying model, the hypothesis to be tested, and the dummy ANOVA table for 4 × 4 Latin square design. How would you compare two treatment effects pairwise in this design?

22.5. The yield data (kg/plot) of a wheat varietal trial are:

B 425	C 442	D 540	A 349
D 384	A 512	B 490	C 408
C 506	D 508	A 536	B 600
A 451	B 568	C 499	D 347

   a. What is the ANOVA model for this investigation?
   b. What is the hypothesis of main interest in the trial?
   c. Is there any real difference among the wheat varieties?
   d. Which mean separation procedure do you follow in this case?
   e. Find the standard error for the mean. What is the CV for the experiment?

22.6. Give two examples of Latin square design.

22.7. When the numbers of treatments are many, Latin square is not advisable. Why?

22.8. Give the formula for estimating a missing value in the Latin square design. What is the change that you will make in the ANOVA table?

22.9. What are the advantages and disadvantages of Latin square design?

22.10. Suppose an experimenter wants to study four irrigation levels (including one standard, i.e., no control) in Latin square design, and is interested in comparing the standard with other treatments. What advice will you offer the experimenter?

22.11. What is the ANOVA model for the Latin square design when $t = 8$?

22.12. How will you estimate if there is one missing value in one treatment? What modifications will you make in the analysis of variance table? What is the standard error of difference for comparing the mean of the estimated value with those of others?

22.13. Suppose an experimenter wants to compare three treatments in Latin square design. What suggestions will you make for layout of design and analysis? Precision is important in this case.

22.14. A 2×3 factorial experiment is to be conducted in Latin square design. Give the layout plan and show how you will partition the treatment sum of squares in the ANOVA. Write the standard errors for the different treatments.

REFERENCES

Anderson, V.L. and R.A. McLean (1974). *Design of experiments—A realistic approach.* New York: Marcel Dekker Inc.

Cochran, W.G. and G.M. Cox (1957). *Experimental design,* Second edition. New York: Wiley.

Fisher, R.A. and F. Yates (1963). *Statistical tables for biological, agricultural and medical research,* Sixth edition. New York: Hafner.

Kempthorne, O. (1952). *The design and analysis of experiments.* New York: Wiley.

Pearce, S.C. (1985). *Agricultural experimentation in a developing country—A celebration of statistics:* The ISI Centenary Volume. New York: Springer-Verlag.

Chapter 23

Factorial Experiments

Experiments in which only one factor is included are called single-factor experiments or univariate experiments or simple experiments. In univariate experiments, we compare treatments such as different drugs, methods of treatment for a disease, varieties of different crops, irrigation methods, fertilizers, effect of vitamins on growth of children, effect of smoking intensities on physical activities of human beings, etc. These are called factors.

LEVELS

When analyzing vitamins, for instance, we study several vitamins' effects on the growth of children. We study the efficacy of different fertilizers like ammonium sulfate, urea, ammonium nitrate, calcium ammonium nitrate, etc. Different vitamins and different fertilizers are called levels. Similarly in studying the effects of smoking, smoking in different intensities, i.e., number of cigarettes per day, cigarettes are called levels. Thus, factor is a treatment and each factor may give rise to several levels depending upon the nature of the experiment. Thus, level is also a treatment within a factor. If two or more factors and more than one level in each factor are studied together in an experiment, it is called a factorial experiment. Such factorial experiments are gaining popularity in many fields such as industry, medicine, social sciences, agriculture, engineering, biotechnology, animal sciences, etc. Some examples of factors and levels are shown in Exhibit 23.1.

Examples of Factorial Experiments

1. Different vaccine preparations can be studied with different amounts of additives to test the effect in terms of antibody response. In this experiment, types of vaccines and amounts of additives are the two factors. The first factor can occur at two levels and the second at five levels.

2. Five different breeds and four different diets are studied in an experiment on scurvy in guinea pigs. Breeds and diets are the two factors, which are studied with five levels of breed and four levels of diet.
3. Two types of leukemia can be studied with four different methods of treatment. Here leukemia is in two levels and methods of treatment are in four levels. It is a 2×4 factorial experiment.
4. Two different types of teaching methods and students in three age groups (5-6, 7-8, 9-10) may form a factorial experiment to evaluate the effect of teaching methods on different age groups of students. Teaching methods are in two levels and age groups are in four levels.
5. Different amounts of protein and carbohydrates may be studied relative to the growth of children with different selected levels.
6. Sometimes, levels of irrigation to a crop are dependent upon different amounts of manure. The two factors are not independent but interact with each other. In single experiments, we cannot find the optimum combination of the two.

EXHIBIT 23.1. Examples of factors and levels.

Factor	Level
Agriculture	
Planting dates	July 15, August 15, September 15
Planting season	Summer, winter, cold
Fertilizer	Super phosphate, diammonium phosphate
Nitrogen fertilizer	0, 60, 120, 180 N kg/ha
Soil	Alkaline, saline, loamy
Breed	Different breeds of an animal
Medicine	
Analgesic drug	Aspirin, placebo, new drug
Age group	5-10 years, 10-15 years, 15-20 years
Sex	Male, female
Drug	Different doses
Industry	
Washing machines	Different types of machines
Laundry detergent	Different types of laundry detergent
Temperature	50, 60, 70, 80, 90°C
Nutrition	
Diet	Protein, nonprotein, or different combinations of nutrient diets
Milk	Cow milk, buffalo milk, goat milk
Protein	Animal protein, vegetable protein

Factorial Experiment

A factorial experiment is one in which different levels of a given factor are combined with levels of every other factor in the experiment (Hicks, 1973).

Reasons for Using Factorial Experiments

In agriculture, crop yields are invariably recorded and based on yields; the treatments are evaluated. The aim of the experimenter is to increase production, which is affected by several factors. For example, increased production may be affected by application of two or more factors simultaneously.

In factorial experiments, the treatments are obtained by combining the different levels of one factor with the different levels of another factor. Such treatments are called factorial treatments. For example, two drugs (A, B) were evaluated by administering them to four patients. The reaction time was noted at three different times (morning, afternoon, and evening). This is a well-designed 2×3 factorial experiment. The factor is one, the effect of which we wish to assess. Factors occur in experiments at several levels. These levels are specific fixed values or states of the factor. For example, in agriculture, season is a factor and it may occur at two levels, winter and summer.

Concept

Factorial experiments should not be referred to as experimental designs since for example, CRD, RCBD, and LSD designs are used for the factorial experiments. Treatments determine whether it is a factorial experiment or not.

Consider that we have to test four varieties of corn using six replications. If we use RCBD we require 24 plots. Also consider that we have to compare two levels of nitrogen in seven replications and in this case we require 14 plots. If we conduct two separate experiments, we need $24 + 14 = 38$ plots. Comparisons are made based on the mean yields of six plots and seven plots, respectively, in each case. If we conduct a factorial experiment with four levels of variety and two levels of nitrogen in four replications, we require only 32 plots instead of 38. The source of variation and degrees of freedom in each case are shown in Table 23.1.

In this factorial experiment, we get the mean yields of varieties from 8 (4 $\times$ 2) plots and of nitrogen from 16 plots. In addition, we get information on

TABLE 23.1. Variation and degrees of freedom.

Variety trial		Nitrogen trial		Factorial experiment	
SV	df	SV	df	SV	df
Replication	5	Replication	6	Replication	3
Varieties	3	Nitrogen	1	Treatments	7
Error	15	Error	6	Variety	(3)
				Nitrogen	(1)
				Var x Nitro-gen	(3)
				Error	21
Total	23		13		31

interaction of these two factors. Thus, the precision of the experiment is en-hanced in factorial experiments. The yield of a crop is the result of a number of factors. Some of the factors are dependent on others for their effects and this phenomenon is referred to as interaction. The traditional method of varying only one factor at a time will work only if the interaction is absent. The factorial experiment allows us to study the interaction of the various factors. By using factorial experiments, we can also gain time and effort in arriving at inferences. The results have wide application since each factor has been studied with varying levels of the other factors.

Notation Used in Factorial Experiments

1. *Factors:* Factors are denoted by capital letters as A, B, C, etc.
2. *Levels:* Levels are denoted by small (lower-case letters) letters such as a_0, a_1, a_2, or a_1, a_2, and a_3 where 0, 1, 2, etc., indicate levels in the in-creasing order, if the factor is quantitative.
3. *Factorial treatments:* If there are two factors A and B, and in each there are two levels namely a_0, a_1, and b_0, b_1, respectively, then the factorial treatments are a_0b_0, a_0b_1, a_1b_0, and a_1b_1. These treatments are also written as (1), a, b, and ab, respectively.
4. *Designation of factorial experiments:* Suppose p denotes the number of levels in factor A and q in factor B, then the experiment is called $p \times q$ factorial experiment. A 2×4 factorial experiment will mean that there are two levels in the first factor and four levels in the second fac-tor. The levels of the different factors in an experiment may be equal or unequal.

SINGLE-FACTOR EXPERIMENT VERSUS FACTORIAL EXPERIMENT

Suppose an agronomist intends to study the nitrogen (N) and phosphorous (P) requirements for cotton crop, and selects two levels of N (n_0, n_1), and two levels of P (p_0 and p_1). N and P are two major nutrients to the crop. The problem can be approached in two ways:

1. By studying the effect of N and P in two separate experiments.
2. By combining the two levels of N with two levels of P (factorial experiment).

First, the agronomist evaluates n_0 or n_1 of N in the first experiment by choosing say p_0 as common factor. If randomized complete block design is used with six replications, the ANOVA will be as shown in Table 23.2.
In such a study 12 plots are required. The standard error of the mean is given by $s/\sqrt{6}$.

Suppose the agronomist infers that n_1 level is superior to n_0, then p_0, and p_1 levels of P are evaluated using n_1 level in the same design and same number of replications. The standard error of phosphorous mean is again given by $s/\sqrt{6}$. Thus to conduct the two single-factor experiments, 24 plots are required, and the effect of N and P are estimated with the same precision. Again, we do not get the following information:

1. the effect of P at n_0 ($n_o p_1 - n_o p_o$) level of N, and
2. the effect of interaction.

If the experimenter utilizes the same 24 plots, the four factorial treatments can be studied with greater precision in six replications. The field plan without randomization will be as follows:

TABLE 23.2. ANOVA results.

Source of variation	df	MS	F	Table *F* 5 percent	1 percent
Replication	5	RMS			
Treatments	1	TMS			
Error	5	EMS (s^2)			
Total	11				

n_0p_0	n_0p_0	n_0p_0	n_0p_0	n_0p_0	n_0p_0
n_0p_1	n_0p_1	n_0p_1	n_0p_1	n_0p_1	n_0p_1
p_1n_0	n_1p_0	n_1p_0	n_1p_0	n_1p_0	n_1p_0
n_1p_1	n_1p_1	n_1p_1	n_1p_1	n_1p_1	n_1p_1
Rep 1	Rep 2	Rep 3	Rep 4	Rep 5	Rep 6

From the plan, the effect of each n_0, n_1, p_0, and p_1 is estimated in two plots in each replication. This is sometimes called hidden replication. The standard error of each mean is obtained by $s / \sqrt{12}$. Thus, the treatment effect is more precisely estimated. Furthermore, the precision could be estimated in three (12 plots) replications in a factorial experiment instead of six (24 plots) in single-factor experiments. Another advantage in factorial experiment is that we get information on interaction between the factors.

TWO-FACTOR EXPERIMENTS

Simple Effect, Main Effect, and Interaction

Consider a two-factor experiment with two levels of each factor (nitrogen and phosphorous in cotton crop). The two levels are denoted by p_0 and p_1 for the factor phosphorus (P), and n_0 and n_1 for the factor nitrogen (N). The average response (kg/plot) for the factorial treatments namely n_0p_0, n_0p_1, n_1p_0, and n_1p_1 are shown as follows (see Table 23.3):

TABLE 23.3. The simple and main effects of the two factors N and P.

		Phosphorous (P)				
		p_0	p_1	Total	Mean	Response
Nitrogen (N)	n_0	n_0p_0	n_0p_1			
		13	17	30	15	+4.0 (17–13)
	n_1	n_1p_0	n_1p_1	37	18.5	
		15	22			+7.0 (22–15)
Total		28	39	67	16.75	
Mean	14.0	19.5				
Response to N	+2.0	+5.0				
	(15–13)	(22–17)				

Simple Effect

The effect of factor P at level of n_0 of factor N is given by

$n_0p_1 - n_0p_0 = 17 - 13 = +4.$

This is called simple effect of P at n_0 level.
The effect of factor P at level of n_1 of factor N is given by

$n_1p_1 - n_1p_0 = 22 - 15 = +7.$

This is called simple effect of P at n_1 level.
The average of these two simple effects is defined as the main effect of P, i.e.,

Main effect of $P = [1/2 \{(n_0p_1 - n_0p_0) + (n_1p_1 - n_1p_0)\}]$
$= \frac{1}{2}[(17-13) + (22-15)] = \frac{1}{2} (4 + 7) = 5.5$

In the same fashion, the two simple effects of N at p_0, and p_1 levels and the main effects are

$(n_1p_0 - n_0p_0), (n_1p_1 - n_0p_1)$ and $\frac{1}{2}[\{(n_1p_0 - n_0p_0) + (n_1p_0 - n_0p_1)\}]$, respectively.

In figures, $n_1p_0 - n_0p_0 = 15 - 13 = 2$
$n_1p_1 - n_0p_1 = 22 - 17 = 5$

$\frac{1}{2}[\{(n_1p_0 - n_0p_0) + (n_1p_1 - n_0p_1)\}] = \frac{1}{2} (2+5) = 3.5$

Independent

P is said to be independent of N when the response of P is the same at n_0 and n_1 levels of N. The two simple effects $(n_0p_1 - n_0p_0)$ (4) and $(n_1p_1 - n_1p_0)$ (7) do not differ significantly and the difference that is observed is due to experimental error. Thus, the average of two simple effects can be taken to give the *main effect of p* (5.5). Similarly, if the response of N is the same at p_0 $(n_1p_0 - n_0p_0)$ and p_1 $(n_1p_1 - n_0p_1)$, we say N and P are independent. That is the two simple effects (2 and 5) do not show significant difference and the average (3.5) will give the *main effect of n*. Thus, when N and P are independent, we can infer that application of P at p_1 level increases the yield by 5.5 kg/plot, and application of N at n_1 level over the zero level increases the yield by 3.5 kg/plot.

Interaction

When N and P are not independent, then we can assume that N affects P or P affects N. When N does affect P, the difference $[(n_1p_1 - n_1p_0) - (n_0p_1 - n_0p_0)]$ $[7 - 4] = 3$ is considered to be *interaction effect*. This shows the difference between two simple effects, i.e.,

$$NP = [(n_1p_1 - n_1p_0) - (n_0p_1 - n_0p_0)] = 7 - 4 = 3.$$

Similarly, the interaction of P with N is defined as PN =

$$[(n_1p_1 - n_0p_1) - (n_1p_0 - n_0p_0)] = 5 - 2 = 3.$$

Thus, the interaction of N with P or P with N conveys the same meaning. If the statistical test shows that this difference is significant, then conclude there is interaction. Otherwise N and P are said to be independent. These ideas can be extended to higher order factorial experiments.

Numerical example: 2×2 factorial experiment in randomized complete block design. In an experiment, the effect of N at two levels, n_0 and n_1, and P at p_0 and p_1 levels was conducted in wheat crop in a randomized complete block design with four replications. The data are given here.

Replications	n_0p_0 (1)	n_1p_0 (n)	n_0p_1 (p)	n_1p_1 (np)	Total
			Treatments		Total
I	4	4	5	6	19
II	4	4	5	5	18
III	3	3	4	6	16
IV	2	4	3	5	14
Total	13	15	17	22	67

Calculations

1. Find the correction factor = $67^2/16 = 280.6$
2. Replication SS = $(19^2 + 18^2 + 16^2 + 14^2)/4 - CF = 284.25 - 280.6 = 3.7$
3. Treatment SS = $(13^2 + 15^2 + 17^2 + 22^2) 4 - CF = 291.75 - 280.6 = 11.2$
4. Total SS = $4^2 + \ldots + 5^2 - CF = 299 - 280.6 = 18.4$
5. Form two-way table for N and P.
6. *PSS* = $(28^2 + 39^2)/8 - CF = 7.5$

P

N		p_0	p_1	Total
	n_0	13	17	30
	n_1	15	22	37
	Total	28	39	67

7. $NSS = (30^2 + 37^2)/8 - CF = 3.0$

PN (interaction) SS = Table SS $- CF - NSS - PSS = 291.7 - 280.6 - 3.0 - 7.5 = 0.6$.

Sometimes the treatment mean square may not show significance. But when partitioned into main effects and interaction, we can get significant results. Hence, in factorial experiments, the treatment sum of squares must be partitioned into main effects and interaction. Here the nitrogen sum of squares (NSS) corresponds to a comparison between N_o and N_1, and the phosphorous sum of squares (PSS) corresponds to a comparison between P_o and P_1.

8. Form the analysis of variance table. See Table 23.4.

Comparison of N Effect

$$LSD_{0.05} = t\sqrt{2s^2 / rn} = 2.262\sqrt{(2 \times 0.4)/4 \times 2} =$$
$$2.262 \times 0.3 = 0.6786$$

Treatments	n_1	n_o
Mean values	4.625	3.75

The difference $(4.625 - 3.75)$ is greater than LSD, and hence the effect of N_1 treatment is significantly greater than N_o.

Comparison of P effects

LSD $= 0.6786$. Conclusion: The P_1 mean (4.875) is significantly greater than P_o mean (3.5).

TABLE 23.4. ANOVA results.

Source of variation	df	SS	MS	F	Table F 5 percent	Table F 1 percent
Replications	3	3.7	1.3	3.3		
Treatments	3	11.2	3.7	9.3**	3.86	6.99
Nitrogen (N)	(1)	3.0	3.0	7.5**	8.12	10.56
Phosphorous (P)	(1)	7.5	7.5	18.8**	8.12	10.56
NP	(1)	0.7	0.7	1.8ns	8.12	10.56
Error	9	3.5	0.4			
Total	15		18.4			

** = significant at $P = 0.01$; ns = not significant.

Interaction: N × P interaction effect is not significant. It indicates N_1 and N_2 means do not differ significantly at P_1 and P_2 levels. It means the difference between N_1 and N_2 is independent of P or we have approximately the same difference between N_0 and N_1 regardless of the levels of P. The mean data are given here for more information:

P		p_0	p_1	Difference
N	n_0	3.25	4.25	1.00
	n_1	3.75	5.50	1.75
	Difference	0.50	1.25	

MEANING OF SIGNIFICANT INTERACTION

Significant interaction is a condition in which the effect of one independent variable on the dependent variable is different at different levels of the second independent variable. Interaction can be studied by graphs. The significant interaction is suggested when the lines are not parallel. Normally main effects are not interpreted when interaction is significant. When interaction is not significant, interpreting the results of a two-factor analysis of variance is very simple. Interactions are easy to understand in a figure. When interaction is present, at least two line segments are not parallel to each other. A significant interaction means that the effect of N at one level of P is different from its effect at the other level of P. The lines are not parallel. When interaction is significant, testing and comparison of main effects are not recommended.

Numerical example (to explain the interaction): Grain density in rice crop is defined as the number of grains per centimeter of the panicle length. In an experiment with two varieties (IR22 and IR 662) (say factor A at a_0 and a_1 levels, and factor B at 3 levels b_0, b_1, and b_2), the grain density was evaluated in a 2 × 3 factorial experiment (Palaniswamy, 1990). The results are given here in the ANOVA table (see Table 23.5).

Mean Values

N (kg/ha)	No. of grains/cm panicle length		Mean
	IR 22	IR 662	
0	4.329	3.540	3.935
60	4.449	3.600	4.025
120	4.252	3.524	3.888
Mean	4.343	3.555	

TABLE 23.5. ANOVA results for experiments at IRRI.

Source of variation	df	SS	MS	F
Treatments	5	924.46914	184.89353	298.01**
Varieties (V)	(1)	898.97658	898.97658	1448.97**
Nitrogen (N)	(2)	14.22340	7.11170	11.46**
Interaction (VN)	(2)	11.26916	5.63458	9.08**
Error	5914	3672.28310	0.62042	
Total	5919	4596.75220		

Conclusion
1. Comparison of varieties: IR 22 IR662
2. Comparison of N levels 60 *0 120*
3. Interaction
   a. Comparison of N levels at each variety:
     IR   22:   60   0   120
     IR   662:   *60   0*   120
   b. Comparison of varieties at each N level:
      0:    IR22   IR662
     60:    IR22   IR662
    120:    IR22   IR662

Example 2. Palaniswamy (1998) conducted an experiment with purslane (*Portulaca oleracea* L.) in growth chambers under controlled conditions with two varieties and four forms of N application in RCBD in pots during 1990 and reported the following results. The varieties are cultivated type (V1) and the wild (V2), and forms of nitrogen applied were (F1 = application of N all in NO_3 form), F2 = 3:1 ratio (3 parts in NO_3 form and 1 part in NH_4 form), F3 = 2:2 (2 parts in NO_3 form and 2 parts in NH_4 form) and F4 = 1:3 (1 part in NO_3 form and 3 parts in NH_4 form). The results are given here:

| | V1 | | | | | V2 | | | | Grand |
F1	F2	F3	F4	Total	F1	F2	F3	F4	Total	total
40.0	37.0	36.0	37.0	150.0	26.0	25.0	24.5	21.5	97.0	247.0
42.5	41.0	42.0	39.2	164.7	25.5	27.0	27.0	23.7	103.2	267.9
41.4	50.0	47.5	42.5	181.4	31.0	30.0	27.0	27.5	115.5	296.9
47.0	46.5	37.0	44.0	174.5	28.0	27.0	27.5	22.5	105.0	279.5
38.0	42.0	39.0	41.0	160.0	26.0	24.0	25.0	23.0	98.0	258.0
208.9	216.5	201.5	203.7	830.6	136.5	133.0	131.0	118.2	518.7	1349.3

Form variety × nitrogen table

Varieties		Nitrogen forms			Total
	F1	F2	F3	F4	
V1	208.9	216.5	201.5	203.7	830.6
V2	136.5	133.0	131.0	118.2	518.7
Total	345.4	349.5	332.5	321.9	1349.3

Computations

$CF = 1349.3^2/40 = 45515.26$
$Rep. SS = 247.0^2 + \ldots + 258.0^2 / 8 - CF = 45701.66 - CF = 186.40$
$VSS = (830.6^2 + 518.7^2)/20 - CF = 47947.30 - CF = 2432.04$
$FSS = (345.4^2 + \ldots + 321.9^2/10 - CF = 45562.73 - CF = 47.47$
$V \times F$ (interaction SS) $= (208.9^2 + \ldots + 118.2^2/5 - CF - VSS - FSS$
$= 48012.18 - 45515.26 - 2432.04 - 47.47 = 2496.02 - 2432.04 -$
$47.47 = 17.41$
See Table 23.6.

LSD for comparing varieties $= t \times SEd = t_{28,0.05} = 2.048 \times =$ $\sqrt{(2 \times 5.26)/(5 \times 4)} = 2.048 \times 0.725 = 1.48$

$SEm = 0.51$
LSD for comparing forms: $2.048 \times = 2.048 \times 1.03 = 2.048 \times 1.03 = 2.11$
Conclusion

1. Comparison of varieties: V1 versus V2—V1 records significantly higher yield.

TABLE 23.6. ANOVA results.

Source of variation	df	SS	MS	F	Table F 5 percent	Table F 1 percent
Replication	4	186.40	46.60	8.86	2.71	4.07
Treatment	7	2496.92	356.57	67.79	2.36	3.36
Variety (V)	(1)	2432.04	2432.04	462.37**	4.20	7.64
Fertilizer (F)	(3)	47.47	15.82	3.01*	2.95	4.57
V x F	(3)	17.41	5.80	1.10ns	2.95	4.57
Error	28	147.21	5.26			
Total	39		2830.53			

2. Comparison of forms:

F2	F1	F3	F4
(34.95)	34.54	33.25	32.19

F2, F1, and F3 give similar results. F2 and F1 are better than F4. There is no difference between F3 and F4.
   3. Interaction: Not significant.

Form the variety and nitrogen form means.

	F1	F2	F3	F4	Mean
V1	41.78	43.30	40.30	40.74	41.53
V2	27.30	26.60	26.20	23.64	25.94
Mean	34.54	34.95	33.25	32.19	

The mean data can be used to draw graphs to study the presence or absence of the interaction. From these data, you will find the lines parallel to each other.
   Example 3: Palaniswamy (1990) conducted a 2 × 5 factorial experiment involving two varieties of rice (IR 20 and IR22) and five nitrogen levels (0, 50, 100, 150, and 200 kg/ha) with three replications in a CRD. Palaniswamy collected 60 grains from each plot, recorded length, breadth, and width measurements using screw gauge (least count = 0.001cm) and reported the data as shown in Table 23.7.

TABLE 23.7. Length, breadth, and thickness in different nitrogen levels in varieties of rice.

Variety	N Levels				
	0	50	100	150	200
Length					
IR22	0.7685	0.7752	0.7793	0.7833	0.7632
IR 20	0.7770	0.7820	0.7873	0.7846	0.7777
Breadth					
IR 22	0.2357	0.2317	0.2339	0.2380	0.2346
IR 20	0.2385	0.2387	0.2373	0.2385	0.2349
Thickness					
IR 22	0.1568	0.1549	0.1590	0.1552	0.1563
IR 20	0.1580	0.1567	0.1589	0.1555	0.1562

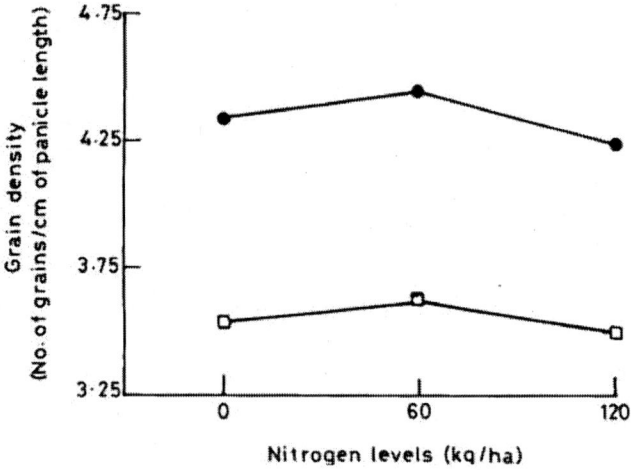

FIGURE 23.1. Effect of different nitrogen treatments on grain density of two rice varieties.

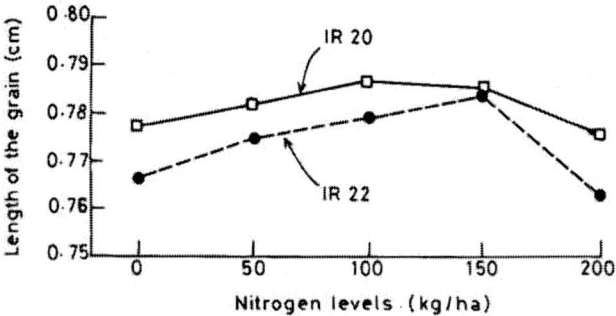

FIGURE 23.2. Graphical representation of interaction between grain length and nitrogen fertilizer (see lack of parallelism of the fitted lines).

The data on length were plotted in a graph and studied (see Figure 23.1). The lines are not parallel, which is an indication of the presence of interaction between varieties and nitrogen levels. The graph is shown in Figure 23.2 in respect to the character grain length. Readers are advised to draw similar graphs in respect of grain breadth and grain thickness in order to study the nature of interaction.

THREE-FACTOR EXPERIMENTS

A 2^3 factorial experiment means that three factors each at two levels are studied in an experiment. Suppose we have three factors namely N, P, and K, and each at 0 and 20 lb/ac are included in the experiment. We get a 2^3 factorial experiment. The treatment combinations are (1), n, p, np, k, nk, pk, and npk. We will denote effects by capital letters and treatment combinations by small letters. It is explained with a numerical example: A 2^3 factorial experiment was conducted in potato crop with three factors (N, P, and K) and each at two levels (0 and 20 lb/ac) in a randomized complete block design with four replications and with eight plots each. The plot yields were recorded. The following yield data were obtained:

Treatment	Replications				Total
	1	2	3	4	
1	18	19	16	17	70
n	20	19	19	19	77
p	21	21	20	21	83
np	23	23	22	22	90
k	22	22	21	22	87
nk	22	20	21	21	84
pk	25	22	24	23	94
npk	26	27	25	25	103
Total	177	173	168	170	688

Calculations

Step 1. Calculate block totals and grand total.
Step 2. Calculate sum of squares due to blocks and total sum of squares.
Step 3. Calculate sum of squares due to treatments.

There are several ways of calculating main effects and interactions. We will follow the Yates method.

1. Correction factor (CF) = $GT^2/rxt = 688^2/32 = 473344/32 = 14792$
2. Block totals: 177, 173, 168, 170
3. Block SS = $(177^2 + 173^2 + 168^2 + 170^2)/8 - CF = 14797.74 - 14792.00$
   = 5.74
4. Total SS = $\Sigma_{yij}^2 - CF = 18^2 + \ldots 25^2 - CF = 14990 - 14792 = 198$
5. Treatment SS: $\Sigma T_i^2/r - CF = (70^2 + 77^2 + 83^2 + 90^2 + 87^2 + 84^2 + 94^2 + 103^2)/4 - CF = 14972 - 14792 = 180$

6. Error $SS = 198 - 5.74 - 80 = 12.26$
7. Form the Yates table (see Table 23.8)

Procedure

Tr	Yield (total)	1	2	3
1	y1	y1 + y2	(y1+y2)+(y3+y4)	{[(y1+y2)+(y3+y4)]+[(y5+y6)+(y7+y8)]}
n	y2	y3 + y4	(y5+y6)+(y7+y8)	{[(y2-y1)+(y4-y3)]+[(y6-y5)+(y8-y7)]}
p	y3	y5 + y6	(y2-y1)+(y4-y3)	{[(y3+y4)-(y1+y2)]+[(y7+y8)-(y5+y6)}
np	y4	y7 + y8	(y6-y5)+(y8-y7)	{[(y4-y3)-(y2-y1)]+[(y8-y7)-(y6-y5)]}
k	y5	y2 - y1	(y3+y4)-(y1-y2)	{[(y5+y6)+(y7+y8)]-[(y1+y2)+(y3+y4)]}
nk	y6	y4 - y3	(y7+y8)-(y5+y6)	{[(y6-y5)+(y8-y7)]-[(y2-y1)+(y4-y3)]}
pk	y7	y6 - y5	(y4-y3)-(y2-y1)	{[(y7+y8)-(y5+y6)]-[(y3+y4)-(y1-y2)]}
npk	y8	y8 - y7	(y8-y7)-(y6-y5)	{[(y8-y7)-(y6-y5)]-[(y4-y3)-(y2-y1)]}

SS due to each effect is obtained as $(\text{col } 3)^2 / r \times 2^n$ where $r = 4$, $r = 3$ (hence $r \times 2^n = 32$)

Each main effect is obtained as $(\text{col } 3)/r \times 2^{n-1}$ where $r = 4$ and $n = 3$ and hence $r \times 2^{n-1} = 16$ (see Table 23.9).

Conclusion: All main effects and interactions except the two-factor interaction namely NK and PK are significant.

1. SE per plot $= \sqrt{\text{error } MS / r} = \sqrt{0.58 / 4} = 0.38$. The LSD may be calculated and treatment means compared in the usual manner.
2. Method: By following the ordinary method, we can also compute the required SS.

TABLE 23.8. Yates table.

Tr.	yield	1	2	3	4: (col 3)^2/ $2^n r$ (32)	5: (col 3)/ $2^{n-1} r$ (16)
1($n_0 p_0 k_0$)	70	147	320	688	14792	43.00
n($n_1 p_0 k_0$)	77	173	368	20	12.50	1.25
p($n_0 p_1 k_0$)	83	171	+14	52	84.50	3.25
np($n_1 p_1 k_0$)	90	197	+6	12	4.50	0.75
k($n_0 p_0 k_1$)	87	+7	+26	48	72.00	3.00
nk($n_1 p_0 k_1$)	84	+7	+26	-8	0.00	-0.50
pk($n_0 p_1 k_1$)	94	-3	0	0	0.00	0
npk($n_1 p_1 k_1$)	103	+9	+12	12	4.50	0.75

TABLE 23.9. ANOVA results.

Source of variation	df	SS	MS	F	Table F 5 percent	Table F 1 percent
Blocks	3	5.74	1.91	3.29	3.07	4.87
Treatments	7	180.00	25.70	44.31**	2.49	3.65
Main effects						
N	1	12.50	12.50	21.55**	4.32	8.02
P	1	84.50	84.50	145.69**	4.32	8.02
K	1	72.00	72.00	124.13**	4.32	8.02
Two factors						
NP	1	4.50	4.50	7.76**	4.32	8.02
NK	1	2.00	2.00	3.45ns	4.32	8.02
PK	1	0.00	0.00		4.32	8.02
Three factors						
NPK	1	4.50	4.50	7.76	4.32	8.02
Error	21	12.26	0.58			
Total	31		198.00			

Form the three-way table as shown here:

	k_0		k_1	
	p_0	p_1	p_0	p_1
n_0	$n_0p_0k_0$ 70	$n_0p_1k_0$ 83	$n_0p_0k_1$ 87	$n_0p_1k_1$ 94
n_1	$n_1p_0k_0$ 77	$n_1p_1k_0$ 90	$n_1p_0k_1$ 84	$n_1p_1k_1$ 103

Form the two-way tables and calculate the *SS* in the usual way.

N × P table	p_0	p_1	Total
n_0	157	177	334
n_1	161	193	354
Total	318	370	688

N × K table	k_0	k_1	Total
n_0	153	181	334
n_1	167	187	354
Total	320	368	688

P × K table	k_0	k_1	Total
p_0	147	171	318
p_1	173	197	370
Total	320	369	688

We can also directly obtain the main effects and interactions from the contrast itself by following the rule of even versus odd.

	1	n	p	np	k	nk	pk	npk
N	−	+	−	+	−	+	−	+
P	−	−	+	+	−	−	+	+
K	−	−	−	−	+	+	+	+

NP	+	−	−	+	+	−	−	+
NK	+	−	+	−	−	+	−	+
PK	+	+	−	−	−	−	+	+
NPK	−	+	+	−	+	−	−	+

Using the treatment totals, the *SS* may be computed, ANOVA table formed, and the treatments tested for their significance. The Table 23.10 will be useful for interpretation of results:

Detailed methods of analysis can be found in Cochran and Cox (1967), Snedecor and Cochran (1967), and Federer (1967).

EXERCISES

23.1. What is a factorial experiment? How does it differ from randomized complete block design? Describe an experiment in which its use would be appropriate.

23.2. Consider a 2×2 factorial experiment and explain clearly the concept of interaction between two factors.

23.3. What do you understand by simple effect, main effect, and interaction effect?

23.4. An experiment is conducted to investigate differences among three methods (P1, P2, and P3), three machines (M1, M2, and M3), and three operators (O1, O2, and O3); the data are available for analysis. Show the layout of the experiment. Assuming the experiment is replicated *r* times, show the dummy ANOVA, and show how to test the various treatments.

23.5. Explain briefly a factorial experiment from your field of specialization.

TABLE 23.10. Summary table.

Factor	Mean response	Response with					
		N		P		K	
		absent	present	absent	present	absent	present
N	1.25	—	—	0.50 (N−NP)	2.00 (N+NP)	1.75 (N−Nk)	0.75 (N+NK)
P	3.25	2.45 (P−NP)	4.00 (P+NP)	—	—	3.25 (P−PK)	3.25 (P+PK)
K	3.00	3.5 (K−NK)	2.5 (K+NK)	3.0 (K−PK)	3.0 (K+PK)	—	—

23.6. Define factorial experiments. Why are these considered better than the experiments in which the factors are tried one by one?

23.7. There are four soybean varieties S1, S2, S3, and S4. Two fertilizers (F1 and F2) are to be tried in the experiment with two replications. The data are as follows:

	S1	S2	S3	S4
F1	110	100	90	90
	125	95	90	80
F2	120	97	85	80
	115	92	86	81

a. Is there significant difference among the varieties and fertilizers?

b. Any interaction?

23.8. An experiment involves four factors, A, B, C, D, each with 3, 2, 2, 3 levels, respectively. There are four replications. Show the sources of variation and degrees of freedom in ANOVA table. How would you determine if there is interaction in different combinations of factors and also main effects?

23.9. In an experiment, seed rates with four levels (S1, S2, S3, and S4) in two varieties (V1 and V2) of chilies were studied. There were three blocks and $4 \times 2 = 8$ treatments. Form the analysis of variance table, and show how you will test the differences in the varieties and seed rates and their interaction. Analyze the data and write down your conclusions.

V1	Replications			V2	Replications		
	1	2	3		1	2	3
S1	4.20	4.94	4.45	S1	2.82	3.14	3.80
S2	4.36	3.50	4.17	S2	3.74	4.43	2.92
S3	5.40	4.50	5.75	S3	4.82	3.90	4.50
S4	5.15	4.40	3.97	S4	4.57	5.32	4.35

23.10. We have a factorial experiment with three factors (A, B, C) each at two levels conducted in RCBD with ten replications. Form the ANOVA table. What pooled sum of squares would be used as an estimate of experimental error and how many degrees of freedom would that sum of squares have?

23.11. Consider that we have three different varieties of corn C1, C2, and C3, and two types of fertilizers F1 and F2. Test whether there is any significant difference in the corn and fertilizers. We have four replications. Each replication has been divided into 3×2 plots to provide the factorial treatments. From the following data, can you conclude that there is (a) significant difference in corn? (b) fertilizer?, and (c) their interaction effect?

Replication	Fertilizer levels	Corn		
		C1	C2	C3
1	F1	9	8	6
	F2	7	9	9
2	F1	11	10	8
	F2	12	11	13
3	F1	22	14	21
	F2	21	17	23
4	F1	31	21	24
	F2	25	27	22

23.12. A multiple choice factorial experiment is conducted on a series of single-factor experiments. Which one of the following three statements is correct?

 a. To know the interaction
 b. To reduce the cost
 c. To estimate the effects precisely

23.13. We have two factors, A and B, each at two levels. There are eight replications. The following data were obtained. Construct the ANOVA table.

Treatments	Replication							
	1	2	3	4	5	6	7	8
a_1b_1	12	10	13	13	12	11	10	7
a_1b_2	9	12	14	11	14	11	12	9
a_2b_1	14	13	8	12	12	8	7	10

23.14. Describe briefly the terms: *factor, factorial experiment, main effect, simple effect, interaction effect between two factors, level of a factor.*

23.15. When will you expect a significant two-factor interaction in a 2×2 experiment?

23.16. Analyze the data from a $p \times q$ factorial experiment involving $p = 4$ and $q = 3$, conducted in four replications. Analyze the data and draw your conclusions.

Rep	P1			P2			P3			P4		
	q_1	q_2	q_3	q_1	q_2	q_3	q_1	q_2	q_3	q_1	q_2	q_3
1	33	49	60	19	16	30	31	30	30	40	40	29
2	40	29	81	38	61	40	76	31	60	50	93	60
3	27	49	57	51	93	71	17	49	62	29	60	60
4	18	18	21	70	41	84	18	60	71	29	29	38
5	40	27	37	38	50	93	60	73	50	40	49	31

23.17. The following analysis table was reported by a scientist working on three factors namely A, B, and C.

SV	df	SS
A	1	23763
B	2	423056
C	5	564689
AB	2	18456
AC	5	85901
BC	10	240115
ABC	10	151394
Error	153	1501642

a. Indicate the mathematical model for the study
b. Show a possible layout for the experiment.
c. Complete the analysis table and comment on significant results.

23.18. A factorial experiment in soybean crop in Latin square design was conducted with two factors each at two levels. The first factor was method of sowing broadcast (a_1) and placement (a_2) and the second factor was fertilizer, fertilizer 1 (b_1) and fertilizer 2 (b_2). The treatments are A (a_1b_1), B (a_1b_2), C (a_2b_1), and D (a_2b_2). The data [plot yield (kg)] collected were as follows.

D 21	A 68	B 55	C 40
B 31	C 42	A 65	D 38
A 43	D 35	C 31	B 40
C 26	B 48	D 35	A 39

Analyze the data and interpret the results. Partition the treatment sum of squares as (a) AB versus CD, (b) AC versus BD, and (c) AD versus BD.

REFERENCES

Cochran, W. and G.M. Cox (1967). *Experimental design,* Second edition. New York: Wiley.

Federer, W.T. (1967). *Experimental design—Theory and application.* New Delhi: Oxford and IBH Publishing Co.

Hicks, C.R. (1973). *Fundamental concepts in the design of experiments.* New York: Holt, Rinehart and Winston.

Palaniswamy, K.M. (1990). Design of statistical (field) experiments and crop forecast in agriculture with special reference to rice. Doctoral thesis, University of Calicut, India.

Palaniswamy, U. (1998). Nitrogen management studies to enhance the essential fatty acids in purslane (*Portulaca oleracea* L). Part of doctoral thesis submitted to the University of Connecticut.

Snedecor, G.W. and Cochran, W.G. (1967). *Statistical methods,* Sixth edition. Ames, IA: The Iowa State Univ. Press.

Chapter 24

Split Plot Design

In a randomized complete block design, all the treatments are randomly allotted in each block separately and independently. Under certain situations, it is not possible to randomize all combinations of factorial experiment over the entire block or replication and hence, a randomized complete block design is not suitable. Split plot design is used to accommodate the factorial treatments.

For example, in crop cultivation studies, treatments such as irrigation methods, application of bulky manures, evaluation of agricultural implements, treatments involving spraying of insecticides, etc., require larger size plots. Other treatments, e.g., varieties, seed rate, spacing, etc., require smaller plots. For example, when the experimenter wants to study four irrigation methods and three different seed rates, irrigation methods can be accommodated in larger sized plots and seed rates in smaller plots. Each block or replication may be divided into four larger plots called main plots to which the irrigation methods are randomly allotted. Each main pot is split or subdivided into three subplots and the three seed rates randomly allotted. The layout so obtained is called a split plot design. (See Figure 24.1 for the plan for one replication.) Here irrigation levels are applied in main plots and each main plot is split into three subplots and seed rates applied in a random manner. The same procedure is followed in other replications separately and independently.

REASONS FOR ADOPTING SPLIT PLOT DESIGN

1. In split plot design, we get increased precision for the effects of subplot treatments and interaction between main plot and subplot treatments equally. Therefore, the split plot design is used when we require increased precision on subplot treatments and interaction effect between main plot treatments and subplot treatments. Thus, when two factors are studied, one factor is considered less important than the other factor and split plot design is used. In such situations, the less

important factor is allotted to main plots and the other factor, whose precision is more important, is allotted to subplots.

2. Economy and nature of the experimental material may cause researchers to use two or more levels of experimental units. One factor may be assigned to a larger plot (e.g., a whole animal) and subdivisions of the animal (skin) to the different levels of the second factor.

3. An entomologist wants to evaluate five chemicals with two different methods of application in two varieties of soybean crop. Since the main interest is in chemicals and methods of application, the varieties will be main plot treatments; subplot treatments are 5 × 2 treatments of chemical and methods of application (see Figure 24.2). In soybean trials, farms are taken as blocks and different fertilizers as main plots and different varieties in each main plot as subplot treatments.

4. Sometimes, between two factors, we can expect larger differences in one factor and smaller differences in the other factor. The factor in which larger differences are expected is used for main plots and the factor with smaller differences for subplots. Examples: (a) In tillage studies, larger plots are needed for tillage treatments and hence this may be applied in main plots. In smaller plots, treatments such as varieties, crops, etc., may be studied; (b) in insecticidal or fungicidal trials, insecticides or fungicides may be applied to main plots and the crops or varieties to subplots.

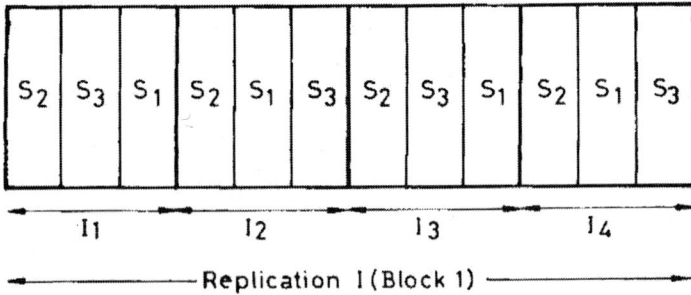

I_1, I_2, I_3 = Irrigation levels in main plots (main plot treatments)

S_1, S_2, S_3 = Seed rate levels in subplots (subplot treatments)

FIGURE 24.1. Field layout plan for a split plot design (only one replication is shown).

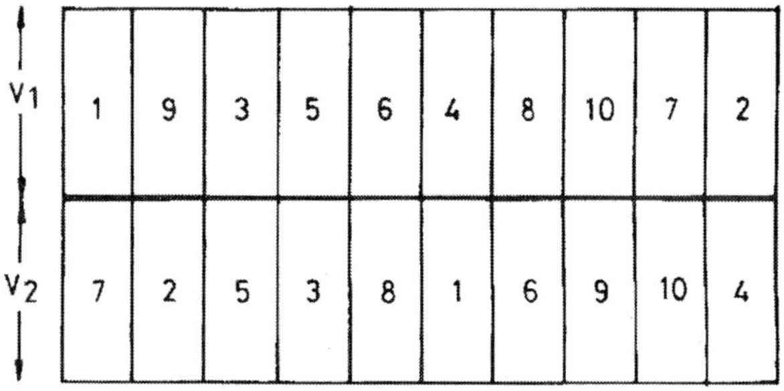

FIGURE 24.2. A split plot design for a 2 × 5 factorial experiment. Factorial experiments are in subplots. Two varieties are in main plots.

5. Sometimes, we have already investigated the effect of the factor, A, and we are not interested in this effect. Hence, it may be allotted to main plots and the factor B, whose effect is more important, may be allotted to the subplots.

Advantages

1. Split plot design is suitable when one factor requires larger plot sizes than the second factor. The levels of the first factor may be accommodated in whole plots and the second factor in smaller plots.
2. Sometimes main effect of a factor is important as compared to the second factor and the interaction. We sacrifice the accuracy of the first factor by allotting it to the main plots.
3. The subplot treatments and the interaction are more precisely estimated in split plot design than in randomized complete block design.
4. We can introduce another factor in the split plot design already started and the experiment can be continued in split-split-plot design model.
5. The main plot treatments may be arranged in Latin square design and overall precision increased.

Disadvantages

1. Main plot treatments are estimated with less precision than they are in a randomized complete block design.
2. When a missing value occurs, the analysis is complicated in split plot design whereas in randomized complete block design, it is not so.

PROCEDURE FOR LAYOUT OF SPLIT PLOT DESIGN

1. There will be minimum of two factors, e.g., A and B. Consider that A is less important than B. Let there be r replications. The levels of A are allotted to main plots. The main plots may be arranged in any design, i.e., completely randomized design, randomized complete block design, or Latin square design. Allot the levels of A at random in each replication.
2. The levels of the second factor are arranged as subplots in each main plot and allotted to the subplots in a random manner.
3. The experimental error for testing the main plot treatments is usually larger than the error used to compare the subplot treatments, i.e., the subplot error is smaller.

The numerical model for split plot design is

$$Y_{ijk} = \mu + \alpha_i + \beta_j + (\alpha\beta)_{ij} + \lambda_k + (\alpha\lambda)_{ik} + \varepsilon_{ijk}$$

where α, λ, and $\alpha\lambda$ are fixed treatment effects; β_j is a random block effect; and $(\alpha\beta)_{ij}$ and ε_{ijk} are random sources of error.

The analysis of variance for the model follows:

SV	df
Blocks	$r-1$
Main plot treatment *(A)*	$a-1$
Error *(a)*	$(r-1)(a-1)$
Sub plot treatment *(B)*	$b-1$
Main plot x sub plot *(AB)*	$(a-1)(b-1)$
Error *(b)*	$a(r-1)(b-1)$
Total	$rab-1$

Method of Analysis

Let r = number of replications, p = number of main plot treatments per replication, and q = number of subplot treatments per main plot. The factor A is assigned to the main plot and the factor B to subplots. The analysis for A will be in the form of a randomized complete block design.

Example: Let $r = 3$, $p = 5$, $q = 4$, then $rpq = 3 \times 5 \times 4 = 60$. The ANOVA for main plot analysis will be as follows.

SV	df
Replication	2
A (MPT)	4
Error *(a)*	8

The levels of *B* are compared within the main plots. The degrees of freedom within the main plots will be $(4 - 1) \times 3 \times 5 = 45$. Out of 45, we have 3 for levels of *B* and 12 for interaction of *AB*. The ANOVA for subplots is as follows.

SV	df
B	3
AB	12
Error *(b)*	30

The degrees of freedom for error *(b)* can be obtained from $(3 + 12) \times 2$ where 3 +12 comes from the treatment degrees of freedom for *B* and *AB*, and 2 from the degrees of freedom for replications. The complete ANOVA will be as follows:

SV	df	SS	MS
Replication	$(r–1) = (3–1) = 2$	SS_R	MS_R
Main plot treatment *(A)*	$(a–1) = (5–1) = 4$	SS_A	MS_A
Error *(a)*	$(r–1)(a–1)\ (3–1)(5–1) = 8$	EaSS	EaMS
Subplot treatment *(B)*	$(b–1)\ (4–1) = 3$	SS_B	MS_B
Interaction *(AB)*	$(a–1)(b–1) = (5–1)(4–1) = 12$	SS_{AB}	MS_{AB}
Error *(b)*	$a(b–1)(r–1)\ 5(4–1)(3–1) = 30$	EbSS	EbMS
Total	$rab–1\ (3 \times 5 \times 4)\text{-}1 = 59$		

COMPARISON OF SPLIT PLOT DESIGN WITH RCBD

1. Increase in accuracy is achieved for subplot treatments, and main plot and subplot interaction over the main plot treatments as the variation within the subplot treatments is smaller because of smaller plots.
2. The degrees of freedom available for main plot treatments is smaller than the degrees of freedom available for subplot treatments.
3. In randomized complete block design, the error *SS* is obtained by summing the sums of squares and degrees of freedom for error *(a)* and error *(b)*. In split plot design, the error (available in randomized block design) is split into two parts: error *(a)* which is larger than error *(b)*.

4. The error degrees of freedom in RCBD is more than in split plot design and hence RCBD is superior to split plot design.

STANDARD ERRORS IN SPLIT PLOT DESIGN

Four different types of comparisons are made. *Ea* refers to the mean square for error *(a)*, and *Eb* to the mean square for error *(b)*.

1. Standard error of difference for comparing two means of *A* over all levels of *B:*

$$\bar{a}_1 - \bar{a}_2 = \sqrt{(2E_a \mathbin{/} rq)}$$

2. Standard error of difference for comparing two means of *B* over all levels of *A:*

$$\bar{b}_1 - \bar{b}_2 = \sqrt{(2E_b \mathbin{/} rP)}$$

3. Standard error of difference for comparing two means of *B* for each level of *A:*

$$\bar{a}_1\bar{b}_1 - \bar{a}_1\bar{b}_2 : \sqrt{(2E_b \mathbin{/} r)}$$

4. Standard error of difference for comparing two means of A for each level of *B:*

$$\bar{a}_1\bar{b}_1 - \bar{a}_2\bar{b}_1 = \sqrt{(^2 2q - 1)E_b + E_a} \mathbin{/} rq$$

Since E_a and E_b are based on different degrees of freedom, $t_{0.05}$ for each case is different. The weighted t is therefore calculated as follows. The weighted t is used for computing LSD.

$$t = [(q-1)E_b t_b + E_a t_a] \mathbin{/} [(q-1)E_b + E_a], \text{ where}$$
$$t_b = t_{0.05} \text{ for } E_b \text{ and } t_a = t_{0.05} \text{ for } E_a.$$

Analysis of variance table for split plot design in Latin square

SV	df
Main plots	
Rows	$a-1$
Columns	$a-1$
A	$a-1$
Error *(a)*	$(a-1)(a-2)$
Subplots	
B	$b-1$
AB	$(a-1)(b-1)$
Eb	$a(a-1)(b-1)$
Total	$a^2 (b-1)$

Example: A layout plan (see Figure 24.3) of split plot design with main plot treatments in Latin square is shown for $r = 3, A = 3, B = 4$ [A = varieties *(V)* and B = fertilizers *(F)*].

The *SV* and degrees of freedom will be replications = 2; positions = 2; v = 2; *(Ea)* = 2; $F = 3$; $FV = 6$; $Eb = 18$; and total 35.

Numerical example: An experiment was conducted in a split plot design with age of seedlings as main plot treatments in ragi crop *(Eleusine coracana)* and strains as subplot treatments. There are six levels of age of seedlings (20, 25, 30, 35, 40, and 45 days) and three levels of strains (Co 1, Co 2, and Co 7). The number of replications was four. The character measured was number of fingers/plant. The data are (see Table 24.1):

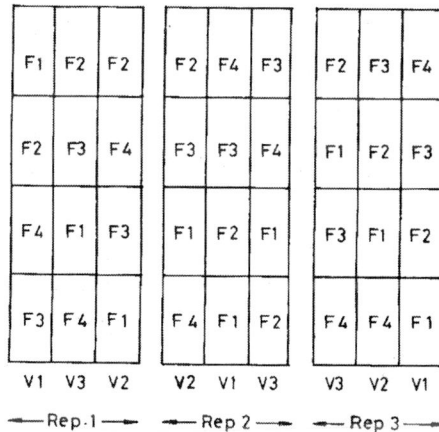

FIGURE 24.3. Field layout plan for a split plot design with three main plots (V1, V2, and V3) and four subplots (F1, F2, F3, and F4).

TABLE 24.1. Replication × age table.

Replications	20	25	30	35	40	45	Total
			Ages of seedlings (days)				
I	21.7	21.6	20.9	20.0	18.0	15.1	117.3
II	20.2	21.5	21.2	19.7	16.9	14.1	113.6
III	20.5	20.3	19.9	20.4	16.3	15.2	112.6
IV	21.8	20.7	21.8	19.3	18.1	14.9	116.6
Total	84.2	84.1	83.8	79.4	69.3	59.3	460.1

Computations

$CF = 460.1^2/72 = 2940.17$
Table $SS/(21.7^2 + ... + 14.9^2)/3 - CF = 46.77$
Replication $SS = (117.3^2 + ... + 116.6^2)/18 - CF = 2941.03 - 2940.17 = 0.86$
Ages $SS = (84.2^2 + ... + 59.3^2)/12 - CF = 2984.02 - 2940.17 = 43.85$
Error (a) = Table SS – Replication SS – Ages $SS = 46.77 - 0.86 - 43.85 = 2.06$

See Table 24.2.

Ages × Strains Table $SS = (25.1^2 + ... + 15.7^2)/4 - CF = 3008.53 - 2940.17 - CF = 68.36$
Strains $SS = (144.3^2 + 164.5^2 + 151.3^2)/24 - CF = 2948.93 - 2940.17 = 8.76$
Interaction SS = Table SS – Ages SS – Strains $SS = 68.36 - 43.85 - 8.76 = 15.75$
Total $SS = 3019.69 - 2940.17 = 79.52$
See Table 24.3.

Calculation of Standard Errors

1. SE for ages mean = $\sqrt{0.137/12} = 0.1068$

   LSD $_{0.05} = 2.131 \times \sqrt{2} \times 0.1068 = 0.322$

(See Appendix, Table A.5 for $t = 2.131$ for 15 df at 5 percent probability level.)

TABLE 24.2. Form ages × strains table.

Strains	20	25	30	35	40	45	Total
			Ages of seedlings (days)				
Co1	25.1	24.7	24.9	23.8	23.2	22.6	144.3
Co2	29.7	29.0	30.3	30.3	24.2	21.0	164.5
Co7	29.4	30.4	28.6	25.3	21.9	15.7	151.3

TABLE 24.3. ANOVA results.

Source of variation	df	SS	MS	F	Table F 5 percent	1 percent
Rep (Blocks)	3	0.86	0.287			
Ages (MPT)	5	43.85	8.770	64.01**	2.90	4.56
Error (a)	15	2.06	0.137			
Strains (SPT)	2	8.76	4.438	10.19**	3.26	5.25
Ages × strains	10	15.75	1.575	6.88**	2.10	2.86
Error (b)	36	8.24	0.228			
Total	71	79.52				

2. SE for strains means: $\sqrt{0.229/24} = 0.0976$. LSD = $2.0282 \times \sqrt{2} \times 0.0976 = 0.280$

Ages × Strains

a. SEd between ages at each strain level:

$$\sqrt{[2(2 \times 0.229) + (0.137)]/(4 \times 3)} = 0.229$$

$t = \{(2 \times 0.229 \times 2.0382) + (0.137 \times 2.1310)\}$
$\sqrt{[(2 \times 0.0229) + (0.137)]} = 1.22086/0.595 = 2.052$

LSD = $2.052 \times 0.2962 = 0.608$

b. SEd between strains at each level of age: $\sqrt{[(2 \times 0.229)/4]} = 0.3383$

LSD = t × SEd = $2.028 \times 0.3383 = 0.686$.

Form the following means table for ages and strains for comparison purpose:

			Ages of seedlings (days)				
Strains	20	25	30	35	40	45	Mean
Co1	6.275	6.175	6.225	5.950	5.800	5.650	6.013
Co2	7.425	7.250	7.575	7.575	6.050	5.250	6.854
Co7	7.350	7.600	7.150	6.325	5.475	3.925	6.304
Mean	7.017	7.008	6.983	6.617	5.775	4.942	

Comparisons
1. Comparison of ages means: <u>1 2 3</u> 4 5 6
2. Comparisons of strains: Co2 Co 7 Co1
3. Comparisons of ages at each level of strain:
     Co1: <u>1 3 2</u> 4 5 6
     Co2: <u>3 4 1</u> 2 5 6
     Co3: <u>3 1 2</u> 4 5 6
4. Comparison of strains at each level of age:
1: <u>Co2 Co7</u> Co1   2: <u>Co 7 Co 2</u> Co1
3: <u>Co2 Co7</u> Co1   4: Co 2 <u>Co 7 Co1</u>
5: <u>Co2 Co1 Co7</u>   6: <u>Co1 Co2</u> Co7
The trial was continued in the second year also and the results are furnished
(see Table 24.4):
Computations

$CF = 436.1^2/72 = 2641.43$
Table SS $= (20.8^2 + \ldots + 16.4^2)/3 - CF = 2676.84 - 2641.43 = 35.41$
Rep. SS $= (110.2^2 + \ldots + 111.1^2)/18 - CF = 2642.20 - 2641.43 =$
0.77
Ages SS $= (83.7^2 + \ldots + 6.09^2)/12 - CF = 2673.58 - 2641.43 = 32.15$
Error *(a)* = Table SS – Rep. SS – Ages SS = 35.41 – 0.77 – 32.15 =
2.49 (see Table 24.5).

Ages × strains table *SS* = $26.1^2 + \ldots + 14.6^2/4 - CF = 2710.00 -$
$2641.48 = 68.57$

TABLE 24.4. Replication × ages table.

	Ages of seedlings (days)						
Rep	20	25	30	35	40	45	Total
I	20.8	19.6	18.3	18.6	17.7	15.2	110.2
II	21.1	20.0	18.6	17.6	16.0	15.3	108.6
III	21.1	20.9	18.3	16.7	15.2	14.0	106.2
IV	20.7	20.3	18.9	18.5	16.3	16.4	111.1
Total	83.7	80.8	74.1	71.4	65.2	60.9	436.1

TABLE 24.5. Form ages × strains table.

| Strains | \multicolumn{6}{c}{Ages of seedlings (days)} | Total |
	20	25	30	35	40	45	
Co1	26.1	24.9	23.7	25.0	24.1	24.2	148.0
Co2	31.1	30.5	28.0	26.6	22.7	22.1	161.3
Co7	26.5	25.4	22.4	19.8	18.4	14.6	127.1

Strains $SS = (148.0^2 + 161.3^2 + 127.1^2)/24 - CF = 2665.81 - 2641.43 = 24.38$

Interaction SS = Table SS – Ages SS – Strains $SS = 68.57 - 32.15 - 24.38 = 12.04$

Total $SS = 2720.41 - 2641.43 = 79.98$ (see Table 24.6).

Form the following means table for ages and strains for comparison purpose.

| Strains | \multicolumn{6}{c}{Ages of seedlings (days)} | Mean |
	20	25	30	35	40	45	
Co1	6.525	6.225	5.925	6.250	6.025	6.050	6.167
Co2	7.775	7.625	7.000	6.650	5.675	5.525	6.708
Co7	6.625	6.350	5.600	4.950	4.600	3.650	5.296
Mean	6.975	6.738	6.175	5.950	5.433	5.075	

Comparisons

1. SE for every age: $\sqrt{(0.166 / 12)} = 0.1176$

   $$LSD = 0.1176 \times \sqrt{2} \times 2.131 = 0.3504$$

2. SE for means of strains = $0.199/24 = 0.091$

   $$LSD = 0.091 \times \sqrt{2} \times 2.0282 = 0.261 \text{ (table value of } t_{0.05,36} = 2.0282)$$

   See Appendix, Table A.5.

3. *SEd* for strains at each age = $\sqrt{\{(2 \times 0.199) / 4\}} = 0.3154$

   $$LSD = 2.0282 \times 0.3154 = 0.640$$

TABLE 24.6. ANOVA results.

Source of variation	df	SS	MS	F	Table F 5 percent	Table F 1 percent
Rep (Blocks)	3	0.77	0.257			
Ages (MPT)	5	32.15	6.430	38.74**	3.06	4.89
Error *(a)*	15	2.49	0.166			
Strains (SPT)	2	24.38	12.190	61.26**	3.26	5.25
Ages × strains	10	12.04	1.204	6.05**	2.10	2.86
Error *(b)*	36	7.15	0.119			
Total	71	78.98				

4. *SEd* for ages at each strain $= \sqrt{[2 \times (0.199) + 0.166]/(4 \times 3)} = 0.307$

$t = [(2 \times 0.199 \times 2.0282) + (0.166 \times 2.131)]/[(2 \times 0.199 + 0.166] = 1.161/0.564 = 2.058$

$LSD = 0.307 \times 2.058 = 0.631$

Comparisons
1. Comparison of ages means: <u>1 2 3 4</u> 5 6
2. Comparisons of strains: Co2 <u>Co1 Co7</u>
3. Comparisons of ages at each level of strain: Co1: <u>1 4 2 6</u> 5 3
   Co2: <u>1 2 3 4 5 6</u>
   Co3: <u>1 2 3 4 5</u> 6
4. Comparison of strains at each level of age:

1: <u>Co2 Co7</u> Co1	2: Co2 <u>Co7 Co1</u>
3: <u>Co2 Co1</u> Co7	4: <u>Co2 Co1</u> Co7
5: <u>Co1 Co2</u> Co7	6: <u>Co1 Co2</u> Co7

Combined Analysis in Split Plot Design

To arrive at a general inference applicable for different years, a combined analysis is to be done. The procedure is given here:

1. Form Table 24.7 of ages × strains × seasons.
Computations

$CF = 896.2^2/144 = 5577.60$
Table $SS = (3008.53 + 2710.00) - 5576.60 = 140.93$
Seasons $SS = (460.1^2 + 436.1^2)/72 - CF = 5581.6 - 5577.60 = 4.00$

TABLE 24.7. Ages × strains × seasons.

Ages/ strains	I (Season)				II (Season)				Grand total
	Co1	Co2	Co7	Total	Co1	Co2	Co 7	Total	
20	25.1	29.7	29.4	84.2	26.1	31.1	26.5	83.7	167.9
25	24.7	29.0	30.4	84.1	24.9	30.5	25.4	80.8	164.9
30	24.9	30.3	28.6	83.8	23.7	28.0	22.4	74.1	157.9
35	23.8	30.3	25.3	79.4	25.0	26.6	19.8	71.4	150.8
40	23.2	24.2	21.9	69.3	24.1	22.7	18.4	65.2	134.5
45	22.6	21.0	15.7	59.3	24.2	22.1	14.6	60.9	120.2
Total	144.3	164.5	151.3	460.1	148.0	161.0	127.1	436.1	896.2

2. Form the seasons × ages table.

Seasons	Ages of seedlings (days)						Total
	20	25	30	35	40	45	
1	84.2	84.1	83.8	79.4	69.3	59.3	460.1
2	83.7	80.8	74.1	71.4	65.2	60.9	436.1
Total	167.9	164.9	157.9	150.8	134.5	120.2	896.2

Seasons $SS:$ = $(460.1^2+436.1^2)/72$ – CF = $5581.60 - 5577.60 = 4.00$
Table $SS:$ = $84.2^2+ \ldots +60.9^2)/6$ – CF = $5657.60 - 5577.60 = 80.00$
Ages $SS:$ = $(167.9^2+ \ldots +120.2^2)/24$ – CF = $5649.74 - 5577.60 =$
72.14
Seasons × ages $SS:$ Table SS – Seasons SS – Ages SS
= $80.00 - 72.14 - 4.00 - 72.14 = 3.86$

3. Form the strains × ages table.

Strains	Ages of seedlings (days)						Total
	20	25	30	35	40	45	
Co 1	51.2	49.6	48.6	48.8	47.3	46.8	292.3
Co 2	60.8	59.5	58.3	56.9	46.9	43.1	325.5
Co 7	55.9	55.8	51.0	45.1	40.3	30.3	278.4
Total	167.9	164.9	157.9	150.8	134.5	120.2	896.2

Table $SS:$ $(51.2^2+ \ldots +30.3^2)/8$ – CF = $5699.85 - 5577.60 = 122.25$
Strains $SS:$ $(292.3^2 + 325.5^2 + 278.4^2)/48$ – CF = $5602.00 - 5577.60$
= 24.40

Strains × ages *SS:* Table *SS* – Strains *SS* – Ages *SS* = 122.25 – 72.14 – 24.40 = 25.71

4. Form seasons × strains table.

		Strains		
Season	Co1	Co2	Co7	Total
1	144.3	164.5	151.3	460.1
2	148.0	161.0	127.1	436.1
Total	292.3	325.5	278.4	896.2

Tables *S:* $(144.3^2 + \ldots + 127.1^2)/24 - CF = 5614.74 - CF = 5614.74 - 5576.60 = 37.14$
Seasons *SS* = 4.00 strains *SS* = 24.40
Seasons × strains *SS* = Table *SS* – Seasons *SS* – Strains *SS* = 37.14 – 4.00 – 24.00 = 8.74

Now calculate the pooled sum of squares (see Table 24.8):

	Replication	Error (a)	Error (b)	Grand total
	0.86	2.06	8.24	3019.69
	0.77	2.49	7.15	2720.41
Total	1.63	4.55	15.39	5740.10 – CF
				5740.10 – 5577.60 = 162.50

TABLE 24.8. ANOVA results.

Source of variation	df	SS	MS	F	Table F 5 percent	Table F 1 percent
Replication (pooled)	6	1.63	0.2717			
Seasons	1	4.00	4.0000	26.37**	4.17	7.56
Ages	5	72.14	14.4280	45.11**	2.53	3.70
Seasons × ages	5	3.86	0.7720	5.09**	2.53	3.70
Error (a) (pooled)	30	4.55	0.1517			
Strains	2	24.40	12.2000	57.06**	3.13	4.92
Seasons × strains	2	8.74	4.3700	20.44**	3.13	4.92
Ages × strains	10	25.71	2.5710	12.03**	1.97	2.59
Ages × seasons × strains	10	2.08	0.2080	$<1^{ns}$		
Error (b)(pooled)	72	15.39	0.2138			
Total	143	162.50				

We require mean tables for the comparisons of various means and study of interactions.

Mean Tables

See Tables 24.9, 24.10, and 24.11.

Comparisons

1. Comparison of seasons

SE of the mean of seasons: $\sqrt{0.1517 / 72} = 0.0459$

LSD $= 0.0459 \times \sqrt{2} \times 2.042 = 0.1325$ (*t* value from table at $t_{0.05,30} = 2.042$)

TABLE 24.9. Seasons × ages.

Seasons	Ages of seedlings (days)						Mean
	20	25	30	35	40	45	
1	7.0167	7.0083	6.9833	6.6167	5.7750	4.9417	6.3903
2	6.9750	6.7333	6.1750	5.9500	5.4333	6.0750	6.0569
Mean	6.9958	6.8708	6.5792	6.2833	5.6042	5.0083	

TABLE 24.10. Ages × strains.

Seasons	Ages of seedlings (days)						Mean
	20	25	30	35	40	45	
Co1	6.4000	6.2000	6.0750	6.1000	5.9125	5.8500	6.0896
Co2	7.6000	7.4375	7.2875	7.1125	5.8625	5.3875	6.7813
Co7	6.9875	6.9750	6.3750	5.6375	5.0375	3.7875	5.8000
Mean	6.9958	6.8708	6.5792	6.2833	5.6042	5.0083	

TABLE 24.11. Seasons × strains.

Season	Strains			Mean
	Co1	Co2	Co 7	
1	6.0125	6.8542	6.3042	6.3903
2	6l1667	6.7083	5.2858	6.0569
Total	6.0896	6.7813	5.8000	

2. Comparison of means of ages

SE of the mean of ages: $\sqrt{0.1517/24} = 0.0795$
LSD $= 0.0795 \times \sqrt{2} \times 2.042 = 0.2295$ (t value from t table at $t_{0.05,30} = 2.042$)

3. Comparison of means of strains:

SE of the means of strains: $\sqrt{0.2138/48} = 0.0667$
LSD $= 0.0667 \times \sqrt{2} \times 2.00 = 0.1886$

4. Seasons × strains interaction:
   a. Between seasons at each strain

SEd $= \sqrt{[2 \times (2 \times 0.2138) + 0.1517]/(24 \times 3)} = 0.1268$
$t_3 = [(2 \times 0.2138 \times 1.993) + (0.1517 \times 2.042)]/[(2 \times 0.2138) + 0.1517] = 2.0058$
LSD $= 0.268 \times 2.0058 = 0.2543$

   b. Between strains at each season

SE $= \sqrt{0.2138/24} = 0.0943$ LSD$_{0.05} = 0.0943 \times \sqrt{2 \times 2} = 0.2667$

5. Ages × strains.
   a. Between ages at each strain:

SEd $= \sqrt{2[(2 \times 0.2138) + 0.1517]/8 \times 3} = 0.2048$, LSD $= 0.2197 \times 2.0058 = 0.4407$
($t_3 = 2.0058$)

   b. Between strains at each age

SEd $= \sqrt{0.2138/8} = 0.1634$ LSD $= 0.1634 \times \sqrt{2 \times 2.0058} = 0.4634$

Conclusions

1. Seasons: 1 2
2. Ages: 1 2 3 4 5 6
3. Strains: Co2 Co1 Co7

4. Seasons × ages

Age	Seasons	
1	1	2
2	1	2
3	1	2
4	1	2
5	1	2
6	2	1

Seasons	Ages					
1	1	2	3	4	5	6
2	1	2	3	4	5	6

5. Seasons × strains

Seasons	Strains		
1	Co 2	Co7	Co1
2	Co2	Co1	Co7

Strains	Seasons	
Co1	2	1
Co2	1	2

6. Ages × strains

Ages	Strains		
1	Co2	Co7	Co1
2	Co2	Co7	Co1
3	Co2	Co7	Co1
4	Co2	Co1	Co7
5	Co2	Co1	Co7
6	Co1	Co2	Co7

Strains	Ages					
Co1	1	2	4	3	5	6
Co2	1	2	3	4	5	6
Co7	1	2	3	4	5	6

Method of Testing Homogeneity of Variances— Bartlett's Test of Homogeneity

It is explained with reference to the split plot errors: Error (a) and Error (b).

Error *(a)*: Season 1 $(s_1^2) = 0.137$; $df = 15$;
Season 2 $(s_2^2) = 0.166$; $df = 15$;
Pooled $= 0.1515$ (s^2); $df = 30$

$$X^2_{n-1} = \log_e 10[(\Sigma k_i) \log_{10} \bar{s}^2 - \Sigma(k_i) \log_{10} \bar{s}_i^2]$$
$$= 2.30259[(30 \times 1.1810) - 15(1.1367 + 1.2201)]$$
$$= 2.30259[(30 \times (-0.8190)) - 15(-0.8633 - 0.7799)]$$
$$= 2.30259[-24.5700 - 24.6480]$$
$$= 2.30259 \times 0.0780 = 0.1796^{ns}.$$

Hence errors are homogeneous.

Testing Error (b)

$$2.30259(72 \times 1.3304) - 36(1.3598 + 1.2989)$$
$$=2.30259(-48.2112 + 48.2868)$$
$$= 2.30259 \times 0.0756 = 0.1741^{ns}.$$

Hence errors are homogeneous.

Variations in Split Plot Design

1. Split plot design (A × B × C factorial experiment in split plot design): Let there be three factors each at two levels: Factor V = 2 (V1 and V2); Factor N = 2 (N1 and N2); Factor T = 2 (T1 and T2). The main plot treatments may be the combinations of V and N. The subplot treatments will then be T1 and T2 (See Figure 24.4 for the layout plan). Here we consider T1 and T2 are more important than the V and N.
2. Suppose there are three factors, i.e., classes (A in four levels), teaching methods (M in two levels), and workbooks (C in three levels). Here we can take A as main plot treatments and combinations of M and C in subplots.

Trend Analysis in a Split Plot Design

In an experiment, a rice breeder used a combination of two factors: nitrogen at five levels (n_0, n_1, n_2, n_3, and n_4), and phosphorous at three levels (p_0, p_1, p_2) as main plot treatments and two varieties as subplot treatments in a split plot design with four replications. The plot yields (kg/plot) ware studied. The following analysis of variance is provided in Table 24.12.

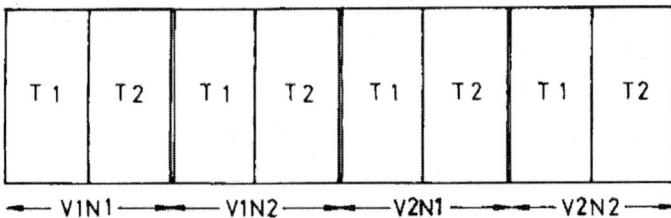

FIGURE 24.4. A split plot design with three factors each at two levels. V × N factorial combinations are in main plots and two levels of factor T (T1 and T2) are in subplots.

TABLE 24.12. ANOVA results.

Source of variation	df	SS	MS	F	Table F 5 percent	Table F 1 percent
Replications	(r–1) 3	277.33	92.44	4.58*		
Main plot treatments	(a–1) 14	754.89	53.92	2.67*	1.94	2.54
Error (a)	(r–1)(a–1) 42	845.93	20.14			
Sub plot treatments	(b–1) 1	2.05	2.05	0.21	4.06	7.24
Interaction (MPTXSPT)	(a–1)(b–) 14	460.11	32.85	3.49*	1.92	2.52
Error (b)	a(r–1)(b–1) 45	423.31	9.40			

The following inference was drawn based on Table 24.12. Replications, main plot treatments, and interaction were found to be significant. From the ANOVA table, we see that the main plot treatments differed significantly. We want to find out whether N is significant or P is significant or NP is significant. An NP analysis was done and the following data obtained.

N	P p_0	p_1	p_2	Total
n_0	222.24	238.05	245.80	706.09
n_1	242.10	248.00	270.19	760.29
n_2	237.74	224.12	221.86	683.72
n_3	213.31	229.11	209.30	651.72
n_4	198.12	194.62	205.31	598.05
Total	1113.51	1133.90	1152.46	3399.87

$$\text{NSS} = \Sigma Ni^2 \, / \, 24 - \text{CF} = \left(706.19^2 + \ldots + 598.05^2\right) / \, 24 - \text{CF}$$
$$= 96936.65 - 96325.95 = 610.70$$

$$\text{PSS} = \Sigma Pi^2 \, / \, 40 - \text{CF} = \left(1113.51^2 + \ldots + 1152.46^2\right) / \, 40 - \text{CF}$$
$$= 96344.94 - 96325.95 = 18.99$$

NP SS = Table SS – NSS – PSS = 754.89 – 610.70 – 18.99 = 125.20
(See Table 24.13.)

TABLE 24.13. ANOVA results.

SV	df	SS	MS	F (5 percent)
N	4	610.70	152.67	7.5*
P	2	18.70	9.49	<1
NP	8	125.20	15.65	<1
Error (a)	42	845.93	20.14	

The effect of N only is found to be significant. We therefore want to know which component (linear, etc.) is significant. We use orthogonal polynomials for this purpose. The following coefficients are obtained from the orthognal polynomial. (See as follows.)

	ξ_1	ξ_2	ξ_3	ξ_4	Total
N_0	-2	+2	-1	+1	706.09
N_1	-1	-1	+2	-4	760.19
N_2	0	-2	0	+6	683.72
N_3	+1	-1	-2	-4	651.72
N_4	+2	+2	+1	+1	598.05

$$SS_{Linear} = [(-2)(706.09) + ... + (2)(598.05)]^2 / (24 \times 10)$$
$$= 105397.6225 / 240 / 439.15$$

$$SS_{Quadratic} = [(2)(706.09) + ... + (2)(598.05)]^2 / 24 \times 14$$
$$= (-171.17)^2 / 235 = 87.1999$$

$$SS_{Cubic} = [(-1)(706.09) + ... + (1)(598.05)]^2 / 24 \times 10$$
$$= 109.10^2 / 240 = 49.5950$$

$$SS_{Quartic} = [(+1)(706.09) + ... + (1)(598.05)]^2 / 24 \times 70$$
$$= 58360.8964 / 1680 = 34.7386$$

Now we form the following analysis of variance table for testing the trend (see Table 24.14).

Thus the trend analysis gives additional information that the linear trend is significant in the factor nitrogen.

TABLE 24.14. ANOVA results.

Source of variation	df	SS	MS	F	Table F 5 percent	1 percent
SS due to linear	1	439.16	439.16	21.80**	4.07	7.27
SS due to quadratic	1	87.20	87.20	4.33*	4.07	7.27
SS due to cubic	1	49.60	49.60	2.46ns	4.07	7.27
SS due to quartic	1	34.74	34.74	1.72ns	4.07	7.27
Error *(a)*	42	845.93	20.14			

Efficiency of Split Plot Design over RCBD

Estimate of error variance (σ^2) for an equivalent RCBD is

(estimated σ^2) = [Ea + (b–1)Eb]/b = [20.14 + (2–1)(9.40)]/2 = 14.77

1. Efficiency of split plot design compared with RCBD when only sub-plot treatments are compared.

   (estimated σ^2 – Eb)/ Eb = [14.77–9.40]/(9.40) = 5.37/(9.40) $\times$ 100 = 57 percent).

2. Efficiency of split plot design compared with RCBD when only the main plot treatments are considered.

   [(estimated s^2–Ea)/(Ea)] $\times$ 100 = [(94.77–20.14)/(20.14)] $\times$ 100
   [(–5.37)/20.14] $\times$ 100 = –27 percent.

Split-Split-Plot Design

In split plot design, the main plot treatment(s) are arranged in a randomized complete block design in *r* replications. A second factor, (B), and several treatments (subplot treatments) are assigned in each of the main plot. If we want to include another factor, (C), in the split plot design, we can further split the subplot and assign the sub-subplot treatments at random. Such a design is called split-split-plot design.

Example: A, B, C are the factors to be studied in an experiment. Consider that we have to study factor A with three levels as main plot treatments, factor B with two levels as subplot treatments, and factor C with three levels as sub-subplot treatments in four replications in a split-split-plot design. (See Figure 24.5 for the layout plan for one replication.)

There are $3 \times 2 \times 3 \times 4 = 72$ plots (a × b × c × r) in the experiment. The analysis is carried out in the usual manner. The sources of variation, degrees of freedom, mean square, and the F tests for various factors and interaction are shown in Table 24.15.

The standard errors for comparison of various treatments and interactions can be calculated using the following formulas.

SE for Comparing

C means =

$$\sqrt{Ec\ /\ rab}.$$

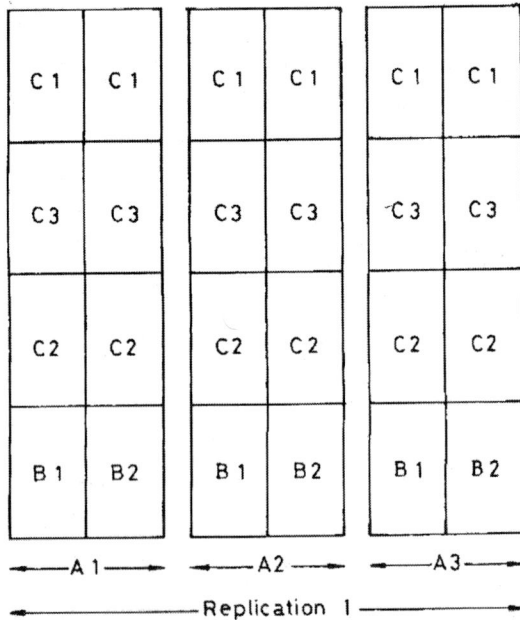

FIGURE 24.5. Split-split design. A = main plot with three levels, B = subplot with two levels, and C = sub-subplot with three levels (randomization not shown).

TABLE 24.15. ANOVA results.

SV	df	SS	MS	F
Replication	3	RSS	RSS/3=MS$_R$	
A	2	SSA	SSA/2=MS$_A$	MS$_A$/E$_a$
E(a)	6	EaSS	EaSS/6=E$_a$	
B	1	SSB	SSB/1=MS$_B$	MS$_B$/E$_b$
AB	2	SSAB	SSAB/2=MS$_{AB}$	MS$_{AB}$/E$_b$
E(b)	9	EbSS	EbSS/9=E$_b$	
C	2	SSC	SSC/2=M$_{SC}$	M$_{SC}$/E$_c$
AC	4	SSAC	SSAC/4=MS$_{AC}$	MS$_{AC}$/E$_c$
BC	2	SSBC	SSBC/2=MS$_{BC}$	MS$_{BC}$/E$_c$
ABC	4	SSABC	SSABC/4=MS$_{ABC}$	MS$_{ABC}$/E$_c$
E(c)	36	EcSS	EcSS/36=E$_c$	
Total	71	Total SS		

The value of t_c is to be used for computing LSD.
C means for same level of A:

$$\sqrt{[E_c \ / \ rb]}.$$

C means for same level of B:

$$\sqrt{E_c \ / \ ra}.$$

The value of t_c is to be used for LSD) B means for same or different levels of C:

$$\sqrt{[(c-1)E_c + E_b]/(rac)}.$$

The t_{bc} is to be calculated as

$$t_{bc} = [(c-1)E_c t_c + E_b t_b]/[(c-1)E_c + E_b].$$

A means for same or different levels of C =

$$\sqrt{[(c-1)E_c + E_a]/(rbc)}.$$

The *t* for LSD is to be calculated as

$$t_{ac} = [(c-1)E_c t_c + E_a t_a] / [(c-1)E_c + E_a].$$

C means for same A and B = $\sqrt{E_c / r}$. For this t_c is to be used.
  B means for same A and same or different C =

$$\sqrt{[(c-1)E_c + E_b] / (rc)}.$$

Here we have to use the *t* to be calculated as

$$t_{bc} = [(c-1)E_c t_c + E_b t_b + E_a t_a] / [(c-1)E_c + E_b].$$

A means for same or different B and C:

$$\sqrt{[b(c-1)E_c + (b-1)E_b + E_a] / (rbc)}.$$

Here the *t* is to be calculated as

$$t_{abc} = [b(c-1t_c) + (b-1)E_b t_b + E_a t_a] / [b(c-1) + (b-1)E_b + E_a].$$

Comparison of Split Plot Design with Strip Plot Design

Let us consider two factors, A and B. A at four levels and B at three levels
and we have five replications. See Table 24.16.
  Note: In a split plot design, two different plot sizes are involved. Main
plot treatments are confounded and are not compared precisely. In strip plot
design (split block design) three different plot sizes and three errors are involved, Ea, Eb and Ec. Ea is used to test treatments allotted to column strips,
Eb to test row strips, and Ec is used to test the plots formed by the interaction of the two factors A and B. Generally Ec < Eb < Ea.

EXERCISES

24.1. What is split plot design? State the circumstances for use of this de-
      sign.
24.2. Give some examples in which split plot design is suitable.
24.3. What are advantages and disadvantages of split plot design?
24.4. Compare split plot design with randomized complete block design.
24.5. How will you calculate the efficiency of split plot design with that of
      randomized complete block design?

TABLE 24.16. ANOVA results.

Split plot design		Strip plot design	
SV	df	SV	df
Replications	$(r-1) = 4$	Replications	$(r-1) = 4$
Main Plot treatments	$(a-1) = 3$	Main plot treatments	$(a-1) = 3$
Error *(a)*	$(r-1)(a-1) = 12$	E*(a)*	$(r-1)(a-1) = 12$
Subplot treatments	$(b-1) = 2$	Subplot treatments	$(b-1) = 2$
MPPT X SPT	$(a-1)(b-1) = 6$	Error *(b)*	$(r-1)(b-1) = 8$
Error *(b)*	$[(r-1)(b-1)+(a-1)(b-1)(r-1)]$ $(4\times2)+(3\times2\times4) = 32$	MPT × SPT	$(a-1)(b-1) = 6$
		Error *(c)*	$(r-1)(a-1))(b-1) = 24$
Total	$r\times a\times b-1 = 59$	Total	$r\times a\times b-1 = 59$

24.6. A company wants to conduct an investigation of the effect of its three incentives (I1, I2, and I3) on its four products (P1, P2, P3, and P4) in two cities. Write down the analysis of variance for split plot design. Researcher selects cities as replications. Researcher is not interested in the products.

24.7. In a regional trial, a breeder has ten outstanding varieties and wants to test the performance of these varieties in four locations and for two seasons using five levels of nitrogen. The breeder already knows the variability due to levels of N.

 a. What is the total degrees of freedom, and what design should be used and why?

 b. If the first, second, and third degree interactions are significant based on F test, how will you interpret your results?

24.8. A soybean breeder has 33 promising lines, 11 are short, 11 are medium, and 11 are tall. His available area is good for only four replications. He wants to test the performance of the lines with respect to levels of nitrogen fertilizer at 0, 30, 60, 90, and 120 kg/ha levels. His main interest is the response of these lines to the increase in N levels or the interaction between lines and N levels.

 a. What design will you recommend and why?

 b. Set up the ANOVA and partition the degrees of freedom.

 c. Set up the field layout plan and show how will you randomize the treatments.

24.9. State the differences between split plot design and strip plot design.

24.10. There are four replications, four varieties of soybean, and two levels of fertilizers in a design. The experimenter is interested in response of the fertilizer to the varieties. The data that he collected are shown here:

Variety		1	Replications 2	3	4
V1	F1	198	207	183	178
	F2	146	101	97	98
V2	F1	214	146	188	206
	F2	106	74	87	73
V3	F1	302	338	254	223
	F2	121	74	95	60
V4	F1	281	252	282	308
	F2	132	115	100	102

Identify the design and analyze the data. Construct a table of means with appropriate standard errors and write a short note on the conclusions to be drawn from the experiment.

Chapter 25

Split Block (Strip Plot)

The following are reasons for selecting split block or strip plot designs:

1. In factorial experiments, the treatment combinations are formed by combining the different levels of one factor with the different levels of another factor. In some experiments, the treatments cannot be randomized at subplot levels. For example, suppose we want to evaluate agricultural implements, irrigation methods, tillage, insecticides, etc., which require larger-sized plots. Let A and B be the two factors. At the first stage, the a_i (A levels) treatments are laid out in horizontal strips in randomized complete block design. At the second stage, the b_i (B levels) treatments are also laid out in strips vertically in the same replication. The resulting design is called a strip plot or split block design.

2. Sometimes, experimenters require more precise estimates on interactions. Split-plot design sacrifices precision in both treatments in order to provide higher precision on the *interaction* between the two treatments, which will generally be more accurately determined than in a randomized compete block design or a split plot design.

LAYOUT

Let factor A be agricultural implements with four levels a_1, a_2, a_3, and a_4, and factor B be the four tillage practices denoted by b_1, b_2, b_3, and b_4. The layout plan for this experiment with four replications is shown in Figure 25.1. The plots for levels of A and levels of B are fairly larger whereas for their interaction treatments, $a_i b_i$ etc., the plot sizes are small. Thus, there are three different plot sizes, one for factor A, the second for factor B, and the third for their interaction. Since there are three different plot sizes involved in this design, we get three different experimental errors called Error (a), Error (b), and Error (c), respectively.

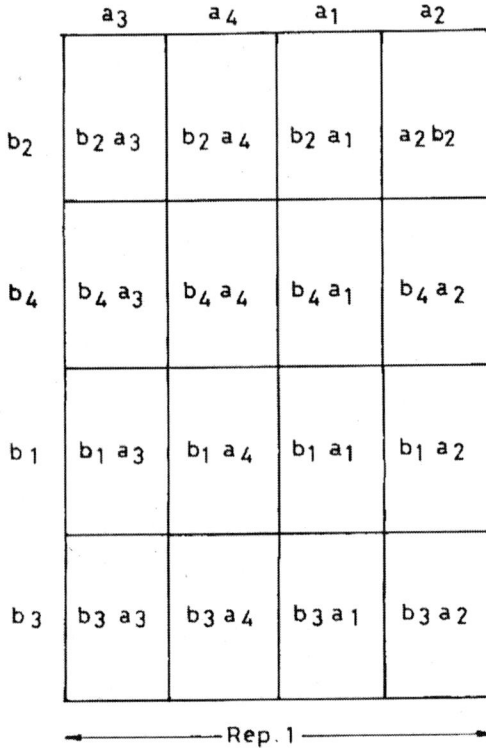

FIGURE 25.1. Strip plot design with two factors. Four levels of factor A allotted in vertical strips and four levels of B allotted in horizontal strips. Shown for one replication only.

TABLE 25.1. ANOVA results.

SV	df	SS	MS	F
Replication	$r-1$	SS_R	MSR	
A	$a-1$	SS_A	MS_A	MS_A/EaMS
Error (a)	$(r-1)(a-1)$	EaSS	EaMS	
B	$b-1$	SS_B	MS_B	MS_B/EbMS
Error (b)	$(r-1)(b-1)$	EbSS	EbMS	
AB	$(a-1)(b-1)$	AB_{SS}	MS_{AB}	MS_{AB}/EcMS
Error(c)	$(a-1)(b-1)(r-1)$	EcSS	EcMS	
Total	$rab-1$			

The ANOVA table for strip plot design is shown as Table 25.1.
For more detailed information, see Cochran and Cox (1957), Leonard and
Clark (1939).

Standard Errors of Differences for Comparing Differences Among the Various Means

1. Standard error of difference for comparing A treatment means:

$$a_1 - a_0 = \sqrt{(2Ea / rb)}$$

2. Standard error of difference for comparing B means:

$$b_1 - b_0 = \sqrt{(2Eb / ra)}$$

3. Standard error of difference for comparing A treatments at the same or different level of B:

$$a_1 b1 - a_o b_1 = \sqrt{2\{[(b - 1)Ec + Ea] / (rb)\}}$$

4. Standard error of difference for comparing B means at the same level of A:

$$a_1 b_1 - a_1 b_0 = \sqrt{2\{[(a - 1)Ec + Eb] / a\}}$$

5. Standard error of difference for comparing AB at the same level:

$$a_1 b_1 - a_0 b_0 = \sqrt{2Ec / r}$$

Numerical example: An experiment with five varieties and six N levels
was conducted in a strip plot design with three replications. The plot size
was 4 m × 4 m, spacing 45 cm × 45 cm in chili crop (*Capsicum annuum* L).
The character studied was pod weight/plant (g). The layout plan (see Figure
25.2) and the yields recorded are given as follows. The nitrogen levels used
were $N_0 = 0$ kg N/ha, $N_1 = 30$ kg N/ha, $N_2 = 60$ kg N/ha, $N_3 = 90$ kg N/ha, $N_4 = 120$ kg N/ha, and $N_5 = 150$ kg N/ha.

Rep. 1

N\V	V1	V2	V3	V4	V5
N2	62	60	75	60	40
N1	40	97	92	50	52
N0	36	30	35	35	38
N3	74	112	55	46	69
N5	72	159	96	62	43
N4	55	71	96	50	74

Rep. 2

N\V	V3	V2	V1	V4	V3
N0	38	28	39	45	39
N4	65	76	132	85	92
N2	55	63	61	49	72
N5	71	33	60	28	63
N1	57	30	51	34	44
N3	62	38	46	50	36

Rep. 3

N\V	V2	V5	V1	V3	V4
N5	33	45	90	69	56
N2	67	64	62	38	57
N3	61	67	75	91	80
N0	59	47	48	64	46
N4	70	74	56	83	61
N1	20	23	54	31	57

FIGURE 25.2. Layout plan for strip plot design with two factors V and N. The V and N treatments are allotted in vertical and horizontal strips, respectively.

498

Analysis

1. Form the N × V table.

		N₀				N₁				N₂		
Variety	R1	R2	R3	Total	R1	R2	R3	Total	R1	R2	R3	Total
V1	36	39	48	123	40	51	54	145	62	61	62	185
V2	30	28	59	117	97	30	20	147	60	63	67	190
V3	35	39	64	138	92	44	31	167	75	72	38	185
V4	35	45	46	126	50	34	57	141	60	49	57	166
V5	38	38	47	123	52	57	23	132	40	55	64	159
Total	174	189	264	627	331	216	185	732	297	300	288	885

		N3				N4				N5		
Variety	R1	R2	R3	Total	R1	R2	R3	Total	R1	R2	R3	Total
V1	74	46	75	195	55	132	56	243	72	60	90	222
V2	112	38	61	211	71	76	70	217	159	33	33	225
V3	55	36	91	182	96	92	83	271	96	63	69	228
V4	46	50	80	176	50	85	61	196	62	28	56	146
V5	69	62	67	198	74	65	74	213	43	71	45	159
Total	356	232	374	962	346	450	344	1140	432	255	293	980

2. Form the replication × varieties table.

	Varieties					
Replications	V1	V2	V3	V4	V5	Total
I	339	529	449	303	316	1936
II	389	268	346	291	348	1642
III	385	310	376	357	320	1748
Total	1113	1107	1171	951	984	5326

3. Form the varieties × N table.

	Varieties					
Nitrogen	V1	V2	V3	V4	V5	Total
N₀	123	117	138	126	123	627
N₁	145	147	167	141	132	732
N₂	185	190	185	166	159	885
N₃	195	211	182	176	198	962
N₄	243	217	271	196	213	1140
N₅	222	225	228	146	159	980
Total	1113	1107	1171	951	984	5326

Computations

1. Correction factor (CF) = $5326^2 / 90 = 315180.84$
2. Replications $SS = (1936^2 + 1642^2 + 1748^2) / 30 - CF = 316658.80 - 315180.84 = 1477.96$
3. Variety $SS = (1113^2 + 1107^2 + 1171^2 + 951^2 + 984^2) / 18 - CF = 317117.56 - 315180.84 = 1936.72$
4. Table $SS = (339^2 + \ldots + 320^2)/6 - CF = 325370.67 - 315180.84 = 10189.83$
5. E(a) SS = Table SS – Rep. SS – Var. SS = $10189.83 - 1477.56 - 1936.72 = 6775.55$
6. Nitrogen $SS = (627^2 + \ldots + 980^2)/15 - CF = 326508.13 - 315180.84 = 11327.29$
   E(bSS) = Table SS – RSS – NSS = $21973.96 - 1477.96 - 11327.29 = 9168.71$
7. Var. × Nitrogen table $SS = (123^2 + \ldots + 159^2)/3 - CF = 330595.33 - 315180.84 = 15414.49$
8. VN SS = Table SS – VSS – NSS = $15414.49 - 1936.72 - 11327.29 = 2150.48$
9. Total $SS = 362956.00 - CF = 362956.00 - 315180.84 = 47775.16$
10. Error (c) = total SS – all others = 14938.85
11. Form the ANOVA table (see Table 25.2).

$$CV(Ea) = \sqrt{846.89 / 59.18} = 49.2 \text{ percent}$$
$$CV(Eb) = \sqrt{916.87 / 59.18} = 51.1 \text{ percent}$$
$$CV(Ec) = \sqrt{373.47 / 59.18} = 32.7 \text{ percent}$$

TABLE 25.2. ANOVA results.

SV	df	SS	MS	F	Table F (0.05)
Replication	2	1477.56	738.98		6.94
Varieties	4	1936.72	484.18	$484.18/846.89 = 0.58^{ns}$	3.84
Error (a)	8	6775.55	846.94		
Nitrogen (N)	5	11327.29	2265.46	$2265.46/916.87 = 2.47^{ns}$	3.33
Error (b)	10	9168.71	916.87		
VN	20	2150.48	107.52	$107.52/373.47 = 0.29^{ns}$	1.84
Error (c)	40	14939.45	373.47		
Total	89	47775.16			

COMBINED ANALYSIS IN STRIP PLOT DESIGN

Comparison of mean values (see Table 25.3).

Treatment comparisons

R = number of replications (r = 3)
A = number of levels of factor A (varieties = 5)
b = number of levels of factor B (N levels = 6)
1. Comparison of varieties
   Standard error of difference for comparison of variety means = SEd =
   $\sqrt{2 \times Ea / rb} = \sqrt{2 \times 846.89 / 3 \times 6} = 9.70$
Conclusion: Differences among the means are not significant.
2. Comparison of N levels:
   Standard error of difference for comparison of N levels = SEd =
   $\sqrt{2 \times Eb / ra} = \sqrt{2 \times 916.87 / 3 \times 5} = 11.06$
   $LSD_{0.05,10} = t \times SEd = 2.228 \times 11.06 = 24.64$
Conclusion: Arrange the means in descending order

N4	N5	N3	N2	N1	N0
76.0	65.3	64.1	59.0	48.8	41.8

Note: In ANOVA, N factor just failed to reach the level of significance.
3. Comparison of varieties at the same or different levels of N
   $SEd = \sqrt{2[(b-1)Ec + Ea] / rb}$
   $= \sqrt{2[(6-1)373.47 + 846.89] / 3 \times 6} = 17.37$
   $LSD_{0.05,40} = t \times SEd = 2.021 \times 17.37 - 35.10$
Variety means do not show statistical difference.

TABLE 25.3. Means table for N levels and varieties.

	Varieties					
N levels	V1	V2	V3	V4	V5	Mean
N0	41.0	39.0	46.0	42.0	41.0	41.8
N1	48.3	49.0	55.7	47.0	44.0	48.8
N2	61.7	63.3	61.7	55.3	53.0	59.0
N3	65.0	70.3	60.7	58.7	66.0	64.1
N4	81.0	72.3	90.3	65.3	71.0	76.0
N5	74.0	75.0	76.0	48.7	53.0	65.3
Mean	61.8	61.5	65.1	52.8	54.7	59.2

4. Comparison of N means at each V level

$$SEd = \sqrt{2[(a-1)Ea + Eb]/ra}$$
$$= \sqrt{2[(5-1)373.47 + 916.87]/3 \times 5} = 17.93$$
$$LSD_{0.05,40} = t \times SEd = 2.021 \times 17.93 = 36.24$$

V1: <u>N4   N5   N3   N2   N1</u>   N0

V2: <u>N5   N4   N3   N2   N1</u>   N0

V3: <u>N4   N5   N2   N3   N1</u>   N0

V4: All are on par

V5: All are on par

Trend Analysis

The effect of N has just failed to reach the level of significance. We can, however, further analyze for significance of the components such as linear, quadratic, etc., and get additional information. We use orthogonal polynomial method.

N levels	ξ_1	ξ_2	ξ_3	ξ_4	ξ_5	Treatment total
N0	−5	+5	−5	+1	−1	627
N1	−3	−1	+7	−3	+5	732
M2	−1	−4	+4	+2	−10	885
N3	+1	−4	−4	+2	+10	962
N4	+3	−1	−7	−3	−5	1140
N5	+5	+5	+5	+1	+1	980
	70	64	180	28	252	5326

Find the sum of squares:

SS linear $= [(-5)(627) + (-3)(732) + (-1)(885) + (1)(962) + (3)(1140)$
$+ (5)(980)]^2 / 15 \times 70 = [(-3135 - 2196 - 885$
$+ 962 + 3420 + 4900)]^2 / 15 \times 70 = 8952.72$

SS Quadratic $= [(-5)(627) + (-1)(732 + (-4)(885) + (-4)(962) + (-1)$
$(1140) + (5)(980)]^2 / 15 \times 84 = [(-3135 - 732 - 3540$
$-3848 - 1140 + 4900)]^2 / 15 \times 84 = 1225^2$
$/15 \times 84 = 1190.97$

$$\text{SS cubic} = [(-5)(627) + (7)(732) + (4)(885) + (-4)(962) + (-7)(1140)$$
$$+(5)(980)]^2 / 15 \times 180 = [(-3135 + 5124 + 3540 - 3848$$
$$-7980 + 4900)]^2 / 15 \times 180 = 1399^2 / 15 \times 180 = 724.89$$

$$\text{SS quartic} = [(+1)(627) + (-3)(732) + (2)(885) + (2)(962) + (-3)(1140)$$
$$+(1)(980)]^2 / 15 \times 28 = [(627 - 2196 + 1770 + 1924 - 3420$$
$$+980)]^2 / 15 \times 28 = 315^2 / 15 \times 28 = 236.25$$

$$\text{SS quintic} = [(-627) + 3600 - 8850 + 9620 - 5700 + 980]^2 / 15 \times 252$$
$$= 917^2 / 15 \times 252 = 222.46$$

Form the ANOVA (see Table 25.4).

Combined Analysis in Strip Plot Design

Often experimenters conduct their experiments over several years or locations and in such cases, they wish to combine the data and obtain general conclusions. Suppose there are 3 years, 5 varieties, 6 N levels, and 3 replications. The ANOVA will be as follows.

SV	df
Replications (Pooled)	6
Years (Y)	$(3-1) = 2$
Varieties (V)	$(5-1) = 4$
$V \times Y$	$(3-1)(5-1) = 8$
Error(a)(Pooled)	24
N levels (N)	$(6-1) = 5$
YN	$(3-1)(6-1) = 10$
Error (b)(Pooled)	30
VN	$(5-1)(6-1) = 20$
YVN	40
Error (c)(Pooled)	120
Total	269

TABLE 25.4. ANOVA

SV	Df	SS	MS	F
Treatments	5	11327.29	2265.46	2.47
SS linear	(1)	8952.72	8952.72	9.76**
SS quadratic	(1)	1190.97	1190.97	1.30^{ns}
SS cubic	(1)	724.89	724.89	< 1
SS quartic	(1)	236.25	236.25	< 1
SS quintic	(1)	222.46	222.46	< 1
Error	10	9168.71	916.87	

EXERCISES

25.1. What is strip plot design?

25.2. Compare the strip plot design with that of split plot design.

25.3. Write down the randomization procedure for the treatments in strip plot design.

25.4. Suggest a few examples for strip plot design.

25.5. Is trend analysis useful when your treatments are in quantitative measures particularly when they are equally spaced?

REFERENCES

Cochran, W.G. and G.M. Cox (1957). *Experimental design,* Second edition. New York: John Wiley and Sons.

Leonard, W.H. and A.G. Clark (1939). *Field plot technique.* Minneapolis: Burgers Publishing Company.

Chapter 26

Completely Confounded Design

In 2^n factorial experiments, the number of treatments increases with increases in numbers of factors. For example, in 2^3, 2^4, 2^5, 2^6, and 2^7 experiments, there will be 8, 16, 32, 64, and 128 treatment combinations, respectively. A large number of combinations of treatments leads to large-sized blocks. As blocks become larger and larger, they are likely to encounter greater and greater soil heterogeneity within blocks. Consequently, the error would be increased. Such a difficulty may be overcome by a device called confounding. Confounding enables the experimenter to carry out the study with blocks containing lesser numbers of treatments than a full replication of the treatment combinations, thus increasing the accuracy of the experiment. Each block is an incomplete block in the sense that a block does not contain all the treatments of the experiment. When treatments are assigned to incomplete blocks, we require several incomplete blocks depending upon the number of treatment combinations. For example, if there are eight treatment combinations, four can be allotted to one block and another four to the second block, i.e., a replication is subdivided into two blocks each containing four plots so that the plots within each block become less heterogeneous. The treatments are then allotted to the incomplete blocks. By doing so, some treatment differences are merged (confounded) with blocks and they are not separable from the blocks.

Example: Consider a 2^2 experiment with two factors, say N and P, each at two levels. We get four treatment combinations (1), n, p, and np. We have two levels of N, namely n_0 and n_1, and two levels of P, namely p_0 and p_1. The treatment combinations will be n_0p_0, n_1p_0, n_0p_1, and n_1p_1 or (1), n, p, and np. Observe the following table.

	(1)	n	p	np
N	–	+	–	+
P	–	–	+	+
NP	+	–	–	+

The plus (+) sign indicates the presence of N and the minus sign (–) indicates absence of N in row 1 in the table, i.e., N is present in the treatments n and np and absent in the treatments (1) and p. Similarly, the plus and minus are written for P in row 2. For NP, the corresponding signs of N and P are multiplied and placed against the NP treatment in row 3.

LAYOUT OF THE EXPERIMENT

NP Confounding

When there is heterogeneity in replication it is not advisable to allot the four treatments in one block in order to have more accuracy in the comparison of the treatments. Hence, to make the block size smaller, divide the replication into two smaller blocks each having two equally sized plots. Consider the NP treatment. The + signs of NP treatment, namely (1) and np, are allotted to one block and the – sign treatments, namely n and p, to the second block in the replication. The layout plan will be as shown in Figure 26.1.

Let the plot yields be 8, 4, 6, and 13 for treatments n, (1), p, and np, respectively (see Figure 26.1). As per assumption of the model, each observed

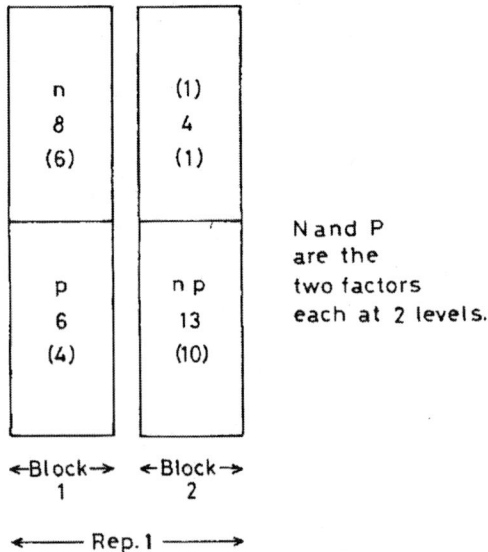

```
┌────────┬────────┐
│   n    │   (1)  │       
│   8    │    4   │       
│  (6)   │   (1)  │       
├────────┼────────┤   N and P
│        │        │   are the
│   p    │   n p  │   two factors
│   6    │   13   │   each at 2 levels.
│  (4)   │  (10)  │       
└────────┴────────┘       
  ←Block→   ←Block→
     1         2

    ←──── Rep. 1 ────→
```

FIGURE 26.1. Layout plan for two factor NP confounding in a 2^2 factorial experiment.

value is the additive effect of treatment effect, block effect, and random effect. Assume that the block effect in block 1 is 2, and the block effect in block 2 is 3. Then the figures in parentheses indicate the treatment effect after removing the block effect. In block 1, the observed values of 8 and 6 for n and p include the block effect and treatment effect. If block effect is 2, the treatment effects for n and p are $8 - 2 = 6$, and $6 - 2 = 4$. Similarly, in block 2, the observed values of 4 and 13 for (1) and np treatments include the block effect and treatment effect. If the block effect in block 2 is 3, then the treatment effects are $4 - 3 = 1$, and $13 - 3 = 10$, respectively. Here 8, 4, 6, and 13 are the observed values of the treatments n, (1), p, and np, respectively. The block effects are assumed as the actual block effects are unknown. As per the model, we assume that each observed value is the additive effect of treatment effect, block effect, and random effect. The treatments in block 1 are favored by 2, and in block 2 by 3. The block difference is $B2 - B1 = 3 - 2 = 1$. We know that no such difference exists within the plots in the same block.

The simple effect of N at p_1 level is $np - p = 13 - 6 = 7$, which contains simple effect as well as the block effect difference. Thus, the actual simple effect is $7 - 1 = 6$. Since we do not know the effect of block differences, the block difference could not be separated from the total difference. Thus, we say that the treatment difference is mixed or confounded with block difference.

Block 2 – Block 1 = NP Interaction Effect

This interaction effect is included with the block difference or it is said to be completely confounded with the blocks.

Calculation of Main Effects and Interaction

1. Main effect, N:

$$N = \text{mean of two simple effects}$$
$$= 1/2 \, [(np - p) + (n - 1)]$$
$$= 1/2 \, [(13 - 6) + (8 - 4)]$$
$$= 1/2 \, [(10 + 3) - (4 + 2) + (6 + 2) - (1 + 3)]$$
$$= 1/2 \, [(10 - 4 + 6 - 1) + (3 - 2 + 2 - 3)]$$
$$= 1/2 \, (11 + 0) = 5.5 + (\text{block effect, which is zero})$$

2. Main effect of P:

$$
\begin{aligned}
P &= \text{mean of two simple effects} \\
&= 1/2\,[(np - n) + (p - 1)] \\
&= 1/2\,[(13 - 8) + (6 - 4)] \\
&= 1/2\,[10 + 3) - (6 + 2) - (4 + 2) - (1 + 3)] \\
&= 1/2\,(10 - 6 + 4 - 1) + (3 - 2 + 2 - 3) \\
&= 1/2\,[7 + (0)] \\
&= 1/2\,(\text{treatment effect} + \text{block effect}) \\
&= 1/2\,(7 + 0) = 3.5 + \text{block effect which is zero.}
\end{aligned}
$$

3. Interaction effect, NP (difference between two simple effects):

$$
\begin{aligned}
NP &= 1/2\,[(np) - (p) - (n) + (1)] \\
&= 1/2\,(13 - 6 - 8 + 4) \\
&= 1/2\,[(10 + 3) - (4 + 2) - (6 + 2) + (1 + 3)] \\
&= 1/2\,[(10 - 4 - 6 + 1) + (3 - 2 - 2 + 3)] \\
&= 1/2\,(\text{Treatment effect} + \text{block effect}) \\
&= 1/2\,(1 + 2) \\
&= 1/2\,(+3) = 1.5
\end{aligned}
$$

The real interaction effect is 0.5 but it is estimated as 1.5 because the interaction effect includes block effect of 1.0. Both the effects are mixed up, i.e., confounded, with the blocks. Thus in such an arrangement of factorial experiments, NP is confounded and information on NP is lost or sacrificed for the sake of the precise estimates of the effects of N and P.

N confounding: If we want to confound N, then the layout plan will be as in Figure 26.2.

P confounding: If we want to confound P, then the layout plan will be as in Figure 26.3.

If the same arrangements of treatments are followed in all the replications, then the design is said to be a completely confounded design. If different treatments are confounded in different replications, then the design is called a partially confounded design.

A brief description of 2^3 factorial experiment in completely confounded design is explained. This design is generally used. It is preferred when block heterogeneity is suspected. Let N, P, and K be the factors each at two levels (0 and 1). Then the treatment combinations can be listed as shown in Table 26.1.

In this experiment, there are three main effects, three first order interactions, and one second order interaction.

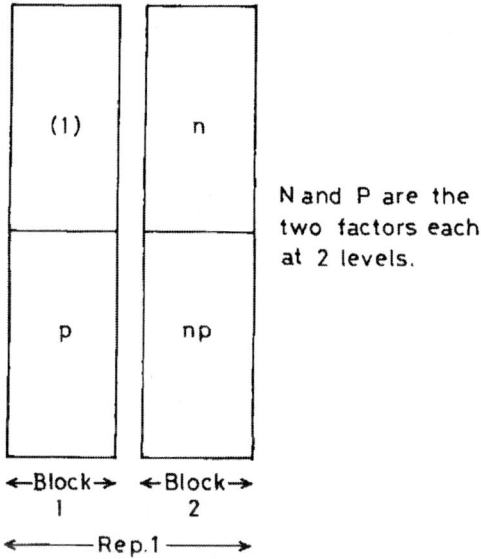

FIGURE 26.2. Layout plan for the main effect, N confounding in a 2² factorial experiment.

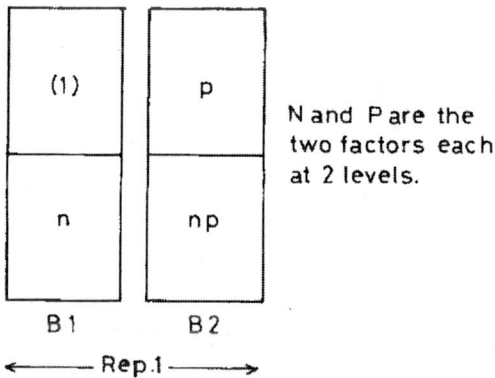

FIGURE 26.3. Layout plan for the main effect, P confounding in a 2² factorial experiment.

TABLE 26.1. A 2^3 factorial experiment in completely confounded design.

	(1) $n_0p_0k_0$	n $n_1p_0k_0$	p $n_0p_1k_0$	np $n_1p_1k_0$	k $n_0p_0k_1$	nk $n_1p_0k_1$	pk $n_0p_1k_1$	npk $n_1p_1k_1$
N	−	+	−	+	−	+	−	+
P	−	−	+	+	−	−	+	+
NP	+	−	−	+	+	−	−	+
K	−	−	−	−	+	+	+	+
NK	+	−	+	−	−	+	−	+
PK	+	+	−	−	−	−	+	+
NPK	−	+	+	−	+	−	−	+

Main Effects

$$N = (n - 1)(p + 1)(k + 1) = (npk + nk + np + n) - (pk + k + p + 1)$$
$$P = (n + 1)(p - 1)(k + 1) = (npk + np + pk + p) - (nk + n + k + 1)$$
$$K = (n + 1)(p + 1)(k - 1) = (npk + nk + pk + k) - (np + n + p + 1)$$

Interaction Effects

$$NP = (n - 1)(p - 1)(k + 1) = (npk + np + k + 1) - (nk + n + pk + p)$$
$$NK = (n - 1)(p + 1)(k - 1) = (npk + p + nk + 1) - (np + pk + n + k)$$
$$PK = (n + 1)(p - 1)(k - 1) = (npk + n + pk + 1) - (np + nk + p + k)$$
$$NPK = (n - 1)(p - 1)(k - 1) = (npk + n + p + k) - (np + nk + pk + 1)$$

In each of the effects, four treatments have the positive (+) sign and the remaining four the negative (–) sign. The positive signs are allotted to one block and the negative signs to the other block. We allot four treatments in the first block and another four in the second block. Thus two blocks will make up one replication. Suppose we decide to confound NPK interaction effect, allot the treatments npk, n, p, and k having + signs, to one block and another four treatments having – signs, namely np, nk, pk, and (1), to another block. NPK is now confounded and we cannot get information on this effect. However, we get precise estimates on the other effects. The experimenter should decide which treatment is to be confounded.

CHOICE OF CONFOUNDING INTERACTIONS

1. Normally, second-order interactions, i.e., higher-order interactions, are difficult to interpret. Therefore, one should confound three-factor interactions. In this case it is NPK.
2. Sometimes, the experimenter may consider the main effect, say P, is not important because of already having sufficient information. In such a situation the experimenter can confound the P effect. In this case, the experimenter would to assign the treatments npk, np, pk, and p having + signs to one block and other treatments namely (1), n, k, and nk having – signs to the second block.

Randomization

When assigning the treatments to each block, random procedure is to be followed. Separate random numbers should be used while allotting treatments in each block.

STATISTICAL ANALYSIS

As in an unconfounded design, the analysis is carried out and there will be no NPK effect. The ANOVA will be as follows:

SV	Df
Blocks	7
Treatments	6
N	(1)
P	(1)
NP	(1)
K	(1)
NK	(1)
PK	(1)
Error	18
Total	31

The treatment means may be compared in the usual manner.

Numerical example (2^4 completely confounded design): An experiment on nitrogen fixing blue-green algae in rice soils was carried out.

Treatments

g = green leaf at 5590kg/ha
a = ammonium sulfate at 168 kg/ha
s = super phosphate at 168 kg/ha
l = algae at 7.4 kg/ha each at 1 (presence) and 0 (absence) levels
Variety: Co 29
Spacing = 6 inches × 6inches
Plot size = 6.5 feet × 5.5 feet (net), gross 7.5 feet × 6 feet
Design: 2^4 factorial design confounded in blocks of eight plots each.
Replications = 3
(See Table 26.2).

Identify the confounded interaction and analyze the data.

IDENTIFICATION OF CONFOUNDED INTERACTION

There are two blocks per replication and eight treatments per block. The block size is 2^{n-k}, where n = 4, k = 1, i.e., eight treatments per block. Then we choose $n - k$, i.e., three independent treatment combinations. They are to

TABLE 26.2.

	Replication 1				Replication 2				Replication 3			
	Block 1		Block 2		Block 1		Block 2		Block 1		Block 2	
gasl	gasl	Yield	gasl	Yield	gasl	Yield	gasl	Yield	gasl	Yield	gasl	Yield
	1101	1172	1010	885	0111	991	1111	1224	1011	1007	1001	919
	1000	947	0101	956	0100	907	0000	667	0010	775	1010	887
	1110	983	0000	690	0001	684	0011	667	0111	1005	1100	1164
	0010	780	1001	1036	1101	1060	0101	875	1110	1220	0011	707
	0001	677	1100	1090	1011	936	1010	912	0100	1021	1111	1225
	0100	1027	0011	953	1000	959	0110	896	1101	1066	0000	676
	1011	945	1111	1194	0010	709	1001	947	0001	740	0101	952
	0111	1030	0110	1033	1110	1132	1100	1163	1000	1006	0110	1024
Total	–	7561	–	7837	–	7378	–	7351	–	7840	–	7554

513

be chosen from the key block. The key block is the one which contains all treatment combinations in the first level, i.e., 0000 (gasl). From the key block 3, independent treatment combinations are taken. In this case it is gs, al, gl. The interaction, which has the even number of letters common with gs, al, gl, is confounded. In this case gasl is confounded.

Procedure

1. Calculate block total, treatment total, and grand total.
   Grand total (GT): 45521
   Block totals: Total of block 1 + block 2t . . . . . . + block 6 = 7561 + 7837 + 7378 + 7351 + 7840 + 7554
   Check: block totals = grand total = 45521
   Treatment totals given in Table 26.3
2. Calculate CF = GT^2 /r t = $45521^2/(3 \times 16)$ = 2072161441/48 = 43170030
3. Block SS = 7561^2 + 7837^2 +....+ $7554^2/8$ − CF = 43198486 − 43170030 = 28456
4. Total SS = 1172^2 + 947^2 +...+ 952^2 + 1024^2 / CF = 44411329 − 43170030 = 1241299
5. Treatment SS = Sum of all the treatments SS and interactions SS, except SS due to gasl because gasl is confounded = 1107950.7
6. Error SS = Total SS − Treatment SS − Block SS = 1241299 − 1107950.7− 28456 = 10489
7. The treatment SS is computed by following Yates method as follows: See Table 26.4.

The $F_{1, 28, 0.05}$ = 4.20 (from table of F; see Appendix, Table A.6)

$$\text{SE of main response} = \sqrt{(3764.1 / 12)} = 17.67[\sqrt{s^2 e / (r \times 2^{n-2})}]$$
$$\text{SE of differential response} = \sqrt{(3746.1) / 6} = 24.9[\sqrt{s^2 e / (r \times 2^{n-3})}]$$

CONCLUSION

The treatments show significant difference. The main effects G and A are significant, and S and L are not. Except SL interaction, all others are not significant.

We get more information if we form a summary table as given (see Table 26.5):

TABLE 26.3. Treatment SS using Yates method.

Treatment	Yield (0)	(1)	(2)	(3)	Total effect (4)	$(Col. 4)^2/$ $r \times 2^n$ $r2^n = 48$ (SS)	$(Col. 4)/$ $r \times (2^{n-1})$ $(r \times 2^{n-1} = 24)$ Main effect	Effect
1	2033	4945	11317	22553	45521	43170090.00	1896.71	
g	2912	6372	11236	22968	4637	447953.50	193.21	G
a	2955	4948	11084	2143	5299	584987.50	220.79	A
ga	3417	6288	11884	2494	-685	9775.52	-28.54	GA
s	2264	5003	1341	2767	719	10770.02	29.96	S
gs	2684	6081	802	2532	-677	9548.52	-28.21	GS
s	2953	5215	1316	-455	289	1740.02	12.04	AS
gas	3335	6669	1178	-230	721	10830.02	30.04	GAS
Total		45521	50158	54772	55824			
1	2101	879	1427	-81	415	3588.02	17.29	1
gl	2902	462	1340	800	351	2566.68	14.62	GL
sl	2783	420	1078	-539	-235	1150.52	-9.79	AL
gal	3298	382	1454	-138	225	1054.68	9.37	GAL
sl	2327	801	-417	-87	881	16170.02	36.71	SL
gsl	2888	515	-38	376	401	3350.02	16.71	GSL
asl	3026	561	-256	379	463	4466.02	19.29	ASL
gasl	3643	617	+56	342	-37	28.52	-1.54	GASL

515

TABLE 26.4. ANOVA results.

SV	df	SS	MS	F	Table F (5 percent)
Between blocks	5	28456.00	569.21		
Between treatments	(16-1-1) = 14	1107950.70	79139.3	21.1**	2.06
Main effect					
G	(1)	447953.50	447953.5	119.58**	4.20
A	(1)	584987.50	584987.5	156.16**	
S	(1)	10770.02	10770.02	2.87ns	4.20
L	(1)	3588.02	3588.02	<1	
Two-factor interactions					
GS	(1)	9848.52	9848.52	2.55ns	
GL	(1)	2566.68	2566.68	<1	
AS	(1)	1740.02	1740.02	<1	
AL	(1)	1150.52	1150.52	<1	
SL	(1)	16170.02	16170.02	4.32*	
Three-factor interactions					
GAS	(1)	10830.02	10830.02	2.89ns	
GAL	(1)	1054.68	1054.68	<1	
GSL	(1)	3350.02	3350.02	<1	
ASL	(1)	4466.02	4466.02	1.12ns	
Error	28	104892.30	3746.1		
Total (by subtraction)	47	1241299.0			

For detailed information on confounding, refer to Federer (1955), Yates (1933), and Snedecor (1956).

EXERCISES

26.1. What do you understand by confounding? What is completely confounding?

26.2. In a 2^3 factorial experiment, you have to confound one of the main effects. Suggest a layout plan.

26.3. State the reasons for preferring confounded design.

26.4. What are main effect, simple effect, and interaction effect? Explain with an example.

TABLE 26.5.

Factor	Mean response	Response with G: Present	G: Absent	A: Present	A: Absent	S: Present	S: Absent	L: Present	L: Absent
G	193.20	—	—	164.66 (G+AG)	221.74 (G–AG)	164.99 (G+GS)	221.40 (G–GS)	207.82 (G+GL)	178.83 (G–GL)
A	220.79	192.25 (A+AG)	249.33 (A–AG)	—	—	232.83 (A+AS)	208.75 (A–AS)	211.0 (A–AL)	230.58 (A+A1)
S	29.95	1.75 (S+SG)	58.15 (S–SG)	41.99 (S+SA)	17.91 (S–SA)	— 54.0	—	66.65 (S+SL)	–6.75 (S–SL)
L	17.29	31.91 (L+LG)	2.62 (L–LG)	7.50 (L+LA)	27.08 (L–LA)	53.75 (L+LS)	19.42 (L–LS)	—	—

REFERENCES

Federer, W.T. (1955). *Experimental design.* New Delhi: Oxford and IBH Publishing Co.

Snedecor, G.W. (1956). *Statistical methods,* Fifth edition. Ames, IA: The Iowa State University Press.

Yates, F. (1933). The principles of orthogonality and confounding in replicated experiments. *J. Agri . Sci* 23:108-145.

Chapter 27

Analysis of Covariance

In analysis of covariance, the techniques of analysis of variance and regression analysis are combined. Reliable data must be obtained from a suitable design in order to make treatment comparisons with greatest precision. In any design, in addition to treatments, the observations are affected by several unknown factors. These factors are the main reasons for experimental error. To some extent, they are controlled or minimized by proper selection of the experimental design, and by the application of the basic principles of experimental design, namely, replication, randomization, and local control. The precision of the estimates can further be enhanced by using confounding, split plot, strip plot, and incomplete block techniques.

Principles in Covariance Analysis

In any design, the significance test depends upon the experimental error term. Any method or procedure that can be adopted to reduce error directly increases the sensitivity of the experiment in detecting treatment effects. The method in which reduced estimate of error could be obtained is called the covariance method. The covariance analysis has become an important tool in many areas of research (Cochran, 1957). In this method, the errors affecting the treatment comparison may further be reduced by involving another variable called supplementary variable, X. Controlling the error using supplementary variables is called statistical control.

The variance of a treatment mean in a design is given by σ^2/r, where σ^2 is the error mean square (EMS) and r is the number of replications. This error can be reduced by using covariance analysis. To perform covariance analysis (ANOCOVA), we require additional information on supplementary or ancillary or concomitant variables. If the concomitance between Y and X is closely related, substantial reduction in error can be achieved and consequently a real difference between the treatment means can be established. If there are t treatments $y_1, y_2, y_3 \ldots y_t$ and $\bar{y}_1, \bar{y}_2, \bar{y}_3 \ldots \bar{y}_t$ are their corresponding means, $\bar{y}$ values should be adjusted in such a way that all the treatment means are obtained on the basis of the same $\bar{x}$. The $\bar{y}$ values are adjusted using the regression coefficient. For example, the adjusted $\bar{y}_i$th treatment mean is given by adjusted $\bar{y}_i = \bar{y}_i - b\,(\bar{x}_1 - \bar{x})$, where b is the regression coefficient, $\bar{x}$ is the general mean, $\bar{x}_i$ mean of the ith treatment of X, $\bar{y}_i$ = mean of the ith treatment of Y. The principle is that the influence of the covariate (x) on the main variate (y) is eliminated. If the experimental error of Y before the adjustment is σ^2_E, then the experimental error after adjustment is $\sigma^2_E\,(1-\rho^2)$.

REASONS FOR COVARIANCE ANALYSIS

Suppose a researcher evaluates the yield potential of soybean varieties. The differences in yields may or may not show significance. If the experimental plots vary greatly with respect to the covariate, say x, the number of plants/plot at the time of harvest of the crop, the experimenter comes to the conclusion that the observed yield differences would have been affected by differences in x values. The experimenter should therefore adjust the Y values, and then carry out the analysis and interpret the results. The nature of the association between Y and X should be examined, and the influence of X on Y should be eliminated during analysis. Furthermore, the differences in X values in the different experimental units within the same block will naturally contribute to the experimental errors. These differences cannot be controlled by experimental design. Thus, the objective of the covariance analysis is to obtain reduced experimental error by taking into account the regression of Y on X. If there is correlation between Y and X, then the ANOCOVA will give a smaller estimate of experimental error than would be obtained from the analysis of variance of Y alone. If the experimental error before adjustment is σ^2_e (without considering X values) and r is the correlation between Y and X, then the experimental error after adjustment is $\{[\sigma^2_e\,(1-\rho^2)]\,[f_e/f_e-1]\}$, where f_e is the degrees of freedom associated with σ^2_e.

USES OF COVARIANCE ANALYSIS

The analysis of covariance is very useful in many areas of research. A few situations in which covariance analysis can be used are given as follows.

1. In agricultural experiments, the plant population, i.e., the number of plants/treatment, in a replication must be the same. The plant population may vary due to death of the plants, unexpected pests and diseases, or for other reasons. The plant population and yield of the crop are highly corre-

lated. The yields of the different treatments can be adjusted according to the same number of plants, and then data are analyzed and interpreted.

2. A researcher is interested in the effects of different nutrition (treatments) on the growth of children. It is difficult to select children of the same weight before the start of the experiment. Children with different weights will respond differently to the diet. The final weights may be due to differences in initial weights. Using covariance analysis, the initial weights can be adjusted to the same weight and then analyzed.

3. Several methods of teaching English are to be evaluated in a class. The final test scores are analyzed. The students' final scores and their IQs are related. The initial IQs must be the same to evaluate the methods. Corrections to the initial IQs are made by covariance analysis and then analyzed.

4. One wants to study the effects of various stress conditions on blood pressure. Initial blood pressure before the experiment may be considered as a covariate.

5. Grain and straw yields: Grain yields may be adjusted considering straw yields as a concomitant variate in agricultural experiments.

6. In assessing the relative yields of different corn varieties in a corn varietal trial, one must consider the intrinsic variation in the soil in which the experiment is conducted. Soil fertility status may be considered as X variable. According to the differences in soil fertility, the Y values may be converted before an analysis is carried out.

7. There are two types of diet, say diet I and diet II. Their effects on weight loss in women are to be studied. One group of women with an initial weight of 70 kg loses 10 kg weight after the study period. Another group with an initial weight of 60 kg loses 5 kg of weight. We cannot compare 10 kg and 5 kg, as loss in weight depends upon the initial weight. Thus some adjustment in the initial weight is necessary before we compare the diets with respect to the weight loss. If the weight loss *(Y)* is related to initial weight *(X)*, it is necessary to make adjustments in Y for the initial weight differences before analyzing Y.

PRINCIPLES IN ANOCOVA WITH REFERENCE TO TWO TREATMENTS

Let there be two treatments A and B; and $\bar{y}_A$ and $\bar{y}_B$ are the means of A and B. The difference between $\bar{y}_A$ and $\bar{y}_B$ may be large and this difference may be attributable to the differences of values. Suppose we adjust $\bar{y}_A$ and $\bar{y}_B$ for the same $\bar{x}$ values for A and B, then the difference is no longer present. The difference between $\bar{y}_A$ and $\bar{y}_B$ may not be seen but real difference may appear if each Y is adjusted to the same $\bar{x}$. The model is $Y_{ij} = \alpha_i + \beta(x_{ij}-$

$\bar{x}$), where β $(x_{ij}-\bar{x})$ is the regression effect of Y on X. This adjusted $Y_{ij} = Y_{ij}-\beta(x_{ij}-\bar{x})$ is the value of Y_{ij} after adjustment by removing the regression effect. The ANOVA performed with adjusted Y_{ij} values is free from regression effect, i.e., the linear relation to the covariate is removed.

Numerical Example

Three varieties of rice crop were tested for the yielding capacity in a randomized complete block design with four replications. At the time of harvest, the researcher noted an uneven number of plants in the experimental units. He suspected the yield of the varieties would have been affected, and hence he wanted to carry out the covariance analysis, making use of the information available on the supplementary variable, namely the number of plants per plot (x). At the time of harvest of the crop, he recorded x values in each experimental plot. The data are given in Table 27.1. The object of the analysis is (1) to adjust for the chance differences among the treatments using the information on the supplementary variable, namely the number of plants/plot *(x)*, and (2) to recuce error.

Analysis of Variance for Y

$$CF = (\Sigma y)^2 \, / \, n = 210^2 \, / \, 12 = 3675.00$$
$$\text{Block SS} = (64^2 + 45^2 + 52^2 + 49^2) \, / \, 3 - CF = 11226 \, / \, 3 - CF$$
$$= 3742 - 3675 = 67$$
$$\text{Treatment SS} = (68^2 + 82^2 + 60^2) \, / \, 4 - CF = 14948 \, / \, 4 - CF$$
$$= 3732 - 3675 = 62$$
$$\text{Total SS} = 15^2 + 20^2 + ... + 19^2 + 12^2 - CF = 4022 - 3675 = 347$$

TABLE 27.1. Yield *(y)* and number of plants/plot *(x)* of the three rice varieties conducted in the experiment in randomized complete block design with four replications *(y* = yield/plot in kg and *x* = number of plants/plot).

	Varieties						Replication total	
	V1		V2		V3			
Replication	X	Y	X	Y	x	Y	x	y
I	660	15	900	33	601	16	2161	64
II	815	20	705	12	580	13	2100	45
III	680	16	801	17	610	19	2091	52
IV	603	17	810	20	569	12	1982	49
Total	2758	68	3216	82	2360	60	8334	210
Mean	689.5	17	804	20.5	590	15	694.5	17.5

ANOVA (for Y variable)

SV	df	SS	MS	F
Blocks	3	67	22.33	0.33ns
Treatment	2	62	31.00	0.86ns
Error	6	218	36.33	
Total	11	347		

GM = 17.5 kg/plot   CV = 34.4 percent

The "F" test indicates that the differences in treatment mean yields is not significant, i.e., the different varieties produce the same yields. The blocks also did not show significant difference, i.e., homogeneity in the plots were seen. Now we perform analysis of variance for the x character.

Analysis for X

$$CF = (\Sigma x)^2 / n = 8334^2 / 12 = 5787963.00$$
Block SS: $(2161^2 + 2100^2 + 2091^2 + 1982^2) / 3 - CF$
$= 17380526/3 - CF = 5793508.6 - 5787963.0 = 5545.6$
Treatment SS: $(2758^2 + 3216^2 + 2360^2) / 4 - CF = 5879705.00$
$-5787963.00 = 91742$
Total SS: $660^2 + ... + 569^2 - CF = 5924022.00 - 5787963.00$
$= 136059.00$

ANOVA (for x variable)

SV	df	SS	MS	F
Blocks	3	5545.60	1848.53	2.53ns
Treatment	2	91742.00	45871.00	7.10*
Error	6	38771.40	6461.90	
Total	11	136059.00		

GM = 694.5   CV = 11.6 percent

The X variable shows significant difference. If F is not significant, the experimenter is to infer that the adjusted yield differences are due to treatments. If F is significant, the adjusted yield differences in y variable may be attributable to differences in X, i.e., the number of plants/plot. If X is expected to show differences in adjusted yields (y_s), the covariance analysis is to be done to eliminate bias due to differences in plant number/plot in estimated yield differences.

Analysis for Sum of Products

Form the XY table (see Table 27.2).

$CF = (\Sigma x)(\Sigma y) / n = (8334 \times 210) / 12 = 145845$
Total sum of products of x and y (Sxy) $[(660 \times 15)+.....+(569 \times 12)]$
$-CF = 150882 - 145845 = 5037$
Block sum of products $= \{(2161 \times 64)+...+(1982 \times 49)\}|/3 - CF$
$= 146218 - 145845 = 373$.
Treatment sum of products: $\{(2758 \times 68) + (3216 \times 82) + (2360 \times 60)\}$
$/4 - CF = 148214 - 145845 = 2369$.

Form the sum of products of x and y and sum of squares table for analysis of covariance and test of significance for adjusted treatment means (see Table 27.3).
Error Sum of Products = Total SP – Rep. SP – Tr. SP = 5037 – 373 – 2369 = 2295
The F test is not significant and this shows that the variety means do not differ significantly. Estimation of regression coefficient of y on x for adjustment for error SS and test of significance:

TABLE 27.2.

Replications	Varieties			Total
	V1	V2	V3	
I	9900	29700	9616	49216
II	16300	8460	7540	32300
III	10880	13617	11590	36087
IV	10251	16200	6828	33279
Total	47331	67977	35574	150882

TABLE 27.3.

SV	df	Sum of squares and sum of products			Sums of estimates	
		Sx^2	Sxy	Sy^2	MS	F
Total	rt–1 = 11	136059.0	5037	347		
Rep.	r–1 = 3	5545.6	373	67		
Treatments	t–1 = 2	91742.0	2369	62	31.00	0.89ns
Error (unadj)	(r–1)(t–1) = 6	38771.4	2295	218	36.33	
Trt. + Error	8	130513.4	4664	280		

Regression coefficient = byx = Sxy/Sx^2 = 2295/38771.40 = 0.0592
Regression sum of squares (Reg. SS) = (Sxy)^2/Sx^2 =
2295^2/38771.40 = 135.85

or

Reg. SS = b Sxy = 0.0592 × 2295 = 135.85
Deviation from reg. = total SS – Reg. SS = 218 – 135.85 = 82.15

See Table 27.4
As F is significant, it is worthwhile to adjust the variety means for the variability in x values in the different plots. The degrees of freedom for error is t(r – 1) – 1, i.e., 3(4 – 1) – 1 = 8.

t Test for the Regression Coefficient

$t = b/SEb$ $SEb = \sqrt{Se^2 / Sx^2} = \sqrt{11.736/38771.40} = 0.0174$
$t = 0.0592/0.0174 = 3.4023**$
Verification: $t^2 = F(3.4023^2 = 11.58)$

The t value from the table for P=0.05 and for 7 df is 2.365.
Computation of correlation coefficient between y and x and test if significance

$r = Sxy / \sqrt{Sx^2 Sy^2} = 2295 / \sqrt{38771.40 \times 218}$
$= 2295 / 2907.3 = 0.7894$
t test: $t = \{r / (\sqrt{1 - r^2})\} \times \sqrt{n - 2} = \{0.7894 / (\sqrt{1 - 0.7894^2})\}\sqrt{12 - 2}$
$= (0.7894 / 0.6138) \times 3.162 = 4.07**$
Correlation between y and x is significant.

TABLE 27.4. ANOVA

Source of variation	df	SS	MS	F	Table F 5 percent	1 percent
Reg. SS	1	135.85	135.85	11.58*	5.59	12.25
Deviation from Reg.	7	82.15	11.736(Se^2)			
Total	8 t(r–1) – 1	218.00				

Now draw a scatter diagram for *y* and *x* values. It shows linear relationship between yield/plot (y) and number of plants/plot (x) indicating that the effect of number of plants/plot on yield must be removed by analysis of covariance.

Adjusted analysis of variance to test the differences in the varieties due to variation in *x* values.

In the sum of squares, sum of products, Sx^2, Sy^2, and Sxy, and degrees of freedom for treatment and error table, compute the last line namely treatment + error (see Table 27.5).

(Treatment + error) adjusted SS is computed as $280 - 4664^2/130513.4 = 280 - 166.67 = 113.33$ (with 7 degrees of freedom). Adjusted error SS = total SS $-$ Reg. SS (in the error line) $= Sy^2 - (Sxy)^2/Sx^2 = 218 - 2295^2/38771.40 = 218 - 135.85 = 82.15$. The adjusted error SS will have the degrees of freedom 5,1 less than the 6 for unadjusted error. Treatment SS (adjusted) for the regression of *y* on *x* is calculated as the difference between treatment + error and error as $113.33 - 82.15 = 31.18$ with 2 degrees of freedom. It is important to note that the adjusted sum of squares for treatment cannot be obtained directly by applying the formula $Sy^2 - (Sxy)^2Sx^2$ to the treatment line (Little and Hills, 1978). The table value of $F_{0.05,2.5} = 3.79$.

F Test for Regression

Adjusted error Mean Square $= 82.15/5 = 16.43$.
Regression Mean Square $= 135.85$. $F = 135.85/16.43 = 8.27*$ with 1 and 5 degrees of freedom.
(Table value of F for 1,5 at $P = 0.05$ is 6.61.)
Calculation of adjusted treatment (variety) means.
The adjusted mean yield is: adj $\bar{y}_i - \bar{y}_i - b(\bar{X}_i - \bar{x})$ where

$\bar{y}_i$ = mean yield for ith treatment
$\bar{x}_i$ = mean number of plants/plot in the ith treatment
$\bar{x}$ = general mean

Form the following table (see Table 27.6).

TABLE 27.5. ANOVA

Source of variation	df	SS	MS	F	Table F 5 percent	Table F 1 percent
Treatment (adj)	2	31.18	15.59	0.95ns	3.79	
Error (adj)	5	82.15	16.43			
Treat + error	7	113.33				

527。15656272

TABLE 27.6.

Tr	$\bar{X}_1$	Deviation $(X_1-x) = x$	Product of $b(X_1-x)$	Unadjusted mean yield	Adjusted mean yield
V1	689.5	689.5–694.5 = –5.00	–0.2960	17.0	17.00–(–.296) = 17.2960
V2	804.0	804.0–694.5 = 109.5	6.4824	20.5	20.50–6.4824 = 14.0176
V3	590.0	590.0–694.5 = –104.5	–6.1864	15.0	15.00–(–.1864 = 21.1864
Total		0	0	17.5	17.5

The adjusted treatment mean is obtained after subtracting $b(\bar{X}_1 - \bar{x})$ from the unadjusted treatment mean. Comparing the adjusted treatment means with the unadjusted treatment means ($\bar{y}$ is), we observed that the ranks in the unadjusted means are V2, V1, and V3 whereas in the unadjusted treatment means the ranks are V3, V1, and V2, which indicates the necessity of covariance analysis.

Effects of Correlation on Increase in Precision of the Experiment

The increase in precision of the covariance analysis depends upon the within treatments correlation between Y and X. Thus the correlation coefficient (within groups) is $Sxy/\sqrt{Sx^2 \, xSy^2} = r = 2295/\sqrt{38771.4 \times 218} = 0.7894$. Error variance before covariance analysis is ESS/df = 218/6 = 36.33. Error variance after covariance analysis is 82.15/5 = 16.43. Thus through the analysis of covariance, we have reduced the error variance from 36.33 to 16.43, i.e., we have increased the precision of the experiment 36.33/16.43 by 2.2 times. Because of high correlation and the effect of covariate, the precision is increased by 2.2 times.

COMPARISONS INVOLVING ADJUSTED MEANS

Since the variance of adjusted treatment means is larger than the correct adjusted treatment mean square, the usual least significant difference test is not appropriate for comparing adjusted treatment means. A different standard error of difference must be calculated for each pair of means (Little and Hills, 1978).

Standard Error of Difference Between Any Pair of the Adjusted Treatment Means and Test of Significance

$SEd = \sqrt{[se^2\{(2/r + (\bar{x}_i - \bar{x}_j)^2 / S_x{}^2\}]}$, where se^2 = mean square in the error line of covariance table. From our example it is $82.15/5 = 16.43$; SSx = SS for X in the same line, r = number of replications, $\bar{x}_i - \bar{x}_j$ = difference between two means in X corresponding to $\bar{y}_i$ and $\bar{y}_i$ treatment means.

SEd (V1 – V2) =
$\overline{\sqrt{[\{16.43\{2/4 + (689.5 - 804)^2 / 38771.4\}] = 3.7109}}$
$t = (V1 - V2) / SEd = (17.296 - 14.0176) / 3.7109 = 3.7109$
$= 3.2784 / 3.7109 = 0.883^{ns}$ (with df. $t[r - 1] - 1 = 3[4 - 1] - 1 = 10$).

Conclusion: V1 is on par with V2.

SEd (V1 – V3) =
$\overline{\sqrt{[\{16.43.(2/4 + (689.5 - 590)^2 / 38771.40\}] = 3.5228}}$
$t = (17.296 - 21.1864) / 3.5228 = -3.8904 / 3.5228 = -1.10$

Conclusion: V1 and V3 do not differ.

SEd (V2 – V3) = $\sqrt{[\{16.43\{2/4 + (804 - 590)^2 / 38771.40\}] = 5.2556}}$
$t = (V2 - V3/SEd = (14.0176 - 21.1864) / 5.2556 =$
$-7.1688 / 5.2556 = -1.36^{ns}$

Conclusion: V2 and V3 do not differ.

The variety V2 ranks first before adjustment for the number of plants/plot and ranks last after adjustment for the number of plants per plot. But in the case of V3, its rank has become first after adjustment. ANOCOVA did not show significance differences among the treatment means contrary to original analysis.

The regression equation within the treatments is

$\hat{y} = \bar{y} + b(x - \bar{x})$ i.e., $\hat{y} = 17.5 + 0.0592(x - 694.5) = -23.6144 + 0.0592x$.

The means of the varieties adjusted for the equal number of plants/plot did not show significant differences attributable to the different yield potential of the varieties. The equation is plotted in the graph. The points are close to the line indicating that the number of plants/plot is related with the grain yields of the varieties.

Standard errors for the variety means. It is given by the formula =

$$\sqrt{[se^2[\{1/r+(\bar{x}i-\bar{x})^2\}/Sx^2]}$$ where se^2 = reduced MS for error, r is the number of replications, Sx^2 is the SS for the covariate, x.

SE for V1: $\sqrt{[\{16.43\{1/4+(689.5-694.5)^2\}/38771.40]}=2.03$

SE for V2: $\sqrt{[\{16.43\{1/4+(804.0-694.5)^2\}/38771.40]}=3.03$

SE for V3: $\sqrt{[\{16.43\{1/4+(590.0-694.5)^2\}/38771.40]}=2.96$

Computation of average variance and comparison of adjusted treatment means.

When the variation in squares of the mean quantities of the number of plants/plot for different pairs of varieties, i.e., $(\bar{x}i-\bar{x}j)^2$ is not large in comparison with error sum of squares, we can compute average variance and least significant difference for comparison of the treatment means. The formula for the computation of average variance is

$(2\times se^2)/r[(1+1/r-1)\times$ (treatment SS for x/ error sum of squares for x)]

i.e., $(2\times16.43)/4[\{1+1/4-1\}\times(91742.0/38771.4)]=14.6945$

and hence

$SEd=\sqrt{14.6945}=3.8333$. Least significant difference is computed $t\times SEd=2.571\times3.8333=7.284$.

The difference between any pair of variety means does not exceed the least significant difference, we conclude that the differences among the treatment means are not significant. In the present example, there is no need to calculate LSD as the F test for the treatments is not significant.

TESTING HOMOGENEITY OF REGRESSION

If the treatments' regression coefficients differ, standard errors for treatment differences cannot be used in the test of significance and hence test of homogeneity of regression is necessary (Federer, 1967). Furthermore, the individual regressions must be the estimates of the same regression. The test of homogeneity of regressions is explained by following the method given by Steel and Torrie (1960).

1. Compute the following for each treatment:

V1 $\Sigma x = 2758 \Sigma x^2 = 1925834 Sx^2 = 1925834 - 2758^2 / 4 = 1925834$
$-1901641 = 24193$
V2 $\Sigma x = 3216 \Sigma x^2 = 2604726 Sx^2 = 2604726 - 3216^2 / 4 = 2604726$
$-2585664 = 19062$
V3 $\Sigma x = 2360 \Sigma x^2 = 1393462 Sx^2 = 1393462 - 2360^2 / 4 = 1393462$
$-1392400 = 1062$
V1 $\Sigma y = 68 \Sigma y^2 = 1170 Sy^2 = 1170 - 68^2 / 4 = 1170 - 1156 = 14$
V2 $\Sigma y = 82 \Sigma y^2 = 1922 Sy^2 = 1922 - 82^2 / 4 = 1922 - 1681 = 241$
V2 $\Sigma y = 60 \Sigma y^2 = 930 Sy^2 = 930 - 60^2 / 4 = 930 - 900 = 30$
V1 $\Sigma xy = 47331 Sxy = \Sigma xy - (\Sigma x)(\Sigma y) / n = 47331 - 2758 \times 68 / 4 =$
$47331 - 46886 = 445$
V2 $\Sigma xy = 67977 Sxy = \Sigma xy - (\Sigma x)(\Sigma y) / n = 67977 - 3216 \times 82 / 4$
$= 67977 - 65928 = 2049$
V3 $\Sigma xy = 35574 Sxy = \Sigma xy - (\Sigma x)(\Sigma y) / n = 35574 - 2360 \times 60 / 4$
$= 35574 - 35400 = 174$

Form the following table (Table 27.7).

Regression SS = SS of Y due to regression on x
Residual SS = Sy^2 – Regression SS
Total residual SS = 5.8120 + 20.7325 + 1.4988 = 28.0433 with
$\Sigma(ni-2)$, i.e., 2 + 2 + 2 = 6 df
Total Regression SS = $2668^2/44317$ = 160.6206
Deviation from regression SS = 285 – 160.6206 = 124.3794 with
$\Sigma(ni-1) - 1 = 9 - 1 = 8$
(Total Error SS = 285 – $2668^2/44317$ = 285 – 160.6206 = 124.3794,
i.e., Σyd^2 = 124.3794
SS in y due to differences among the regression coefficients:
124.3794 – 28.0433 = 96.3361

ANOVA

SV	df	SS	MS	F
Difference	2 (t–1)	96.3361	48.1681	10.3058*
Treatment	6 (Σni–2)	28.0433	4.6739	
Total	8 (Σni–t–1)	124.3794		

TABLE 27.7.

Variety	df	SS in x (SSx = Sx²)	SP = (Sxy)	Total SS (Sy²)	Reg. coeff. (byx)	Reg. SS {(byx) x Sxy}	Σyd²	df
A	3	36.8	28.8	22.8	0.7826	22.5389	0.2611	2
B	3	36.8	28.6	23.2	0.7772	22.2279	0.9721	2
C	3	33.2	17.6	10.8	0.5301	9.3298	1.4702	2
Total	9	106.8	75.0	56.8	0.7022	52.6650	4.135	8

The total reduced sum of squares accounts for eight degrees of freedom, the reduced sum of squares for each variety accounts for six degrees of freedom. The remaining two degrees of freedom represents differences between the three regression coefficients. The "F" is significant. Hence, the regression coefficients are not homogeneous that is they are having different slopes. Under such circumstances, the covariance analysis is not recommended. If the F is not significant, the regression lines for the various treatments are assumed to have a common slope.

ASSUMPTIONS IN COVARIANCE ANALYSIS

1. The regression coefficients within each treatment are homogeneous. Any difference in regression coefficients is due to chance only.
2. The relationship between variate and covariate is linear. Adjustments will not be effective if the line is in some other form, say curvilinear.
3. The errors are normally independently distributed with mean zero and common variance within each treatment.
4. The treatment, block, and regression effects are additive.
5. The x values are measured without error.
6. The distribution of Y_{ij} at each X_{ij} is normal.
7. The adjusted y values within the treatments are normal.
8. The treatments that we apply to the experimental units should not influence x values.

Uses of Covariance Analysis

1. To estimate the true regression of y on x on the assumption that the regression coefficients obtained in each treatment are homogeneous.
2. To use the regression coefficient to adjust or to correct the y means (treatment means) on the basis of some x values.

3. After adjusting the *y* values, the *x* adjusted treatment means are compared and the real treatment effects are assessed.
4. Though computation is tedious, the gain in precision justifies the use of covariance analysis.

About the Regression Coefficient

It is the main criterion in the analysis of covariance. It is used to predict the *y* values from knowledge of the corresponding *x* values.

About Selection of Concomitant Covariables

The important consideration in the selection of supplementary covariable *(x)* is that it must have a high correlation with the dependent variable. Information in X is obtained either prior to the start of the experiment or prior to the harvest of the crop depending upon the nature of the experiment.

About Regression Coefficients

The assumption of homogeneity of regression coefficients within treatments is important. It is on the basis of this assumption that the treatments are adjusted.

EXERCISES

27.1. In an experiment, the *(X)* concomitant and *Y* (variable of interest) data of a trial was conducted in a completely randomized design with three varieties and eight replications (*X* = number of plants/plot and *Y* = grain yield/plot [kg/plot]).

	Varieties					
	V1		V2		V3	
Replication						
	x	y	x	y	x	y
I	10	15	4	6	7	12
II	6	1	8	13	8	9
III	5	4	8	5	7	16
IV	8	6	8	18	3	7
V	9	10	6	9	6	13
VI	4	0	11	7	8	18
VII	9	7	10	15	6	13
VIII	12	13	9	15	8	6

27.2. What do you understand by statistical control?
27.3. What is covariate? Under what criteria will you select the covariate for covariance analysis?
27.4. What are the reasons for the selection of covariance analysis?
27.5. What are the uses of covariance analysis?
27.6. What is the important principle involved in covariance analysis?

REFERENCES

Federer, W.T. (1967). *Experimental design: Theory and application.* New Delhi: Oxford & IBH Publishing Co.

Little, T.M. and F.J. Hills (1978). Agricultural experimentation: Design and analysis. New York: John Wiley and Sons.

Steel, R.G.D. and J.H. Torrie (1960). *Principles and procedures of statistics with special reference to biological sciences.* New York: McGraw-Hill Book Co.

Chapter 28

Transformation of Experimental Data

In biological studies and in several other areas, experimental data are analyzed using F and Student's t tests of significance. These tests are based on certain basic assumptions: additivity (the treatments' effects must be assumed to be additive (sometimes additivity is referred to as linearity); normality (data are obtained randomly from a normal population); homogeneity of variances (equal variances, i.e., the variances are homoscedastic); and independence of errors (errors are noncorrelated). If the assumptions are not met, the basic probabilities may change and lead to faulty conclusions. When we test at $P = 0.05$ level of significance, we actually will be testing at 8 percent level, which will lead to too many significant results. The normal distribution is one important assumption in the analysis of variance (Cochran, 1947). The experimental data may exhibit violations of the assumptions. In the ANOVA, if the treatment effects are not really additive, then the sums of squares attributable to such effects do not represent the true effects (Anderson and Bancroft, 1952). Nonadditivity may exist when the true effects are multiplicative in nature and when aberrant observations are present (Ostle, 1988; Zar, 1974). Additivity effects imply a model that does not contain interaction terms. Experimenters commonly use design models, e.g., completely randomized design, randomized complete block design, Latin square design, factorial experiments, etc., which are all based on the assumption of additivity. The use of transformation in order to obtain additivity has received more attention recently than in the past. Use of Tukey's nonadditivity has been used as a guide for deciding between alternative possible transformation.

Certain procedures are available to normalize the observations. If the data do not show homogeneity of variances, we can make the variances homogeneous by using transformation of the experimental data. Furthermore, the homogeneity of variances, i.e., $\sigma_1^2 = \sigma_2^2 = \ldots = \sigma_k^2$ can be tested by Bartlett's test. If the treatment variances differ significantly, the standard deviation of the mean will differ from treatment mean to treatment mean. The errors must be independent, i.e., noncorrelated. Noncorrelation with the errors can be obtained by adopting a randomization procedure (Fisher, 1947; Cochran, 1947; Yates, 1933).

TRANSFORMATION

Transformation means changing the data from their original form (x values) to a different form (let us call them x' values). By doing so, we are changing the original shape of the observations. Skewed distribution may change to normal distribution, i.e., the shape of the original data may change. Normality can be achieved from the transformed data. The transformation affects the mean and standard deviation but not the order of the original means. Also, adding of a constant to all x values (observations) does not affect the relative position of any observation (Hopkins and Glass, 1978). For example, (1) if x_i (3.1) is the original observed value, $\log (x_i + 1)$ ($x' = 0.61278$) is the transformed value in the log scale; (2) if 17 is the value of the original observation, its transformed value in the square root form is 4.1231 (x'); and (3) when $x = 17$, its transformed value in the inverse sin or angular transformation is 22.30 (x'). There are several types of transformations available. Based on the analysis of the transformed data, the inferences are drawn and reported. Though the transformed data are not as accurate as the original values, the results are still acceptable. Transformation may bring improvements in one assumption and, sometimes, results in the improvement of other assumptions also. For example, the same transformation will sometimes normalize the distribution as well as make the variances more homogeneous (Winer, 1971).

The appropriate transformation is to be decided via trial and error. The transformation, which gives the best straight line when the cumulative frequencies of the transformed data are plotted, is the appropriate one. When assumptions are not satisfied, one need not discard the data. The treatment group having different variances can be discarded and the ANOVA performed. Also, one can resort to nonparametric methods of analysis. More information on transformation can be obtained from Bartlett (1947), Box and Cox (1964), Dolby (1963), Kendall and Stuart (1966), and Draper and Hunter (1969). The most commonly employed transformations are described in this chapter. After transforming the data, a statistical analysis may be performed. Then we can express descriptive statistics in terms of the original data. For example, after analyzing the data in the transformed scale, $x' = \sqrt{(x + 1/2)}$, the original values can be obtained by squaring them and then by subtracting 0.5 although the resultant statistics are slightly biased (Zar, 1974).

Common Types of Transformations

Logarithmic Transformation

In this transformation, the original data are converted to $\log_{10} x$, and then analysis is done. Relationship between mean and standard deviation is studied, and if there is evidence for relationship, this transformation is suggested. This transformation is also recommended when the data show a skewed distribution. Skewed distribution results when the experimental data consist of small and large values. In log transformation, the amount of differences among the variances is reduced leading to stability of variances. Suppose zero values are found, $\log (x + c)$ values, where c is a constant. By doing so, we are avoiding zero as well as negative values. For example: Let there be two treatments in an experiment that contains five observations. The values of the original and $\sqrt{x + 0.5}$ values are shown in Table 28.1. We find that mean and variance are no longer proportional, and the s_1^2 and s_2^2 are homogeneous in the transformed scale. Bartlett (1947) recommended log transformation and $\log (x + 1)$ transformation when zero x value occurs. If the distribution of the x' ($x' = \log x$) is normal, then the distribution of x is said to be log normal.

Square Root Transformation

Sometimes the data follow Poisson distribution. This distribution occurs when the data are in counts. It is applicable when the group variances are proportional to means. Such situations occur in biological data belonging to Poisson distribution. In such a situation, one will get a skewed distribution. The original values (x') are converted to $\sqrt{x}$, $\sqrt{x + 0.5}$ (most commonly used, Barlett, 1936) or $\sqrt{x + 1}$ (Freeman and Tukey, 1950; Bartlett, 1936). In

TABLE 28.1. Values in logarithmic transformation.

	Original values		Transformed values	
	Treatment 1	Treatment 2	Treatment 1	Treatment 2
	2	6	1.58	1.58
	2	6	1.58	2.55
	0	10	0.71	3.24
	4	12	2.12	3.54
	3	6	1.87	2.55
Mean	2.2	8.0	1.57	2.89
Variance	2.2	8.0	0.28	0.22

Poisson distribution, the treatment mean is equal to variance, in two treatments we observe $\mu_1 = \sigma_1^2$ and $\mu_2 = \sigma^2$. The transformed data result in normal distribution. Anscombe (1948) showed the value $X' = \sqrt{x + 3/8}$ as a good transformation. Tharp et al. (1941) added a constant value of 2 to each square root value of the observation.

Arcsine Transformation

Often, when data belong to percentages or proportions, they may show deviations from normal distribution. Under such circumstances, $x = \arcsin \sqrt{p}$ transformation is recommended where x is the transformed value and p is the original value. Such transformation values may be obtained from the tables given by Fisher and Yates (1949), and Snedecor and Cochran (1967). The notation $\sin^{-1}$ (read "inverse sine") is equivalent to the notation "arcsin." (Here "arcsine" is abbreviated as "arcsin.") The arcsine transformation is referred to as angular transformation.

Generally, this transformation is used when the data belong to binomial distribution. If the values lie between 30 and 70, transformation is not necessary (Goulden, 1952). If all percentages are less than 20 or more than 80, the square root transformation is recommended. By transforming the binomial variable to that of angular transformation, we can achieve normal distribution.

Reciprocal Transformation

Reciprocal transformation is recommended when standard deviation is proportional to the square of the mean ($s \propto m^2$). The original x is replaced by $1/x$.

Example 1: Palaniswamy (1975) studied the intensity of stem borer infestation in rice *(Oryza sativa)* crop under different nitrogen rates (0, 60, 120, 180, 240, and 300 kg/ha^{-1}) in a randomized complete block design using five replications. The data were collected in counts and also in percentages. The data were then converted to $\arcsin \sqrt{\text{equation}}$ and $\sqrt{x}$, and analysis of variance was performed individually and the most suitable transformation to the data was investigated. The results obtained in various transformations are reported briefly.

Analysis of Original Data in Counts. When the original observations, i.e., the number of infested tillers in the different treatments were analyzed, correlation between mean and standard deviation was significant at $P = 0.01$. Bartlett's test of homogeneity of variances was also significant (see Table 28.2, Figures 28.1 and 28.2).

TABLE 28.2. Data on the number of infested tillers in different treatments (five levels of nitrogen application) original data in counts and the analysis of variance.

Treatment N (kg·/ha⁻¹)	I	II	III	IV	V	Mean	Std. dev	Total
0	17	14	3	15	18	13.4	6.1	67
60	30	29	16	12	34	24.3	9.6	121
120	44	59	38	42	39	44.4	8.5	222
180	88	52	50	44	40	54.8	19.2	274
240	84	71	90	40	63	69.6	19.7	348
300	76	97	135	98	28	86.8	39.0	434
	339	322	332	251	222	48.9		1466

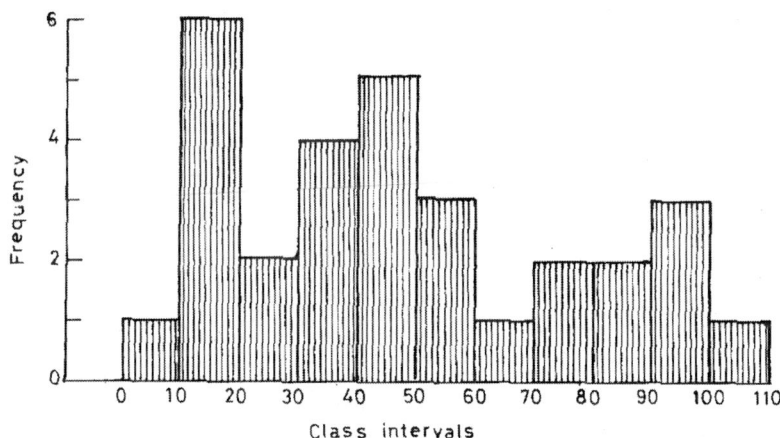

FIGURE 28.1. Histogram for the original data without transformation (count data).

Correlation coefficient between mean and standard deviation $= 0.910**$
χ^2 value for homogeneity of variances $= 43.7**$ (see Table 28.3).
$SE_m = 8.98$ $CV = 41.1$ percent

The analysis reveals that the homogeneity of variances test (Steel and Torrie, 1960) and the Tukey's test of nonadditivity (Snedecor and Cochran, 1967) were also significant. These tests confirmed the violations underlying ANOVA. Thus the data with original observations cannot be analyzed via the analysis of variance technique and conclusions cannot be drawn.

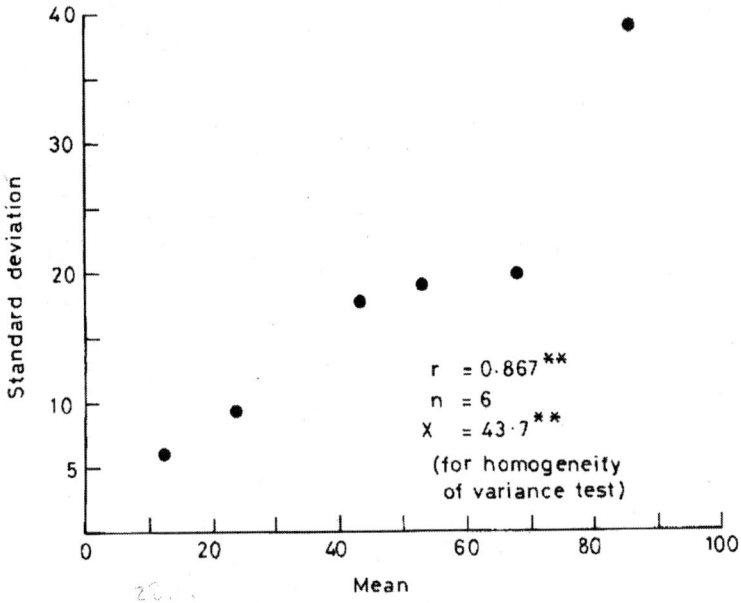

FIGURE 28.2. Relation between mean and standard deviation (original data).

TABLE 28.3. ANOVA results.

Source of variation	df	SS	MS	F	Table F 5 percent	Table F 1 percent
Replication	4	1880.5	470.125	1.17ns	2.87	4.43
Treatments	5	18951.5	3790.300	9.40**	2.71	4.10
Error	20	8063.5	403.18			
Total	29	28895.5				
Nonadditivity	1	2227.1	2227.1	7.25*	4.38	8.10
Residual	19	58364.4	291.4			
Total	29	28895.5				

ns = not significant; *significant at $P = 0.05$; ** = significant at $P = 0.01$.

Analysis of Percent Data. The intensity of infestation was also expressed in percentage [(number of infested tillers/total number of tillers) × 100], and then the percentage data were analyzed. In this case also, significant results with respect to correlation between mean and standard deviation, and homogeneity of variances were reported. Tukey's test for nonadditivity was also observed (see Table 28.4, Figures 28.3 and 28.4).

TABLE 28.4. Data on the number of infested tillers with stem borer in different treatments (five levels of nitrogen application) expressed in percentage and the analysis of variance.

Treatment	Replications						Std
N (kg·/ha⁻¹)	I	II	III	IV	V	Mean	dev.
0	14.41	9.72	2.91	12.71	16.36	11.2	5.2
60	25.21	22.65	11.18	9.16	25.00	18.6	7.8
120	30.55	36.41	28.57	31.11	27.46	30.8	3.5
180	72.13	33.12	29.41	29.33	27.02	38.2	19.1
240	45.40	42.26	48.29	29.24	34.42	39.9	17.9
300	29.68	50.00	58.95	39.51	20.00	39.6	31.0
Total	217.38	194.16	179.31	151.06	150.26	29.72	

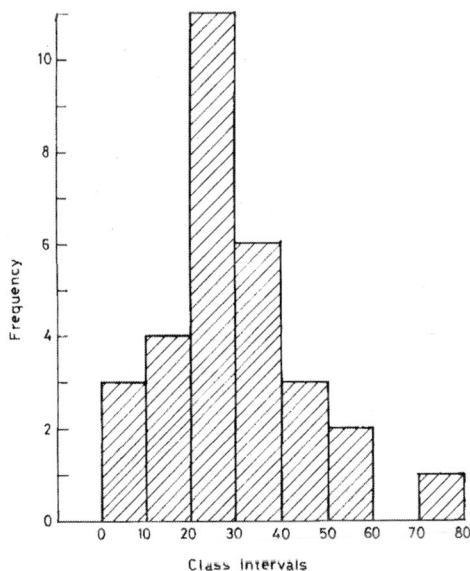

FIGURE 28.3. Histogram for the data expressed in percent.

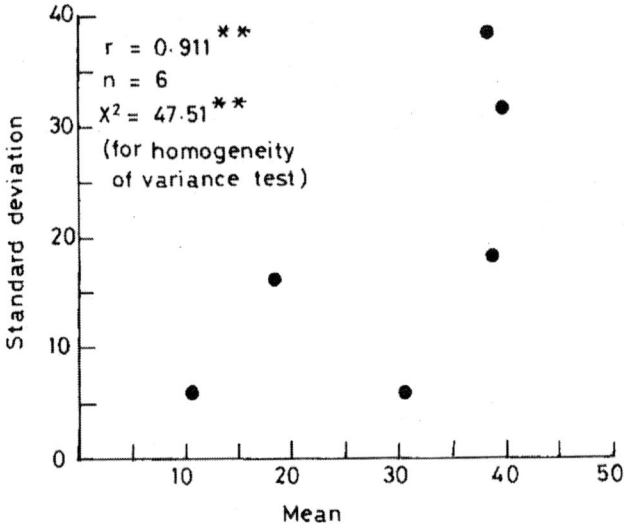

FIGURE 28.4. Relation between mean and standard deviation (original data in percentage).

Correlation coefficient between mean and standard deviation = 0.5722
χ^2 value for homogeneity of variances = 47.51** (see Table 28.5).

Tukey's test for nonadditivity is not significant.
Analysis of Arcsin Data. In this transformation, the test for correlation between mean and standard deviation, and Tukey's test for nonadditivity were not significant, but Bartlett's test for homogeneity of variances was reported. Thus, this transformation is not advisable (see Figures 28.5 and 28.6, and Table 28.6).

Correlation coefficient between mean and standard deviation = 0.4084[ns].
χ^2 value for homogeneity of variances 19.2** (see Table 28.7).

Tukey's test for nonadditivity is not significant.
Analysis of $\sqrt{x}$ Transformed Data. In this transformation, nonsignificance results for (1) homogeneity of variances test, (2) correlation between mean and standard deviation, and (3) Tukey's test for nonadditivity (see Figures 28.7 and 28.8, and Table 28.8).

TABLE 28.5. ANOVA results.

Source of variation	df	SS	MS	F	Table F 5 percent	1 percent
Replication	4	554.72	138.68	1.10[ns]	2.87	4.43
Treatments	5	3717.2	743.44	5.90**	2.71	4.10
Error	20	2520.3	126.02			
Total	29	6792.2				
Nonadditivity	1	221.6	221.6	1.82[ns]	4.38	8.10
Residual	19	2298.7	120.98			

[ns] = not significant; ** significant at $P = 0.01$; $SEm = 5.0$; $CV = 37.8$ percent.

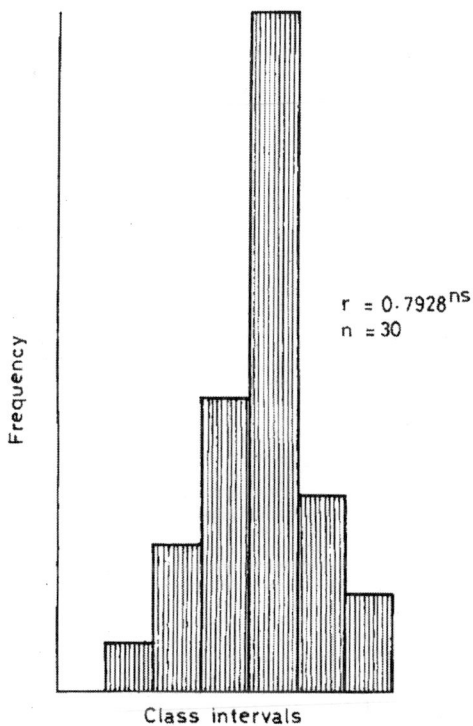

FIGURE 28.5. Histogram for the arcsin percentage transformed data.

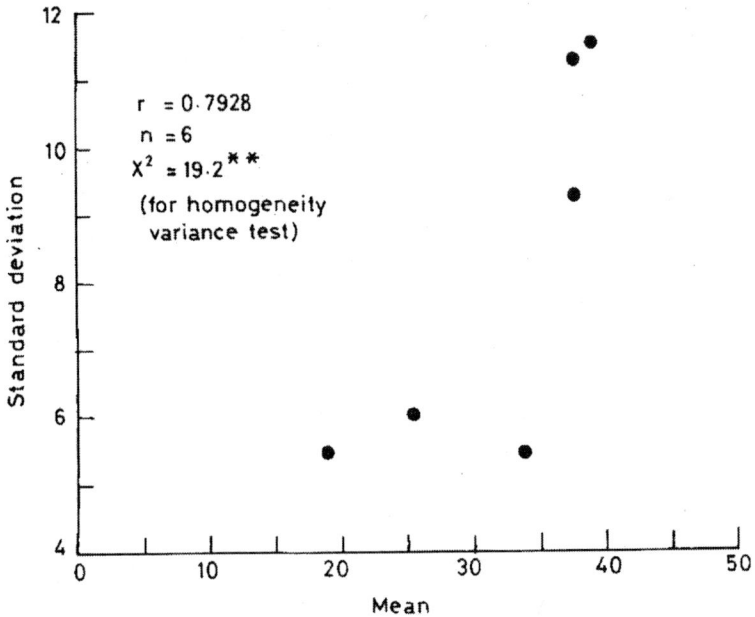

FIGURE 28.6. Relation between mean and standard deviation (arcsin percentage transformed data).

TABLE 28.6. Data on the number of infested tillers with stem borer in different treatments (five levels of nitrogen application) expressed in sin $\sqrt{}$ percentage.

Treatment N (kg·/ha^{-1})	Replications					Mean	Std dev.
	I	II	III	IV	V		
0	22.30	18.15	9.81	20.88	23.89	19.0	5.5
60	30.13	28.45	19.55	17.66	30.00	25.2	6.0
120	33.58	38.88	32.33	33.89	31.63	34.1	2.8
180	58.12	35.24	32.83	32.77	31.31	38.1	11.3
240	42.36	40.57	44.03	29.87	35.91	38.6	5.7
300	33.07	45.00	50.18	38.94	26.56	38.8	9.4
Total	219.56	206.29	188.73	174.01	179.30	32.3	

TABLE 28.7. Analysis of variance and Tukey's test for nonadditivity for the number of tillers infested with stem borer in rice for arcsin $\sqrt{}$percentage transformed data.

Source of variation	df	SS	MS	F	Table F 5 percent	Table F 1 percent
Replication	4	241.20	60.3	1.14ns	2.87	4.43
Treatments	5	1722.50	344.5	6.53**	2.71	4.10
Error	20	1054.20	52.7			
Total	29	3017.9				
Nonadditivity	1	35.9	35.9	0.66ns	4.38	8.10
Residual	19	1018.30	53.59			

ns = not significant; ** significant at $P = .01$; $SEm = 3.3$; $CV = 22.5$ percent.

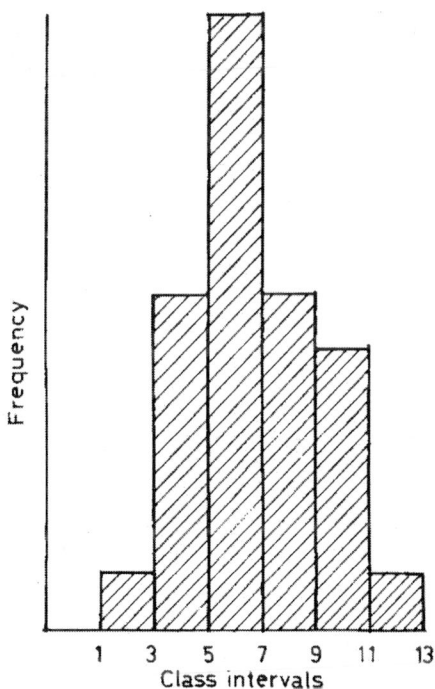

FIGURE 28.7. Histogram for the $\sqrt{X}$ transformed data

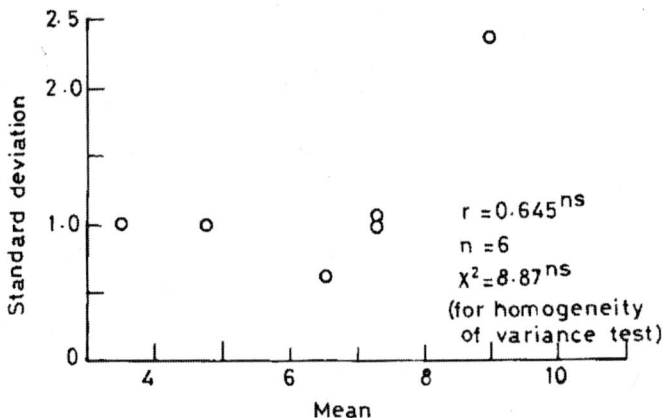

FIGURE 28.8. Relation between mean and standard deviation ($\sqrt{X}$ transformed data).

TABLE 28.8. Data on the number of infested tillers in different treatments (five levels of nitrogen application) expressed as $\sqrt{x}$ transformation and the analysis of variance.

Treatment N (kg·/ha^{-1})	Replications						Std
	I	II	III	IV	V	Mean	dev.
0	4.1231	3.7417	1.7321	3.8730	4.2426	3.5425	1.0309
60	5.4772	5.3852	4.0000	3.4641	5.8310	4.8315	1.0348
120	6.6332	7.6811	6.1644	6.4807	6.2450	6.6409	0.6106
180	9.3808	7.2111	7.0711	6.6332	6.3245	7.3241	1.2026
240	9.1652	8.4261	9.4868	6.3245	7.9373	8.2680	1.2450
300	8.7178	9.8489	11.6189	9.8995	5.2915	9.0753	2.3554
Mean	7.2496	7.0490	6.6789	6.1123	5.9786	6.6137	

Correlation coefficient between mean and standard deviation = 0.645[ns]. χ^2 value for testing homogeneity of variances = 8.87[ns].

Among the transformations, the square root transformation appears to be suitable as all four assumptions are satisfied. From the analysis using the transformation, it is suggested that $\sqrt{x}$ transformation as the most appropriate one (see Table 28.9). Furthermore, the coefficient of variation is the lowest in the transformed data, which is an indication in the improvement in the precision of the experiment. Bliss and Owen (1958) studied transforma-

TABLE 28.9. Analysis of variance and Tukey's test for nonadditivity for the data expressed as $\sqrt{x}$ transformation.

Source of variation	df	SS	MS	F	Table F 5 percent	1 percent
Replication	4	7.5152	1.8788	1.0ns	2.87	4.43
Treatments	5	109.5331	21.9066	11.93**	2.71	4.10
Error	20	36.7094	1.8358			
Total	29	153.757				
Nonadditivity	1	3.6482	3.6482	2.10ns	4.38	8.10
Residual	19	33.0612	1.7401			

ns = not significant; ** significant at $P = .01$; $SEm = 0.606$; $CV = 20.6$ percent.

tion of various sets of insect count data to bring about additivity. The readers may refer to LeClerg et al. (1962) for more information about the transformation of experimental data with suitable examples.

EXERCISES

28.1. State the four important assumptions underlying the analysis of variance.

28.2. State two reasons for recommending transformation of experimental data.

28.3. What is transformation?

28.4. State the most important types of transformations.

28.5. In an experiment involving five treatments, A, B, C, D, and E, were evaluated. Weed counts were made at the time of the harvest of the crop. Apply square root transformation and analyze the data and discuss. The data are given here (Design: RCB, replications = 5):

	Treatments A	B	C	D	E	Transformed values A	B	C	D	E
	28	7	6	177	184	5.3	2.6	2.4	13.3	13.6
	22	11	9	151	146	4.7	3.3	3.0	12.3	12.1
	54	30	26	110	131	7.3	3.5	5.1	10.5	11.4
	19	6	7	117	110	4.4	2.4	2.6	10.8	10.5
	32	11	7	135	134	5.7	3.3	2.6	11.6	11.6
Mean	31	13	11	138	141	5.48	3.42	3.14	11.7	11.8
Variance	152.8	76.4	57.2	584.8	596.8	1.03	1.21	1.00	1.04	1.04

28.6. The following data were collected from an experiment conducted in randomized complete block design in potato crop.

Replications	Treatments				
	A	B	C	D	E
I	73	46	36	40	14
II	61	50	32	22	23
III	69	44	46	44	31
IV	60	40	47	34	28

Perform analysis of variance on the original data and on the transformed data using inverse sin transformation to determine whether the treatments differ significantly.

28.7. What two suggestions will you make, if the data are not following the assumptions underlying the analysis of variance, to analyze the data to obtain maximum information?

REFERENCES

Anderson, R.L and T.A. Bancroft (1952). *Statistical theory in research.* New York: McGraw-Hill.

Anscombe, F.J. (1948). The transformation of Poisson, binomial and negative binomial data. *Biometrika* 35:246-254.

Bartlett, M.S. (1936). Square root transformation in analysis of variance. *J Royal Stat. Soc. Supplement* 3:68-78.

Bartlett, M.S. (1947). The use of transformation. *Biometrics* 3:39-52.

Bliss, C.I. and A.R.G. Owen (1958). Negative binomial distributions with a common *k*. *Biometricka.* 45:37-58.

Box, G.E.P. and D.R. Cox (1964). An analysis of transformations. *J Royal Stat. Soc.* 6:211-214.

Cochran, W.G. (1947). Some consequences when the assumptions for the analysis of variance are not satisfied. *Biometrics* 3:22-38.

Dolby, J.L. (1963). A quick method for choosing a transformation. *Tecnometrics* 5: 317-326.

Draper, N.R. and W.G. Hunter. (1969). Transformations: Some examples revisited. *Technometrics,* 11:23-40.

Fisher, R.A. (1947). *Design of experiments,* Fourth edition. Edinburgh and London: Oliver and Boyd Ltd.

Fisher, R.A. and F.T. Yates (1949). *Statistical tables,* Third edition. Edinburgh and London: Oliver and Boyd.

Freeman, M.F. and J.W. Tukey (1950). Tansformation relates to the angular and the square root. *Annals of Mathematical Statistics* 21:607-611.

Goulden, C.H. (1952). *Methods of statistical analysis,* Second edition. New York: John Wiley and Sons Inc.

Hopkins, K.D and G.V. Glass (1978). *Basic statistics for behavioral sciences.* Englewood Cliffs, NJ: Prentice Hall, Inc.

Kendall, M.G. and A. Stuart (1966). *The advanced theory of statistics,* Volume 3. New York: Hafner Publishing Co., Inc.

LeClerg, E.L., W.H. Leonard, and A.G. Clark (1962). *Field plot technique,* Second edition. Minneapolis, MN: Burgess Publ. Co.

Ostle, B. (1988). *Statistics in research—Basic concepts and techniques for research workers.* Ames, IA: The Iowa State University Press.

Palaniswamy, K.M. (1975). A note on the use of data transformation in the analysis of variance as applicable to some rice experimental data, *Oryza* 12(2):105-106.

Snedecor, G.W. and W.G. Cochran (1967). *Statistical methods,* Sixth edition. Ames, IA: Iowa State University Press.

Steel, R.G.D. and J.H. Torrie (1960). *Principles and proceedures of statistics.* New York: McGraw-Hill.

Tharp, W.H., C.H. Wadleigh, and H.D. Barker (1941). Some problems in handling and interpreting plant disease data in complex factorial designs. *Phytopath* 31:26-48.

Winer, B.J. (1971). *Statistical principles in experimental design,* Second edition. New York: McGraw-Hill.

Yates, F. (1933). The formation of latin squares for use in field E experiments. *Empire J Exptl. Agr* 1:238-244.

Zar, J.H. (1974). *Biostatistical analysis.* Englewood, NJ: Prentice Hall, Inc.

Chapter 29

Quality Control

Agriculture is the mainstay of many countries and hence it must be developed. To develop agriculture, improved agricultural implements must be put into use for various operations such as plowing, sowing, planting, intercultivation, weeding, spraying, irrigation, fertilizer application, harvesting, cleaning, packaging, etc. Industry plays a significant role in the production of implements on a large scale to meet the demands. Almost all industries such as electronics, textiles, chemicals, fertilizers, food processing, automobiles, laboratory equipment, medical instruments, fruit-related beverages, soft drinks, cereals, pesticides, and pharmaceuticals produce products continuously on a mass scale employing machines, labor, and materials.

Johnson and Tetley (1962) studied the use of quality control in the estimation of the percentage of available nitrogen in the fertilizer packed in sacks by taking six small samples from every third sack in a series of sacks ordered according to time of packing. They studied the fluctuations of this percentage comparing the mean nitrogen content of 6 percent (μ). They considered the ordering factor, not the time, the number allotted to the sacks. Lewis (1971) emphasizes the use of a quality control chart in the study of the estimation of nitrogen fixation in plants or the rate of protein changes in the serum of animals by taking known concentrations of ammonium sulfate as standard or true mean μ and the concentration of each unknown as measured at each time as x values. Thus, quality control methods are useful not only in industry but also in laboratories.

If industries are to thrive, they must produce quality products, and avoid production of items not satisfying the prescribed specifications. For example, if the specification of a product is prescribed as 0.995 ± 0.005 inches, then all the items that lie in between the measurements, namely 0.990 and 1.000, are said to be satisfying the prescribed specifications. It is a well-known fact that no machine or machines can produce two items exactly alike. Also, one cannot construct two machines that can produce exactly identical units of output. Grant (1964) states that measurements made on products are always subject to a certain amount of variation as a result of chance. Certain specifications are prescribed prior to the start of production processes. The quality of the product is judged or evaluated by comparing the measurements to those of the specifications. Thousands of units are produced continuously and they must be checked for quality. The main objective of any production process is not only maintaining quality control but also maintaining the quality standards.

QUALITY CONTROL AND STATISTICAL QUALITY CONTROL

Controlling the quality of production according to specifications is referred to as quality control. In quality control, the production trend is studied with sufficient accuracy and if any deviation from the specifications is noticed, necessary action must be taken to prevent future occurrences (Goulden, 1956). The use of statistical methods in industries and scientific laboratories to check on quality in successive batches of materials, and components in the manufacturing process is called statistical quality control (SQC) (Johnson and Tetley, 1962). When known causes are not present, the process is said to be under stable condition or statistical control.

CAUSES OF VARIATION

Because manufactured products show variation, one has to identify the sources of variation in order to eliminate or minimize them. There are two types of variation, one relates to assignable (known) causes and the other chance (unknown) causes.

Chance Causes

Measurements are recorded in the characteristics of interest. Measurements exhibit variation and follow normal distribution. Based on the assumption of normal distribution, variability is investigated to find out the factors for out-of-control conditions and to take action to rectify them.

Variation is studied graphically. In 1924, Dr. W. A. Shewart (1931) found a solution to detect and maintain quality using an application of statistical tools. He devised a quality control chart method to watch and maintain quality. We can ascertain the quality of any characteristic one at a time using the control chart method.

Control Chart

During the process of manufacturing, samples are taken and measurements on the desired characteristics are made. These individual measurements follow normal distribution. About 99.73 percent of the observations from a normal distribution will include $\pm 3\sigma$ limits and 0.27 percent outside the limits. Since we cannot judge the quality from a single observation, we resort to samples. We know that if the standard deviation of the individual observations in the population is σ, then the estimated standard deviation of the means of samples is $\sigma/\sqrt{n}$. Even if the individual observations fail to follow normal distribution, the means of samples follow normal distribution. In quality control, we take samples and use sample means and ranges. Shewart (1931) suggested the sample size of four or five observations in each sample, and about 25 to 30 samples as sufficient in quality control programs. The sampling distribution of sample means is studied and its variability assessed and reported as standard deviation of the means (or standard error of the mean) in quality control. The $\bar{x} \pm 3\sigma/\sqrt{n}$ limits are used as good criteria for determining the nonexistence of known causes. The mean, standard deviation, and ranges of successive samples vary, even if the process is in control due, to sampling variation. The extent of sampling variation must be known in order to know the lack of control in the process. The control chart based on variability is more sensitive than the R chart. The s-chart, where s is the standard deviation from the sample is not preferred as it is more tedious to calculate (Johnson and Leone, 1964).

STABLE AND UNSTABLE CONDITIONS

In a process, if the mean and standard deviation remain constant, we say the process is in stable condition. On the other hand, if the mean and standard deviation change, we consider the process unstable or out of control. The process of manufacturing is said to be stable or in control if variations between individuals are caused by chance only (Kennedy and Neville, 1964).

Suppose we take 25 samples each size equals 5. Here we assume the observations in a sample as a random selection. Calculate mean and range for each sample. We do not know the population standard deviation, σ'. Hence we estimate σ' by a sample and denote it by s. We estimate the standard error of the mean using the relationship, $SEm = \sigma/\sqrt{n}$.

Along the x-axis, take sample (subgroup) number and along the y-axis the characters measured. The mean and range are plotted in a graph for each sample number. The mean of the ith sample $\bar{x}_i = 1/n\sum x_i$. The means ($\bar{x}_I$'s) are normally independently distributed with mean μ and standard deviation σ/$\sqrt{n}$. Calculate the mean of means

$$\overline{\overline{X}} = \Sigma\overline{x}_1 / n.$$

At $\overline{\overline{X}}$, the center line (CL) (solid line) is drawn parallel to the x-axis horizontally. Then calculate $\overline{\overline{X}} \pm 3$ s/$\sqrt{n}$. The upper control (UCL) is given by $\overline{\overline{X}} + 3$ s/$\sqrt{n}$ and the lower control limit (LCL) by $\overline{\overline{X}} - 3$ s/$\sqrt{n}$. These two lines are drawn above and below the center line. Plot the means of samples in the graph (see Figure 29.1). If a point ($\overline{x}_i$) falls within the control limits, the variation is attributed to chance and conclude that the process is in control. When a point ($\overline{x}_i$) falls outside the control limits, it indicates something is going wrong which must be identified and corrected.

A plot of the sample means with central line and control limits is referred to as $\overline{x}$ chart in quality control. If all the points lie within the control limits, it indicates that the variation is attributed to chance causes and no action is warranted. In such a situation, one can achieve quality products. When the sample is quite small, the range can be an adequate measure of variation. It is used widely in industrial quality control to keep a close check on the consistency of raw materials or products or on the uniformity of the process (Freund, 1979). There are two types of control charts, one deals with measurement data (variable) and the other with count data (attribute). It is described as follows:

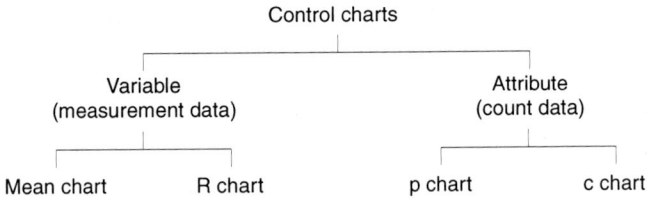

Control charts

Variable (measurement data): Mean chart, R chart

Attribute (count data): p chart, c chart

Examples for measurement data are length, width, height, diameter, tensile strength, moisture content, percentage impurities in chemical substances, breaking strength, filled weight of cans, and thickness of cloth. Examples for count data include number of defective articles in a safety box or screw box. In general, the control charts are the screening mechanisms for detecting assignable causes of variation in the process. Each type of control chart is described briefly with suitable examples in this chapter. It is not the intention of this chapter to give full perspective of the quality control topic. Hence the readers are advised to refer to Burr (1953), Schrock (1950), Volk

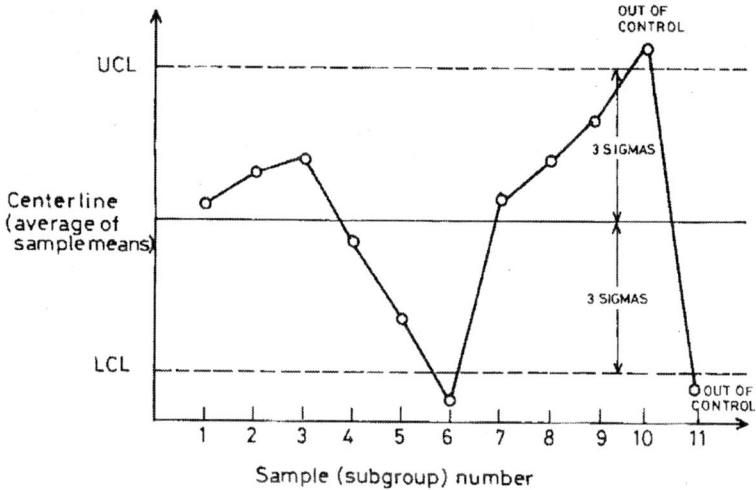

FIGURE 29.1. Outline of a control chart with points within and outside control limits.

(1969), Kennedy and Neville (1964), and Grant and Leavenworth (1985) for detailed information.

MEAN CHART

Mean chart is a valuable and useful statistical tool. Since Shewart (1931) proposed this chart, it is called Shewart control chart for mean. By using this chart during the industrial process, one can find out whether the process is in control or out of control (see Figure 29.2).

Procedure

1. Collect 20 to 25 samples of size 4 or 5 each (Shewart, 1931) from the process over a period of time and record the measurements in each sampling unit in the samples. The samples are also called subgroups. They are numbered as 1, 2, 3 . . . *k*. It is assumed that the variability within subgroup is attributed to chance and one can expect homogeneity among the observations within a subgroup. They are considered as random observations (Tippett, 1950).
2. Compute the means. We get $x_1, \bar{x}_2, \ldots \bar{x}_k$.
3. Calculate the mean of means $(x)[1/n(\Sigma \bar{x}_1)]$.

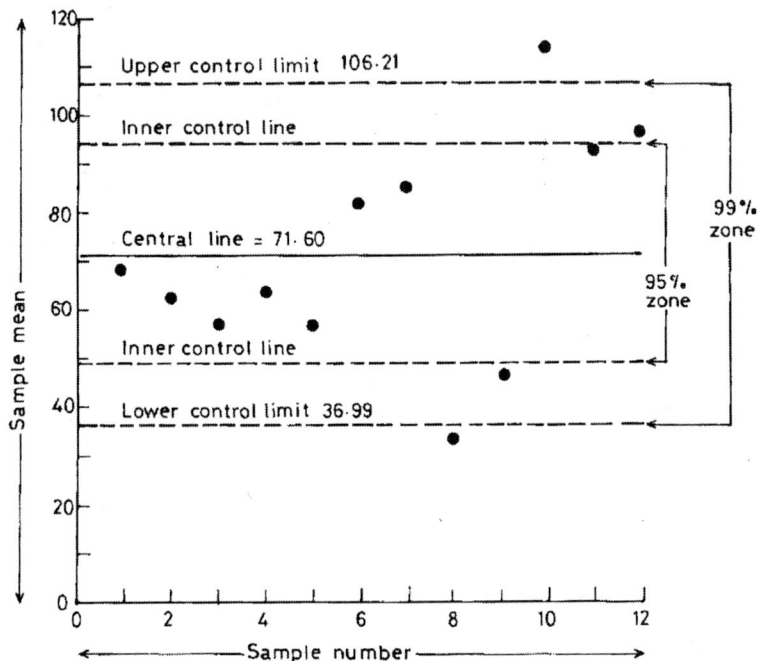

FIGURE 29.2. The $\overline{X}$ control chart with inner control lines and upper control limits.

4. Take sample numbers along the x-axis and the measurements in the y-axis.
5. Plot the sample means in the graph against the sample numbers.
6. Draw the center line (CL) at $\overline{x}$ horizontally parallel to the x-axis considered as estimate of population, mean μ.
7. Calculate standard deviation for the means as $s / \sqrt{n_1}$, where $s =$ estimate of the population standard deviation, σ (method of calculation of s is given as follows).
8. Calculate upper control limit (UCL) $= \overline{\overline{X}} + 3s / \sqrt{n}$ and lower control limit (LCL) $= \overline{\overline{X}} - 3s\sqrt{n}$.
9. Draw UCL and LCL above and below center line. These two lines are shown in dotted lines. Note: In the control chart, the sampling numbers, or time, is taken in x-axis and $\overline{X}_i$ values as ordinates.
10. Study the distribution of $\overline{x}_{is}$. If the points lie within UCL and LCL, we infer that the process is supposed to be in control. If the points lie

outside the control limits, we infer that the process is out of control, and one has to stop the process and look for the reasons for the trouble (see Figure 29.2). Two lines if drawn at a distance of $\sqrt{n} \pm 1.96$ $s/\bar{x}$ are called warning limits (Kennedy and Neville, 1964).

Method of Computation of Standard Deviation from the Range

Ordinarily, the standard deviation is calculated using the formula

$$\sqrt{\{\Sigma x_i^2 - (\Sigma x)^2 / n\} / (n-1)}$$

using sum of squares method. Here, it is estimated using coefficient d. The sampling distribution of the range is difficult to devise mathematically but data sufficient for the purpose of quality control have been tabulated. The population standard deviation, σ, can be estimated from the mean range, $\bar{R}$ of a number of samples by using the relationship $\sigma = \bar{R} \times d$, where the value of d can be obtained from Table 29.1.

Example: The data obtained in a quality control program are given in Table 29.2.

X (mean of the sample means) = $1/5(39.5+ \ldots +34.00) = 166.72/5 =$ 33.25

TABLE 29.1. Values of d for different sample sizes for conversion of range to standard deviation.

n	d
2	0.8862
3	0.5908
4	0.4857
5	0.4299
6	0.3946
7	0.3698
8	0.3512
9	0.3367
10	0.3249
11	0.3152
12	0.3069
13	0.2998

Source: Adapted from Lindley and Miller, *Cambridge elementary statistical tables,* page 7.

TABLE 29.2. Data collected in a quality control program (quality characteristic is in dimension).

Sample no.	1	2	3	4	5
1	47	33	34	12	35
2	32	33	34	21	23
3	44	34	31	24	38
4	35	34	34	47	40
Total	158	134	133	104	136
Range R	15	1	3	35	17
Mean $\bar{x}$	39.5	33.5	33.25	26	34

Mean range $= \bar{R} = 1/5(15+ \ldots +17) = 1/5(71) = 14.20$
From the table for $n = 4$, the value of d is 0.4857.

Method 1. Estimated standard deviation =

$$\bar{R}d = 14.20 \times 0.4857 = 6.896.$$

Method 2.

$$d_2 = \bar{R}/s, \text{ i.e., } s = \bar{R}/d_2 = 14.20/2.059 = 6.896.$$

d_2 is obtained from the table of factor for control charts (Gupta, 1983). Example: For construction of $\bar{X}$ and R charts (see Table 29.3). Method 1. Calculation: $n = 5$

Overall mean $= \bar{\bar{X}} = 1/12 \Sigma \bar{X}_i = 859.2/12 = 71.60$
$\bar{R} = 1/12 \Sigma R_i = 715/12 = 59.58$
Factor for control limits for $\bar{x}$ when $n = 5$
$A_2 = 0.58$ (from the table of factor for control charts) (Gupta, 1983).
Formula: UCL $\bar{x} = \bar{\bar{X}} + A_2\bar{R} = 71.60 + 0.58 \times 59.58 = 106.16$
LCL $\bar{x} = \bar{\bar{X}} - A_2\bar{R} = 71.60 - 0.58 \times 59.58 = 37.04$
Center line (CL) for $\bar{\bar{X}} = 71.60$.

In the example given, the mean and standard deviation were based on 12 samples. In Figure 29.2, the $\bar{x}$ chart is explained. It shows that the process is out of control since the points corresponding to the eighth and tenth samples lie outside the control limits (see Figure 29.2).

TABLE 29.3. Data obtained in a factory on fuse production (sample size = 5; number of samples = 12).

Sample number	Sample observations					Total	Sample mean	Sample range
1	42	65	75	78	87	347	69.4	45
2	42	45	68	72	90	317	63.4	48
3	19	24	80	81	81	285	57.0	62
4	36	54	69	77	84	320	64.0	48
5	42	51	57	59	78	287	57.4	36
6	51	74	78	74	132	410	82.0	80
7	60	60	72	95	138	425	85.0	78
8	18	20	27	42	60	167	33.4	42
9	15	30	39	62	84	230	46.0	69
10	69	109	113	118	153	562	112.4	84
11	64	90	93	109	112	468	93.6	48
12	61	78	94	109	136	478	95.6	75
					Total	4296	71.6	59.5

Note: One can also plot the individual observations in a graph and find the distribution as normal, and compare the scatter among the individual observations and to those of the means. You will find less scatter amongst the mean values. The means give normal distribution with the same mean and standard error of the mean $\sigma/\sqrt{n}$).

Method 2. Standard deviation =

$$\overline{R} \times d = 59.58 \times 0.4299 = 25.61$$

We can now compute the control limits:

$$71.60 \pm 3\sigma / \sqrt{n} = 71.60 \pm 3(25.61 / \sqrt{5})$$
$$= 71.60 \pm 34 = 105.96 - 37.24$$

We can also calculate the warning limits at 95 percent as follows:

$$95 \text{ percent control limits} = 71.60 \pm 1.96 \, \sigma / \sqrt{n}$$
$$= 71.60 \pm 1.96(25.61) / \sqrt{5} = 71.60 \pm 22.45 = 94.05 - 49.15$$

QUALITY CONTROL CHART FOR RANGE (R CHART)

In addition to mean chart, one has to investigate the amount of variability in the process also. For this, we consider ranges, generally the R chart is studied along with the $\bar{x}$ chart, i.e., one accompanies the other. The range does not vary unless some serious fault develops. Hence mean and range charts are studied together. The range values of each subgroup are plotted against the sample numbers as dots. The range is computed as $X_{max} - X_{min}$. The average range $(\bar{R})$ is computed as $\Sigma R_i/n$ (sum of the ranges of the subgroups divided by the number of subgroups).

The $\bar{R}$ is represented as the central line with 3 σ limits. The calculations of 3 σ limits are as follows:

$$UCL_R = D_4\bar{R} = \bar{R} + 3\sigma$$
$$LCL_R = D_3\bar{R} = \bar{R} - 3\sigma$$

The values of D_3 and D_4 are taken from the table of control charts for constants (Gupta, 1983).

R Chart Calculations

Example. Data given in Table 29.3 are used for calculating R chart also. In the y-axis, the statistical measures are plotted and in the x-axis the subgroup numbers are plotted. If our data relate to dates, time, or lot numbers, they may be plotted in the horizontal scale (x-axis). The points may or may not be connected. The central line of the $\bar{R}$ should be drawn as a solid horizontal line. The upper control limits should be drawn as dotted lines. If the subgroup size is 6 or less, the lower control limit for R is zero. Calculations:

Upper control limit = $D_4 \times \bar{R} = 2.11 \times 59.58 = 125.71$

See control chart (Gupta, 1983) for D_4 value = 2.11, for subgroup = 5.

Central line = 59.58
Lower control limit = 0
The details are shown in Figure 29.3.

Since all the sample points fall within control limits, this chart shows that the process is in control. However, because $\bar{x}$ chart shows lack of control, the process is considered as lack of control. First, the variability is to be controlled and the quality variation is to be brought under control.

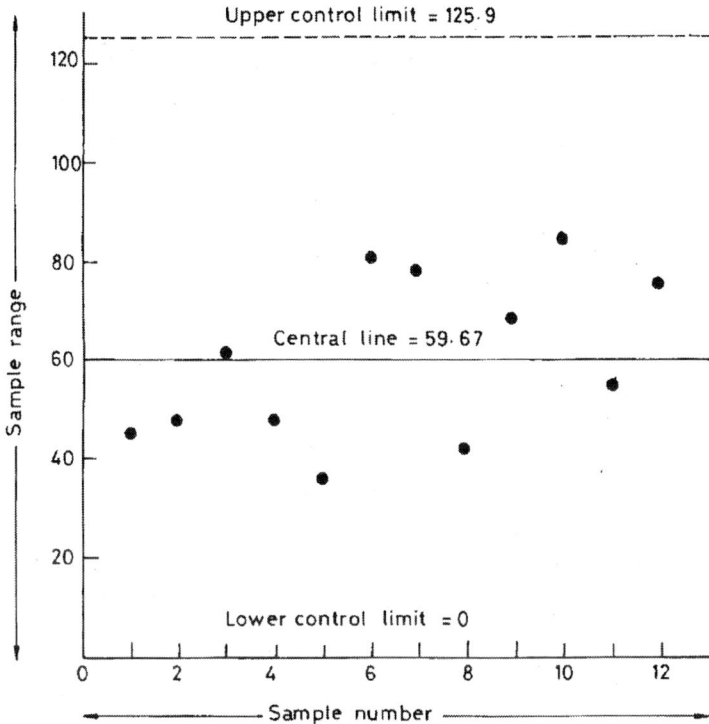

FIGURE 29.3. The range chart (*R* chart) for the data given in Table 29.3.

P CHART

Sometimes, the articles manufactured are not classified by measurements but considered as good or bad, conforming or not conforming, acceptable or not acceptable. If the articles comply with the specifications, they are considered as sound (not defective) and if they do not comply with the specifications, they are considered as defective or not acceptable. The process is said to be in statistical control if all the samples or subgroups have the same population proportion, P. The quality is judged by the sample proportion, p. The fraction rejected is defined as the ratio of the number of nonconforming articles in any inspection or series of inspections to the total number of articles actually inspected. If d is the number of defectives in a sample of size n, then the sample proportion defective $p = d/n$. d is a binomial variable with parameters n and p.

Then $E(d) = nP$ and the Var $(d) = nPQ$ and $Q = 1 - P$.
The $E(p) = E(d/n) = 1/n\ E(d) = P$.
Var $(p) = \text{Var}(d/n) = 1/n^2\ nPQ = PQ/n$.

The quality is the proportion, p. A graphical representation of the p chart is shown in Figure 29.4. It contains a center line, which is drawn at the average of the sample proportion. The proportions are plotted in the graph. The Shewart control chart for fraction rejected is described as p chart (Grant and Leavenworth, 1985). The procedure for the calculation of p value and the control limits are as follows:

1. The quality character to be investigated is to be decided.
2. The sample size is to be determined.
3. The number of samples to be taken is to be fixed. About 20 to 30 samples have to be collected.
4. For each sample, compute the defective proportion (p) which is given as $p = d/n$, where d is the number of defective items and n = number of items inspected.

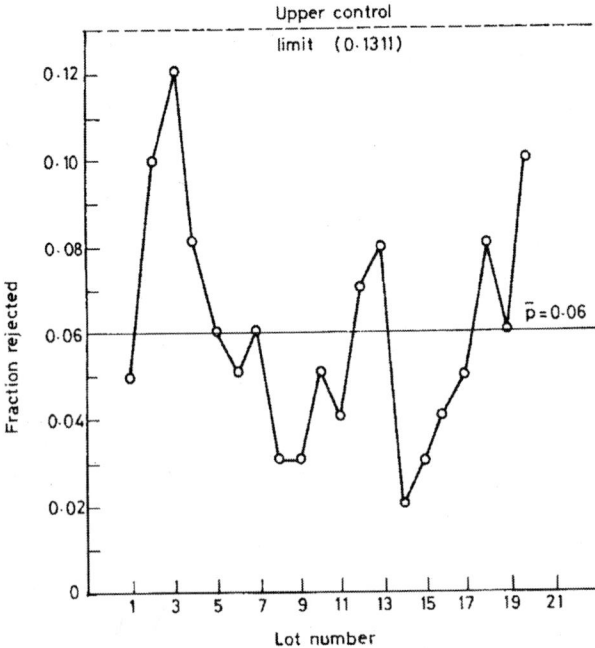

FIGURE 29.4. Control chart (p chart) for fraction rejected for the spark plug data given in Table 29.4.

5. Compute the average proportion p ($\bar{p}$) as $\Sigma p/k$, where p = sample proportion and k = number of subgroups or samples. Calculate the center line and the control limits as follows:

Central line is drawn at p.
Upper control line = UCL = $\bar{p} + 3\sqrt{pq/n}$.
Lower control line = LCL = $\bar{p} - 3\sqrt{pq/n}$.
Draw the UCL and LCL in the p chart.

When the observed defective fraction exceeds these lines, it is suspected that the manufacturing process has broken down to some degree and the p began to increase. If all the points fall within the control limits, we infer that the process is in control. The defective proportion may be expressed in percent, and p chart drawn for easy understanding. The p chart is explained with a numerical example.

Example: In a spark plug factory, 20 lots, each lot containing 100 numbers, were inspected, and the number of rejected items are given in Table 29.4.

Total 120: $\bar{p} = (120/20)/100 = 0.06$

Calculations

$\bar{p} = (120)/(20 \times 100) = 0.06$
$UCLp = \bar{p} + 3\sqrt{pq/n} = 0.06 + 3\sqrt{(0.06 \times 0.94)/100}$

TABLE 29.4. Data collected on the number of defectives from a factory producing spark plugs.

Sample number	Number rejected	Fraction rejected	Sample number	Number rejected	Fraction rejected
1	5	0.050	11	4	0.040
2	10	0.100	12	7	0.070
3	12	0.120	13	8	0.080
4	8	0.080	14	2	0.020
5	6	0.060	15	3	0.030
6	5	0.050	16	4	0.040
7	6	0.060	17	5	0.050
8	3	0.030	18	8	0.080
9	3	0.030	19	6	0.060
10	5	0.050	20	10	0.100

$$= 0.06 + 3 \ (0.0237) = 0.1311$$
$$\text{LCLp} = \bar{p} - 3\sqrt{pq \ / \ n} = 0.06 - 3 \times 0.0237 = -0.011$$

Since the fraction rejected cannot be negative, the LCLp is taken as zero (see Figure 29.4). The control chart shows that all the points are within the control limits and hence variation is random; the process is in control. Instead of p chart, the percentage p chart can also be plotted for easy understanding.

C CHART

The Shewart control chart for the number of defects (nonconformities) per unit is defined as c chart (Grant and Leavenworth, 1985). The word defective refers to articles that do not conform to one or more of the specifications. A defective article may contain one or more defects. The c chart applies to the number of nonconformities. The probability of occurrence or nonconformities in each subgroup is the same. The distribution is assumed to follow Poisson distribution (Johnson and Leone, 1964), and on this assumption the control limits are calculated. Examples in which c chart are applicable include: (1) number of typographical errors on the printed page, (2) number of rust spots on steel sheets. The control limits are calculated as follows:

$$\text{UCL} = \bar{c} + 3\sqrt{\bar{c}}$$
$$\text{LAC} = \bar{c} - 3\sqrt{\bar{c}}$$

Example (see Figure 29.5): From a bundle of cloth, 20 samples of cloth each having equal length and width were examined before it was released to the public. The number of nonconformities observed per sample are given in Table 29.5.
Now we have to draw a control chart for the c values, and find out whether the process is in control or out of control.

Calculations

1. Find $\bar{c} = \Sigma c \ / \ n = (1 + 4 + \ldots + 6 + 4) \ / \ 20 = 93 \ / \ 20 = 4.65$
2. $\text{UCL} = \bar{c} + 3\sqrt{\bar{c}} = 4.65 + 3\sqrt{4.65} = 11.12$
3. $\text{LCL} = \bar{c} - 3\sqrt{\bar{c}} = 4.65 - 3\sqrt{4.65} =$ a negative value for c hence taken as zero. The c chart is shown in Figure 29.5.

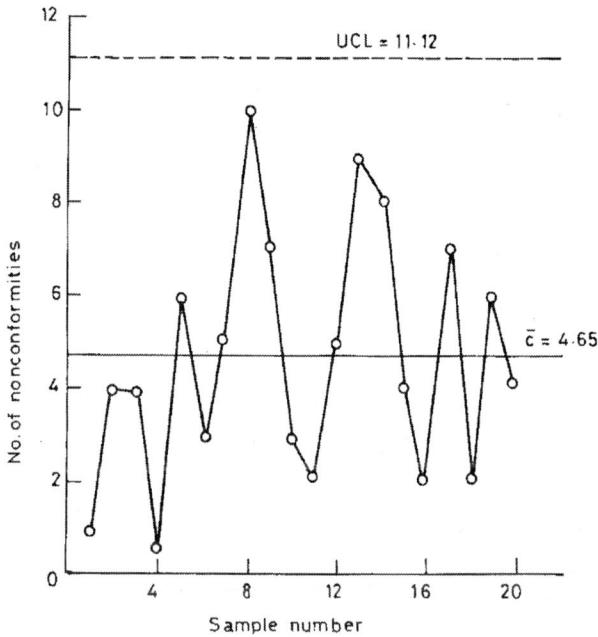

FIGURE 29.5. C chart for the data given in Table 29.5.

TABLE 29.5. Number of nonconformities recorded in 20 samples of cloth collected from a textile mill.

Sample no.	No. of errors	Sample no.	No. of errors
1	1	11	2
2	4	12	5
3	4	13	9
4	1	14	8
5	6	15	4
6	3	16	2
7	5	17	7
8	10	18	2
9	7	19	6
10	3	20	4

TABLE 29.6. Assumptions and formulas for the most commonly used control charts.

Chart	Assumed distribution	Center line	Upper control limit (UCL)	Lower control limit (LCL
$\bar{x}$	Normal	$\bar{\bar{X}}$	$\bar{\bar{X}} + A_2\bar{R}$	$\bar{\bar{X}} - A_2\bar{R}$
R	Normal	$\bar{R}$	$D_4\bar{R}$	$D_3\bar{R}$
p	Binomial	$\bar{p}$	$\bar{p} + 3\sqrt{\bar{p}(1-\bar{p})/n}$	$\bar{p} - 3\sqrt{\bar{p}(1-\bar{p})/n}$
c	Poisson	$\bar{c}$	$\bar{c} + 3\sqrt{\bar{c}}$	$\bar{c} - 3\sqrt{\bar{c}}$

Source: Ostle, 1988.

From the distribution of the points in the figure, it is evident that the process is in control (see Table 29.6).
The X and $\bar{R}$ charts deal with measurement data and P and C charts deal with attribute (enumeration) data.

Advantages of Quality Control

Quality control is used in almost all industries and it is a useful statistical tool to maintain the quality of product. If a fault is found during the manufacturing process, it can be rectified immediately and excessive defects can be prevented. It can also be used in laboratories where equipment, which are subject to wear and tear, are in use. It also helps to increase the precision of the results.

EXERCISES

29.1. Explain the terms quality control, statistical quality control, stable condition, unstable condition.
29.2. What are control limits and specification limits?
29.3. The mean of an article is 20 mm and standard deviation is 0.30 mm. Find the upper control limit and lower control limit. Given $n = 4$. Use 3 sigmas. (See Gupta, 1983, for tables.)
29.4. The following data give the values of means and ranges of sample observations.

Sample number	Mean	Range	Sample number	Mean	Range
1	4.17	0.14	11	4.35	0.62
2	4.15	0.30	12	4.54	0.23

3	4.08	0.20	13	4.28	0.38
4	4.26	0.26	14	4.50	0.22
5	4.13	0.19	15	4.07	0.09
6	4.22	0.24	16	4.33	0.17
7	4.33	0.65	17	4.61	0.20
8	4.25	0.17	18	4.32	0.20
9	4.54	0.58	19	4.57	0.40
10	4.54	0.22	20	4.31	0.28

    a. Find the estimate of σ' from the data.
    b. Construct control charts for $\overline{X}$ and R.
    c. If the specifications are 4.35 ± 0.5, what do you say about the process?

29.5. Explain warning limits in the quality control program.

29.6. The mean = 0.5230 inches and standard deviation = 0.003 in a quality control problem with $n = 5$. Find the inner and outer limits.

29.7. You have five samples, each with four observations. Compute the mean and standard deviation to the data given as follows. (Data are thousandths of an inch above 1.000.)

		Sample number		
1	*2*	*3*	*4*	*5*
10	2	9	5	6
2	7	7	11	2
7	8	3	2	1
7	5	0	8	8

29.8. In a soil testing laboratory, soil samples were collected from a place and analyzed for available nitrogen (ml) in 10 different consecutive days. There were five analyses in each subgroup. Draw control chart for mean and range, and comment on your results.

29.9. What is a p chart?

				Sample number					
1	*2*	*3*	*4*	*5*	*6*	*7*	*8*	*9*	*10*
7.6	7.7	8.1	10.6	7.5	8.7	6.0	8.0	7.9	8.0
7.8	7.9	7.7	11.0	8.5	8.6	6.0	7.9	7.6	6.0
6.2	8.6	6.5	8.8	6.3	9.1	6.6	7.8	7.0	6.8
5.8	8.4	6.7	10.8	6.1	3.0	7.2	7.7	6.1	8.4
6.2	7.5	7.9	11.8	5.7	4.5	8.2	7.4	6.9	8.4
Mean 6.72	8.02	7.38	11.2	6.82	6.78	6.80	7.76	7.10	7.52
Range 2.00	1.10	1.60	6.00	2.80	6.10	2.20	0.60	1.80	2.40

29.10. The number of defective rubber belts were noted from each of 22 lots, each lot containing 2,000 rubber belts. Data are as follows:

451	425	216	225	280	337	356	216	126	193	280
420	430	341	322	306	305	402	264	409	326	388

Draw the control chart for the fraction of defective belts and comment on the state of control of process.

29.11. The diameter of rods in a sample was observed as 5.613, 5.682, 5.636, 5.671, 5.652, 5.629, 5.677, 5.645, 65.639, and 5.668. Estimate the variance using the sum of squares method. Also estimate using range. Do you observe any difference?

29.12. In an electronics factory, samples of 100 ohm electrical resistors were tested for quality control during production. Random samples of five resistors were drawn and the resistances measured. Resistances are given above 80.

				Sample number					
1	*2*	*3*	*4*	*5*	*6*	*7*	*8*	*9*	*10*
21	32	21	33	18	13	21	27	19	16
28	32	28	25	14	21	23	17	19	25
25	22	25	20	26	21	19	19	19	24
21	20	21	23	22	26	19	20	26	23
25	23	21	23	20	26	17	20	26	23

				Sample number					
11	*12*	*13*	*14*	*15*	*16*	*17*	*18*	*19*	*20*
25	19	25	23	21	24	22	21	19	21
10	20	20	22	23	24	27	16	23	21
22	18	20	32	17	19	27	23	18	19
22	20	22	20	19	23	18	23	24	26
13	17	19	20	23	22	16	23	19	24

Set up a quality control chart for mean and range with 3σ control limits.

29.13. What is c chart? When you will use it?

29.14. From a factory producing articles, 20 articles were collected from every 100th article. The number of defectives per sample was noted. In 20 samples, the number of defectives found was as follows: 3, 2, 2, 1, 0, 3, 1, 4, 2, 0, 2, 2, 1, 1, 2, 3, 0, 2, 1, 4. Obtain the control chart for the mean number of defectives (c-chart).

29.15 Give an example of your choice for statistical quality control.

29.16. Given here are the values of sample mean $\bar{x}$ and sample range R for ten samples of size 5 each. Draw the appropriate control charts and comment on the state of control, or otherwise, of the process.

Sample no.	1	2	3	4	5	6	7	8	9	10
$\bar{x}$	43	49	37	44	45	37	51	46	43	47
R	5	6	5	7	7	4	8	6	4	6

29.17. A plant produces newsprint and rolls of paper are inspected for defects. The results of inspection on 25 rolls for the number of defects are 20, 10, 8, 12, 13, 15, 25, 7, 13, 18, 16, 14, 6, 5, 4, 2, 3, 6, 8, 9, 15, 18, 20, 10, 5. Using a control chart, determine whether roll production is under statistical control.

29.18. The following are the figures for the number of defectives of 20 lots each containing 2,300 rubber belts.

430	425	221	346	230	327	285	311	342	308
456	394	285	331	398	414	131	269	221	407

Draw the control chart for the defective fraction. Plot the points on it. Comment on the state of the process.

29.19. The mean strength and mean range of timber of a certain variety is 7020 and 780 k N/m^{-2}. Taking sample size as five, set up a mean chart with control limits.

REFERENCES

Burr, I.W. (1953). *Statistics and quality control.* New York: McGraw-Hill.

Freund, J.E. (1979). *Modern elementary statistics,* Fifth edition. Englewood Cliffs, NJ: Prentice Hall, Inc.

Goulden, C.H. (1956). *Methods of statistical analysis.* London: John Wiley and Sons Inc.

Grant, E.L. (1964). *Statistical quality control,* Third edition. New York: McGraw-Hill Book Company.

Grant, E.L. and R.S. Leavenworth (1985). *Statistical quality control,* Fifth edition. London: McGraw-Hill Book Company.

Gupta, R.C. (1983). *Statistical quality control book,* Third edition. Delhi: Khanna Publishers.

Johnson, N.L. and F.C. Leone (1964). *Statistics and experimental design in engineering and the physical sciences,* Volume 1. New York: John Wiley and Sons, Inc.

Johnson, N.L. and H. Tetley (1962). *Statistics—An intermediate textbook.* Cambridge. London: Institute of Actuaries at the University Press.

Kennedy, J.B. and A.M. Neville (1964). *Basic statistical methods for engineers and scientists,* Second edition. Harper International Edition. New York: Thomas Y. Crowell Company, Harper and Row.

Lewis, A.E. (1971). *Biostatistics.* New Delhi: Affiliated East-West Press, Pvt. Ltd.

Ostle, B. (1988). *Statistics in research,* Second edition. Ames, IA: The Iowa State University Press.

Schrock, E.M. (1950). *Quality control and statistical methods.* New York: Reinhold Publishing Corporation.

Shewart, W.A. (1931). *Economic control of quality of manufactured product.* Princeton, NJ: Van Nostrand.

Tippett, L.H.C. (1950). *Technological applications of statistics.* New York: John Wiley and Sons Inc.

Volk, W. (1969). *Applied statistics for engineers,* Second edition. New York: McGraw-Hill Book Company.

APPENDIX:
STATISTICAL TABLES

TABLE A.1. Table of random numbers.

02711	08182	75997	79866	58095	83319	80295	79741	74599	84379
94873	90935	31684	63952	09865	14491	99518	93394	34691	14985
54921	78680	06635	98689	17306	25170	65928	87709	30533	89736
77640	97636	37397	93379	56454	59818	45827	74164	71666	46977
61545	00835	93251	87203	36759	49197	85967	01704	19634	21898
17147	19519	22497	16857	42426	84822	92598	49186	88247	39967
13748	04742	92460	85801	53444	65626	58710	55405	17173	69776
87455	14813	50373	28037	91182	32786	65261	11173	34376	36408
08999	57409	91185	10200	61411	23392	47797	56377	71635	08601
78804	81333	53809	32471	46034	36306	22498	19239	85428	55721
82173	26921	28472	98958	07960	66124	89731	95069	18625	92405
97594	25168	89178	68190	05043	17407	48201	83917	11413	72920
73881	67176	93504	42636	38233	16154	96451	57925	29667	30859
46071	22912	90326	42453	88108	72064	58601	32357	90610	32921
44492	19686	12495	93135	95185	77799	52441	88272	22024	80631
31864	72170	37722	55794	14636	05148	54505	50113	21119	25228
51574	90692	43339	65689	76539	27909	05467	21727	51141	72949
35350	76132	92925	92124	92634	35681	43690	89136	35599	84138
46943	36502	01172	46045	46991	33804	80006	35542	61056	75666
22665	87226	33304	57975	03985	21566	65796	72915	81466	89205
39437	97957	11838	10433	21564	51570	73558	27495	34533	57808
77082	47784	40098	97962	89845	28392	78187	06112	08169	11261
24544	25649	43370	28007	06779	72402	62632	53956	24709	06978
27503	15558	37738	24849	70722	71859	83736	06016	94397	12529
24590	24545	06435	52758	45685	90151	46516	49644	92686	84870

TABLE A.1 (continued)

48155	86226	40359	28723	15364	69125	12609	57171	86857	31702
20226	53752	90648	24362	83314	00014	19207	69413	97016	86290
70178	73444	38790	53626	93780	18629	68766	24371	74639	30782
10169	41465	51935	05711	09799	79077	88159	33437	68519	03040
81084	03701	28590	70013	63794	53169	97054	60303	23259	96196
69202	20777	21727	81511	51887	16175	53746	46516	70339	62727
80561	95787	89426	93325	86412	57479	54194	52153	19197	81877
08199	26703	95128	48599	09333	12584	24374	31232	61782	44032
98883	28220	39358	53720	80161	83371	15181	11131	12219	55920
84568	69286	76054	21615	80883	36797	82845	39139	90900	18172
04269	35173	95745	53893	86022	77722	52498	84193	22448	22571
10538	13124	36099	13140	37706	44562	57179	44693	67877	01549
77843	24955	25900	63843	95029	93859	93634	20205	66294	41218
12034	94636	49455	76362	83532	31062	69903	91186	65768	55949
10524	72829	47611	93315	80875	28090	97728	52560	34937	79548
68935	76632	46984	61772	92786	22651	07086	89754	44143	97687
89450	65665	29190	43709	11172	34481	95977	47535	25658	73898
90696	20451	24211	97310	60446	73530	62865	96574	13829	72226
49006	32047	93086	00112	20470	17136	28255	86328	07293	38809
74591	87025	52368	59416	34417	70557	86746	55809	53628	12000
06315	17012	77103	00968	07235	10728	42189	33292	51487	64443
62386	09184	62092	46617	99419	64230	95034	85481	07857	42510
86848	82122	04028	36959	87827	12813	08627	80699	13345	51695
65643	69480	46598	04501	40403	91408	32343	48130	49303	90689
11084	46534	78957	77353	39578	77868	22970	84349	09184	70603

TABLE A.2. Areas of the unit normal (z) distribution.

$\frac{x-\bar{x}}{\sigma}$	0.09	0.08	0.07	0.06	0.05	0.04	0.03	0.02	0.01	0.00
-3.5	0.00017	0.00017	0.00018	0.00019	0.00019	0.00020	0.00021	0.00022	0.00022	0.00023
-3.4	0.00024	0.00028	0.00026	0.00027	0.00028	0.00029	0.00030	0.00031	0.00033	0.00034
-3.3	0.00035	0.00038	0.00038	0.00039	0.00040	0.00042	0.00043	0.00045	0.00047	0.00048
-3.2	0.00050	0.00052	0.00054	0.00056	0.00058	0.00060	0.00062	0.00054	0.00055	0.00059
-3.1	0.00071	0.00074	0.00076	0.00079	0.00082	0.00085	0.00087	0.00090	0.00094	0.00097
-3.0	0.00100	0.00104	0.00107	0.00111	0.00114	0.00118	0.00122	0.00126	0.00131	0.00135
-2.9	0.0014	0.0014	0.0015	0.0015	0.0016	0.0016	0.0017	0.0017	0.0018	0.0019
-2.8	0.0019	0.0020	0.0021	0.0021	0.0022	0.0023	0.0023	0.0024	0.0025	0.0028
-2.7	0.0028	0.0027	0.0028	0.0029	0.0030	0.0031	0.0032	0.0033	0.0034	0.0035
-2.6	0.0038	0.0037	0.0038	0.0030	0.0040	0.0041	0.0043	0.0044	0.0045	0.0047
-2.5	0.0048	0.0049	0.0051	0.0052	0.0054	0.0055	0.0057	0.0059	0.0060	0.0062
-2.4	0.0064	0.0056	0.0068	0.0059	0.0071	0.0073	0.0075	0.0078	0.0080	0.0082
-2.3	0.0084	0.0087	0.0089	0.0091	0.0094	0.0096	0.0099	0.0102	0.0104	0.0107
-2.2	0.0110	0.0113	0.0116	0.0119	0.0122	0.0125	0.0129	0.0132	0.0136	0.0139
-2.1	0.0143	0.1146	0.0150	0.0154	0.0158	0.0162	0.0166	0.0170	0.0174	0.0179
-2.0	0.0183	0.0185	0.0192	0.0197	0.0202	0.0207	0.0212	0.0217	0.0222	0.0228
-1.9	0.0233	0.0238	0.0244	0.0250	0.0256	0.0262	0.0268	0.0274	0.0281	0.0287
-1.8	0.0294	0.0301	0.0307	0.0314	0.0322	0.0329	0.0336	0.0344	0.0351	0.0359
-1.7	0.0367	0.0375	0.0384	0.0392	0.0401	0.0409	0.0418	0.0427	0.0436	0.0448
-1.6	0.0455	0.0465	0.0475	0.0485	0.0495	0.0505	0.0516	0.0526	0.0537	0.0548
-1.5	0.0559	0.0571	0.0582	0.0594	0.0606	0.0638	0.0630	0.0643	0.0655	0.0668
-1.4	0.0681	0.0694	0.0708	0.0721	0.0735	0.0749	0.0764	0.0778	0.0793	0.0808
-1.3	0.0823	0.0838	0.0853	0.0859	0.0885	0.0901	0.0918	0.0934	0.0951	0.0958

TABLE A.2 (continued)

x−xσ	0.09	0.08	0.07	0.06	0.05	0.04	0.03	0.02	0.01	0.00
−1.2	0.0985	0.1003	0.1020	0.1038	0.1057	0.1075	0.1093	0.1112	0.1131	0.1151
−1.1	0.1170	0.1190	0.1210	0.1230	0.1251	0.1271	0.1292	0.1314	0.1335	0.1357
−1.0	0.1379	0.401	0.1423	0.1446	0.1469	0.1492	0.1515	0.1539	0.1582	0.1587
−0.9	0.1611	0.1635	0.1660	0.1685	0.1711	0.1736	0.1762	0.1788	0.1814	0.1841
−0.8	0.1867	0.1891	0.1922	0.1949	0.1977	0.2003	0.2033	0.2061	0.2090	0.2119
−0.7	0.2146	0.2177	0.2207	0.2236	0.2280	0.2297	0.2327	0.2358	0.2380	0.2420
−0.6	0.2451	0.2483	0.2514	0.2546	0.2578	0.2611	0.2643	0.2676	0.2709	0.2743
−0.5	0.2776	0.2810	0.2843	0.2877	0.2912	0.2948	0.2981	0.3015	0.3050	0.3085
−0.4	0.3621	0.3156	0.3192	0.3228	0.3264	0.3300	0.3330	0.3372	0.3409	0.3445
−0.3	0.3483	0.3520	0.3557	0.3594	0.3632	0.3600	0.3707	0.3745	0.3783	0.3821
−0.2	0.3859	0.3897	0.3936	0.3974	0.4013	0.4052	0.4090	0.4129	0.4168	0.4207
−0.1	0.4247	0.4286	0.4325	0.4364	0.4404	0.4443	0.4485	0.4522	0.4562	0.4602
−0.0	0.4541	0.4581	0.4721	0.4761	0.4801	0.4840	0.4880	0.4920	0.4960	0.5000
−0.0	0.5000	0.5040	0.5080	0.5120	0.5160	0.5120	0.5239	0.5279	0.5319	0.5359
−0.1	0.5398	0.5436	0.5478	0.5517	0.5557	0.5596	0.5636	0.5675	0.5704	0.5753
−0.2	0.5793	0.5832	0.5871	0.5910	0.5948	0.5087	0.6026	0.6064	0.6103	0.6141
−0.3	0.6179	0.6217	0.6255	0.6293	0.6331	0.6368	0.6406	0.6443	0.6480	0.6517
−0.4	0.6554	0.6591	0.6628	0.6554	0.6700	0.6736	0.6772	0.6808	0.6844	0.6879
−0.5	0.6915	0.6950	0.6985	0.7019	0.7054	0.7088	0.7123	0.7157	0.7190	0.7224
−0.6	0.7257	0.7291	0.7324	0.7357	0.7389	0.7422	0.7454	0.7456	0.7517	0.7549
−0.7	0.7580	0.7611	0.7642	0.7673	0.7704	0.7734	0.7764	0.7794	0.7823	0.7852
−0.8	0.7881	0.7910	0.7939	0.7967	0.7995	0.8023	0.8051	0.8079	0.8106	0.8133
−0.9	0.8159	0.8186	0.8212	0.8238	0.8264	0.8289	0.8315	0.8340	0.8365	0.8389
−1.0	0.8413	0.8438	0.8461	0.8485	0.8508	0.8531	0.8554	0.8577	0.8599	0.8621
−1.1	0.8543	0.8665	0.8686	0.8708	0.8720	0.8749	0.8770	0.8790	0.8810	0.8830

-1.2	0.8849	0.8869	0.8888	0.8907	0.8925	0.8944	0.8962	0.8980	0.8997	0.9015
-1.3	0.9032	0.9049	0.9056	0.9092	0.9099	0.0135	0.9131	0.9147	0.9162	0.9177
-1.4	0.9192	0.9207	0.9222	0.9236	0.9251	0.9265	0.9279	0.9292	0.9306	0.9319
-1.5	0.9332	0.9345	0.9357	0.9370	0.9382	0.9394	0.9406	0.9418	0.9429	0.9441
-1.6	0.9452	0.9463	0.9474	0.9484	0.9495	0.9505	0.9515	0.9525	0.9535	0.9545
-1.7	0.9554	0.9584	0.9573	0.9582	0.9591	0.9599	0.9608	0.9616	0.9625	0.9633
-1.8	0.9641	0.9649	0.9656	0.9664	0.9671	0.9678	0.9686	0.9693	0.9699	0.9706
-1.9	0.9713	0.9719	0.9726	0.9732	0.9738	0.9744	0.9750	0.9758	0.9761	0.9767
-2.0	0.9773	0.9778	0.9783	0.9788	0.9793	0.9798	0.9803	0.9808	0.9812	0.9817
-2.1	0.9821	0.9825	0.9830	0.9834	0.9838	0.9842	0.9846	0.9850	0.9854	0.9857
-2.2	0.9861	0.9854	0.9868	0.9871	0.9875	0.9878	0.9881	0.9884	0.9887	0.9890
-2.3	0.9893	0.9895	0.9898	0.9901	0.9904	0.9906	0.9909	0.9911	0.9913	0.9916
-2.4	0.9918	0.9920	0.9922	0.9925	0.9927	0.9929	0.9931	0.9932	0.9934	0.9936
-2.5	0.9938	0.9940	0.9941	0.9943	0.9945	0.9946	0.9948	0.9949	0.9951	0.9952
-2.6	0.9953	0.9955	0.9956	0.9957	0.9959	0.9960	0.9961	0.9952	0.9963	0.9964
-2.7	0.9965	0.9965	0.9957	0.9958	0.9969	0.9970	0.9971	0.9972	0.9973	0.9974
-2.8	0.9974	0.9975	0.9976	0.9977	0.9977	0.9978	0.9979	0.9979	0.9980	0.9981
-2.9	0.9981	0.9982	0.9983	0.9983	0.9984	0.9984	0.9985	0.9985	0.9986	0.9986
-3.0	0.99865	0.99869	0.99674	0.99878	0.99882	0.99885	0.99889	0.99893	0.99898	0.99200
-3.1	0.99903	0.99905	0.99910	0.99913	0.99915	0.99915	0.99921	0.99924	0.99926	0.99929
-3.2	0.99931	0.99934	0.99936	0.99938	0.99940	0.99942	0.99944	0.99946	0.99948	0.99950
-3.3	0.99952	0.99953	0.99955	0.99957	0.99958	0.99960	0.99961	0.99952	0.99964	0.99985
-3.4	0.99965	0.99957	0.99969	0.99970	0.99971	0.99972	0.99973	0.99974	0.99975	0.99975
-3.5	0.99977	0.99978	0.99978	0.99979	0.99981	0.99981	0.99981	0.99982	0.99983	0.99983

Source: R. C. Gupta, Statistical Quality Control Book, Third Edition, 1983. Khanna Publishers, Nai Sarak, Delhi, India. Reprinted with permission.

TABLE A.3. Ordinates and areas of the normal curve (in terms of σ units).

x σ	Area	Ordinate	x σ	Area	Ordinate	x σ	Area	Ordinate
00	.0000	.3989	.50	.1915	.3521	C	.3413	.2420
.01	.0040	.3989	.51	.1950	.3503	1.01	.3438	.2396
.02	.0080	.3989	.52	.1985	.3485	1.02	.3461	.2371
.03	.0120	.3988	.53	.2019	.3467	1.03	.3485	.2347
.04	.0160	.3986	.54	.2054	.3448	1.04	.3508	.2323
.05	.0199	.3984	.55	.2088	.3429	1.05	.3531	.2299
.06	.0239	.3982	.56	.2123	.3410	1.06	.3554	.2275
.07	.0279	.3980	.57	.2157	.3391	1.07	.3577	.2251
.08	.0319	.3977	.58	.2190	.3372	1.08	.3599	.2227
.09	.0359	.3973	.59	.2224	.3352	1.09	.3621	.2203
.10	.0398	.3970	.60	.2257	.3332	1.10	.3643	.2179
.11	.0438	.3965	.61	.2291	.3312	1.11	.3665	.2155
.12	.0478	.3961	.62	.2324	.3292	1.12	.3686	.2131
.13	.0517	.3956	.63	.2357	.3271	1.13	.3708	.2107
.14	.0557	.3951	.64	.2389	.3251	1.14	.3729	.2083
.15	.0596	.3945	.65	.2422	.3230	1.15	.3749	.2059
.16	.0636	.3939	.66	.2454	.3209	1.16	.3770	.2036
.17	.0675	.3932	.67	.2486	.3187	1.17	.3790	.2012
.18	.0714	.3925	.68	.2517	.3166	1.18	.3810	.1989
.19	.0753	.3918	.69	.2549	.3144	1.19	.3830	.1965
.20	.0793	.3910	.70	.2580	.3123	1.20	.3849	.1942
.21	.0832	.3902	.71	.2611	.3101	1.21	.3869	.1919
.22	.0871	.3894	.72	.2642	.3079	1.22	.3888	.1895
.23	.0910	.3885	.73	.2673	.3056	1.23	.3907	.1872
.24	.0948	.3876	.74	.2703	.3034	1.24	.3925	.1849

x			x			x		
.25	.0987	.3867	.75	.2734	.3011	1.25	.3944	.1826
.26	.1026	.3857	.76	.2764	.2989	1.26	.3962	.1804
.27	.1064	.3847	.77	.2794	.2966	1.27	.3980	.1781
.28	.1103	.3836	.78	.2823	.2943	1.28	.3997	.1758
.29	.1141	.3825	.79	.2852	.2920	1.29	.4015	.1736
.30	.1179	.3814	.80	.2881	.2897	1.30	.4032	.1714
.31	.1217	.3802	.81	.2910	.2874	1.31	.4049	.1691
.32	.1255	.3790	.82	.2939	.2850	1.32	.4066	.1669
.33	.1293	.3778	.83	.2967	.2827	1.33	.4082	.1647
.34	.1331	.3765	.84	.2995	.2803	1.34	.4099	.1626
.35	.1368	.3752	.85	.3023	.2780	1.35	.4115	.1604
.36	.1406	.3739	.86	.3051	.2756	1.36	.4131	.1582
.37	.1443	.3725	.87	.3078	.2732	1.37	.4147	.1561
.38	.1480	.3712	.88	.3106	.2709	1.38	.4162	.1539
.39	.1517	.3697	.89	.3133	.2685	1.39	.4177	.1518
.40	.1554	.3683	.90	.3159	.2661	1.40	.4192	.1497
.41	.1591	.3668	.91	.3186	.2637	1.41	.4207	.1476
.42	.1628	.3653	.92	.3212	.2613	1.42	.4222	.1456
.43	.1664	.3637	.93	.3238	.2589	1.43	.4236	.1435
.44	.1700	.3621	.94	.3264	.2565	1.44	.4251	.1415
.45	.1736	.3605	.95	.3289	.2541	1.45	.4265	.1394
.46	.1772	.3589	.96	.3315	.2516	1.46	.4279	.1374
.47	.1808	.3572	.97	.3340	.2492	1.47	.4292	.1354
.48	.1844	.3555	.98	.3365	.2468	1.48	.4306	.1334
.49	.1879	.3538	.99	.3389	.2444	1.49	.4319	.1315
.50	.1915	.3521	1.00	.3413	.2420	1.50	.4332	.1295

Source: J.E. Wert, 1938. *Educational Statistics,* New York: McGraw-Hill.

TABLE A.3. Ordinates and areas of the normal curve (continued).

x σ	Area	Ordinate	x σ	Area	Ordinate	x σ	Area	Ordinate
1.50	.4332	.1295	2.00	.4772	.0540	2.50	.4938	.0175
1.51	.4345	.1276	2.01	.4778	.0529	2.51	.4940	.0171
1.52	.4357	.1257	2.02	.4783	.0519	2.52	.4941	.0167
1.53	.4370	.1238	2.03	.4788	.0508	2.53	.4943	.0163
1.54	.4382	.1219	2.04	.4793	.0498	2.54	.4945	.0158
1.55	.4394	.1200	2.05	.4798	.0488	2.55	.4946	.0154
1.56	.4406	.1182	2.06	.4803	.0478	2.56	.4948	.0151
1.57	.4418	.1163	2.07	.4808	.0468	2.57	.4949	.0147
1.58	.4429	.1145	2.08	.4812	.0459	2.58	.4951	.0143
1.59	.4441	.1127	2.09	.4817	.0449	2.59	.4952	.0139
1.60	.4452	.1109	2.10	.4821	.0440	2.60	.4953	.0136
1.61	.4463	.1092	2.11	.4826	.0431	2.61	.4955	.0132
1.62	.4474	.1074	2.12	.4830	.0422	2.62	.4956	.0129
1.63	.4484	.1057	2.13	.4834	.0413	2.63	.4957	.0126
1.64	.4495	.1040	2.14	.4838	.0404	2.64	.4959	.0122
1.65	.4505	.1023	2.15	.4842	.0395	2.65	.4960	.0119
1.66	.4515	.1006	2.16	.4846	.0387	2.66	.4961	.0116
1.67	.4525	.0989	2.17	.4850	.0379	2.67	.4962	.0113
1.68	.4535	.0973	2.18	.4854	.0371	2.68	.4963	.0110
1.69	.4545	.0957	2.19	.4857	.0363	2.69	.4964	.0107
1.70	.4554	.0940	2.20	.4861	.0355	2.70	.4965	.0104
1.71	.4564	.0925	2.21	.4864	.0347	2.71	.4966	.0101
1.72	.4573	.0909	2.22	.4868	.0339	2.72	.4967	.0099
1.73	.4582	.0893	2.23	.4871	.0332	2.73	.4968	.0096
1.74	.4591	.0878	2.24	.4875	.0325	2.74	.4969	.0093

z			z			z		
1.75	.4599	.0863	2.25	.4878	.0317	2.75	.4970	.0091
1.76	.4608	.0843	2.26	.4881	.0310	2.76	.4971	.0088
1.77	.4616	.0833	2.27	.4884	.0303	2.77	.4972	.0086
1.78	.4625	.0818	2.28	.4887	.0297	2.78	.4973	.0084
1.79	.4633	.0804	2.29	.4890	.0290	2.79	.4974	.0081
1.80	.4641	.0790	2.30	.4893	.0283	2.80	.4974	.0079
1.81	.4649	.0775	2.31	.4896	.0277	2.81	.4975	.0077
1.82	.4656	.0761	2.32	.4898	.0270	2.82	.4976	.0075
1.83	.4604	.0748	2.33	.4901	.0264	2.83	.4977	.0073
1.84	.4671	.0734	2.34	.4904	.0258	2.84	.4977	.0071
1.85	.4678	.0721	2.35	.4906	.0252	2.85	.4978	.0069
1.86	.4686	.0707	2.36	.4909	.0246	8.86	.4979	.0067
1.87	.4693	.0694	2.37	.4911	.0241	2.87	.4979	.0065
1.88	.4699	.0681	2.38	.4913	.0235	2.88	.4980	.0063
1.89	.4706	.0669	2.39	.4916	.0229	2.89	.4981	.0061
1.90	.4713	.0656	2.40	.4918	.0224	2.90	.4981	.0060
1.91	.4711	.0644	2.41	.4920	.0219	2.91	.4982	.0058
1.92	.4726	.0632	2.42	.4922	.0213	2.92	.4982	.0056
1.93	.4732	.0620	2.43	.4925	.0208	2.93	.4983	.0055
1.94	.4738	.0608	2.44	.4927	.0203	2.94	.4984	.0053
1.95	.4744	.0596	2.45	.4929	.0198	2.95	.4984	.0051
1.96	.4750	.0584	2.46	.4931	.0194	2.96	.4985	.0050
1.97	.4756	.0573	2.47	.4932	.0189	2.97	.4985	.0048
1.98	.4761	.0562	2.48	.4934	.0184	2.98	.4986	.0047
1.99	.4767	.0551	2.49	.4936	.0180	2.99	.4986	.0046
2.00	.4772	.0540	2.50	.4938	.0175	3.00	.4987	.0044

Source: J.E. Wert, 1938. *Educational Statistics,* New York: McGraw-Hill.

TABLE A.4. Critical values of F distribution.

Denomi-nator df	a	Numerator df								
		1	2	3	4	5	6	7	8	9
1	0.100	39.86	49.50	53.59	55.83	57.24	58.20	58.91	59.44	59.86
	0.050	161.4	199.50	215.7	224.6	230.2	234.0	236.8	238.9	240.5
	0.025	647.8	799.50	864.2	899.6	921.8	937.1	948.2	956.7	963.3
	0.010	4052	4999.5	5403	5625	5764	5859	5928	5982	6022
2	0.100	8.53	9.00	9.16	9.24	9.29	9.33	9.35	9.37	9.38
	0.050	18.51	19.00	19.16	19.25	19.30	19.33	19.35	19.37	19.38
	0.025	38.51	39.00	39.17	39.25	39.30	39.33	39.36	39.37	39.39
	0.010	98.50	99.00	99.17	99.25	99.30	99.33	99.36	99.37	99.39
3	0.100	5.54	5.46	5.39	5.34	5.31	5.28	5.27	5.25	5.24
	0.050	10.13	9.55	9.28	9.12	9.01	8.94	8.89	8.85	8.81
	0.025	17.44	16.04	15.44	15.10	14.88	14.73	14.62	14.54	14.47
	0.010	34.12	30.82	29.46	28.71	28.24	27.91	27.67	27.49	27.35
4	0.100	4.54	4.32	4.19	4.11	4.05	4.01	3.98	3.95	3.94
	0.050	7.71	6.94	6.59	6.39	6.26	6.16	6.09	6.04	6.00
	0.025	12.22	10.65	9.98	9.60	9.36	9.20	9.07	8.98	8.90
	0.010	21.20	18.00	16.69	15.98	15.52	15.21	14.98	14.80	14.66
5	0.100	4.06	3.78	3.62	3.52	3.45	3.40	3.37	3.34	2.32
	0.050	6.61	5.79	5.41	5.19	5.05	4.95	4.88	4.82	4.77
	0.025	10.01	8.43	7.76	7.39	7.15	6.98	6.85	6.76	6.68
	0.010	16.26	13.27	12.06	11.39	10.97	10.67	10.46	10.29	10.16
6	0.100	3.78	3.46	3.29	3.18	3.11	3.05	3.01	2.98	2.96
	0.050	5.99	5.14	4.76	4.53	4.39	4.28	4.21	4.15	4.10
	0.025	8.81	7.26	6.60	6.23	5.99	5.82	5.70	5.60	5.52
	0.010	13.75	10.92	9.78	9.15	8.75	8.47	8.26	8.10	7.98
7	0.100	3.59	3.26	3.07	2.96	2.88	2.83	2.78	2.75	2.72

df	prob									
	0.050	5.59	4.74	4.35	4.12	3.97	3.87	3.79	3.73	3.68
	0.025	8.07	6.54	5.89	5.52	5.29	5.12	4.99	4.90	4.82
	0.010	12.25	9.55	8.45	7.85	7.46	7.19	6.99	6.84	6.72
8	0.100	3.46	3.11	2.92	2.81	2.73	2.67	2.62	2.59	2.56
	0.050	5.32	4.46	4.07	3.84	3.69	3.58	3.50	3.44	3.39
	0.025	7.57	6.06	5.42	5.05	4.82	4.65	4.53	4.43	4.36
	0.010	11.26	8.65	7.59	7.01	6.63	6.37	6.18	6.03	5.91
9	0.100	3.36	3.01	2.81	2.69	2.61	2.55	2.51	2.47	2.44
	0.050	5.12	4.26	3.86	3.63	3.48	3.37	3.29	3.23	3.18
	0.025	7.21	5.71	5.08	4.72	4.48	4.32	4.20	4.10	4.03
	0.010	10.56	8.02	6.99	6.42	6.06	5.80	5.61	5.47	5.35
10	0.100	3.29	2.92	2.73	2.61	2.52	2.46	2.41	2.38	2.35
	0.050	4.96	4.10	3.71	3.48	3.33	3.22	3.14	3.07	3.02
	0.025	6.94	5.46	4.83	4.47	4.24	4.07	3.95	3.85	3.78
	0.010	10.04	7.56	6.55	5.99	5.64	5.39	5.20	5.06	4.94
11	0.100	3.23	2.86	2.66	2.54	2.45	2.39	2.34	2.30	2.27
	0.050	4.84	3.98	3.59	3.36	3.20	3.09	3.01	2.95	2.90
	0.025	6.72	5.26	4.63	4.28	4.04	3.88	3.76	3.66	3.59
	0.010	9.65	7.21	6.22	5.67	5.32	5.07	4.89	4.74	4.63
12	0.100	3.18	2.81	2.61	2.48	2.39	2.33	2.28	2.24	2.21
	0.050	4.75	3.89	3.49	3.26	3.11	3.00	2.91	2.85	2.80
	0.025	6.55	5.10	4.47	4.12	3.89	3.73	3.61	3.51	3.44
	0.010	9.33	6.93	5.95	5.41	5.06	4.82	4.64	4.50	4.39

Source: M. Merrington and C.M. Thompson, "Tables of Percentage Points of the Inverted Beta (F) Distribution." Biometrika, 1943, 33:73–78. Reprinted with permission of Oxford University Press.

TABLE A.4. Critical values of F distribution (continued).

Denominator df	∂	Numerator of C/F									
		10	12	15	20	24	30	40	60	120	∞
1	0.100	60.19	60.71	61.22	61.74	62.00	62.26	62.53	62.79	63.06	63.33
	0.050	241.9	243.9	245.9	248.0	249.1	250.1	251.1	252.2	253.3	254.3
	0.025	968.6	976.7	984.9	993.1	997.2	1001	1006	1010	1014	1018
	0.010	6056	6106	6157	6209	6235	6261	6287	6313	6339	6366
2	0.100	9.39	9.41	9.42	9.44	9.45	9.46	9.47	9.47	9.48	9.49
	0.050	19.40	19.41	19.43	19.45	19.45	19.46	19.47	19.48	19.49	19.50
	0.025	39.40	39.41	39.43	39.45	39.46	39.46	39.47	39.48	39.49	39.50
	0.010	99.40	99.42	99.43	99.45	99.46	99.47	99.47	99.48	99.49	99.50
3	0.100	5.23	5.22	5.20	5.18	5.18	5.17	5.16	5.15	5.14	5.13
	0.050	8.79	8.74	8.70	8.66	8.64	8.62	8.59	8.57	8.55	8.53
	0.025	14.42	14.34	14.25	14.17	14.12	14.08	14.04	13.99	13.95	13.90
	0.010	27.23	27.05	26.87	26.69	26.60	26.50	26.41	26.32	26.22	26.13
4	0.100	3.92	3.90	3.87	3.84	3.83	3.82	3.80	3.79	3.78	3.76
	0.050	5.96	5.91	5.86	5.80	5.77	5.75	5.72	5.69	5.66	5.63
	0.025	8.84	8.75	8.66	8.56	8.51	8.46	8.41	8.36	8.31	8.26
	0.010	14.55	14.37	14.20	14.02	13.93	13.84	13.75	13.65	13.56	13.46
5	0.100	3.30	3.27	3.24	3.21	3.19	3.17	3.16	3.14	3.12	3.10
	0.50	4.74	4.68	4.62	4.56	4.53	4.50	4.46	4.43	4.40	4.36
	0.025	6.62	6.52	6.43	6.33	6.28	6.23	6.18	6.12	6.07	6.02
	0.010	10.05	9.89	9.72	9.55	9.47	9.38	9.29	9.20	9.11	9.02
6	0.100	2.94	2.90	2.87	2.84	2.82	2.80	2.78	2.76	2.74	2.72
	0.050	4.06	4.00	3.94	3.87	3.84	3.81	3.77	3.74	3.70	3.67
	0.025	5.46	5.37	5.27	5.17	5.12	5.07	5.01	4.96	4.90	4.85
	0.010	7.87	7.72	7.56	7.40	7.31	7.23	7.14	7.06	6.97	6.88

7	0.100	2.70	2.67	2.63	2.59	2.58	2.56	2.54	2.51	2.49	2.47
	0.050	3.64	3.57	3.51	3.44	3.41	3.38	3.34	3.30	3.27	3.23
	0.025	4.76	4.67	4.57	4.47	4.42	4.36	4.31	4.25	4.20	4.14
	0.010	6.62	6.47	6.31	6.16	6.07	5.99	5.91	5.82	5.74	5.65
8	0.100	2.54	2.50	2.46	2.42	2.40	2.38	2.36	2.34	2.32	2.29
	0.050	3.35	3.28	3.22	3.15	3.12	3.08	3.04	3.01	2.97	2.93
	0.025	4.30	4.20	4.10	4.00	3.95	3.89	3.84	3.78	3.73	3.67
	0.010	5.81	5.67	5.52	5.36	5.28	5.20	5.12	5.03	4.95	4.86
9	0.100	2.42	2.38	2.34	2.30	2.28	2.25	2.23	2.21	2.18	2.16
	0.050	3.14	3.07	3.01	2.94	2.90	2.86	2.83	2.79	2.75	2.71
	0.025	3.96	3.87	3.77	3.67	3.61	3.56	3.51	3.45	3.39	3.33
	0.010	5.26	5.11	4.96	4.81	4.73	4.65	4.57	4.48	4.40	4.31
10	0.100	2.32	2.28	2.24	2.20	2.18	2.16	2.13	2.11	2.08	2.06
	0.050	2.98	2.91	2.85	2.77	2.74	2.70	2.66	2.62	2.58	2.54
	0.025	3.72	3.62	3.52	3.42	3.37	3.31	3.26	3.20	3.14	3.08
	0.010	4.85	4.71	4.56	4.41	4.33	4.25	4.17	4.08	4.00	3.91
11	0.100	2.25	2.21	2.17	2.12	2.10	2.08	2.05	2.03	2.00	1.97
	0.050	2.85	2.79	2.72	2.65	2.61	2.57	2.53	2.49	2.45	2.40
	0.025	3.53	3.43	3.33	3.23	3.17	3.12	3.06	3.00	2.94	2.88
	0.010	4.54	4.40	4.25	4.10	4.02	3.94	3.86	3.78	3.69	3.60
12	0.100	2.19	2.15	2.10	2.06	2.04	2.01	1.99	1.96	1.93	1.90
	0.050	2.75	2.69	2.62	2.54	2.51	2.47	2.43	2.38	2.34	2.30
	0.025	3.37	3.28	3.18	3.07	3.02	2.96	2.91	2.85	2.79	2.72
	0.010	4.30	4.16	4.01	3.86	3.78	3.70	3.62	3.54	3.45	3.36

Source: M. Merrington and C.M. Thompson, "Tables of Percentage Points of the Inverted Beta (F) Distribution." *Biometrika*, 1943, 33:73-78. Reprinted with permission of Oxford University Press.

TABLE A.4. Critical values of F distribution (continued).

Denominator df	a	Numerator of df 1	2	3	4	5	6	7	8	9
13	0.100	3.14	2.76	2.56	2.43	2.35	2.28	2.23	2.20	2.16
	0.050	4.67	3.81	3.41	3.18	3.03	2.92	2.83	2.77	2.71
	0.025	6.41	4.97	4.35	4.00	3.77	3.60	3.48	3.39	3.31
	0.010	9.07	6.70	5.74	5.21	4.86	4.62	4.44	4.30	4.19
14	0.100	3.10	2.73	2.52	2.39	2.31	2.24	2.19	2.15	2.12
	0.050	4.60	3.74	3.34	3.11	2.96	2.85	2.76	2.70	2.65
	0.025	6.30	4.86	4.24	3.89	3.66	3.50	3.38	3.29	3.21
	0.010	8.86	6.51	5.56	5.04	4.69	4.46	4.28	4.14	4.03
15	0.100	3.07	2.70	2.48	2.36	2.27	2.21	2.16	2.12	2.09
	0.050	4.54	3.68	3.29	3.06	2.90	2.79	2.71	2.64	2.59
	0.025	6.20	4.77	4.15	3.80	3.58	3.41	3.29	3.20	3.12
	0.010	8.68	6.36	5.42	4.89	4.56	4.32	4.14	4.00	3.89
16	0.100	3.05	2.67	2.46	2.33	2.24	2.18	2.13	2.09	2.06
	0.050	4.49	3.63	3.24	3.01	2.85	2.74	2.66	2.59	2.54
	0.025	6.12	4.69	4.08	3.73	3.50	3.34	3.22	3.12	3.05
	0.010	8.53	6.23	5.29	4.77	4.44	4.20	4.03	3.89	3.78
17	0.100	3.03	2.64	2.44	2.31	2.22	2.15	2.10	2.06	2.03
	0.50	4.45	3.59	3.20	2.96	2.81	2.70	2.61	2.55	2.49
	0.025	6.04	4.62	4.01	3.66	3.44	3.28	3.16	3.06	2.98
	0.010	8.40	6.11	5.18	4.67	4.34	4.10	3.93	3.79	3.68
18	0.100	3.01	2.62	2.42	2.29	2.20	2.13	2.08	2.04	2.00
	0.050	4.41	3.55	3.16	2.93	2.77	2.66	2.58	2.51	2.46
	0.025	5.98	4.56	3.95	3.61	3.38	3.22	3.10	3.01	2.93

	0.010	8.29	6.01	5.09	4.58	4.25	4.01	3.84	3.71	3.60
19	0.100	2.99	2.61	2.40	2.27	2.18	2.11	2.06	2.02	1.98
	0.050	4.38	3.52	3.13	2.90	2.74	2.63	2.54	2.48	2.42
	0.025	5.92	4.51	3.90	3.56	3.33	3.17	3.05	2.96	2.88
	0.010	8.18	5.93	5.01	4.50	4.17	3.94	3.77	3.63	3.52
20	0.100	2.97	2.59	2.38	2.25	2.16	2.09	2.04	2.00	1.96
	0.050	4.32	3.49	3.10	2.87	2.71	2.60	2.51	2.45	2.39
	0.025	5.87	4.46	3.86	3.51	3.29	3.13	3.01	2.91	2.84
	0.010	8.10	5.85	4.94	4.43	4.10	3.87	3.70	3.56	3.46
21	0.100	2.96	2.57	2.36	2.23	2.14	2.08	2.02	1.98	1.95
	0.050	4.32	3.47	3.07	2.84	2.68	2.57	2.49	2.42	2.37
	0.025	5.83	4.42	3.82	3.48	3.25	3.09	2.97	2.87	2.80
	0.010	8.02	5.78	4.87	4.37	4.04	3.81	3.64	3.51	3.40
22	0.100	2.95	2.56	2.35	2.22	2.13	2.06	2.01	1.97	1.93
	0.050	4.30	3.44	3.05	2.82	2.66	2.55	2.46	2.40	2.34
	0.025	5.79	4.38	3.78	3.44	3.22	3.05	2.93	2.84	2.76
	0.010	7.95	5.72	4.82	4.31	3.99	3.76	3.59	3.45	3.35
23	0.100	2.94	2.55	2.34	2.21	2.11	2.05	1.99	1.95	1.92
	0.050	4.28	3.42	3.03	2.80	2.64	2.53	2.44	2.37	2.32
	0.025	5.75	4.35	3.75	3.41	3.18	3.02	2.90	2.81	2.73
	0.010	7.88	5.66	4.76	4.26	3.94	3.71	3.54	3.41	3.30
24	0.100	2.93	2.54	2.33	2.19	2.10	2.04	1.98	1.94	1.91
	0.050	4.26	3.40	3.01	2.78	2.62	2.51	2.42	2.36	2.30
	0.025	5.72	4.32	3.72	3.38	3.15	2.99	2.87	2.78	2.70
	0.010	7.82	5.61	4.72	4.22	3.90	3.67	3.50	3.36	3.26

Source: M. Merrington and C.M. Thompson, "Tables of Percentage Points of the Inverted Beta (F) Distribution." *Biometrika*, 1943, 33:73-78. Reprinted with permission of Oxford University Press.

TABLE A.4. Critical values of F distribution (continued).

Denomi- nator df	a	Numerator of df									
		10	12	15	20	24	30	40	60	120	z
13	0.100	2.14	2.10	2.05	2.01	1.98	1.96	1.93	1.90	1.88	1.85
	0.050	2.67	2.60	2.53	2.46	2.42	2.38	2.34	2.30	2.25	2.21
	0.025	3.25	3.15	3.05	2.95	2.89	2.84	2.78	2.72	2.66	2.60
	0.010	4.10	3.96	3.82	3.66	3.59	3.51	3.43	3.34	3.25	3.17
14	0.100	2.10	2.05	2.01	1.96	1.94	1.91	1.89	1.86	1.83	1.80
	0.050	2.60	2.53	2.46	2.39	2.35	2.31	2.27	2.22	2.18	2.13
	0.025	3.15	3.05	2.95	2.84	2.79	2.73	2.67	2.61	2.55	2.49
	0.010	3.94	3.80	3.66	3.51	3.43	3.35	3.27	3.18	3.09	3.00
15	0.100	2.06	2.02	1.97	1.92	1.90	1.87	1.85	1.82	1.79	1.76
	0.050	2.54	2.48	2.40	2.33	2.29	2.25	2.20	2.16	2.11	2.07
	0.025	3.06	2.96	2.86	2.76	2.70	2.64	2.59	2.52	2.46	2.40
	0.010	3.80	3.67	3.52	3.37	3.29	3.21	3.13	3.05	2.96	2.87
16	0.100	2.03	1.99	1.94	1.89	1.87	1.84	1.81	1.78	1.75	1.72
	0.050	2.49	2.42	2.35	2.28	2.24	2.19	2.15	2.11	2.06	2.01
	0.025	2.99	2.89	2.79	2.68	2.63	2.57	2.51	2.45	2.38	2.32
	0.010	3.69	3.55	3.41	3.26	3.18	3.10	3.02	2.93	2.84	2.75
17	0.100	2.00	1.96	1.91	1.86	1.84	1.81	1.78	1.75	1.72	1.69
	0.50	2.45	2.38	2.31	2.23	2.19	2.15	2.10	2.06	2.01	1.96
	0.025	2.92	2.82	2.72	2.62	2.56	2.50	2.44	2.38	2.32	2.25
	0.010	3.59	3.46	3.31	3.16	3.08	3.00	2.92	2.83	2.75	2.65
18	0.100	1.98	1.93	1.89	1.84	1.81	1.78	1.75	1.72	1.69	1.66
	0.050	2.41	2.34	2.27	2.19	2.15	2.11	2.06	2.02	1.97	1.92
	0.025	2.87	2.77	2.67	2.56	2.50	2.44	2.38	2.32	2.26	2.19

df	p										
	0.010	3.51	3.37	3.23	3.08	3.00	2.92	2.84	2.75	2.66	2.57
19	0.100	1.96	1.91	1.86	1.81	1.79	1.76	1.73	1.70	1.67	1.63
	0.050	2.38	2.31	2.23	2.16	2.11	2.07	2.03	1.98	1.93	1.88
	0.025	2.82	2.72	2.62	2.51	2.45	2.39	2.33	2.27	2.20	2.13
	0.010	3.43	3.30	3.15	3.00	2.92	2.84	2.76	2.67	2.58	2.49
20	0.100	1.94	1.89	1.84	1.79	1.77	1.74	1.71	1.68	1.64	1.61
	0.050	2.35	2.28	2.20	2.12	2.08	2.04	1.99	1.95	1.90	1.84
	0.025	2.77	2.68	2.57	2.46	2.41	2.35	2.29	2.22	2.16	2.09
	0.010	3.37	3.23	3.09	2.94	2.86	2.78	2.69	2.61	2.52	2.42
21	0.100	1.92	1.87	1.83	1.78	1.75	1.72	1.69	1.66	1.62	1.59
	0.050	2.32	2.25	2.18	2.10	2.05	2.01	1.96	1.92	1.87	1.81
	0.025	2.73	2.64	2.53	2.42	2.37	2.31	2.25	2.18	2.11	2.04
	0.010	3.31	3.17	3.03	2.88	2.80	2.72	2.64	2.55	2.46	2.36
22	0.100	1.90	1.86	1.81	1.76	1.73	1.70	1.67	1.64	1.60	1.57
	0.050	2.30	2.23	2.15	2.07	2.03	1.98	1.94	1.89	1.84	1.78
	0.025	2.70	2.60	2.50	2.39	2.33	2.27	2.21	2.14	2.08	2.00
	0.010	3.26	3.12	2.98	2.83	2.75	2.67	2.58	2.50	2.40	2.31
23	0.100	1.89	1.84	1.80	1.74	1.72	1.69	1.66	1.62	1.59	1.55
	0.050	2.27	2.20	2.13	2.05	2.01	1.96	1.91	1.86	1.81	1.76
	0.025	2.67	2.57	2.47	2.36	2.30	2.24	2.18	2.11	2.04	1.97
	0.010	3.21	3.07	2.93	2.78	2.70	2.62	2.54	2.45	2.35	2.26
24	0.100	1.88	1.83	1.78	1.73	1.70	1.67	1.64	1.61	1.57	1.53
	0.050	2.25	2.18	2.11	2.03	1.98	1.94	1.89	1.84	1.79	1.73
	0.025	2.64	2.54	2.44	2.33	2.27	2.21	2.15	2.08	2.01	1.94
	0.010	3.17	3.03	2.89	2.74	2.66	2.58	2.49	2.40	2.31	2.21

Source: M. Merrington and C.M. Thompson, "Tables of Percentage Points of the Inverted Beta (F) Distribution." Biometrika, 1943, 33:73-78. Reprinted with permission of Oxford University Press.

TABLE A.5. Critical values of *t*.

df	Level of significance for one-tailed test						
	.10	.05	.025	.01	.005	.0005	
	Level of significance for two-tailed test						
	.20	.10	.05	.02	.01	.001	
1	3.078	6.314	12.706	31.821	63.657	636.619	
2	1.886	2.920	4.303	6.965	9.925	31.598	
3	1.638	2.353	3.182	4.541	5.841	12.941	
4	1.535	2.132	2.776	3.747	4.604	8.610	
5	1.476	2.015	2.571	3.365	4.032	6.859	
6	1.440	1.943	2.447	3.143	3.707	5.959	
7	1.415	1.895	2.365	2.998	3.499	5.405	
8	1.397	1.860	2.306	2.896	3.355	5.041	
9	1.383	1.833	2.262	2.821	3.250	4.781	
10	1.372	1.812	2.228	2.764	3.169	4.587	
11	1.363	1.796	2.201	2.718	3.106	4.437	
12	1.356	1.782	2.179	2.681	3.055	4.318	
13	1.350	1.771	2.160	2.650	3.012	4.221	
14	1.345	1.761	2.145	2.624	2.977	4.140	
15	1.341	1.753	2.131	2.602	2.947	4.073	
16	1.337	1.746	2.120	2.583	2.921	4.015	
17	1.333	1.740	2.110	2.567	2.898	3.965	

18	1.330	1.734	2.101	2.552	2.878	3.922
19	1.328	1.729	2.093	2.539	2.861	3.883
20	1.325	1.725	2.086	2.528	2.845	3.850
21	1.323	1.721	2.080	2.518	2.831	3.819
22	1.321	1.717	2.074	2.508	2.819	3.792
23	1.319	1.714	2.069	2.500	2.807	3.767
24	1.318	1.711	2.064	2.492	2.797	3.745
25	1.316	1.708	2.060	2.485	2.787	3.725
26	1.315	1.706	2.056	2.479	2.779	3.707
27	1.314	1.703	2.052	2.473	2.771	3.690
28	1.313	1.701	2.048	2.467	2.763	3.674
29	1.311	1.699	2.045	2.462	2.756	3.659
30	1.310	1.697	2.042	2.457	2.750	3.646
40	1.303	1.684	2.021	2.423	2.704	3.551
60	1.296	1.671	2.000	2.390	2.660	3.460
120	1.289	1.658	1.980	2.358	2.617	3.373
∞	1.282	1.645	1.960	2.326	2.576	3.291

Source: Table III from Statistical Tables for Biological, Agricultural and Medical Research, Sixth Edition, by R.A. Fisher and F. Yates, Pearson Education Limited, copyright 1963. Reprinted with permission.

TABLE A.6. Critical values of Chi-square (x^2) distribution.

df	.99	.98	.95	.90	.80	.70	.50	.30	.20	.10	.05	.02	.01	.001
1	.00016	.00063	.0039	.016	.064	.15	.46	1.07	1.64	2.71	3.84	5.41	6.64	10.83
2	.02	.04	.10	.21	.45	.71	1.39	2.41	3.22	4.60	5.99	7.82	9.21	13.82
3	.12	.18	.35	.58	1.00	1.42	2.37	3.66	4.64	6.25	7.82	9.84	11.34	10.27
4	.30	.43	.71	1.06	1.65	2.20	3.35	4.88	5.99	7.78	9.49	11.67	13.28	18.46
5	.55	.75	1.14	1.61	2.34	3.00	4.35	6.06	7.29	9.24	11.07	13.39	15.09	20.52
6	.87	1.13	1.64	2.20	3.07	3.83	5.35	7.23	8.56	10.64	12.59	15.03	16.81	22.46
7	1.24	1.56	2.17	2.83	3.82	4.67	6.35	8.38	9.80	12.02	14.07	16.62	18.48	24.32
8	1.65	2.03	2.73	3.49	4.59	5.53	7.34	9.52	11.03	13.36	15.51	18.17	20.09	26.12
9	2.09	2.53	3.32	4.17	5.38	6.39	8.34	10.66	12.24	14.68	16.92	19.68	21.67	27.88
10	2.56	3.06	3.94	4.86	6.18	7.27	9.34	11.78	13.44	15.99	18.31	21.10	23.21	29.59
11	3.05	3.61	4.58	5.58	6.99	8.15	10.34	12.90	14.63	17.28	19.68	22.62	24.72	31.26
12	3.57	4.18	5.23	6.30	7.81	9.03	11.34	14.01	15.81	18.55	21.03	24.05	26.22	32.91
13	4.11	4.76	5.89	7.04	8.63	9.93	12.34	15.12	16.98	19.81	22.36	25.47	27.69	34.53
14	4.66	5.37	6.57	7.79	9.47	10.82	13.34	16.22	18.15	21.08	23.68	26.87	29.14	36.12
15	5.23	5.98	7.26	8.55	10.31	11.72	14.34	17.32	19.31	22.31	25.00	28.26	30.58	37.70
16	5.81	6.61	7.96	9.31	11.15	12.62	15.34	18.42	20.46	23.54	26.80	29.63	32.00	39.29
17	6.41	7.26	8.67	10.08	12.00	13.53	16.34	19.51	21.62	24.77	27.59	31.00	33.41	40.75

df														
18	7.02	7.91	9.39	10.86	12.86	14.44	17.34	20.60	22.76	25.99	28.87	32.35	34.80	42.31
19	7.63	8.57	10.12	11.65	13.72	15.35	18.34	21.69	23.90	27.20	30.14	33.69	36.19	43.82
20	8.26	9.24	10.85	12.44	14.58	16.27	19.34	22.78	25.04	28.41	31.41	35.02	37.57	45.32
21	8.90	9.92	11.59	13.24	15.44	17.18	20.34	23.86	26.17	29.62	32.67	36.34	38.93	46.80
22	9.54	10.60	12.34	14.04	16.31	18.10	21.34	24.94	27.30	30.81	33.92	37.66	40.29	48.27
23	10.20	11.29	13.09	14.85	17.19	19.02	22.34	26.02	28.43	32.01	35.17	38.97	41.64	49.73
24	10.86	11.99	13.85	15.66	18.06	19.94	23.34	27.10	29.55	33.20	36.42	40.27	42.98	51.18
25	11.52	12.70	14.61	16.47	18.94	20.87	24.34	28.17	30.68	34.38	37.65	41.57	44.31	52.62
26	12.20	13.41	15.38	17.29	19.82	21.79	25.34	29.25	31.80	35.56	38.88	42.86	45.64	54.05
27	12.88	14.12	16.15	18.11	20.70	22.72	26.34	30.32	32.91	36.74	40.11	44.14	46.96	55.48
28	13.56	14.85	16.93	18.94	21.59	23.65	27.34	31.39	34.03	37.92	41.34	45.42	48.28	56.89
29	14.26	15.57	17.71	19.77	22.48	24.58	28.34	32.49	35.14	39.09	42.56	46.69	49.59	58.30
30	14.95	16.31	18.49	20.60	23.36	25.51	29.34	33.53	36.25	40.26	43.77	47.96	50.89	59.70

Source: Table IV from *Statistical Tables for Biological, Agricultural and Medical Research*, Sixth Edition, by R. A. Fisher and F. Yates, Pearson Education Limited, copyright 1963. Reprinted with permission.

TABLE A.7. Significant studentized ranges for Duncan's new multiplication range test.

/k df/	2	3	4	5	6	7	8	9	10	11	12	13	14	15	16	17	18	19
2	14.04																	
3	8.261	8.321																
4	6.512	6.677	6.740															
5	5.702	5.893	5.989	6.040														
6	5.243	5.439	5.549	5.614	5.655													
7	4.949	5.145	5.260	5.334	5.383	5.416												
8	4.746	4.939	5.057	5.135	5.189	5.227	5.256											
9	4.596	4.787	4.906	4.986	5.043	5.086	5.118	5.142										
10	4.482	4.671	4.790	4.871	4.931	4.975	5.010	5.037	5.058									
11	4.392	4.579	4.697	4.780	4.841	4.887	4.924	4.952	4.975	4.994								
12	4.320	4.504	4.622	4.706	4.767	4.815	4.852	4.883	4.907	4.927	4.944							
13	4.260	4.442	4.560	4.644	4.706	4.755	4.793	4.824	4.850	4.872	4.889	4.994						
14	4.210	4.391	4.508	4.591	4.654	4.704	4.743	4.775	4.802	4.824	4.843	4.859	4.872					

15	4.168	4.347	4.463	4.547	4.610	4.660	4.700	4.733	4.760	4.783	4.803	4.820	4.834	4.846				
16	4.131	4.309	4.425	4.509	4.572	4.622	4.663	4.696	4.724	4.748	4.768	4.786	4.800	4.813	4.825			
17	4.099	4.275	4.391	4.475	4.539	4.589	4.630	4.664	4.693	4.717	4.738	4.756	4.771	4.785	4.797	4.807		
18	4.071	4.246	4.362	4.445	4.509	4.560	4.601	4.635	4.664	4.689	4.711	7.729	4.745	4.759	4.772	4.783	4.792	
19	4.046	4.220	4.335	4.419	4.483	4.534	4.575	4.610	4.639	4.665	4.686	4.705	4.722	4.736	4.749	4.761	4.771	4.780
20	4.024	4.197	4.312	4.395	4.459	4.510	4.552	4.587	4.617	4.642	4.664	4.684	4.701	4.716	4.729	4.741	4.751	4.761
24	3.956	4.126	4.239	4.322	4.386	4.437	4.480	4.516	4.546	4.573	4.596	4.616	4.634	4.651	4.665	4.678	4.690	4.700
30	3.889	4.056	4.168	4.250	4.314	4.366	4.409	4.445	4.477	4.504	4.528	4.550	4.569	4.586	4.601	4.615	4.628	4.640
40	3.825	3.988	4.098	4.180	4.244	4.296	4.339	4.376	4.408	4.436	4.461	4.483	4.503	4.521	4.537	4.553	4.566	4.579
60	3.762	3.922	4.031	4.111	4.174	4.226	4.270	4.307	4.340	4.368	4.394	4.417	4.438	4.456	4.474	4.490	4.504	4.518
120	3.702	3.858	3.965	4.044	4.107	4.158	4.202	4.239	4.272	4.301	4.327	4.351	4.372	4.392	4.410	4.426	4.442	4.456
∞	3.643	3.796	3.900	3.78	4.040	4.091	4.135	4.172	4.205	4.235	4.261	4.285	4.307	4.327	4.345	4.363	4.379	4.394

Source: H. Leon Harter, "Critical Values for Duncan's New Multiple Range Test." *Biometrics*, 1960, 16:671–685. Reprinted with permission of the International Biometric Society.

TABLE A.7. Significant studentized ranges for Duncan's new multiplication range test (continued).

/k df/	2	3	4	5	6	7	8	9	10	11	12	13	14	15	16	17	18	19
2	6.085																	
3	4.501	4.516																
4	3.927	4.013	4.033															
5	3.635	3.749	3.797	3.814														
6	3.461	3.587	3.649	3.680	3.694													
7	3.344	3.477	3.548	3.588	3.611	3.622												
8	3.261	3.399	3.475	3.521	3.549	3.566	3.575											
9	3.199	3.339	3.420	3.470	3.502	3.523	3.536	3.544										
10	3.151	3.293	3.376	3.430	3.465	3.489	3.505	3.516	3.522									
11	3.113	3.256	3.342	3.397	3.435	3.462	3.480	3.493	3.501	3.506								
12	3.082	3.225	3.313	3.370	3.410	3.435	3.459	3.474	3.484	3.491	3.496							
13	3.055	3.200	3.289	3.348	3.389	3.419	3.442	3.458	3.470	3.478	3.484	3.488						
14	3.033	3.178	3.268	3.329	3.372	3.403	3.426	3.444	3.457	3.467	3.474	3.479	3.482					
15	3.014	3.160	3.250	3.312	3.356	3.389	3.413	3.432	3.446	3.457	3.465	3.471	3.476	3.478				
16	2.998	3.144	3.235	3.298	3.343	3.376	3.402	3.422	3.437	3.449	3.458	3.465	3.470	3.473	3.477			

17	2.984	3.130	3.222	3.285	3.331	3.366	3.392	3.412	3.429	3.441	3.451	3.459	3.465	3.469	3.473	3.475		
18	2.971	3.118	3.210	3.274	3.321	3.356	3.383	3.405	3.421	3.435	3.445	3.454	3.460	3.465	3.470	3.472	3.474	
19	2.960	3.107	3.199	3.264	3.311	3.347	3.375	3.397	3.415	3.429	3.440	3.449	3.456	3.462	3.467	3.470	3.472	3.473
20	2.950	3.097	3.190	3.255	3.303	3.339	3.368	3.391	3.409	3.424	3.436	3.445	3.453	3.459	3.464	3.467	3.470	3.472
21	2.919	3.066	3.160	3.226	3.276	3.315	3.345	3.370	3.390	3.406	3.420	3.432	3.441	3.449	3.456	3.461	3.465	3.469
30	2.888	3.035	3.131	3.199	3.250	3.290	3.322	3.349	3.371	3.389	3.405	3.418	3.430	3.439	3.447	3.454	3.460	3.466
40	2.858	3.006	3.102	3.171	3.224	3.266	3.300	3.238	3.352	3.373	3.390	3.405	3.418	3.429	3.439	3.448	3.456	3.463
60	2.829	2.976	3.073	3.143	3.198	3.241	3.277	3.307	3.333	3.355	3.374	3.391	3.406	3.419	3.431	3.442	3.451	3.460
120	2.800	2.947	3.045	3.116	3.172	3.217	3.254	3.287	3.314	3.337	3.359	3.377	3.394	3.409	3.423	3.435	3.446	3.457
∞	2.772	2.918	3.017	3.089	3.146	3.193	3.232	3.265	3.294	3.320	3.343	3.363	3.382	3.399	3.414	3.428	3.442	3.454

Source: H. Leon Harter, "Critical Values for Duncan's New Multiple Range Test." *Biometrics*, 1960, 16:671-685. Reprinted with permission of the International Biometric Society.

TABLE A.8. Critical values Dunnett's test for comparing treatment means with a control—one-tailed comparisons.

df error	α	2	3	4	5	6	7	8	9	10
5	.05	2.02	2.44	2.68	2.85	2.98	3.08	3.16	3.24	3.30
	.01	3.37	3.90	4.21	4.43	4.60	4.73	4.85	4.94	5.03
6	.05	1.94	2.34	2.56	2.71	2.83	2.92	3.00	3.07	3.12
	.01	3.14	3.61	3.88	4.07	4.21	4.33	4.43	4.51	4.59
7	.05	1.89	2.27	2.48	2.62	2.73	2.82	2.89	2.95	3.01
	.01	3.00	3.42	3.66	3.83	3.96	4.07	4.15	4.23	4.30
8	.05	1.86	2.22	2.42	2.55	2.66	2.74	2.81	2.87	2.92
	.01	2.90	3.29	3.51	3.67	3.79	3.88	3.96	4.03	4.09
9	.05	1.83	2.18	2.37	2.50	2.60	2.68	2.75	2.81	2.86
	.01	2.82	3.19	3.40	3.55	3.66	3.75	3.82	3.89	3.94
10	.05	1.81	2.15	2.34	2.47	2.56	2.64	2.70	2.76	2.81
	.01	2.76	3.11	3.31	3.45	3.56	3.64	3.71	3.78	3.83
11	.05	1.80	2.13	2.31	2.44	2.53	2.60	2.67	2.72	2.77
	.01	2.72	3.06	3.25	3.38	3.48	3.56	3.63	3.69	3.74
12	.05	1.78	2.11	2.29	2.41	2.50	2.58	2.64	2.69	2.74
	.01	2.68	3.01	3.19	3.32	3.42	3.50	3.56	3.62	3.67
13	.05	1.77	2.09	2.27	2.39	2.48	2.55	2.61	2.66	2.71
	.01	2.65	2.97	3.15	3.27	3.37	3.44	3.51	3.56	3.61
14	.05	1.76	2.08	2.25	2.37	2.46	2.53	2.59	2.64	2.69
	.01	2.62	2.94	3.11	3.23	3.32	3.40	3.46	3.51	3.56
15	.05	1.75	2.07	2.24	2.36	2.44	2.51	2.57	2.62	2.67
	.01	2.60	2.91	3.08	3.20	3.29	3.36	3.42	3.47	3.52
16	.05	1.75	2.06	2.23	2.34	2.43	2.50	2.56	2.61	2.65
	.01	2.58	2.88	3.05	3.17	3.26	3.33	3.39	3.44	3.48
17	.05	1.74	2.05	2.22	2.33	2.42	2.49	2.54	2.59	2.64
	.01	2.57	2.86	3.03	3.14	3.23	3.30	3.36	3.41	3.45
18	.05	1.73	2.04	2.21	2.32	2.41	2.48	2.53	2.58	2.62
	.01	2.55	2.84	3.01	3.12	3.21	3.27	3.33	3.38	3.42
19	.05	1.73	2.03	2.20	2.31	2.40	2.47	2.52	2.57	2.61
	.01	2.54	2.83	2.99	3.10	3.18	3.25	3.31	3.36	3.40
20	.05	1.72	2.03	2.19	2.30	2.39	2.46	2.51	2.56	2.60
	.01	2.53	2.81	2.97	3.08	3.17	3.23	3.29	3.34	3.38
24	.05	1.71	2.01	2.17	2.28	2.36	2.43	2.48	2.53	2.57
	.01	2.49	2.77	2.92	3.03	3.11	3.17	3.22	3.27	3.31
30	.05	1.70	1.99	2.15	2.25	2.33	2.40	2.45	2.50	2.54
	.01	2.46	2.72	2.87	2.97	3.05	3.11	3.16	3.21	3.24
40	.05	1.68	1.97	2.13	2.23	2.31	2.37	2.42	2.47	2.51

df error	α	2	3	4	5	6	7	8	9	10
	.01	2.42	2.68	2.82	2.92	2.99	3.05	3.10	3.14	3.18
60	.05	1.67	1.95	2.10	2.21	2.28	2.35	2.39	2.44	2.48
	.01	2.39	2.64	2.78	2.87	2.94	3.00	3.04	3.08	3.12
120	.05	1.66	1.93	2.08	2.18	2.26	2.32	2.37	2.41	2.45
	.01	2.36	2.60	2.73	2.82	2.89	2.94	2.99	3.03	3.06
∞	.05	1.64	1.92	2.06	2.16	2.23	2.29	2.34	2.38	2.42
	.01	2.33	2.56	2.68	2.77	2.84	2.89	2.93	2.97	3.00

K = number of treatment means, including control.

Source: C.W. Dunnett. "A Multiple Comparison Procedure for Comparing Several Treatments with a Control," *Journal of the American Statistical Association,* 1955, 50(272):1096-1121. Reproduced with permission from the *Journal of the American Statistical Association.* Copyright 1955 by The American Statistical Association. All rights reserved.

TABLE A.8. Critical values Dunnett's test for comparing treatment means with a control—two-tailed comparisons.

df error	a	2	3	4	5	6	7	8	9	10
5	.05	2.57	3.03	3.29	3.48	3.62	3.73	3.82	3.90	3.97
	.01	4.03	4.63	4.98	5.22	5.41	5.56	5.69	5.80	5.89
6	.05	2.45	2.86	3.10	3.26	3.39	3.49	3.57	3.64	3.71
	.01	3.71	4.21	4.51	4.71	4.87	5.00	5.10	5.20	5.28
7	.05	2.36	2.75	2.97	3.12	3.24	3.33	3.41	3.47	3.53
	.01	3.50	3.95	4.21	4.39	4.53	4.64	4.74	4.82	4.89
8	.05	2.31	2.67	2.88	3.02	3.13	3.22	3.29	3.35	3.41
	.01	3.36	3.77	4.00	4.17	4.29	4.40	4.48	4.56	4.62
9	.05	2.26	2.61	2.81	2.95	3.05	3.14	3.20	3.26	3.32
	.01	3.25	3.63	3.85	4.01	4.12	4.22	4.30	4.37	4.43
10	.05	2.23	2.57	2.76	2.89	2.99	3.07	3.14	3.19	3.24
	.01	3.17	3.53	3.74	3.88	3.99	4.08	4.16	4.22	4.28
11	.05	2.20	2.53	2.72	2.84	2.94	3.02	3.08	3.14	3.19
	.01	3.11	3.45	3.65	3.79	3.89	3.98	4.05	4.11	4.16
12	.05	2.18	2.50	2.68	2.81	2.90	2.98	3.04	3.09	3.14
	.01	3.05	3.39	3.58	3.71	3.81	3.89	3.96	4.02	4.07
13	.05	2.16	2.48	2.65	2.78	2.87	2.94	3.00	3.06	3.10
	.01	3.01	3.33	3.52	3.65	3.74	3.82	3.89	3.94	3.99
14	.05	2.14	2.46	2.63	2.75	2.84	2.91	2.97	3.02	3.07
	.01	2.98	3.29	3.47	3.59	3.69	3.76	3.83	3.88	3.93
15	.05	2.13	2.44	2.61	2.73	2.82	2.89	2.95	3.00	3.04
		2.95	3.25	3.43	3.55	3.64	3.71	3.78	3.83	3.88
16	.05	2.12	2.42	2.59	2.71	2.80	2.87	2.92	2.97	3.02
	.01	2.92	3.22	3.39	3.51	3.60	3.67	3.73	3.78	3.83
17	.05	2.11	2.41	2.58	2.69	2.78	2.85	2.90	2.95	3.00
	.01	2.90	3.19	3.36	3.47	3.56	3.63	3.69	3.74	3.79
18	.05	2.10	2.40	2.56	2.68	2.76	2.83	2.89	2.94	2.98
	.01	2.88	3.17	3.33	3.44	3.53	3.60	3.66	3.71	3.75
19	.05	2.09	2.39	2.55	2.66	2.75	2.81	2.87	2.92	2.96
	.01	2.86	3.15	3.31	3.42	3.50	3.57	3.63	3.68	3.72
20	.05	2.09	2.38	2.54	2.65	2.73	2.80	2.86	2.90	2.95
	.01	2.85	3.13	3.29	3.40	3.48	3.55	3.60	3.65	3.69
24	.05	2.06	2.35	2.51	2.61	2.70	2.76	2.81	2.86	2.90
	.01	2.80	3.07	3.22	3.32	3.40	3.47	3.52	3.57	3.61
30	.05	2.04	2.32	2.47	2.58	2.66	2.72	2.77	2.82	2.86
	.01	2.75	3.01	3.15	3.25	3.33	3.39	3.44	3.49	3.52
40	.05	2.02	2.29	2.44	2.54	2.62	2.68	2.73	2.77	2.81

df error	a	2	3	4	5	6	7	8	9	10
	.01	2.70	2.95	3.09	3.19	3.26	3.32	3.37	3.41	3.44
60	.05	2.00	2.27	2.41	2.51	2.58	2.64	2.69	2.73	2.77
	.01	2.66	2.90	3.03	3.12	3.19	3.25	3.29	3.33	3.37
120	.05	1.98	2.24	2.38	2.47	2.55	2.60	2.65	2.69	2.73
	.01	2.62	2.85	2.97	3.06	3.12	3.18	3.22	3.26	3.29
∞	.05	1.96	2.21	2.35	2.44	2.51	2.57	2.61	2.65	2.69
	.01	2.58	2.79	2.92	3.00	3.06	3.11	3.15	3.19	3.22

K = number of treatment means, including control.

Source: C.W. Dunnett. "New Tables for Multiple Comparisons with a Control." *Biometrics,* 1964, 20:482-491. Reprinted with permission of the International Biometric Society.

TABLE A.9. Distribution of q statistic (Tukey's test).

	A	2	3	4	5	6	7	8	9	10	11	12	13	14	15
1	0.5	18.0	27.0	32.8	37.1	40.4	43.1	45.4	47.4	49.1	50.6	52.0	53.2	54.3	55.4
	.01	90.0	135	164	186	202	216	227	237	246	253	260	266	272	277
2	.05	6.09	8.3	9.8	10.9	11.7	12.4	13.0	13.5	14.0	14.4	14.7	15.1	15.4	15.7
	.01	14.0	19.0	22.3	24.7	26.6	28.2	29.5	30.7	31.7	32.6	33.4	34.1	34.8	35.4
3	.05	4.50	5.91	6.82	7.50	8.04	8.48	8.85	9.18	9.46	9.72	9.95	10.2	10.4	10.5
	.01	8.26	10.6	12.2	13.3	14.2	15.0	15.6	16.2	16.7	17.1	17.5	17.9	18.2	18.5
4	.05	3.93	5.04	5.76	6.29	6.71	7.05	7.35	7.60	7.83	8.03	8.21	8.37	8.52	8.66
	.01	6.51	8.12	9.17	9.96	10.6	11.1	11.5	11.9	12.3	12.6	12.8	13.1	13.3	13.5
5	.05	3.64	4.60	5.22	5.67	6.03	6.33	6.58	6.80	6.99	7.17	7.32	7.47	7.60	7.72
	.01	5.70	6.97	7.80	8.42	8.91	9.32	9.67	9.97	10.2	10.5	10.7	10.9	11.1	11.2
6	.05	3.46	4.34	4.90	5.31	5.63	5.89	6.12	6.32	6.49	6.65	6.79	6.92	7.03	7.14
	.01	5.24	6.33	7.03	7.56	7.97	8.32	8.61	8.87	9.10	9.30	9.49	9.65	9.81	9.95
7	.05	3.34	4.16	4.69	5.06	5.36	5.61	5.82	6.00	6.16	6.30	6.43	6.55	6.66	6.76
	.01	4.95	5.92	6.54	7.01	7.37	7.68	7.94	8.17	8.37	8.55	8.71	8.86	9.00	9.12
8	.05	3.26	4.04	4.53	4.89	5.17	5.40	5.60	5.77	5.92	6.05	6.18	6.29	6.39	6.48
	.01	4.74	5.63	6.20	6.63	6.96	7.24	7.47	7.68	7.87	8.03	8.18	8.31	8.44	8.55
9	.05	3.20	3.95	4.42	4.76	5.02	5.24	5.43	5.60	5.74	5.87	5.98	6.09	6.19	6.28
	.01	4.60	5.43	5.96	6.35	6.66	6.91	7.13	7.32	7.49	7.65	7.78	7.91	8.03	8.13
10	.05	3.15	3.88	4.33	4.65	4.91	5.12	5.30	5.46	5.60	5.72	5.83	5.93	6.03	6.11
	.01	4.48	5.27	5.77	6.14	6.43	6.67	6.87	7.05	7.21	7.36	7.48	7.60	7.71	7.81
11	.05	3.11	3.82	4.26	4.57	4.82	5.03	5.20	5.35	5.49	5.61	5.71	5.81	5.90	5.99
	.01	4.39	5.14	5.62	5.97	6.25	6.48	6.67	6.84	6.99	7.13	7.26	7.36	7.46	7.36

12	.05	3.08	3.77	4.20	4.51	4.75	4.95	5.12	5.27	5.40	5.51	5.62	5.71	5.80	5.88
	.01	4.32	5.04	5.50	5.84	6.10	6.32	6.51	6.67	6.81	6.94	7.06	7.17	7.26	7.36
13	0.5	3.06	3.73	4.15	4.45	4.69	4.88	5.05	5.19	5.32	5.43	5.53	5.63	5.71	5.79
	.01	4.26	4.96	5.40	5.73	5.98	6.19	6.37	6.53	6.67	6.79	6.90	7.01	7.10	7.19
14	.05	3.03	3.70	4.11	4.41	4.64	4.83	4.99	5.13	5.25	5.36	5.46	5.55	5.64	5.72
	.01	4.21	4.89	5.32	5.63	5.88	6.08	6.26	6.41	6.54	6.66	6.77	6.87	6.96	7.05
16	.05	3.00	3.65	4.05	4.33	4.56	4.74	4.90	5.03	5.15	5.26	5.35	5.44	5.52	5.56
	.01	4.13	4.78	5.19	5.49	5.72	5.92	6.08	6.22	6.35	6.46	6.56	6.65	6.74	6.82
18	.05	2.97	3.61	4.00	4.28	4.49	4.67	4.82	4.96	5.07	5.17	5.27	5.35	5.43	5.50
	.01	4.07	4.70	5.09	5.38	5.60	5.79	5.94	6.08	6.20	6.31	6.41	6.50	6.58	6.65
20	.05	2.95	3.58	3.96	4.23	4.45	4.62	4.77	4.90	5.01	5.11	5.20	5.28	5.36	5.43
	.01	4.02	4.64	5.02	5.29	5.51	5.69	5.84	5.97	6.09	6.19	6.29	6.37	6.45	6.52
24	.05	2.92	3.53	3.90	4.17	4.37	4.54	4.68	4.81	4.92	5.01	5.10	5.18	5.25	5.32
	.01	3.96	4.54	4.91	5.17	5.37	5.54	5.69	5.81	5.92	6.02	6.11	6.19	6.26	6.33
30	.05	2.89	3.49	3.84	4.10	4.30	4.46	4.60	4.72	4.83	4.92	5.00	5.08	5.15	5.21
	.01	3.89	4.45	4.80	5.05	5.24	5.40	5.54	5.56	5.76	5.85	5.93	6.01	6.08	6.14
40	.05	2.86	3.44	3.79	4.04	4.23	4.39	4.52	4.63	4.74	4.82	4.91	4.98	5.05	5.11
	.01	3.82	4.37	4.70	4.93	5.11	5.27	5.39	5.50	5.60	5.69	5.77	5.84	5.90	5.96
60	.05	2.83	3.40	3.74	3.98	4.16	4.31	4.44	4.55	4.65	4.73	4.81	4.88	4.94	5.00
	.01	3.76	4.28	4.60	4.82	4.99	5.13	5.25	5.36	5.45	5.53	5.60	5.67	5.73	5.79
120	.05	2.80	3.36	3.69	3.92	4.10	4.24	4.36	4.48	4.56	4.64	4.72	4.78	4.84	4.90
	.01	3.70	4.20	4.50	4.71	4.87	5.01	5.12	5.21	5.30	5.38	5.44	5.51	5.56	5.61
∞	.05	2.77	3.31	3.63	3.86	4.03	4.17	4.29	4.39	4.47	4.55	4.62	4.68	4.74	4.80
	.01	3.64	4.12	4.40	4.60	4.76	4.88	4.99	5.08	5.16	5.23	5.29	5.35	5.40	5.45

K = number of treatment groups. Source: H. Leon Harter, Donald S. Clemm, and Eugene H. Guthrie, "The probability integrals of the range and of the studentized range," WADC Tech. Rep. 1959, 2:58-484.

Index

Page numbers followed by the letter "f" indicate figures; those followed by the letter "t" indicate tables.

Order a copy of this book with this form or online at:
http://www.haworthpress.com/store/product.asp?sku=5256

HANDBOOK OF STATISTICS FOR TEACHING AND RESEARCH IN PLANT AND CROP SCIENCE

_____ in hardbound at $99.95 (ISBN-13: 978-1-56022-292-7; ISBN-10: 1-56022-292-1)

_____ in softbound at $79.95 (ISBN-13: 978-1-56022-293-4; ISBN-10: 1-56022-293-X)

Or order online and use special offer code HEC25 in the shopping cart.

COST OF BOOKS_____

POSTAGE & HANDLING_____
(US: $4.00 for first book & $1.50
for each additional book)
(Outside US: $5.00 for first book
& $2.00 for each additional book)

SUBTOTAL_____

IN CANADA: ADD 7% GST_____

STATE TAX_____
(NJ, NY, OH, MN, CA, IL, IN, PA, & SD
residents, add appropriate local sales tax)

FINAL TOTAL_____
(If paying in Canadian funds,
convert using the current
exchange rate, UNESCO
coupons welcome)

☐ **BILL ME LATER:** (Bill-me option is good on US/Canada/Mexico orders only; not good to jobbers, wholesalers, or subscription agencies.)
☐ Check here if billing address is different from shipping address and attach purchase order and billing address information.

Signature_____

☐ **PAYMENT ENCLOSED: $_____**

☐ **PLEASE CHARGE TO MY CREDIT CARD.**

☐ Visa ☐ MasterCard ☐ AmEx ☐ Discover
☐ Diner's Club ☐ Eurocard ☐ JCB

Account #_____

Exp. Date_____

Signature_____

Prices in US dollars and subject to change without notice.

NAME_____

INSTITUTION_____

ADDRESS_____

CITY_____

STATE/ZIP_____

COUNTRY_____ COUNTY (NY residents only)_____

TEL_____ FAX_____

E-MAIL_____

May we use your e-mail address for confirmations and other types of information? ☐ Yes ☐ No
We appreciate receiving your e-mail address and fax number. Haworth would like to e-mail or fax special discount offers to you, as a preferred customer. **We will never share, rent, or exchange your e-mail address or fax number.** We regard such actions as an invasion of your privacy.

Order From Your Local Bookstore or Directly From
The Haworth Press, Inc.
10 Alice Street, Binghamton, New York 13904-1580 • USA
TELEPHONE: 1-800-HAWORTH (1-800-429-6784) / Outside US/Canada: (607) 722-5857
FAX: 1-800-895-0582 / Outside US/Canada: (607) 771-0012
E-mail to: orders@haworthpress.com

For orders outside US and Canada, you may wish to order through your local
sales representative, distributor, or bookseller.
For information, see http://haworthpress.com/distributors

(Discounts are available for individual orders in US and Canada only, not booksellers/distributors.)
PLEASE PHOTOCOPY THIS FORM FOR YOUR PERSONAL USE.
http://www.HaworthPress.com

BOF04